Dispute Resolution
and
Conflict Management
in
Construction

Also available from E & FN Spon

Engineering Law and the ICE Contracts
Fourth edition
M W Abrahamson

Construction Management and Economics
Edited by R Bon and W Hughes

The Construction Net
A Bridges

Understanding JCT Standard Building Contracts
Fifth edition
D M Chappell

Financial Protection in the UK Building Industry
W Hughes, P Hillebrandt and J Murdoch

Post Construction Liability and Insurance
J Knocke

An Introduction to Building Procurement Systems
J W E Masterman

Construction Contracts
J Murdoch and W Hughes

Understanding the Building Regulations
S Polley

Project Management Demystified
G Reiss

Programme Management Demystified
G Reiss

Failures and the Law
H P Rossmanith

Procurement Systems
Edited by S Rowlinson

The Building Regulations Explained
J Stephenson

Building Education and Research
Edited by J Yang and W P Chang

For more information about these and other titles please contact:
The Marketing Department, E & FN Spon, 11 New Fetter Lane, London, EC4P 4EE Tel: 0171 842 2180

Dispute Resolution and Conflict Management in Construction

An international review

Edited by
Peter Fenn, Michael O'Shea
and Edward Davies

E & FN Spon
An imprint of Routledge
London and New York

This edition published 1998
by E & FN Spon, an imprint of Routledge
11 New Fetter Lane, London EC4P 4EE

Simultaneously published in the USA and Canada
by Routledge
29 West 35th Street, New York, NY 10001

Printed and bound in Great Britain by
TJ International Ltd, Padstow, Cornwall

British Library Cataloguing in Publication Data
A catalogue record for this book is available
from the British Library

ISBN 0 419 23700 3

Contents

Contents (continued)

Foreword

There is a great deal of anecdotal evidence of the occurrence of construction disputes and the research carried out by TG15 to gain empirical evidence both of the incidence of disputes and the techniques used to resolve them will be a most useful insight for those involved in the construction industry. It is only too easy to draw exclusively from one's own experience or the traditions of the cultures in which one is working for ideas and techniques of dispute management and resolution. This is too narrow an approach as we all have plenty to learn from each other.

TG15 has carried out much of its work via the Internet. The group first set up its home page in 1994 and the majority of correspondence is e-mail. This book, and indeed this foreword, has been produced by the transfer of electronic files. The effect of Information Technology on global construction is just beginning, and the implications for construction disputes are obvious. This book will be a valuable first stop for anyone considering construction, whether design, construction or advice, in a new street of the "Global Village".

This book gives an interesting insight into the management and resolution of disputes in a number of countries and will be a useful source of background information on the countries concerned. Even if not working in the countries listed readers looking for new ideas or fresh approaches will find inspiration. Although the book reveals a wide range of techniques it also shows, perhaps comfortingly, that many problems are not limited to particular jurisdictions or cultures and that there are inherent features of the industry and its disputes that are common to us all.

John Bishop
Honorary President
Official Referee's
Solicitors Association

March, 1998

Acknowledgements

There are many people and organisations that deserve recognition and thanks.

The editors in their roles as members of Conseil International du Bâtiment pour la Recherche l'Étude et la Documentation (CIB), The International Council for Building Research, Studies and Documentation saw the need for an international study of the area of construction disputes. CIB agreed and formed International Task Group TG15, which was to study construction conflict management and dispute resolution. The first acknowledgement must therefore go to CIB for providing the vehicle for the group in particular the Deputy Secretary General Chris Pollington.

Once the group was formed the many members of TG15 provided valuable guidance in drafting the required guidelines for the authors of the national monographs. The second acknowledgement is to the members of TG15 who have contributed their time to direct the group, attend meetings and provide advice on dispute resolution around the world.

During the period that TG15 has been in existence the administration of the group has been shared; first UMIST then Masons Solicitors carried the burden and expense of running a group with more than 130 members in over 30 countries. The contributions of Masons and UMIST are gratefully acknowledged. The production of this book is further example of the commitment of Masons and UMIST; the time of the editors has been freely given. Peter Fenn was a visiting professor at the University of Kentucky during the book's final stages and the contribution of UK in particular the Chair of the Department of Civil Engineering Dr Donn Hancher must be recorded.

The final two acknowledgements go to the parties who carried out the real hard work.

The monographs were written, for no financial reward, by eminent international experts; their contributions are the essence of this book. We are deeply grateful for their contributions. We only hope that our editing into a complete volume, and our commentaries on their work have not detracted from the quality of their work. Any mistakes that have crept in are the responsibility of the editors and not the contributors. Although we have covered many countries, time and space have limited us. We hope to produce a second edition and would actively encourage potential authors to join TG15 and to contribute to the second edition.

This book was produced from a collection of files by Sue Shepherd at Masons. We hope that the final product is testimony to the excellent work carried out by Sue; her patience despite our hurried e-mails and scribbled notes was beyond the call of duty.

The material in this book is for general background information only. Detailed professional advice should be obtained before taking or refraining from action based on any of it.

Neither the editors or the authors can accept responsibility for any loss or damage occasioned as a result of reliance placed on any part of the contents of this book.

Peter Fenn   Michael O'Shea, Edward Davies
Lexington   Manchester
KY USA   England

About the editors

Peter Fenn is a lecturer at The University of Manchester Institute of Science and Technology. Peter is a Chartered Surveyor, a registered mediator and a Fellow of the Chartered Institute of Arbitrators. During the period of this book's production Peter was Scholar-in-Residence, and a Visiting Professor, at the University of Kentucky.

Michael O'Shea is a solicitor at Masons Solicitors and Privy Council Agents. Michael is a Chartered Surveyor and before qualifying as a solicitor he had previously held posts with construction companies and as a lecturer at the University of Salford.

Edward Davies is a solicitor and a partner in Masons Solicitors and Privy Council Agents. Edward is a graduate of Manchester University, Guildford College of Law and Kings College London. He is a trained mediator and a Visiting Research Fellow at the University of Manchester Institute of Science and Technology.

List of Contributors

Australia	**Vicky Watts** University of Melbourne, Australia
Canada	**Richard Beifuss** Consultant, Ontario, Canada
China	**Deepak Bajaj** Faculty of Design, University of Technology, Sydney, Australia **Rui Zhang** Page Kirkland Partnership, St. Leonards, New South Wales, Australia
England and Wales	**Peter Fenn** University of Kentucky **Michael O'Shea and** **Edward Davies** Masons Solicitors, Manchester, England
Estonia	**Nicholas Gould** Masons Solicitors, Manchester, England
Hong Kong	**Colin Wall** Commercial Mediation & Arbitration Services Limited **Timothy Hill** Masons Solicitors, Hong Kong
Iraq	**Mohammed Dulaimi** University of the West of England, Bristol, England
Ireland	**Nael Bunni and Anne Bunni** Bunni & Associates, Dublin, Eire

List of Contributors (continued)

Italy	**Alberto Feliciani and Carlo Massini** Pavia e Ansaldo, Rome, Italy
Japan	**Fumio Matsushita** Nikken Sekkei, Tokyo, Japan
Malaysia	**Inpamathi Natkunasingam and** **Satkunabalan K Sabaratnam** Rashid and Lee, Kuala Lumpur, Malaysia
Netherlands	**A G Dorée and M H B Nelissen** University of Twente, Twente, The Netherlands
Oman	**Richie Alder** Trowers and Hamlins, Sultanate of Oman
Portugal	**José Filipe Abecasis** A M Pereira Sárragga Leal, Oliveira Martins, Júdice e Associados, Lisbon, Portugal
Quebec	**Daniel Alain Dagenais** Lavery de Billy, Montreal, Quebec
Romania	**Magda Brosteanau** Gheorghe Asachi Technical University, Iasi, Romania
Scotland	**Duncan Cartlidge** The Robert Gordon University, Aberdeen, Scotland

List of Contributors (continued)

Sweden

Jan Bröchner
SBUF
Ulf Franke
Stockholm Chamber of Commerce
Gunnar Lindgren
Nyström & Partners
Lars Ranhem
Ferax Contract Management

Editorial assistance from
Jens Pedersen and Peter Dyer
Advokatfirman Foyen & Co

Switzerland

Roland Hürlimann
Baur Schumacher & Partner, Zurich,
Switzerland

USA

Stuart Nash
Freshfields, London, England
Douglas L Patin
Spriggs & Hollingsworth,
Washington DC, USA
Ralph C Nash
Professor Emeritus of Law,
The George Washington University,
Washington DC, USA

Note

All the monograph authors can be contacted via the publisher, many
contact details are available for the TG15 Home Page url:

**http://picard.umist.ac.uk/UMIST_Building_Engineering/tg15.htm/t
g15.htm**

Introduction and Commentary

Peter Fenn

Michael O'Shea

Edward Davies

Introduction and Commentary

Introduction

The need for empirical data

Following an international conference at UMIST in 1992 Conseil International du Bâtiment pour la Recherche l'Étude et la Documentation (CIB), The International Council for Building Research, Studies and Documentation saw a need for an international study of the area of construction disputes. International Task Group TG15 was formed in 1993 to study construction conflict: management and dispute resolution.

TG15 resolved to carry this out by investigating the techniques and procedures used to manage conflict and resolve disputes on construction projects around the world. This area of construction receives considerable attention from both lawyers and construction professionals. Despite this interest, there is an absence of any empirical data on the nature of such disputes. Anecdote abounds in technical and legal journals, and unsubstantiated theorising is allowed to pass unchallenged.

Construction industries around the world play a large part in national economies. In developed nations construction output on a value-added basis often exceeds 10% of Gross Domestic Product. The received wisdoms claim that construction contracts are bedeviled by disputes, that the frequency of disputes has risen in recent memory and that the frequency continues to rise. The industry is often claimed to perform badly when compared to other industries. Economists in particular debate the problems of the industry against an image of an industry that delivers buildings and structures which are late, over budget and of a poor quality. Ball goes as far as describing the UK construction industry as backward[1].

While the economic debate is grounded in theory and empirical observations, the debate which surrounds construction disputes is almost exclusively a theoretical one. In the UK virtually all the published forms of contract for construction contain arbitration clauses, yet there is little published research into arbitration which includes any statistics on arbitration. The UK empirical work which exists centres on the opinion of arbitrators[2]. This lack of any evidence to support theorising on arbitration appears to be reflected internationally.[3]

TG15 proposed that for effective management it is essential that data be collected to support, or disprove, theory. Research will seldom *prove* but it allows the logic of analysis to be followed. There is an analogy with the rules of evidence followed by courts and arbitrators. The reader is asked to consider the weight which would be given to expert evidence presented with no analysis to support the opinion presented. In the recent debate which surrounded the proposed introduction of Adjudication provisions to all UK construction contracts via *The Housing Grants, Construction and Regeneration Act* 1996 and the *Scheme for Construction*, the Department of the Environment claimed *inter alia*:

> "... there is compelling anecdotal evidence that adjudication would reduce overall project costs".[4]

It is suggested that the phrase "compelling anecdote" is an oxymoron. Anecdote can prove anything. TG15 therefore encouraged the collection and dissemination of empirical data wherever possible, to test the anecdote.

Conflict and dispute

Is there a difference between conflict and dispute? And if so should TG15 concern itself with the difference?[5] Some authors interchange the two terms, others point to conceptual differences, even if they are blurred. The problem would appear to have arisen with the development of international conflict strategies where the term dispute is seldom used. Schelling [6] describes a field of research that is variously described as Theory of Bargaining; Theory of Conflict or Theory of Strategy. He further argues that the correct title is Theory of Games.

This theory of games allows a further split of conflict, and therefore disputes; the researchers fall into two camps:

·   Those who treat conflict and dispute as pathological states and seek to understand the causes and treatment;

·   Those who take conflict for granted and seek to study the behaviour associated with it.

TG15's first task was to explore conflict and dispute and propose a taxonomy. The UK Construction Industry exists within an adversarial society, that is not a criticism although many present it as such; the culture is a western dialectic argument idiom.[7] It follows therefore that conflict is pandemic and exists where there is an incompatibility of interest. Conflict can be managed, and perhaps avoided. Dispute may or may not flow from conflict and is associated with distinct justiciable issues. Disputes require resolution but that is not to say they cannot be managed rather that the process may lend itself to third party intervention.

The proposed taxonomy

In addition to the split between conflict and dispute the subject might further be categorised by the techniques used for managing the conflict or resolving the dispute. This further split may be made between the binding and non-binding processes used to manage the conflict or resolve the dispute. Of course the conflict management techniques may be made binding by agreement, in much the same way that non-binding mediation can be converted to a binding solution by agreement between the parties. TG15 debated this taxonomy and the final result was included in the Guidelines for Authors produced by the group (See Appendix).

The proposed taxonomy allows members and researchers to consider whether their interest is in conflict and its management or disputes and their resolution. The taxonomy agrees with commentators who argue that conflict is an unavoidable fact of organisational life.[8] Conflict has positive aspects to do with commercial risk taking which is the basis of free enterprise and competition. Disputes on the other hand provide the negative aspects that contribute little to the Construction Industry (other than by, perhaps, incentivising their avoidance by compliance with obligations or willingness to compromise) and its performance; disputes can be studied to attempt to understand causes and treatment. The difference between conflict and dispute is therefore important and researchers and practitioners should concern themselves with the difference.

The International Review

Following the production of the taxonomy and the Guidelines for Authors the group commenced the collection of national monographs which would form the International Review. Invitations were issued in 1996, and the end result is this book. The co-ordinators had great difficulty in finding authors in certain parts of the world and as a result the book contains notable absences. It is hoped that this book will contribute to the body of knowledge on construction disputes. A second edition is planned and it is hoped that range of countries can be extended.

Commentary

The concept of this part of the work of TG15 is simple enough; a collection of national monographs describing practice and procedure on construction conflict and disputes. The stated aims of the monographs were the collection of empirical data where possible and commonality to allow meaningful comparison. The national monographs stand on their own, but to reflect on the usefulness and success of the work of TG15 the co-ordinators and editors resolved to provide a commentary. This commentary has two sections:

- A report of the national authors comments on the taxonomy and list of terms;
- A brief comment on the highlights of each national monograph.

The taxonomy and list of terms

The Taxonomy

The taxonomy was developed by the group after much discussion and formed the structure of the monographs. As with any classification system, anomalies emerged and misclassification occurred through the authors' attempts to deal with the rigidity of the system. However it is clear that the taxonomy allowed the authors to develop clear separation between the management of conflict and the resolution of disputes. It is reassuring to note that several contributors have offered the opinion that the distinction between conflict and dispute is necessary; they refer to the two different stages in the relationship between the parties.

Some authors noted that the taxonomy was restrictive and affected their monographs. Where the language of the country is not English it was noted that many terms did not translate; some gave useful national glossaries.

The list of terms

The list of terms was used by many authors as a checklist, there were no clear additions to the list but it is obvious that a useful review could now be carried out. The monographs have identified many new and different techniques for consideration.

National Monographs

Australia

The author reports that the speed of development of alternative developments in Australia has been described as *breathtaking*, and certainly the range of techniques available is impressive. The idea of a toolbox of conflict management and dispute resolution techniques is hinted at in TG15's taxonomy and classification and Australia appears to be some way down that road. Interesting techniques include Reference Out of Court to a Referee and the Domestic Building Tribunal. The concept of dispute prevention, perhaps the logical extension of a study of disputes, is recognised.

Canada

The failings of court based dispute resolution are emphasised by the Canadian author.   Again it is interesting to note the wide range of available techniques, and the author's emphasis is on dispute avoidance. Partnering receives much attention in the national monographs but many ignore the role of the facilitator during Partnering. The Canadian author describes the role in detail and the importance of the Partnering agreement or Charter.

China

A number of different forces and trends such as the introduction of free market economy thinking and the gradual sophistication of the legal system in China are encouraging formalisation of contract and dispute resolution techniques. Given the enormous importance of China in the world economy the continuing development of dispute management resolution in this country will prove interesting for many years.

Estonia

The challenges, and opportunities, for construction as economies are opened up by the fall of communism are widely recognised. Estonia is a prime example of the opportunities available; the author reports the dream of many Estonians for the trading links offered by their ports to create the Hong Kong of the Baltic. The absence of formal dispute resolution as reported may hinder the accomplishment of that dream. Many will feel that the Estonian culture of non-adversarialism is a positive feature but whether this will soothe international fears about the lack of credible dispute resolution remains to be seen.

Hong Kong

Following the transfer of sovereignty to The People's Republic of China the author reports that the uncertainties which have dogged Hong Kong have now dissipated. An impressive list of developments is proposed. Hong Kong shows a remarkable mix of eastern and western cultures and procedures, but the confidence demonstrated by the international community is evidence of Hong Kong's success. Surprisingly the author reports that mediation is not well understood in Hong Kong, and that the use of evaluative mediation has contributed to the failure of some mediations.

Iraq

The current international political situation will no doubt be a major influence on the direction which the construction industry will take. Assuming those political difficulties will in due course be resolved, it will be interesting to see whether a country that has strong directive state control balanced with an Islamic tradition will continue to adopt many aspects of international practice into its construction industry.

Ireland

As the Irish economy strengthens further as the new millennium approaches so too it appears does its resolve to adopt arbitration as the major process for dispute resolution within the construction industry. This is evidenced by the recent publication of a draft new Arbitration Act and the courts' support of Arbitration by the extensive use of its powers to stay court proceedings where a valid arbitration clause exists.

Italy

This monograph details the distinctive differences which operate to regulate construction contracts in the public and private sectors. A specific conflict management mechanism is prescribed for contracts involving public works. The mechanism includes a strictly enforced monthly notice provision for claims in relation to construction projects.

Failure to notify results in the contractor losing its entitlement to pursue such claims in the future. In contrast private construction contracts are not, without the agreement of the parties, subject to any mandatory statutory conflict management mechanisms and the parties are free to agree the dispute resolution process to be adopted.

Japan

An early draft of the Guidelines for Authors asked authors to provide if possible benchmark data for the country. One of the suggestions was that the number of lawyers in relation to the total population might be useful. This was eventually dropped for fear it might prove intimidating should the data not be readily available or precipitate anti lawyer sentiments. The Japanese author reports that in a country of more than 125 Million citizens there exist less than 20,000 attorneys. The explanation for this astonishing figure is that Japanese culture results in a deep reluctance to call for legal redress unless the dispute is desperate. Construction conflict management and dispute resolution in Japan is similarly affected by the Japanese culture. The author reports that no court battle is conceivable between sub-contractors and main or general contractors.

Québec

A national concern with the failings of the civil justice system is described which is echoed in many other countries. The current aim to simplify the procedures and make justice more accessible and less costly for citizens is of course laudable. However the author reminds us that moves towards alternatives cannot be considered in isolation from existing rights and liabilities. The author provides an example which highlights the fact that even if the parties must show goodwill in the application of alternative dispute resolution methods, they must be careful to preserve their contractual rights.

Malaysia

As with many countries Malaysia is undergoing a major review of its justice system in an attempt to address what appear to be the universal problems of both the cost and delay in bringing disputes to trial in court. The early stages of the review have provided some success with the backlog of civil cases, including those related to construction disputes, being reduced at the end of 1996 from 80,000 to 25,000. There is also an emphasis on judges becoming case managers to maintain control over the litigation process. It is interesting to note that in a society which combines many cultures and stresses the observance of courtesy and the "saving of face" non binding forms of ADR such as conciliation and mediation are in their infancy and their promotion has only recently been intensified

The Netherlands

This monograph offers insight into the Dutch construction industry and Dutch practice. Again standard forms of contract are widely used and the normal dispute resolution procedure is arbitration. The author reports that the Dutch construction industry is small in a small country - too small for "hit and run" strategies. There is a tendency for disputes to be settled amicably between the parties and judges often refer the parties in dispute back to the negotiation table.

Oman

The five year development plan which began in January 1996 is leading to steps forward in the law and procedure which will affect the construction industry. At present it seems that litigation and Court supervised arbitration are very dominant and it remains to be seen whether ADR techniques will move into the "mainstream" especially given the high involvement of international contracting organisations.

Portugal

The author sets out the extensive legislation which dictates the operation of construction contracts within the country and the elaborate requirements for public tendering that exists. The Portuguese courts are not subject to the principle of judicial precedent and are completely free to render their own decisions on the basis of the facts presented under the terms of the applicable law. During the course of judicial proceedings the Judge has a duty, under the Civil Procedure Code, to promote the conciliation between the parties to the dispute serving, as the author describes, as a form of mediator.

Romania

Although significant economic problems still exist the Romanian nation is optimistic about its future. There is a widespread redevelopment of the country's structure as a result of the now irreversible liberalisation of the Romanian market. This includes the development of legal systems to improve access to justice and the harmonising and unification of construction legislation enabling it to follow international codes. Despite its problems and unique characteristics the author succinctly summarises what he considers to be the future of construction conflict in Romania namely, "the essence of dispute avoidance is quality". This would seems to have universal application.

Scotland

It is often incorrectly assumed that Scottish law is the same as English and that the system is the same. In fact the country is legally and constitutionally separate from England and its legal system owes more to continental systems than to the English system. However, the two countries are very influential on each other, Scotland seeming to have the advantage of a smaller country in being able to implement change more effectively but maintaining its own traditions whilst selectively adopting those from England which seem appropriate.

Sweden

The authors report that the use of standard forms for construction work is almost total, with the benefits of standardised procedures being widely recognised. What may be of interest to other countries where standard forms exist is that the Swedish construction industry has developed a standard form for residential construction works where the buyer is an individual consumer. This consumer protection is often a problem elsewhere. The normal dispute resolution in the standard forms is arbitration. The distinction between conflict and dispute is recognised.

Switzerland

The Swiss monograph provides an in depth study of construction conflict within the complexities of the legal systems which operates within the 26 "Cantons" provided for by the Swiss Federal Constitution. Each Canton has its own court system and the majority of construction disputes are decided by district courts who have general jurisdiction within a Canton. Final judgments of Cantonal Courts may be appealed to the Federal Supreme Court. The author reports that in recent times construction disputes have amounted to approximately 10% of all cases litigated within the Supreme Court.

USA

The USA boasts almost 1,000,000 lawyers within its overall population of over 265 million. Lawyers operate within the dual legal system of federal law which overlies individual state systems. It is within the federal courts that the majority of construction cases which are litigated and many are heard before a lay jury. Litigation is, however, no longer viewed as the first choice of those parities wishing to resolve construction disputes. As a result of the increasing costs of litigation and the time taken to pursue a matter to trial the use of "ADR" techniques, particularly arbitration, is now prevalent. This is also true for disputes involving government bodies where litigation was always the traditional method of dispute resolution. This initiative has been heavily influenced by the United States government who have recently passed legislation confirming their commitment to ADR.

1	Ball, M. (1990). *Rebuilding Construction*, Routledge, London.

2	See Hoare et al (1992), 'Consumer Reaction to Arbitration in the Construction Industry' Arbitration, November p278-283. And Ndekugri and Hughes (1994), 'Arbitrators in Construction', *International Construction Law Review*, Vol 3 p366-383.

3	But see Farley 1995 and Hart 1995 for Canada; Davidson 1990 for Scotland and Stipanowich and Henderson 1992 for the USA. Davidson (1990), 'The Practice of Arbitration in Scotland' Scottish.
Office: *Central research Unit Papers*, ISSN 0950-2254. Farley (1995), 'Alternative Dispute Resolution and The Ontario Court Experiment', *ADRLJ* 263. Hart (1995), Court Connected ADR in Ontario', *ADRLJ* 274.
Stipanowich and Henderson, (1992) 'Mediation and Mini-Trial of Construction Disputes', in Fenn, P and Gameson, R (Eds) (1992), *Construction Conflict: Management and Resolution*, Chapman Hall, London.

4	Making the Scheme for Construction Contracts: a consultation paper issued by the Department of the Environment November 1996 at page 47.

5	For a detailed coverage of the literature on this topic see Fenn, P. Lowe, D and Speck, C. (1997), 'Conflict and Dispute in Construction, *Construction Management and Economics*, Volume 15 Number 15 October 1997 p513-518.

6	Schelling, T. (1960), *The Strategy of Conflict*, OUP, Oxford.

7	De Bono, E. (1985), *Conflicts*, Harrap, London.

8	See for example Kolb, D. (1992), *Hidden Conflict in Organisations*, Sage Publications, London, or Lax, D. and Sebenius, J. (1986), *The Manager as Negotiator,* New York Free Press, New York.

AUSTRALIA

Vicky Watts
University of Melbourne, Australia

1.   BACKGROUND

1.1   Economic[1]

Australia is a country of 7,680,000 square kilometres having a population of a little over 18 million. More than 70 percent of the population is situated in 14 major cities most of which are in the south-east corner of the continent. Thus despite appearances to the contrary, it is a highly urbanised society.

The Gross Domestic Product (GDP) of Australia for 1996/97 was $AUS510 billion.[2] The building and construction industry has contributed around eight to nine percent of GDP.

The industry comprises the following broad sectors:

Residential	$ 13.0 billion
Alterations and additions to residential	$ 2.6 billion
Non-residential	$ 12.7 billion
Engineering construction	$ 15.2 billion
Total expenditure	**$ 43.5 billion**

The residential sector produced about 130,000 houses in 1996/7, down from 170,000 in 1994/5, including detached houses, semi-detached and row houses and apartments the total being 6.68 million residences. The non-residential sector produces office, retail, industrial and other building stock. Finally, the engineering construction sector is responsible for roads and bridges, other transport construction and infrastructure development.

The industry normally employs about seven percent of the total workforce (currently about 600,000 of a total of 8.4 million persons).[3] The building and construction industry consists of about 100,000 separate enterprises most of which are small house building firms and trade based firms such as plumbers, electricians, carpenters, bricklayers and painters. There are some 4,000 lager building firms in the non-residential and engineering construction sectors.

1.1.1   Trends

Recent trends in construction activity are shown in the attached table at 1989/9 prices ($AUSbillion).

Year	Residential	Non-residential	Engineering	Total	GDP
90/91	12.8	13.6	11.1	37.6	366
91/92	12.9	10.4	10.3	33.6	368
92/93	15.2	9.3	10.6	35.1	380
93/94	16.7	9.1	11.5	37.3	396
94/95	17.4	10.2	11.8	39.4	413
95/96	14.5	11.5	12.9	38.9	430
96/97	14.4	12.8	13.3	40.5	444

Table 1   Recent trends in construction activity, 1989/9 prices ($AUSbillion).

This shows a steady increase in residential construction up to 1994/5 which dropped by 17 percent to 1995/6 and output remains steady in 1996/7. Non-residential work fell during the early 1990s as a result of the general recession but has steadily improved since.[4] Engineering and infrastructure work has remained fairly steady throughout the 1990s but with some recent increases.[5] Most of the volatility in the above figures is in the private sector as public sector building and construction has been constant in recent years.

1.2   Legal[6]

1.2.1   Construction Law

The Australian legal system is governed by the Australian Constitution (1901) which divides powers to legislate between the Federal Government ant the six State and two Territory Governments.

The Federal government has the power under Section 51 of the Constitution to legislate with respect to various national-related matters.[7] The States and Territories may also make laws on all of these subjects but only to the extent that they are not inconsistent with Federal laws. Any area not specifically stated to be a matter on which the Federal Government may legislate, is left to the exclusive domain of the States and Territories. The power to legislate with respect to building and construction law is left to the States and Territories.

The court system of Australia where these laws are administered and challenged is based on a hierarchy. In descending order the courts are: the High Court, Supreme Court; County or District Court; and Magistrates' Court. Each court has jurisdictional limits and powers. Appeals from each court are made to the next court in the hierarchy, the last court of appeal being the High Court of Australia. There is also a Federal Court as well as a limited number of specialist courts and tribunals.[8]

The lack of uniformity in laws and procedures of each State and Territory is one of the most common complaints of national and international companies involved in the Australian construction industry. Although there are some attempts being made to create uniform laws across each jurisdiction.

1.2.1   Regulation of Building Practice

All States and Territories have legislation governing the construction industry.[9]   The Building Code of Australia prescribes in detail the design and construction requirements with which a building must comply. It is universally accepted as the standard of Australian construction and used in all States and Territories. Each State and Territory has set up authorities to monitor and enforce the various rules and regulations governing the construction industry.[10]

Local Government also plays a significant role in the building and construction through such things as issuing planning and building permits.[11]   Building surveyors, whether municipal or private, inspect building work, ensure it complies with the building regulations, and issue final certificates of inspection and occupancy permits.

1.2.2   Construction Contracts

The Australian construction industry very much favours standard form contracts and there are a number to choose from. The JCC suite of contracts are most popular for traditional commercial building projects whilst the Australian Standard contracts such as AS2124 or its successor AS4000 are favoured for civil works or projects with more of an engineering slant.

Industry organisations (e.g. the Housing Industry Association, Master Builders Association and the Royal Australian Institute of Architects) also publish standard form contracts for use in domestic or small commercial projects.

In more recent years parties have been looking at alternative project delivery vehicles and have explored, with some measure of success, more innovative styles of contracting, such as partnering and alliancing. There is also a trend developing of clients looking at the life cycle cost of a project and exploring

Design Construct and Maintain (DC&M) contracts based on the theory that if the contractor also has to maintain the building it will be less likely to select products which are cheaper up-front but have higher maintenance costs.

1.3   The Development of ADR in Australia

1.3.1   General Development

Australia inherited the common law of England and Wales and its courts system, complete with procedures developed in accordance with that country's 19th century culture and experience.[12]   Over time that traditional court system has developed and adapted to the needs of the Australian people. There has been considerable growth in the court system with the addition of specialised courts and tribunals; as well significant expansion of the jurisdiction and powers of lower courts.   More recently, in addition to the formal court system for resolving disputes, there has grown a trust and acceptance of using various alternative methods for resolving disputes.

The success and general acceptance of these alternative methods has been so great that the courts themselves are now modifying their Rules to allow such methods to be incorporated into their range of resolution options.   Astor and Chinkin (1992) comment that there is:

growing pressure for the introduction of more alternative processes within formal court structures . . . current trend of mainstreaming ADR by bringing it within the court system or incorporating it within government decision making . . . closer association of ADR with the formal justice system through the development of court-connected schemes. [13]

In fact methods for resolving disputes are evolving constantly, for example, a new initiative by the Victorian Chapter of the Institute of Arbitrators Australia has been to establish a new service designed for resolving most types of small or single-issue disputes. The aim of the service is to arrive at a resolution within a very short period of time (within days) for disputes which do not require a lot of fact finding; full details of this service are still to be released.

In recent years many organisations have been set up to facilitate the growing ADR environment. These range from the older and more active organisations such as the Australian Institute of Arbitrators, the Australian Commercial Dispute Centre and LEADR (Lawyers Engaged in Alternative Dispute Resolution), through to smaller groups within existing structures such as the Law Societies in various states who are developing boards or committees for those specialising in ADR.[14] Some organisations are specific to the building and construction industry, such as, the Building Dispute Practitioners' Society which runs discussion nights and produces a newsletter, although it only exists in Victoria at present. There is even international membership of some Australian organisations, encouraged by their acclaimed training courses and publications.[15] A detailed list of the major organisations involved in Dispute Resolution and Management in Australia is contained in Section 4.2.

Justice Ronald Sackville (1996) made an observation which summarises the Australian situation:

> *In a sense, the rapidity of the development of alternative dispute resolution is breathtaking. Within the space of a decade and a half a large body of skilled practitioners has emerged to play a critical role in resolving disputes.*[16]

1.3.2   Development within the Building and Construction Industry

In the 1980's the building and construction industry developed the view that there had been an increase in the incidence of contractual claims and disputes. These concerns within the industry led to the formation of a Research Project Group comprising senior management people from the Australian Federation of Construction Contractors, the Australian Institute of Quantity Surveyors and Federal and State Government Construction Authorities.[17] After visiting a number of overseas countries to investigate the level and causes of construction disputation, forms of contract and methods of dispute resolution, a report was published in 1988 called "Strategies for the Reduction of Claims and Disputes in the Construction Industry - A Research Report".

There followed a meeting between Government and Private Construction Industry Organisations in 1989 which agreed that the findings of the Report were of such significance that more should be done to implement the recommendations. This led to the establishment of a Joint Working Party with the aim of developing improved practices, better quality work, with the over-riding aim of achieving a reduction in claims and disputes. Papers produced by the sub-groups of the Joint Working Party were published in 1990 in a document 'No Dispute: Strategies for Improvement in the Australian Building and Construction Industry', commonly known as *'No Dispute'*.

No Dispute addressed dispute resolution in Paper 10 and proposed the following guidelines[18]:

- encourage, facilitate and expedite genuine negotiation
- avoid legal representation
- avoid arbitration and litigation processes
- specify compulsory conferences of senior management of both parties before embarking on formal third party processes
- concentrate on cost of mitigation of the problem area, rather than procrastination about negotiating and resolving the dispute
- be cost-conscious, contemplating end financial implications of resolution processes once genuine negotiation has failed
- encourage use of alternative dispute resolution processes

In the seven years since the publication of *No Dispute* the Australian construction industry has moved a long way towards achieving the goals identified in these early guidelines. Duncombe and Heap (1995) commented upon the impact of these two reports on the Australian construction industry stating:

Both reports recognised that it was essential that there be a change from adversarial culture in the industry and that more cost effective and efficient dispute resolution procedures be introduced. [19]

They go on to observe that subsequent initiatives reflecting this change include: the use by government agencies of ADR processes; the introduction of codes of practice[20]; as well as the widespread use of mediation and the introduction of partnering.

Also in 1990, the Monash University Centre for Commercial Law (which no longer exists), published a Report on a one year research project on ADR usage in commercial disputes. This Report found '*more than 85 percent of identifiable ADR is taking place within the construction/civil engineering industry*'.[21] However, with the rise in usage of mediation for Family Law disputes over recent years such statistics would probably differ somewhat if similar research were undertaken today.

1.4   Usage rating for procedures

1.4.1   Preliminary research

It was decided that it would be useful for the purposes of this Monograph to be able to indicate how much any particular resolution or management procedure was used in Australia. Preliminary research was consequently carried out in early 1997. As a result most procedures addressed in this Monograph are accompanied by an indication of their rate of usage in Australia, described as a 'Usage Rating in Australia'. The scale adopted is from 0 to 10, with a Usage Rating of 10 indicating the procedure is frequently used, while 0.5 or 1 indicates a very slight usage.

1.4.2   A brief summary of research

Over 50 companies responded to a mailout questionnaire sent to over 250 building firms and law firms throughout Australia. These companies were selected from the telephone directory. The response rate was 23 percent however no follow up approaches were made to attempt to increase that response rate. Five percent of questionnaires were returned as 'addressee unknown'. Within the sub-group of responses from building firms 20 percent returned the survey stating that they had not experienced any disputes in the survey period which covered 10 years. For the purposes of this CIB Monograph the results were ranked for each of the building and law firm subgroups and then a combined rating was developed to indicate the frequency of usage of each resolution method in Australia.[22]

2.   CONFLICT MANAGEMENT

Astor and Chinkin (1992) in referring to dispute management explain:

The need for ongoing dispute resolution mechanisms which are established at the time the contract is entered into, or shortly afterwards. This has been especially the case within the construction industry, where the long-term nature and complexity of many projects and the claim-prone nature of the industry means that a variety of both large and small claims can be anticipated throughout the contract performance. What is often desired is a means of resolving these disputes with a minimum of delay and while allowing performance to continue. Where decision-maker(s) can also build up familiarity with the contract and its performance, the mechanism is likely to be more effective. [23]

2.1 Non-binding

2.1.1 Dispute Review Boards (or Dispute Board of Review, or Boards of Dispute Avoidance, as they are also known in Australia)

Usage rating in Australia : Rating 0.5

A Board is set up soon after the contract is awarded and comprises senior executives from client organisation and construction organisation plus a third party neutral. The Board meets regularly during construction phase. When disputes occur they are presented to the Board who then makes a recommendation to the parties.

If the Board is set up under the ACDC model they are called 'Boards of Dispute Avoidance (BODA)' and the neutral expert is appointed by ACDC. ACDC publish a set of Guidelines to assist in establishing a Board of Dispute Avoidance.

Regardless of the name, all these Boards function to prevent or resolve disputes quickly during the course of a lengthy construction project. In 1990 *No Dispute* commented that such Boards are in increasing use. [24] Tryill (1992) also stated that *'there has been good experience of this dispute resolution system in Australia in tunnelling projects and in defence procurement contracts.'* [25]

In the Preliminary Research Survey (see Section 1.4) conducted in early 1997 a very low Usage Rating was obtained for this procedure. However, it is worth noting that, in the preliminary survey it was named as a Dispute Review Board, rather than either Dispute Board of Review or Board of Dispute Avoidance and the choice of terminology may have effected the survey result.

Process

The parties to the contract each select a person to be a member of the Board and then either those two people select a third or the facilitating organisation nominates a third person, who acts as chairperson. The third person selected will ideally have some knowledge of the type of building project being undertaken. All three members of the Board should be impartial and have no conflicts of interest with respect to the contract. The Board members then enter into an agreement with the parties which provides a description of the scope of work to be undertaken by the Board, including establishing procedures for resolving disputes; payment, termination, and liability of Board members etc.

The concept is to prevent disagreements from escalating into disputes. The process does not impinge on the supervisors certifying function. The idea is to provide a non-binding procedure in an attempt to resolve any disagreements before the parties move to other forms of resolution provided for in the contract, such as litigation or arbitration. The Board meets regularly and more frequently if matters require it.

If a matter requires a hearing then the parties will generally submit papers to the Board before the hearing. Legal representation is usually discouraged. After the hearing the Board meet in private to arrive at their recommendations. These recommendations are non-binding, unless the contract between the parties requires the Board to make a formal determination which is binding on the parties. To assist the parties the independent third party may indicate the likely decision that could issue from any arbitration on the dispute.[26]

2.1.2   Dispute Review Advisers

Usage rating in Australia : Rating 0

As for Dispute Review Boards but with an individual adviser instead of a Board. The preliminary research did not reveal any building disputes being resolved by this indicating that it is rarely used in Australia.

2.1.3   Quality Matters[27]

In Australia techniques which centre on management control can be described as follows:

a)   Total Quality Matters - The contract requirements for a project quality plan vary. However there is strong evidence that a commitment to a PQP provides a comprehensive (total) quality commitment by contractors; and

b)   Quality Assurance - Deregulation of inspectorial responsibility by government bodies and now larger contractors is placing an increasing importance on QA to pass down the compliance responsibility to minimise risks.

The approach to standard control began with the Federal and State governments who initiated QA and certification. This has moved to individual contractors who are now requiring suppliers and consultants certification as a basis for claim settlement. ISO9000 is now generally accepted in the construction industry. Client's obligations are not always understood nor effectively implemented.

One initiative to improve the quality of information control has been by VicRoads, who use electronic QA systems for document and quality records which facilitates their earliest certification by

DLIQ, minimises document control issues and maximises understanding at all levels.

2.2   Binding

2.2.1   Partnering

<u>Usage rating in Australia</u> : Rating 0.5

Major players in the project from the client, construction, design and consultant organisations are brought together early in the project, to establish communication, forge partnerships and set common goals in the hope of avoiding disputes. The process also manages any disputes occurring during the construction phase.

Charles Cowan of the United States came to Australia in 1992 to present a series of seminars on partnering. Partnering received further exposure in a seminar conducted in July 1993 by the Commonwealth Department of Administrative Services and the Royal Institute of Public Administration Australia. CIDA estimates that

'By *November 1993 projects valued at over $1billion were being partnered in Australia. Most of the projects involved public sector developments such as dams, roads, schools and waste treatment works'.* [28]

Partnering in Australia tends to be on a project by project basis in both the public and private sectors, known as 'Project Partnering'. This is especially so in relation to public sector projects because of the stringent requirements for competitive tendering where public funding is involved. There are some private organisations which carry out extensive multi-project construction over longer periods of time, however, they often continue to use competitive tendering for each separate project.

Journals, magazines and newsletter published throughout the industry often feature descriptive articles on both partnering and strategic alliances (long term partnering commitments). However, it is hard to locate well documented evidence of such methods being widely used. One report which did indicate the level of use of project partnering was the Public Works 1994-5 Annual Report, which stated:

The Department of Public Works & Services has adopted a flexible approach to partnering tailoring it to suit the needs of individual projects. Partnering was adopted on a trial basis on 20 percent of projects valued at over $AUS 5 million in 1994/95. An assessment of partnering on these projects indicated that improved project outcomes have been achieved on partnered projects, in comparison with non-partnered project over a similar period.

Consequently the Department concluded its Report by stating that is expected to extend the use of partnering in the future.

At present none of the Standard Forms of Contract available in Australia are suitable for use in partnered projects. However, the Department of Public Works & Services are in the process of developing such a contract.[29]

Process

Principal participants in the project meet before the project commences at a partnering workshop to develop a partnering statement/charter which: identifies and records, in writing, their common goals and interests; agrees for the participants to co-operate with each other to achieve these goals; agrees procedures to be adopted to achieve these goals and interests; and promotes co-operative problem solving rather than adversarial confrontation.

An independent person usually acts as facilitator at the partnering workshop. Partnering usually involves sharing information; establishing effective communication; problem solving at the lowest level; a procedure for resolving disagreements early or identifying problem disputes and moving them to a higher level of authority to assist resolution; a mechanism for continuous joint evaluation of performance.

A Construction Industry Institute (CII) Report characterises a partnering agreement as a 'moral' agreement superimposed over a contractual agreement where:

the contract defines the legal relationship, whilst partnering defines the working relationship.[30]

The agreements and representations associated with the partnering process, even if not intended to be legally binding, may impact upon the contractual risk allocation in the following ways, according to Jones (1996):

- the implication of a contractual duty of good faith;

- the creation of fiduciary obligations misleading and deceptive conduct (under *The Trade Practices & Fair Trading* Act);

- promissory estoppel and waiver; and

- confidentiality and 'without prejudice' discussions.[31]

3.   DISPUTE RESOLUTION

3.1   Non-binding

3.1.1   Conciliation

Usage rating in Australia : Rating 1

Conciliation is very similar to mediation however the third party neutral does not always meet together with the parties. The conciliators role is also broader than in mediation as it includes advising the parties on the likely outcome of the dispute if it were resolved in an alternative forum.

A conciliator is selected via agreement or appointment. Several organisations have lists of conciliators and some also have procedural notes and draft agreements for conciliation.[32]   A conciliator has no immunity from action for actions of negligence. In light of Brabazon J's (1997) comments, the function of expressing opinions, as to the likely outcome of a dispute if it were to be arbitrated or litigated, places the conciliator at risk of suit for negligence.

Process:

In conciliation, the process begins with identification of the issues, then the options for resolution are explored, the conciliator advises on likely outcome of dispute in other forums and in light of this the options for resolution are considered; and ideally a consensual agreement is then reached.

In February 1988 the Institute of Arbitrators Australia published "Rules for the Conduct of Commercial Conciliations". The Notes accompanying the Rules outline the functions which may be performed by the conciliators as follows:

- attempt to gain a compromise agreement

- provide an assessment of respective rights and obligations of the parties

- provide an opinion as to the defectiveness of the work

- provide an opinion on the probable outcome of the dispute if it were to be arbitrated or litigated

3.1.2   Expert Tribunal (or 'mini trial') (or Senior Executive Appraisal as it is commonly known in Australia)

Usage rating in Australia : Rating 1

In an American style mini-trial both sides briefly present their case to a panel then the panel negotiates a settlement. The third party acts as a mediator/facilitator of discussions and usually gives an opinion on the likely outcome if the dispute were to be litigated or arbitrated.

Such an American style process has undergone some transition in its application in Australia, where Sir Laurence Street Chief Justice of New South Wales:

has adapted the mini-trial to develop a process for commercial dispute resolution which he has called 'senior executive appraisal'. [33]

The resulting Australian process is a less adversarial, more consensus orientated process than the American mini-trial. [34] It has a focus on the ongoing or future relationships of the parties as an overlay on settlement of the particular dispute.

Tyrill (1992) observed that there has been some use of senior executive appraisal in the construction industry and the advantage of this process is to elevate disputes above lower or middle (site-based) management level, where they often bog down, up to senior management level. It also allows each party to assess the respective strengths of both their own and the other party's case, often for the first time. [35]

The neutral third party in a senior executive appraisal has no immunity from action for actions of negligence. In terms of liability for negligence the third party neutral could be liable for any opinions expressed.

Process:

Street (1989) describes the process as:

(it) involves senior executives of the disputing companies, the parties or their representatives and a neutral advisor. The process comprises the senior executives meeting in conference where they appraise the dispute and attempt to formulate some possible basis for settlement. This conference will be chaired by a neutral consultant chosen by the executives. Before the conference each side will have prepared a brief position paper which will be forwarded to the executives, the neutral consultant and the other party. The parties will also have had the opportunity to prepare a response to the other side's paper and this too will be in the possession of both the executives and the consultant. At the appraisal conference, the parties, or their representatives, may make a short oral presentation and the panel may require elucidation of points as the feel necessary. In a private session following the conference the senior executives and the neutral adviser appraise the commercial aspects of the dispute and attempt negotiate a settlement. The executives may request the neutral adviser to give an opinion on the legal or factual merits of the dispute, or on the likely course and outcome of litigation. The executives may agree in advance of the conference to accept any opinion on the legal or factual merits of the dispute as the basis for settlement. 'The object of the consultant will be to facilitate in every way possible the

settlement of the whole dispute, or failing this, he resolution of some of the issues.[36]

3.1.3   Mediation

<u>Usage rating in Australia</u> : Rating 4

Mediation is a private process where a third party neutral assists the parties come to an agreed outcome.   The third party is impartial and their role is to control the resolution process. The content of the mediation and the resultant outcome are controlled by the parties - the outcome is not a decision of the third party. Where a company is involved the person representing the company must have authority to enter into any agreement which might be reached.

It has been commented that mediation to date in Australia has a high success rate and the costs of the process are small compared to more formal arbitration and litigation processes.[37]   It is a very popular method of resolving disputes across a variety of industries.   Several organisation have draft agreements for mediation; Rules for the conduct of mediations; and panels/published lists of mediators from which parties can select a mediator.[38]   Under the Institute of Arbitrators Australia's' Rules for the Mediation of Commercial Disputes there is provision for the Institute to appoint a mediator where the parties have not made other provision for appointment

A mediator has no immunity from action for actions of negligence. However according to Brabazon (1997):

In a private mediation, it seems that no civil or criminal legal proceeding has been taken against a mediator in this country.
It seems to me that the greatest danger for mediators lies in the giving of advice that turns out to be wrong, and negligent . . .
Mediators should not give advice to disputants . . . the Law Council of Australia's Ethical Standards for Mediators warns of

the dangers of giving advice . . . Even the best form of agreement will not protect a mediator in all cases. There will never be protection against fraud or criminal conduct. Sometimes mediations are conducted by corporations . . . in those cases, section 52 (of the Trade Practices Act) will apply . . . also the Fair Trading Acts of the states protect consumers against misleading and deceptive conduct on the part of individuals, who can not exempt themselves from liability. [39]

Rule 14 of the Institute of Arbitrators Australia's' Rules for the Mediation of Commercial Disputes contains an exclusion of Liability and indemnity clauses. These seek to give the mediator protection for any acts, other than fraud. However the strength of such clauses has not been tested in the courts.

<u>Process:</u>

The well recognised LEADR Model for mediation proceedings in Australia follows the following stages:

a)   Pre-mediation - a preliminary conference is held to explain the process and respective roles; agree to date, time and location for mediation to occur; establish who is to attend and their authority to act and the limit of that authority; schedule exchange of experts reports etc. if required.

b)   Mediators opening statement - explains role of parties, mediator, legal advisors, and experts; explain confidentiality of joint sessions and private sessions; outline mediation process; confirm authority held by parties to settle; answer any queries raised by the parties; set parties to sign a mediation agreement.

c)   Parties-statements - parties explain their side of the dispute; parties are encourage to speak rather than their advisors; parties outline what they hope to achieve from the mediation.

d)   Summarising and identification of issues - each parties' statement is summarised by the mediator; and identifies the issues for discussion in the mediation; an agenda for conducting the mediation is set; the parties are encouraged to assist the mediator with these items.

e)   Exploration of issues - the mediator works through each of the issues identified for discussion assisting the parties to explore options for settlement for each issue.

f)   Private session - each party meets in turn with the mediator to discuss any underlying issues and explore proposed settlement options further; discussions in private sessions with one party are not revealed to the other party by the mediator unless authorised to do so; more than one session with each party may occur.

This ends the Problem Defining stages.

g)   Subsequent joint session - parties attempt to negotiate a settlement of the issues in dispute.

h)   Subsequent private session - each party meets in turn with the mediator to discuss that the evolving agreement is workable and that all the issues in dispute are covered, and that the parties have adequately discussed all issues identified earlier.

i)   Agreement - parties are encouraged to reach a final agreement; any agreement is encouraged to be set out in writing; where final agreement is not reached parties may sign a heads of agreement.

This ends the problem solving stage.[40]

In commercial disputes the written mediation agreement is usually intended to be final and binding, however the status is a matter for the parties, and their legal advisors, to decide upon.

It has been expressed in several texts that the private sessions in mediation pose potential problems regarding confidentiality, however, it is widely acknowledged that these sessions are essential to arriving at a settlement. Sir Laurence Street (1991) commented that private caucusing sessions were *'the heart of the mediation process'.*[41] However, the risks are highlighted by Brabazon (1997) who details the provisions in Queensland concerning this area:

> *s100S provides a criminal penalty for a breach of the obligation of an ADR convenor to keep confidential information*

> and *provides an ADR convenor must not, without reasonable excuse, disclose information coming to the convenors knowledge during an ADR process. Maximum penalty - $3000.*[42]

3.1.4   Negotiation

<u>Usage rating in Australia</u> : Rating 10

In negotiation parties approach each other for discussions to find a mutually acceptable outcome to the dispute. Negotiations may, or may not, involve partisans supporting each of the parties.

Some of the Standard Forms of Contract available in Australia contain negotiation style clauses which require the parties *'to confer at least once'* and only if this attempt at negotiation fails may the dispute then be refereed to arbitration or litigation.[43]

According to Astor and Chinkin (1992):

Some lawyers are beginning consciously to refine their negotiation skills . . . such things as communication, active listening and the importance of understanding the   impact of different personality traits on negotiation styles are currently being taught in training sessions within law firms and in law school courses.[44]

Process:

As negotiation is a consensual process, it requires a willingness by both parties to attempt resolution by this method. If the relationship between the parties has broken down beyond this point then negotiation will not be possible. Negotiation can take many forms ranging from an informal chat or a telephone conversation between the parties or their lawyers to a highly structured and complex process taking place over an extended period of time.[45]

Distinction has been drawn between simple bilateral negotiation which does not involve third parties in the process, the content or the resolution; and supported negotiation where a third party acts in some partisan role for each of the parties. Any third party involvement does not extend to acting as facilitator or dispute resolver.[46]

Whatever the form of the negotiation the process, content and outcome always remain in control of the parties.

3.1.5 Fact-based Mediation

Usage rating in Australia : Rating 1

Relevant facts are supplied to mediator who then prepares for each party an opinion on the likely outcome if the dispute were to be arbitrated, and costs there in and evaluates strengths and weaknesses of their case. This information is provided to the party prior to the mediation so they have an expert opinion of their case prior to the mediation commencing.

Little material exists commenting on the application of this method in Australia.

3.2   Binding

3.2.1   Adjudication

The decision of a third party neutral, named in the contract, is binding upon the parties with respect to any matter in dispute until the contract is complete. At that time the parties may challenge the decision through arbitration or litigation.

Some contracts make provision for this in Australia, however, little material exists in the literature commenting on adjudication.[47]

3.2.2   Arbitration

Usage rating in Australia : Rating 5

3.2.2.1 Traditional Arbitration

Arbitration is a process where a third party who is independent of parties, but may be selected by them, makes an Award determining the dispute. The Award is binding and can be enforced by the Courts. The process is controlled by the Commercial Arbitration Act. The proceedings are private and the Award is not formally published.

Australia inherited the English Arbitration Act 1697 (9 Will III c 15). In 1979 reforms to the English Arbitration Act prompted the various states in Australia to review their own legislation. In

1984 the Standing Committee of Attorneys General prepared and adopted a uniform bill on commercial arbitration. To accommodate the recommended uniformity the various States passed similar, but unfortunately not identical, Commercial Arbitration Acts.[48] Astor and Chinkin comment:

The philosophy of the uniform legislation was 'to recognise and respect party autonomy in choosing a tribunal and procedure suitable for the resolution of their dispute' and to reduce judicial intervention into arbitration.[49]

The Institute of Arbitrators Australia was established in 1975 and provides facilities for arbitration throughout Australia. Amongst other things, it provides training courses, qualifying examinations, and recommended dispute resolution clauses for arbitration which can be inserted into agreements. See Section 4.2 for a detailed list of services provided by the Institute. There remains a shortage of people recognised and accepted by those in the industry as suitable to act as commercial arbitrators in complex or specialist cases despite the training efforts of the Institute.[50] The level of appointment being made by the Institute of Arbitrators has fallen in recent years to approximately 20 per year,[51] compared to 300 in 1989.[52] In part this reduction may be explained by the downturn in the economy in the early 1990's, however construction activity has steadily risen since then. So one is left to hypothesise that the recent lower levels of appointments are because of less arguments occurring and/or by a move away from arbitration to other ADR procedures.

Process

Under the uniform Commercial Arbitration Act, parties may agree to submit a commercial dispute to arbitration, either at the time the dispute arises or by an arbitration clause in their agreement. Arbitration begins with a Preliminary Conference, followed by a hearing over several days or weeks. The arbitrator may publish Interim Awards as well as the Final Award which are binding on the parties.

Many sections of the Act commence with the words ' Unless otherwise agreed in writing', giving the parties the flexibility to make their own arrangements.[53]   Some of the more consequential sections of the Act are described below.

s14   subject to the arbitration agreement s14 allows the Arbitrator *to conduct the proceedings as he or she sees fit'*.   Recent interpretation of this section of the Commercial Arbitration Act by the Courts has varied between the States. In a South Australian the Full Court found that it had the right to supervise arbitrators.[54]   The judgment means that in South Australia 'procedural justice requires that arbitrators should, in long complex arbitrations, follow as nearly as reasonable practicable the pre-trial pleading, discovery and other procedures of the court.' This concept has been strongly rejected in New South Wales[55] and in other States the arbitrator is more free to conduct the proceedings as he sees fit under s14 of the Commercial Arbitration Act. As a result of these two cases arbitrators in South Australia, in large complex cases, need to follow the traditional court format.

s19   the Arbitrator is not bound by the rules of evidence unless otherwise agreed in writing and may inform him/her self in relation to any matter as the arbitrator or umpire thinks fit.

s22(2)   the Arbitrators decision is based on evidence presented by the parties. The award is made on the basis of law, unless the parties have agreed that the arbitrator is to act as an amiable compositeur, when the decision can be based on considerations of general justice and fairness.

s27   The uniform Commercial Arbitrations Acts have given legislative impetus to the trend towards using processes in conjunction with each other by making provision for mediation or conciliation within the arbitral process.

s28   The award is binding on the parties.

s38(2)   Appeal on a point of law is allowed with the consent of all the parties or leave of the Supreme Court, resulting in only limited grounds for jurisdictional review. Some of the States' Arbitration Acts go further in limiting the circumstances upon which leave to appeal may be granted. For example s38(5) in NSW requires there to have been 'manifest error of law on the face of the award' or 'strong evidence that the arbitrator or umpire made an error of law and that the determination of the question may add or may be likely to add substantially to the certainty of commercial law'.

s39   The Court can determine a preliminary point of law

s51   *'An arbitrator or umpire is not liable for negligence in respect of anything done or omitted to be done by the arbitrator . . . but is liable for fraud . . .'* In arbitrations subject to the Act s51 operates to give the arbitrator immunity, but this is limited to actions for negligence.

3.2.2.2 Expedited Arbitration

The Institute of Arbitrators Australia has a standard clause which is recommended for insertion into agreements where the parties wish to resolve disputes by means of expedited arbitration in accordance with the Institutes' Rules. Such streamlined arbitration is still within the terms of the uniform Commercial Arbitration Act. The Institute also publishes Rules and Notes for Expedited Commercial Arbitrations.

If the parties specify in their arbitration agreement that these expedited Rules shall apply to any dispute then the arbitrator is bound to apply the expedited Rules, which modifies the arbitrators power 'to conduct the proceedings as they see fit' under section 14. Hackett (1995) commented that:

the benefit of including the expedited Rules in the arbitration agreement is that it makes it clear to the arbitrator that the parties do not necessarily wish to adopt a Court style procedure.[56]

Process

Expedited arbitration is really a streamlined version of the normal arbitration process and was developed in response to requests from the commercial community. Several of these rules, designed to promote time and cost savings, are detailed below.

Pursuant to Rule 18:

- Pleadings may be limited or omitted altogether,
- Discovery is limited,
- opening and final statements are time limited or omitted altogether,
- pre-hearing submissions are lodged by the parties and are accompanied by sworn witness statements and documents upon which the parties wish to rely; parties have a right to reply and deponents of statements may be cross-examined,
- expert witness numbers are limited,
- experts reports to be relied upon are exchanged at least seven days before the hearing,
- oral evidence is omitted,
- strict time limits are applied to all parts of Rule 18.

and Rule 19:

- judgment may be in terms of the general principles of justice and fairness.

and Rule 21:

- reasons may be omitted from the Award.

Sometimes legal representation may also be prohibited.

According to Tyrill (1992) even when not adopting the Institutes' Rules they might still be useful to the parties by serving:

> *"as a checklist for the parties to develop (perhaps in conjunction with the arbitrator) their own expedited arbitration procedure for the particular dispute"*[57]

3.2.2.3 International Arbitration

International commercial arbitration in Australia is subject to a separate legal regime from that applicable to domestic commercial arbitration. Australia's International Arbitration Act 1974 gave effect to three international laws - the New York Convention, the ICSID or Washington Convention and the UNCITRAL Model Law.[58]

In 1985 the Australian Centre for International Commercial Arbitration (ACICA) was established with the support of the Victorian State Government, the Institute of Arbitrators Australia, the Law Council of Australia, the Australian Bar Association and the Victorian Attorney-General. It is primarily devoted to international arbitration and to acting as an appointing authority for statutory arbitrations. It was recently reported that *'ACICA is a highly regarded organisation which now has co-operating agreements with some 30 centres around the world'.*[59] The Institute of Arbitrators Australia also has a recommended dispute resolution clause for insertion into contracts to provide for international arbitration.

3.2.3   Expert Determination (or Expert Appraisal as it is known in Australia)

Usage rating in Australia : Rating 1

Parties appoint an expert to carry out an independent, inquisitorial style, appraisal of the disputed facts/issues. The experts written opinion/determination may be binding or not depending on the parties agreement when making the appointment.

Several organisations can facilitate the appointment of such Experts.   A sub-committee of the Institute of Arbitrators Australia has recently finalised a set of Rules for Expert Determination which is expected to be published in the near future. A promotional brochure on Expert Determination is also being drafted.[60]   Also, an outline Expert Appraisal agreement is available from ACDC.

There is no immunity from action for actions of negligence for such Experts. In light of Brabazon J's comments (see Section 3.1.3) the function of expressing opinions, as to the likely outcome of a dispute if it were to be arbitrated or litigated, places the Expert at risk of suit for negligence. [61]

Process:

The expert can meet with all or some of the parties, together or separately as he/she chooses; the style of the meetings may be inquisitorial; a picture of the main issues in dispute are outlined by presenting a statement of main points, the main documents, and maybe also brief oral presentations and some evidence so the expert can form an opinion on the dispute; the opinion may be in draft form initially in order to allow the parties some input into its form if it is to become binding.

Tyrill (1992) observed that there has been considerable use of independent expert appraisal in the construction industry in recent years. Further, the independent expert appraisal system has been found to be effective in satisfying management needs for an impartial assessment of liability. It satisfies the need, usually for accountability purposes, of justification for any settlement reached.   In this regard, the system has advantages over mediation, which is an oral system whereby any decision reached is that of the parties. However, expert appraisal is a much more time consuming and tedious process for the independent than mediation in the work required to prepare a written opinion. There is greater rigour involved in written expressions of opinion. Consequently, it can be considerably more expensive than mediation. That additional cost might be justified where there is need for a written third party assessment.[62]

3.2.4   Litigation

<u>Usage rating in Australia</u> : Rating 5

Litigation involves adjudication before a judge in Court, and the resulting judgment is binding. The monetary sum involved in a building dispute determines which tier of the Court system will have jurisdiction over the case.   Larger building disputes generally fall under the jurisdiction of the Supreme Court. Smaller disputes are handled by Magistrates or County Courts or other tribunals. Some disputes which start by going to Court are in turn referred out to Court-annexed Mediation or Court-annexed Arbitration or to a Referee for a Report.

The various court procedures and the alternative procedures available under the umbrella of the court will be further described in the following sub-sections.

3.2.4.1 Building Cases List

Australian Courts do not have any equivalent of the Official Referees Court of England which was set up to exclusively hear construction disputes. A number of State courts have "building lists", which are monitored by a particular judge.[63]   Strictly speaking, this separation of building disputes operates only during the preliminary stages of the litigation and when it comes to trial the case *may* be assigned to *any* judge. In reality though, the building cases usually go to the same small group of judges.

In the New South Wales Supreme Court this specialised list to regulate building disputes is administered by the judges of the Commercial Division.   Rogers J. sets out the guidelines adopted (Practice Note 58) by the court, whose aim is to achieve efficient resolution of building disputes, as follows[64]:

1. In so far as possible, strip the disputes of unnecessary legal technicalities in documentation and presentation.

2. Ensure that the issues between the parties are clearly defined at the earliest possible time.

3. Establish procedures which, with the minimum of delay, will allow for a hearing of those issues.

4. Require all the proposed evidence to be in the form of statements so as to avoid surprise and to cut down on required court time.

5. Send technical issues to a technical expert, designated as a referee, for a report and confine the proceedings in court to the determination of legal and non-technical factual issues" (refer to section 3.2.4.6).

Victoria also has a Building Cases List and, according to Justice David Byrne (1996) [65], the number of cases entered onto it is *'fairly stable at 22 per annum'.*[66]

3.2.4.2 Case Management by Judges

The courts have responded to criticism of being slow and expensive by:

incorporating, within their existing armouries, the methods of so-called Alternative Dispute Resolution and Differential Case Management.   Judges are becoming more active in their management of cases through the interlocutory stages; more alive to the desirability of lifting cases from the ordinary litigation stream into mediation, case appraisal and the like; more willing to design an interlocutory procedure, or hearing format, to suit the particular case ... The Judge's approach, in complex cases especially, is much more of a "hands on" involvement, plainly directed towards having the case resolved at the earliest possible time and without undue expense - whether (and preferably) by agreement, or by adjudication.[67]

For such processes to work the Judge can non longer be non-interventionist as in the traditional adversarial model of adjudication. Pre-trial conferences (also refereed to as mediation conference, conciliation conferences or settlement conferences) which aim to negotiate a settlement or at least narrow the issues in dispute, are widely used now in most State courts.[68]

3.2.4.3 Determination of Separate Issues/Preliminary Questions

According to research carried out over 1989 to 1991, in the States of Victoria and New South Wales, 10 percent of litigated building disputes resulted in an order for summary judgment.[69]
Based on Justice Byrnes statistics for building cases disposed of in the Supreme Court of Victoria, during 1995, almost 20% were ordered for trial of Preliminary Question.[70]

3.2.4.4 Court-annexed Arbitration

Usage rating in Australia : Rating 0.5

In Australia, court-annexed arbitration has been introduced in many States.[71] Much of the legislation is directed towards civil cases rather than commercial cases, where building cases usually fall, however a brief outline of the process will still be included.

The court makes an order to refer the dispute to Arbitration. Such disputes usually do not involve complex legal questions. The arbitrator is selected by the Court; some of the legislation requires the arbitrator to be a barrister or solicitor.

Process:

The arbitrator must first attempt to assist the parties achieve a settlement of their dispute, known as *'an attempt at conciliation'*. Then, subject to any direction from the court, the arbitrator may determine the procedures to be followed. The court retains some control over arbitration allowing the court *'to give such directions for the conduct of the proceedings before the arbitrator as appear best adapted for the just, quick and cheap disposal of the proceedings'.*[72]

In general awards are in writing and take effect as if they were judgments of the court, being subject to review only on the grounds of breach of the rules of natural justice or lack of jurisdiction.

3.2.4.5 Court-annexed Mediation

Usage rating in Australia : Rating 2

Many States now have legislation for referring cases to mediation. For example, building disputes from the Victorian County Court

Building Cases List have been referred to mediation for a period of some 11 years, this can be done with or without the consent of the parties. In Victoria the Supreme Court Building Cases List commenced to refer cases to mediation from about 1992.[73] In New South Wales the legislation was amended in 1994 to allow for court-annexed mediation and neutral evaluation in the District Courts and the Supreme Court.[74]

Court annexed mediation occurs when the Court directs the dispute to Mediation. A mediator acting pursuant to an appointment of the court has statutory immunity as for a judge.

Process:

Building disputes can be referred to mediation with or without the consent of the parties. The mediator is selected by the Court. Orders for referring a case to mediation from a Building Cases List usually occur at the end of the interlocutory stage.[75]

The development of guidelines for the conduct of Court-annexed Mediations are more advanced in some states of Australia than others. The various States need to hold responsibility for the quality, integrity and accountability of consensual dispute resolutions processes used within courts and tribunals, according to the New South Wales Law Reform Commission, which also asserted that:

'a dispute resolution program connected with courts and tribunals must operate in accordance with clear guidelines and adequate resources to ensure the integrity of the process and quality of service'.[76]

3.2.4.6 Reference Out of Court to a Referee

<u>Usage rating in Australia</u> : Rating 2

References out of court permit technical issues, or whole cases, to be determined by a technically qualified referee. The Referee conducts a hearing and writes a report. The Court can adopt, vary, reject, or remit the report in whole or in part. The report is not legally binding until it is adopted by the Court.[77]   The procedure is widely used in New South Wales for the resolution of complex technical cases such as building disputes.   More recently the court Rules have been amended in Victoria to bring the process into line with the New South Wales provisions and thereby to hopefully encourage more use of References Out of Court for building matters. For the purposes of this Monograph the following discussion will focus on New South Wales as it is the leading state in this area.

Referees appointed by the court can be "any person", although the court does have a list of 'approved referees'. The rules also provide for the appointment of two referees, where required. The Court of Appeal has stated :

The power of the Court to appoint a person as a referee is discretionary . . . in many cases referees are appointed by consent and, indeed, the parties often select their own referee. On the other hand, the power exists to appoint a referee without the consent of one party, or for that matter, over the objection of both parties.[78]

It has been held by the Court of Appeal in New South Wales that referees enjoy the same immunity as a judge of the court.[79]   In addition to this common law ruling some other states have legislation providing for referees to have the same immunity as a judge of the court.

Process

The procedure, under a Part 72 reference, commences when the judge makes an order referring the whole of the proceeding or any question or questions arising in the proceeding out to a referee. A question under Part 72 includes:

> "any question or issue arising in any proceedings, whether of fact or law or both and whether raised by pleadings, agreement of the parties or otherwise"

The Referee then holds a hearing. The court has the power to give directions with respect to the conduct of the proceedings under the reference.[80] The referee may conduct the proceedings, subject to any directions given by the court, as he/she thinks fit.[81] Further, the referee is not bound by the rules of evidence; and the referee may inform him/her self as the referee thinks fit. However, this does not mean the referee can ignore the concepts of natural justice.[82]

After the hearing has concluded, the referee is required to prepare a written report for the Court which must contain the referee's opinions with reasons. The referee's report is given to the Court, which then serves it on the parties. The parties have the opportunity to argue in Court for or against the adoption of the report.[83] It is not intended that the hearing pursuant to this rule be a re-trial of the issues already considered by the referee.

The report of a referee has no legal consequence until it has been accepted by the court. The court has the power to deal with the report in a variety of ways.

The court may :

> "on a matter of fact or law or both -

(a)   adopt, vary or reject the report in whole or in part;

(b)   require an explanation by way of report from the referee;

(c)   on any ground, remit for further consideration by the referee the whole or any part of the matter referred for a further report;

(d)   decide any matter on the evidence taken before the referee, with or without additional evidence and shall give judgment or make such order as the Court thinks fit.[84]"

Watts (1994) suggests the benefits of this system are demonstrated by examining two likely situations. Firstly, in the situation where the whole proceeding is referred out, then the process is similar to an arbitration proceeding. However, the court maintains control of the time involved and can influence the process by its power to give directions. Secondly, where only part of a matter is referred out - the judge retains the legal questions for resolution by the court, while the technical issues, are referred to a Referee who is qualified in the appropriate technical area. By only referring out technical issues the judge keeps control over the whole case and in particular the legal issues. The time given to the referee to make his/her report is also set by the judge, so the duration of the proceedings is also under the judge's control.[85]

Giles J.(1993) commented:

"disputants have the dual advantage, never previously available to them, of having legal issues resolved by judges, accompanied by a court considering a report upon technical issues by a court appointed referee having special expertise in the area of the technical dispute"[86]

3.2.4.7 Magistrates Court - Small Claims

Sometimes small value building disputes are handled as 'small claims'. This occurs in South Australia where the Magistrates' Court has a small claims jurisdiction within its civil jurisdiction, with a jurisdictional limit in small claims at $5000. This process:

Takes the form of an inquiry by the court into the matters in dispute between the parties rather than an adversarial contest between them . . . the court is not bound by the Rules of Evidence and the court must act in accordance with equity, good conscience and the substantial merits of the case without regard to technicalities and legal forms.[87]

Domestic building disputes have now been added to a new class of claims which are dealt with as small claims in South Australia. Disputes above $5000 may also be treated as a small claim provided neither party objects. There also exists the power for experts to sit with the Magistrate to assist with technical matters.[88]

3.2.4.8 Domestic Building Tribunal

Domestic building disputes are handled by special tribunals in Victoria and Queensland. This Monograph will only consider the Victorian tribunal.

The Victoria Domestic Building Contracts and Tribunal Act came into effect in April 1996. Not long after the Tribunal commenced it had 12 sessional members, 52 mediators and in excess of 200 matters on its list.[89] An officer of the Tribunal has complete protection for any negligent actions in the conduct of their duty.

The jurisdiction of the Victorian tribunal includes *all* domestic building disputes, (including multiple-unit developments) disputes concerning insurance claims, references to the Tribunal under the House Guarantee Act and a miscellaneous category. Domestic building disputes can include disputes between owners, builders, architects, sub-contractor, insurer and Building Practitioners under the Building Act 1993. The type of dispute includes claims in contract, negligence, nuisance and trespass. Tribunal can send small claims for mediation to attempt settlement prior to the tribunal hearing.

The Victorian tribunal is regulated by practice directions and not by formal rules, some of which are outlined here:

- s77 states that the tribunal is not bound by the rules of evidence and may inform itself on any matter in any manner it thinks fit.
- s77(1)(D) *'must conduct its proceedings with as little formality and technicality and with as   much speed as the requirements of this act and a proper consideration of the matters before it permit'*
- s94 allows the Tribunal to retain experts to advise the tribunal and also to refer matters to an expert for investigation
- s97 gives the tribunal discretion to dismiss an application if the tribunal forms the view that the proceedings *'would be more appropriately dealt with by a court after having regard to such   matters as the size and legal complexity of the claim'*

In addition:

- there are no pleadings (unless circumstances are exceptional)
- issues in dispute are listed in a type of Scott schedule document, refereed to as a 'dispute schedule'
- witness statements and documents are filed
- outlines of legal submissions and chronologies are filed
- onsite meeting with an independent expert retained by the tribunal may also be required

- proceeding may be conducted by conference using telephones, video links or any other system of telecommunications

Process

A streaming process where small claims are separated from other claims, based on the amount in dispute and likely duration of the hearing Reference of a small claim to mediation may then occur, which can last for a maximum of 2 hours and if it is unsuccessful then the hearing takes place immediately.

Cases which involve other claims (ie those not considered to be small claims) can also be sent for immediate mediation. If they are not resolved at the mediation they proceed to a case management conference , where further attempts to reach a settlement are also included. If this type of case remains unsettled it proceeds to a hearing.

Where mediations have failed to produce a settlement the Mediators have to report to the tribunal what issues the mediator and parties agree are in dispute.

Criticism of the *Victorian Domestic Building Contracts and Tribunal Act 1995* has come from the President of the Institute of Arbitrator because it makes the Domestic Building Tribunal chiefly responsible for the resolution of all domestic building disputes rendering void any arbitration clauses in domestic building contracts. The President said :

'It is to be regretted that in a State supposedly dedicated to private enterprise, we see a private system of dispute resolution jeopardised by the rise of yet another government tribunal. It is also another unfortunate restriction upon the ability of parties to choose their own form of dispute resolution in their contracts in an era when the general thrust of legislation is to remove restrictions on freedom to contract'. [90]

3.2.5   Negotiation

Refer to section 3.1.4

3.2.6   Concilio-Arbitration

<u>Usage rating in Australia</u> : Rating 0

The concilio-arbitrator is provided with details about each party's case at the commencement of the process and from this forms an opinion on the likely outcome if the dispute were to be litigated. The parties are informed of this opinion and have the opportunity to revise their case. The concilio-arbitrator then writes a final opinion which is binding after a certain period of time. A cost penalty occurs if the final award is not accepted and litigation is undertaken unsuccessfully.

Preliminary research did not reveal any building disputes being resolved by this indicating that it is rarely used in Australia.

3.2.7   Med-Arb

<u>Usage rating in Australia</u> : Rating 0.5

Section 27 of *the Commercial Arbitration Act* creates what amounts to a quasi med-arb system in Australia.[91] Pursuant to s27 the parties may elect to have a conference to attempt settlement, adopting a form of ADR. If no settlement occurs the arbitration proceeding recommences. A different person may conduct the settlement conference proceeding to avoid problems associated with impartiality if it were to be conducted by the arbitrator.

Astor and Chinkin (1992) explain the potential difficulties which s27 conferences may cause:

From their inception the uniform Commercial Arbitration Acts have allowed an arbitrator to seek the settlement of a dispute submitted to arbitration otherwise than by arbitration . . . If no settlement is reached the arbitrator is not precluded from continuing with the arbitration, although the role played in any conference will have differed significantly from the arbitral role. (27(2)(b) uniform Commercial Arbitration Acts). The potential for a breach in the rules of natural justice lies in the markedly different methods for the exchange of information in mediation and arbitration.[92]

The Working Group on the Working of Uniform Commercial Arbitration Legislation in Australia considered these arguments with respect to s27. It recommended amending the section to allow parties specifically to *contract into* participating in an ADR process such as mediation or conciliation in the course of an arbitration, rather than being required to *contract out* as was the position under the original legislation. This proposal has been progressively adopted in most States.[93]

In essence the revision allows the parties to opt into a s27 conference, the parties can then give the arbitrator written permission to proceed in the s27 conference without the constraints of being bound by natural justice.

Arguing against the need for changing the legislation was Bradbrook (1990), who suggested that the concern regarding the earlier version of s27 conference was unwarranted and came about because of confusion of mediation and conciliation:

While the terms 'conciliation' and 'mediation' are not defined in the Commercial Arbitration Act it is as commonly understood that mediators merely assist the parties by facilitating the process and will not express opinions. It is only when a s27 conference takes the form of a conciliation, where the disputants are offered opinions by the conciliator that the rules of natural justice are put in jeopardy by a subsequent resuming of the role of arbitrator.[94]

4.   THE ZEITGEIST

4.1   Current focus, the future and some concerns

There exists continued recognition of the benefits of preventing disputes. In 1990 *No-Dispute* acknowledged this with a recommendation that further education, by way of post-graduate courses in Contract Administration, be introduced in an effort to reduce contractual disputation.[95] Unfortunately to date there has been little effort made by Universities to act upon the No Dispute recommendation for advanced contract administration training.[96] Although, universities have broadened their undergraduate and some post-graduate courses, to embrace ADR and conflict management.

Professional bodies active in the building and construction area are also directing their efforts more and more towards preventative measures. An example of this an initiative, being trialed in Victoria, by the Australian Institute of Building, where a series of Forums are being conducted on contract administration, ethics, and dispute avoidance. The aim of these Forums is to build on the university knowledge base of recently qualified graduates in the hope of increasing professionalism and thereby reducing the incidence of contractual disputation.[97]

It seems that Australia is fortunate enough to have a wide range of resolution options available - offering experienced arbitration, a variety of established ADR techniques as well as conventional litigation plus a range of innovative alternatives under the umbrella of the courts. There is a procedure available for virtually any dispute resolution requirement the parties may have. As a result our industry is shifting its focus to conflict management techniques as well as addressing prevention through further education of the professionals practising in the industry.

Some concerns have been expressed in literature as to the long term effect of ADR and resolution outside the courts. ADR being private means decisions are not reported. So precedents and common law do not develop. This means the law may not develop at a rate to keep up with development in the commercial building environment. Also, those active in the building industry can not gauge the likely consequence of their actions nor be guided in their conduct, by case law which may not exist to reflect current interpretation of legal matters. These issues point to concern as to what will be the long term impact of ADR upon the common law adversarial system.

4.2   Active Organisations

The organisations listed below are non-profit or Government-sponsored organisations which are currently active in Australia. The organisations are listed in alphabetical order. Journals published by these organisations are listed in bold italic.

Australian Centre for International Commercial Arbitration (ACICA)
Internet: http://www.acica.com.au   ph(613) 96296799
The centre was established in 1985 and is primarily devoted to international arbitration. Services include appointing arbitrators, provision of facilities and administrative support. It also offers support for international mediation and conciliation, plus administration facilities for domestic commercial arbitration

Australian Commercial Disputes Centre (ACDC)
Internet: http://www.austii.edu.au/au/other/acdc/
        ph (61 2) 92671000

An independent, non-profit company established by the NSW State Government in 1986 to introduce procedures to resolve disputes. Governed by a Board of Directors. Services include:

- dispute resolution services: mediation; expert determination; expert determination; ACDC arbitration; boards of dispute avoidance;
- dispute resolution systems - design, management and evaluation of systems to manage conflict;
- consulting services including facilitating partnership agreements;
- training and accreditation; Over 350 panel members covering mediation, experts, assessors and arbitrators
- dispute resolution clauses;

According to Duncombe and Heap[98] in its first five years of operation 'the majority of the mediations arranged by ACDC were in relation to construction industry disputes'.

Australian Dispute Resolution Association Inc (ADRA)
Non-profit organisation formed in 1987, incorporated 1989. Board of Management elected annually. Members principally operate in areas of Family, community and environmental disputes. Services include quarterly newsletter; seminars and meetings; conferences.

Building Dispute Practitioners' Society (Victoria)
Fax (61 3) 59966188
An informal multidisciplinary group of individuals who practice or are interested in the building disputation. The society runs discussion nights, produces a newsletter, and hold regular meetings. There does not appear to be anything similar in other States, at this point in time.

Institute of Arbitrators Australia (IArbA)
Internet: http://www.instarb.com.au   ph (61 3) 96296799
A national non-profit organisation formed in 1975 to provide alternative resolution services for a wide range of disputes,

including construction. There are Chapters in all States. Services include: nomination and appointment of arbitrators; list of arbitrators; list of mediators; publication of Rules for mediation, arbitration and expert determination; publication of quarterly journal *The Arbitrator* ; continuing professional development and training courses in arbitration and mediation; examinations; pupilage program; model dispute resolution contract clauses; library; hearing and conference rooms; publications. Arbitrator training will be conducted through universities commencing in 1998. Expert Determination programmes are currently being developed in collaboration with LEADR and Bond University.

LEADR (Lawyers Engaged in Alternative Dispute Resolution)
Email: leadr@sl.asn.au   ph 1800 651650
Non-profit company limited by guarantee formed in 1989 to promote and facilitate ADR. Elected Board of Directors. Chapters in all States. Membership open to non-lawyers also. Worldwide membership. Services include negotiation, mediation and conciliation training; ADR seminars; annual International ADR conference; panels of mediators; model dispute resolution contract clauses; ADR publications and information including *LEADR Brief.*

National Dispute Centre
ph (61 3) 92231044
Established to develop and promote dispute resolution methods both in conjunction with the courts and outside the stream of conventional litigation. services include arbitration; mediation; court reference; early neutral evaluation; expert appraisal; expert determination.

The organisations listed below are Centres currently operating at Australian tertiary Universities, in alphabetical order:

Centre for Conflict Resolution, Macquarie University, Sydney
ph (61 2) 98509963
The Centre has an advisory board consisting of members from the business, legal and mediation professions. Services include newsletter; seminars; conferences; research and also offers a Diploma in Conflict Resolution. Believed to have been renamed as Centre for Conflict Management

Centre for Dispute Resolution, University of Technology, Sydney
ph (91 2) 95143753
The Centre is jointly sponsored by the Faculties of Law and Legal Practice and Business.   It is governed by a management committee and has an advisory board consisting of members of the business, legal and mediation professions. Services include newsletters; publishing the *Commercial Dispute Resolution Journal*; seminars; conferences; research and development; establishing standards; training and consultancy services in dispute resolution-mediation and negotiation; the Centre also offers a Master of Dispute Resolution.

Dispute Resolution Centre, Bond University, Gold Coast
Internet:
http://www.bond.edu.au/Bond/Schools/Law/centres/DRC/
ph(61 7) 55952039
The Centre was established in 1990. Services include training and consultancy services in dispute resolution- mediation, early neutral evaluation and negotiation; and the Centre also offers a Master of Dispute Resolution.

Proposed Course in Construction Law, University of Melbourne
ph (61 3) 92444000
The Law Faculty in conjunction with the Faculties of
Engineering, and Architecture, Building and Planning are hoping
to introduce a Graduate Diploma in Building and Engineering
Law, in 1999 and possibly develop it as a Masters in the future.

4.3   Additional Reference Texts and Journals

Additional journals to those listed above are:

- *Australian Construction Law Newsletter*, Construction
  Publication, PO Box 298, Avalon, NSW, 2107, Telephone 61
  2 9974 5667, fax 61 2 9974 4405.
- *Australian Construction Law Reporter* (ACLR),
  Construction & Housing Service Australia on behalf of
  Master Builders Association, Construction House, 217
  Northbourne Ave, ACT, 2601

Principal text books and loose leaf services include the following:

- Duncombe and Heap in association with LEADR, 1995,
  Alternative Dispute Resolution, Law Book Company
  Information Services, looseleaf service.
- Astor, H. and Chinkin, C. 1992, *Dispute Resolution in
  Australia*, Butterworths.
- Dorter, J. and Sharkey, J. 1980, *Building & Construction
  Contracts in Australia*, The Law Book Company, looseleaf
  service.
- Jones, D. 1996, *Building and Construction Claims and
  Disputes*, Construction Publications Pty Ltd, Sydney.
- Fitch, R. 1989, *Commercial Arbitration in the Australian
  Construction Industry*, The Federation Press, Annandale,
  NSW.
- Jacobs, M.S. 1990, *Commercial Arbitration Law & Practice*,
  The Law Book Company, looseleaf service.

4.4   Acknowledgements

The author makes the following acknowledgements: Ms Phillipa Ryan of Deacons Graham & James for providing access to their library; Ms Leigh Cunningham of The Institute of Arbitrators Australia for access to their library and other statistics; Mr Roger Quick for providing a copy of his book chapter; and especially the contributors to the Monograph text: Mr Leon Gardiner, Ms Paula Gerber, and Professor Jon Robinson.

1   Sections 1.1 and 1.1.1 were kindly contributed by Professor Jon
         Robinson of the Faculty of Architecture, Building and Planning at
         the University of Melbourne, Parkville, Victoria, 3052, Australia.
2   McLennan, W. (1997), National Income Expenditure and
         Product, Australian Bureau of Statistics, Canberra, p. 5206.0.
3   McLennan, W. (1997), 1997 Yearbook Australia, Australian
         Bureau of Statistics, Canberra, p. 1301.0.
4   McLennan, W. (1997); Building Activity Australia, Australian
         Bureau of Statistics, Canberra, p. 8752.0.
5   McLennan, W. (1997); Engineering Construction Activity
         Australia, Australian Bureau of Statistics, Canberra, p. 8762.0.
6   Sections 1.2.1 and 1.2.2 were kindly contributed by Paula
         Gerber, a Partner in the Construction Group at Maddock Lonie &
         Chisholm, Lawyers, 140 William Street, Melbourne, 3000,
         Australia.
7   National issues include such things as: taxation; defence; banking;
         naturalisation; and marriage and divorce, etc.
8   for example, the Family Court and the Administrative Appeals
         tribunal.
9   for example, in Victoria these include the following:
         Building Act 1993;
         Construction Industry Long Service Leave Act 1983
         Domestic Building Contract and Tribunal Act 1995
         Project Development and Construction Management Act 1994
         Architects Act 1991
         Local Government Act 1989.
10   for example, in Victoria, builders are regulated by the Building
         Practitioners Board and Building Control Commission. Similar
         bodies exist to regulate architects, engineers, electricians etc.
11   In Victoria and South Australia the legislation provides that a
         building permit may also be issued by private building surveyors.
12   The Hon. Mr Justice Fitzgerald, Companions on the Journey,
         1996, The Arbitrator, Nov 1996, pp.151-180.
13   Astor, H. & Chinkin, C. (1992) *Dispute Resolution in Australia*,
         Butterworths, p. 70.

14   For example, the Law Society in New South Wales has a Dispute
        Resolution Committee and in Victoria it offers accreditation for
        solicitors; and in Queensland the Bar Association has a dispute
        resolution scheme.

15   LEADR has full members from many Asian countries, South
        Africa, New Zealand and the United Kingdom as well as
        Australia. They have recently been invited to China and also
        asked to set up a branch in Singapore. Telephone conversation
        with Sue Thaler, Executive Officer, LEADR (see Section 4.2).

16   Address given by Justice Ronald Sackville, 1996, Australian
        Dispute Resolution Journal, May 1996, pp. 145 to 156.

17   A Report by NPWC/NBCC joint working party, May 1990, *No
        Dispute: Strategies for Improvement in the Australian Building
        and Construction Industry,* National Public Works Council,
        Dickson, ACT, Australia, p. 1.

18   *No Dispute,* p. 176.

19   Duncombe & Heap in association with LEADR, (1995), *Dispute
        Resolution in Australia,* looseleaf service, Law Book Company
        Information Services.

20   The codes of practice mentioned above included:
        a) In 1992 the NSW Government introduced a 'Code of Practice
        for the Construction Industry' to promote reform of the industry.
        The Code requires avoidance or minimisation of disputes and the
        adoption of non-adversarial methods of resolution disputes in the
        first instance. Non-compliance with the Code may result in
        reduced tendering opportunities or even being removed from
        tendering lists.
        b) In 1991 the Commonwealth government established the
        Construction Industry Development Agency (CIDA) which
        introduced the 'Australian Construction Industry Pre-
        Qualification Criteria' to establish the suitability of participants
        for projects. The criteria considers the claims history of the firms
        and their willingness to enter partnering processes and embrace
        avoidance of adversarial confrontation.
        c) In 1994 the National Master Builders Association
        Incorporation (MBA) embraced the principals of both the NSW

Code and the CIDA criteria in its own Code called the 'Master Builders Australia National Code of Practice'.

21   Riekert, J. 1990, *Alternative Dispute Resolution in Australian Commercial Disputes: Quo Vadis*, Australian Dispute Resolution Journal, pp. 31-43.

22   A publication, by the author of this monograph, containing further analysis of the results should be available in 1998.

23   Astor & Chinkin, p.147.

24   No Dispute, p. 180.

25   Tyrill, J. 1992, *Construction Industry Dispute Resolution - a brief overview*, Australian Dispute Resolution Journal, August 1992, pp. 168-183, at p. 81.

26   No Dispute, p. 181.

27   Section 2.1.4 was kindly contributed by Leon Gardiner a Director at Davis Langdon Australia, Consultants to Building & Construction, 1st Floor, 79-81 Franklin Street, Melbourne, 3000, Australia.

28   Duncombe & Heap, pp. 3-3102.

29   Jones, D. 1996, *Where are Standard Forms Going?*, Australian Construction Law Newsletter, Issue 47, pp. 15-34.

30   CII Partnering Taskforce Interim Report, 1996, *Partnering: Meeting the Challenges of the Future*, Aug 1996.

31   Jones, D. 1996, *Building and Construction Claims and Disputes*, Construction Publications P/L, Sydney, p.146.

32   see Section 4.2

33   Street, Sir L. 1989, *Senior Executive Appraisal*, Australian Construction Law Newsletter, No 6, July-Aug 1989, pp. 9-11.

34   Collins, R.1989, *Alternative Dispute Resolution- Choosing the Best Settlement Option*, Australian Construction Law Newsletter, No 8, 1989, pp. 17-27.

35   Tyrill, p 180.

36   Street, Sir L. 1989, as quoted in Astor & Chinkin, p.143.

37   No Dispute, p. 111.

38   see Section 4.2

39   Brabazon,C. 1997, *Dispute Resolvers Liability in Negligence*, The Arbitrator, Vol. 15, No 4, pp. 227-237.

40   see Section 4.2

41   Street, Sir L. 1991, *The Courts and Mediation - A Warning*, 2ADRJ203.

42   Brazabon, p. 227.

43   for example AS 2124, clause 47.2 and JCC clause 13.02.

44   Astor & Chinkin, p. 89.

45   Astor & Chinkin, p. 80.

46   Roberts, 1983, *Mediation in Family Disputes*, Modern Law Review, Vol 46, No5, pp. 537-57.

47   Clause 46 of AS2124, 1992, provides for contract adjudication through the appointment of a separate senior person who has no involvement with the day to day running of the Contract as Superintendent to handle formal decisions which have failed resolution.

48   Uniform Commercial Arbitration Acts of the various states were adopted in the following years: 1984- New South Wales and Victoria; 1985- Western Australia and Northern Territory; 1986- south Australia, Tasmania and the Australian Capital Territory; and lastly Queensland in 1990.

49   Astor & Chinkin, p. 117.

50   Astor & Chinkin, pp. 122 - 123.

51   There were 92 appointments made between July 1990 and October 1995 and 21 appointments made between October 1995 to December 1996. Private correspondence from Leigh Cunningham, Executive Officer, Institute of Arbitrators, 1997.

52   Around 300 appointments were made in 1989 according to Monash Study, see Riekert, p. 33.

53   Astor & Chinkin, p. 118.

54   South Australian Superannuation Investment Trust v Leighton Contractors Pty Ltd (1990) 55 SASR 327.

55   Imperial Leatherware Co Pty Ltd v Marci and Marcellimo Pty Ltd (1991) 22 NSWLR 653.

56   Hackett, C. *Expedited rules and mediation in the context of the arbitration process* , The Arbitrator, Nov 1995, Vol. 14, No 3, pp. 173-185.

57   Tyrill, p. 181.

58   Astor & Chinkin (1992) detailed the background of international
           arbitration at page 156:
           The Arbitration (Foreign Awards and Agreements) Act 1974
           (Cth) implemented the 1958 New York Convention on the
           Recognition and Enforcement of Foreign Arbitral Awards. The
           Convention is also incorporated into the uniform Commercial
           Arbitration Acts by sections 56-59 and Schedule 2 to ensure
           uniformity between the Commonwealth and the States in
           recognition and enforcement of foreign arbitral awards. In
           1989 the federal government incorporated the UNCITRAL
           Model Arbitration Law into Australia . . . under this Act the
           Model Law will be applicable to all international commercial
           arbitrations conducted within Australia unless the parties have
           deliberately contracted out in writing.

59   Pryles, M. 1996, *News from ACICA*, The Arbitrator, Nov 1996,
           pp. 193.

60   Easton, G. 1997, *Report from Chairman ADR Committee*,
           National Newsletter of the Australian Institute of Arbitrators,
           April, p. 2.

61   Brazabon, p. 181.

62   Tyrill, pp. 177-178.

63   notably New South Wales, Victoria and Queensland.

64   Cole J. 1990, *Advantages of referring building disputes to the*
           courts, Master Builders Centennial Conference,6/6/1990.

65   Byrne J. 1996, *Mr Justice Byrne's note consequent on his*
           presentation to the Building Dispute Practitioners Society, 21
           February 1996.

66   During 1995, 19 cases were disposed of and 12 of those were
           settled. Of the 19 cases, six were ordered for Special Reference
           (refer to section 3.2.4.6); four for Trial of a Preliminary Question
           (refer to section 3.2.4.3); and 13 were referred to Court-annexed
           Mediation (refer to section 3.2.4.5) where over half the
           mediations resulted in settlements.

67   The Hon. Mr Justice de Jersey, 1996, *Reform of the Arbitration*
           Process: Interlocutory and Hearing Steps: Problems and
           Solutions, The Arbitrator, November 1996, pp. 181-198.

68   for example, in Victoria pre-trial conferences were introduced in the County Courts and Supreme Court in 1984, and in 1986 a pre-trial conference unit was established 'as a composite body to pre-trial matters as directed by the chief Justice and Chief Judge'. In New South Wales all courts have pre-trial procedures designed to promote early identification of the real dispute between the parties. See Astor & Chinkin, pp. 159-160.

69   Watts, V. 1994, *A survey of litigated building disputes*, Master of Building thesis, Department of Architecture and Building, University of Melbourne.

70   Byrne J. 1996.

71   for example, this was as early as 1983 in New South Wales District and Local Courts, and 1989 for the Supreme Court; and since 1989 in the Victorian Magistrates (s102 Victorian Magistrates' Court Act 1989).

72   through section 27(2)(b).

73   Golvan, G. 1996, *The Use of Mediation in Commercial and Construction Disputes*, Australian Dispute Resolution Journal, August 1996, pp. 188-196.

74   Construction Headlines a news sheet from the Allens Arthur Robinson Group, as republished in Australian Construction Law Newsletter, Issue 45, pg 25-27.

75   Cole J. 1990.

76   NSW Law Reform Commission Training and Accreditation of Mediators, 1991, *Report No 67*, p. 6.2.

77   for example, Order 50 Rules of Supreme Court - Victoria.

78   Super Pty Ltd v SJP Formwork (Aust) P/L, (1992) 29 NSWLR 549.

79   Najjar v Haines (1991) 25 NSWLR 224.

80   Part 72 rule 8 (1).

81   Part 72, r8 (2) (a).

82   Hooper Bailie Associated v Natcom Group, unreported, NSW SC, Cole J. , 17/11/1989.

83   Supreme Court Rules, Part 72, r13.

84   Pursuant to Part 72, r13 (1).

85 Watts, V. 1994, *References Out of Court: An examination of recent building cases referred out from the Supreme Court of New South Wales*, in Construction Conflict : Management and Resolution ed by P.Fenn, Counseil International Du Batiment (CIB) The International Council for Building Research, Studies and Documentation, The Netherlands, 1994, pp 188-199 and Australian Construction Law Reporter, Vol 15, No 3, 1996, pp. 83-89.

86 Hooper Bailie Associated v Natcom Group, unreported, NSWSC, Giles J., 9/11/1993.

87 section 38, *South Australian Magistrates Court* Act 1991.

88 Cannon, A. 1996, *Alternative Adjudication - Adjudication by a mixture of mediation and inquiry - a small claims experience*, Law Reform Commission Conference, QUT, Session 2, 8/3/96.

89 Davey, J. 1996, *Of Shoes and Ships and the domestic building tribunal*, The Arbitrator, Nov 1996, pp. 137-146.

90 Fitzgerald, p. 161.

91 Bradbrook, A. 1990, *s27 of the Uniform Commercial Arbitration Act - a new proposal for reform*, The Arbitrator, Vol 9, No 3, pp. 107-21.

92 Astor & Chinkin, p. 145.

93 First in New South Wales as the 1990 amendment to the Commercial Arbitration Act. This revised version of s27 has been adopted in NSW, ACT and NT.

94 Bradbrook, p. 111.

95 In commenting that many of the problems dealt with in No Dispute 'are the product of, or exacerbated by, imperfect understanding of the management, legal and commercial principles and practices which underpin contracts' the Report went on to state that 'there is a need for industry-wide continuing education across the private and public sectors to supplement the inadequate level of Contact Administration training from existing sources', at pg 219 - see footnote 17.

96 Training in contract administration is still learnt largely 'on-the-job'; through irregular seminars or conferences.

97   The Australian Institute of Building, Construction House,217
        Northbourne Ave., Turner, ACT, 2601.
98   Duncombe & Heap, p. 3-3106.

CANADA

Richard Beifuss
Consultant, Ontario, Canada

1.   BACKGROUND

1.1   Economic

The Canadian Construction Industry 1 in 1996 was an 88 billion dollar industry.

1.1.1   Trends

The construction industry continues to be used by governments (federal, provincial and municipal) as a means to create employment. At this time it has become a very useful tool to deal with youth unemployment and replace ageing infrastructure. The forecast for Canada is for continued steady two percent growth for 1998. Unemployment or underemployment of a highly educated young work force remains a serious concern.

1.2   Legal

1.2.1   Construction Law

Canadian contracting is carried out under two legal systems, the Code Civil in the Province of Quebec and Common Law in the remaining nine Provinces and two Territories. There is no specific court procedure established to deal with commercial claims. All commercial claims are dealt with by the same judiciary that is asked to deal with civil and criminal cases. There are, however, no jury trials for commercial cases as is sometimes the case in the US All lawyers are permitted to act as either barrister or solicitor and in the common law jurisdictions advertise themselves under both headings.

1.2.2   Construction Contracts

Canada utilises primarily the design, bid, construct model for the implementation of built-works such as marine, transportation and building related projects. While schedulers and estimators are part of the project team no bill of quantities are employed. The role of the estimator (Quantity Surveyors) is to provide an estimate primarily at the tender stage which the contractor utilises for his bid.

While design-build contracts have been employed with greater frequency in recent years they represent a very small proportion of the construction program. The logic of utilising the design-bid-construct method of construction is that it allows for a much wider participation in the construction program than would be possible under design-build.

2.   CONFLICT MANAGEMENT

Canada has two major ADR Centres the Dispute Resolution Centre in Vancouver, British Columbia and the National and International Arbitration Centre in Quebec (City), Quebec. The Arbitration and Mediation Institute is the largest Alternate Dispute Organisation in Canada.

The Canadian Construction Association (CCA) has been at the forefront of promoting non court non litigious dispute resolution procedures. The CCA has encouraged the contracting community, governments (provincial, federal and municipal) as well as other private sector purchasers of construction services to include stepped dispute resolution clauses in their contract documents. The CCA did finally in 1994 produce contract documents, which can be used by any party, that include stepped ADR clauses. Various levels of

Government in Canada have over the years experimented with ADR procedures. To date no master contract documents have been produced that include comprehensive ADR clauses as promoted by the CCA. Mediation and arbitration procedures are currently being used more and more to avoid the more costly and lengthy court process to resolve disputes. Canada became a signatory to the 1958 New York Convention on May 7, 1985. and adopted the United Nations Foreign Arbitral Awards Convention Act RSC 1985, Chapter 16 Second Supplement effective August 10, 1986. Arbitration Awards are not published and the privacy of participants in the process is respected.

There are at this time experiments ongoing with court annexed mediation in various provincial jurisdictions across Canada as well as individual experiments with commercial based judiciary procedures. By and large the jury is still out insofar as client satisfaction and success with mediation (current success rate is fifty percent which is considered successful (July 1997) by the case file master in Ontario). One thing to note is that the court annexed mediation processes stress knowledge of the legal court process and liability insurance for the practitioners. Quite the message for a process where the practitioner ostensibly only facilitates negotiation between the parties. Insofar as the commercial court goes, it is possible, with stringent judicial control, to deal with a commercial dispute in sixteen months.

2.1   Non-binding

2.1.1   Dispute Review Boards

Dispute Review Boards are used primarily for large construction contracts. These are most efficacious for large projects and have been applied on the Boston Transit way reconstruction. The purported success rate where used is in excess of 98%. Essentially under this model three independent experts are jointly engaged by the contracting parties at the signing of the contract. These individuals are paid a retainer and are expected to keep up to date on the project's progress. If a dispute occurs this is given to the "Board" who review the issues and render a decision. The parties have the option of acceptance or rejection. However, as earlier noted, the parties almost always accept the recommendations. The usual next step is binding arbitration for those cases where the parties are unwilling to accept the recommendation. In the interim however, the parties proceed on the basis of the recommendation.

2.1.2   Dispute Review Advisors (DRA)

Dispute Review advisors have been used in Canada but on an ad hoc basis. There is no formal acceptance of this role within the construction industry. There are Claim Consultants in Canada but they do not exercise this role. The DRA function, would be a welcome tool in our attempt to manage conflict in the construction industry.

2.1.3   Negotiation

Negotiation both interest based and rights based is taught and practised in Canada. Generally speaking, most front line negotiators (e.g. Project Managers) pick up their negotiation skills on the job without benefit of formal training in this area.

2.1.4   Quality Matters

Delivery of quality products remains an ongoing concern in the construction industry. Large owners are responding differently to this problem and various techniques are adopted and promoted from time to time. The Department of Public Works developed a Quality

Control Manual which was to be used by its Design and Construction Branch on all its contracts (both consultant and construction). It has not been implemented to date even though at the time it was considered an extremely thorough practical treatise in this area. It incorporated Total Quality Management (TQM) , Quality Assurance (QA) , Value Engineering (VE) and Value for Money (VDM) concepts in its project delivery concepts. ISO 9000 is being applied on an ad hoc basis but to date there are no national standards that formally adopt this standard. This is also the case for TQM, QA, VE and VFM. Co-ordinated Project Information (through the use of various computer software programs) has assisted project authorities greatly in the delivery of timely quality built works products. The bid depository system is used primarily for the procurement of 99% of construction projects. The volume across Canada is in the hundreds of thousands of contracts annually valued in the billions of dollars.

2.1.5 Partnering

One of the newer claim prevention innovations is partnering. In the design bid construct model this process occurs after the contractor's bid has been accepted. The undertaking to participate in such a partnering session is part of the contract documents and the cost for same is included in the bid price.

A facilitator is engaged to conduct the partnering session. The partnering session can last up to a week and all the project participants (owner, designer, contractor and sub-trades) are present. The facilitator, through the review of the contract documents, works to obtain commitment to the delivery of the common goal. a quality project. At the end of the session a Charter is drafted and signed by all the participants which sets out the goals, communication links, and dispute procedures.

It is crucial that the Charter is:

1)   given prominence on the job site by being prominently posted; and

2)   is revisited on a regular basis to keep the commitment achieved at the partnering session uppermost in the minds of the participants.

The benefits of the process have been twofold:

1)   very few claims have been filed and

2)   insurers have reduced their rates for partnered projects.

2.2   Binding

2.2.1   Partnering

The Partnering concept has not been used in binding the construction partners to anything more than agreeing to follow an agreed communication concept. This process has nevertheless turned out to be a powerful tool in avoiding contracting problems.

3.   DISPUTE RESOLUTION

The primary system for dispute resolution in Canada are still the courts. This venue is as in the US prohibitively lengthy and expensive. As noted earlier the courts are encouraging and supporting the use of ADR techniques primarily arbitration and mediation.

The various ADR techniques adopted/utilised for the prevention and resolution of construction and consultant claims in Canada are:

1) Arbitration
    a) Binding
    b) Non-binding

2) Mediation
3) Dispute Review Boards
4) Adjudication
5) Neutral Fact Finding
6) Referee
7) Contract Disputes Advisory Board
8) Partnering
9) Conciliation
10) Executive Tribunal or Mini Trial

3.1   Non-binding

3.1.1 Conciliation

Conciliation is used primarily in the labour negotiation process and usually precedes a strike. It is not normally used in the construction industry. It is suggested that retired judges have a tendency to practice conciliation when they are engaged as mediators.

3.1.2 Executive Tribunal

There are many instances where this process has been utilised with considerable success. It remains, however, an ad hoc process used at the initiation of the parties usually prior to initiating court proceedings.

3.1.3   Mediation

As with Arbitration a number of models are being implemented. The two extremes are neutral facilitated negotiation (both face to face and long distance) with no lawyer participation and lawyer controlled. In the latter the mediator must address his questions to counsel rather than the party. This is essentially the last ADR model where control of the outcome rests with the goodwill of the parties. Mediation is being used with far greater frequency in Canadian Construction related disputes.

3.1.4   Contract Disputes Advisory Board

This model was adopted by the Department of Public Works in 1987 at the direction of the then Minister, Stuart McInnes. The author was involved from the outset as one of the designers and until 1993 as the Director Claims Administration charged with managing the process. This process was originally designed to provide advice to the Minister so that he could decide whether or not to offer an amount in settlement of a dispute. The process was designed as a Board Room Arbitration (non-binding) at which both sides would explain their position to a neutral Chairperson utilising written and oral representation. The process was designed to run three days. The first day was used by the claimant, the second by the defendant and the third day was set aside for closing statements and to allow the Board Members to review the issues. To ensure that the issues were well articulated both parties were allowed to nominate an individual whose role was to advocate their respective client's position. The parties participated jointly in the selection of the Chairperson. The Chairperson was given two weeks from the close of the Hearing to provide his report to the Minister in which he set out his recommendations. This process dealt with claims up to twenty million dollars at a per case cost of fifteen thousand dollars. A resolution rate up to ninety-five percent was achieved during the writer's mandate.

3.1.6   Non-binding Arbitration

This model is dependent on the claim and rebuttal being reduced to writing and the Arbitrator being asked to render a recommendation following oral representation by the parties. This type of process is normally funded by a large organisation seeking an independent opinion to help it resolve claims and thereby trying to encourage parties to do business with it. The safety valve in this case is that the owner if it is a public concern can verify that the recommended settlement is within the parameters of the contract. Settlement rates for some organisations utilising this methodology are in excess of 90%.

3.1.7   Referee

The organisation that adopted this model in Canada is British Columbia Hydro, a very large Utility on Canada's Pacific Coast. Under this model B.C. Hydro maintains a list of independent experts from which it can engage a referee whenever it has a dispute with a private contractor.

The referee is given the following correspondence:

1)   The Contractor's written Claim.
2)   The written B.C. Hydro offer of settlement.
3)   The Contractor's written response to B.C. Hydro's offer.

The referee reviews the position taken by both parties and forwards his recommendation to B.C. Hydro. B.C. Hydro will forward this recommendation to the Contractor as written. If the Contractor agrees, the claim is settled as B.C. Hydro considers itself to be bound by this recommendation. The success rate for resolution is in excess of 90%.

3.1.8   Neutral Fact Finding

In this case an outside expert is engaged and paid for by one of the parties with the concurrence of the other to examine the project files. The independent is requested to provide a recommendation which the engaging party will act on. The other party is not obliged to accept the solution so developed and can proceed to court or binding arbitration.

3.2   Binding

3.2.1   Adjudication

Under this model the adjudicator can be hired at the outset of the contract or when the dispute arises. The outcome of the adjudicator's deliberations are almost always non-binding with binding arbitration as the next step. In this scenario written documents with oral presentations are made to the adjudicator. There is normally no legal representation or argument. The adjudicator's recommendations are acted on with the proviso of appeal to binding arbitration where one of the parties continues to dispute the decision. At this time Canadian Contracts do not include the Adjudicator in the contract as has been the case in the UK Civil Engineering Contract Documents.

3.2.2   Arbitration

There are various models followed in this area ranging from fact based boardroom to formal International (more rigid than the courts) Arbitration. The model adopted depends on the sophistication and expertise of the parties as well as if they are truly interested in a decision based on business practice. In Canada there are many qualified arbitrators who can function efficiently under any of the above named processes. The arbitration process is governed by various Acts across Canada. For instance the Ontario Arbitration Act envisages a single arbitrator unless the parties otherwise direct and prohibits the arbitrator from acting as a mediator. The Ontario International Arbitration Act envisages a tribunal and permits the arbitrator the role of mediator during the process. The Federal Government has adopted the Commercial Arbitration Act for its contracts. The arbitrator's powers are governed by the various Acts and the power that the parties confer on him or her. The parties can permit the Arbitrator to act ex aequo et bono or as an amiable compositeur. The parties can also decide which elements of the dispute can be heard by the arbitrator and which elements will be forwarded to another jurisdiction such as the court. All of this must be included in the parties joint agreement to arbitrate. The arbitrator's decision/award is enforced by the courts and he or she is immune from prosecution unless he or she has behaved in a criminal manner. As earlier noted, awards are not published. Both the UNCITRAL Model Law for Arbitration has been adopted and the New York convention 1958 has been ratified.

3.2.3   Expert Determination

This has been used in a binding and non-binding manner and is commonly interchanged with the terms Referee and Adjudication .

3.2.4   Litigation

In Canada parties can elect to use arbitration and the courts will enforce both the choice of the parties to select this venue and the resulting award. There are no specialist judges in Canada. The courts particularly in Ontario are forcing parties to engage mediation before proceeding with the court process. But at this time mediation is not an entrenched process. The court process is adversarial in nature both in Canada and the US. The Court can hear appeals from arbitrator s awards but these are limited by the various Acts. The courts as well as arbitrators will enforce without prejudice agreements. Parties can give evidence orally or in writing and opportunities for cross examination exist both in court and in arbitration.

3.2.5   Ombudsman

This process is not in use in the construction industry. Ombudsman have been appointed primarily to deal with complaints from individuals against various institutions and governments.

4.   THE ZEITGEIST

It could be said that ADR is alive and well in Canada and is gaining converts. The legal community is rediscovering itself in this new conciliatory role. Hopefully it will also mean a revisit and redesign of the legal procedures and principles that currently guide the court process both for commercial as well as other disputes.

The economic zeitgeist is officially encouraging. In fact the investment in construction for 1996 was 88 Billion which represents 11% of the Canadian GDP. The forecast for the next five years is even more promising.

Bibliography

Golder Associates Ltd. Report To - FIDIC "Amicable Dispute Resolution Task Committee - The Use of ADR in the Canadian Construction Industry 1991".

Beifuss R. - Report on behalf of Industry Canada "Contract Disputes Advisory Board 1992".

Attorney General of Ontario Discussion Paper - "Too Many Disputes! Too Much Litigation! - Dispute Resolution Opportunities For The Construction Industry 1994"

CCH ADR Manual - Jackman, Stitt, Beifuss etc. 1995

Geddes W. "Demand ADR Now 1990"

Australian Government and Private Sector Commission - "Strategies For The Reduction of Claims and Disputes in The Construction Industry - A Research Report 1988"

Beifuss R. - "Contract Claims Avoidance, Settlement Procedures and Options" - A Professional Development Course (Saudi Arabia, Kuwait 1996, 97 & 98).

Defence Construction Canada- "Partnering 1997"

British Columbia Hydro-Electric -"Referee" Eric Kosty 1988

Arbitration and Mediation Institute of Canada-Arbitration/Mediation Practice Manual(s) 1996

The Dispute Resolution Advisor System in Practice, Wall, C., (1994), CIB
Publication No 171, CIB, The Netherlands.

CHINA

Deepak Bajaj
Faculty of Design, University of Technology, Sydney, Australia
Rui Zhang
Page Kirkland Partnership, St. Leonards,
New South Wales, Australia

1.   BACKGROUND

1.1   Economic

Since the late 1970's, the Chinese government has gradually moved away from a sluggish centrally controlled economy to a new, dynamic, market oriented mechanism. In the past decade China, with its 1.2 billion strong population, has accelerated its economic reforms with an averaged 9% annual growth under the "Open Door Policy" adopted in the 1980s. As a result, the national economy has more than doubled since 1978.

To accommodate this economic growth, China has designated an extraordinary budget for construction and infrastructure development. In 1993, the net value of construction output was 6.7 % of Gross Domestic Product (GDP) and it is predicted that the building industry will grow at an average rate of 12 % per annum from now on to the year 2000 (Davis Langdon & Seah International, 1997). In fact currently the industry is gaining momentum from a new surge of high demand for residential and commercial development. For instance, by the year 2000, the Chinese government plans to provide each urban resident with 8m² of floor space. In addition, many key transportation and other infrastructure development projects are planned and under way to bring the benefits of economic growth from the coastal and urban regions to the remaining 70 % of the population who live inland. These could amount to as much as US $ 200 billion by the year 2000.

1.1.1   Trends

China has achieved extraordinary economic growth and managed to lift the average living standard dramatically in the past two decades with a comparatively small compromise on social-political stability. However, still some major hurdles need to be removed in order to maintain a long lasting modern economy and a healthy economic growth. Firstly, with current unbalanced development and regionally fragmented markets, how to sustain its high growth rate while maintaining a low inflation rate remains the biggest dilemma China has to face. Secondly, social instability caused by an unprecedented high unemployment as a result of the nation wide re-structuring and incorporatisation of State Owned Enterprises (SOEs), which employ the majority of the work force, is damaging and eroding the current economy. Finally, with strong public pressure to reform not only the economic system but also the political one, it is arguable whether it is feasible at all to speed up the market economy without providing political freedom. Therefore, the success of the current government's liberalisation of the economy whilst maintaining its political solidity and social stability will prove to be crucial to China's future development.

1.2   Legal

China's legal system is based on the civil law system to which the common law principle of judicial precedent does not formally apply. Therefore Chinese law mainly consists of enacted legislation, court cases do not form a part of the law (Baker & McKenzie, 1997c). Laws passed by the National People's Congress (NPC), which functions as a parliament, are effective at all levels of government.

The legal system in China was almost dissolved during the Cultural Revolution (mid 1960s until mid 1970s). Only after the formal adoption of the Open Door policy in 1979 was a basic law system devised. Since then the NPC has enacted more than 300 laws and regulations together with more than 4,000 local laws and regulations promulgated by People's Congress at lower government levels.

Apart from some specialised tribunals such as maritime and military tribunals, there are four tiers of courts in China. The Supreme People's Court, responsible for the NPC and its standing committee, is the highest judicial organ and the final appellate court. Higher People's Courts are at the provincial and Directly Administrated Municipal (DAM)[1] level. Intermediate People's Courts are at municipal level and Basic People's Courts are at the county and municipal district level (see Figure 1-1). Depending on the size of a case and the impact in the relevant area, different levels of courts may assume their jurisdiction as courts of first instance. Parties not consenting to a judgment made by the court of first instance may appeal to the court at the next higher level. The ruling of the court of second instance is final. Litigation in China is governed by Civil Procedure Law 1991.

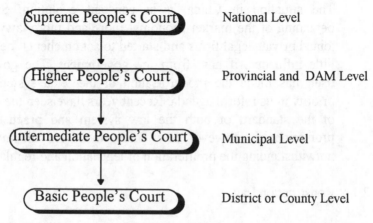

Figure 1-1 Court System in China

Both adversarial and inquisitorial systems are used in China depending on the different nature of the concerning cases. Although the trend is to strengthen the inquisitorial procedure, a distinctive feature of court inquiry characterised by pure prosecution and defence has been maintained. During trials, litigants and advocates are permitted to debate over the evidence presented and legal professionals are usually employed to act on parties' behalf and heated courtroom debates are becoming commonplace.

The legal profession in China is not bifurcated into barristers and solicitors as those in most British Commonwealth countries. They are generally referred to as lawyers and have rights of audience in courts once registered and certified. All lawyers in China used to be government employees thus their impartiality, especially in dealing with cases involving government functionaries, has always been doubted. Similarly, the client-agency relationship between clients and professional legal advisers may not always be taken for granted due to the perceived lack of independence of the legal profession.

The situation in China is in constant evolution. Since the beginning of the market economy, more and more lawyers have joined private legal firms and started to act on clients' behalf with little influence, if any, from the government. The government itself has made the modernisation of the legal system a high priority in its reform agenda. Recent years have seen the elevation of the standard of both the law system and practising legal professionals. However law enforcement is still ineffective, notwithstanding the proliferation of legislation and regulations.

1.2.1   Construction Law

Construction law in China does not form a special realm in its own right but rather is an amalgamation of regulations. Again, they are usually enacted by various government organisations. Contractually, the construction industry is governed by the Law of Economic Contract 1981 and Arbitration Law 1995. Numerous regulations are specifically devised for the administration of the construction industry and the conformation of technical standards. Due to the current booming of the construction industry and escalation in the numbers of construction related conflicts and disputes, specialist legal experts for the industry are emerging, albeit still in an early stage.

1.2.2   Construction Contracts

In order to understand the contractual relationship existing in the construction industry in China, it is essential to review the social and economic history of the industry. From 1949 to 1981, construction companies were assigned work by the state and therefore no tendering and contracting systems were needed. Only simple agreements were drafted between the parties instead of the comprehensive contractual documents normally adopted in the West. Competitive tendering was reintroduced in China in 1981 (it was practised on a limited scale before 1949 in which the People's Republic was formed) in the Shenzhen Special Economic Zone (SEZ). Since then competitive tendering methods have been employed in most areas (over 15% of the total construction work contracted by tendering in 1993 compared with only 2% in 1990). Among them, selective tendering is the most favoured procurement system and negotiation occurs very frequently during both the tendering and construction stage.

The government issued Regulation of Building Construction and Installation Contract in 1983. There are 20 clauses in the regulation which form a guideline for how to draft construction contracts. However, it is not a standard form of construction contract by its own right. Compared with international standards, construction contracts drafted on the basis of the regulation are too simple in many aspects. Lack of legal terms is commonplace and usually a contract contains many implied (even confused) rather than explicit terms. Influenced by traditional Chinese philosophy which emphasises social consensus and harmony rather than individualism, people prefer mutual agreement to legal confrontation. Therefore, most contracts consist of very broad guidelines, and tend to leave anything not addressed to be agreed upon on the basis of "mutual understanding and trust" rather than clearly defined legal obligations (Walker, 1991).

In 1991 a more comprehensive standard form of contract (GF-91-0201) based on the Federation International Des Ingenieurs-Conseils (FIDIC) was issued by the Ministry of Construction. It contains 41 clauses and has since been widely used across the country. Clause 30 outlines the dispute resolution commonly used and incorporated in the standard contract:

1.   mediation through the organisations or individuals appointed in the contract clause;
2.   arbitration conducted by the local governing arbitration committee;
3.   litigation under the jurisdiction of the local People's Court.

It should be noted here that although mediators can be appointed, the parties normally can not select their arbitration committees or People's Court. As usually only the local arbitration committees or People's Court which has jurisdiction over the area where the project is located can make an adjudication, this is particularly in the case of disputes involving only domestic parties. Disputes involving a foreign element are treated differently in China and this is further elaborated in 3.2.2.

Given the large amount of foreign investment in China, international contract forms are often in use albeit subject to modification and amendments. All international contracts to be agreed upon within China are translated into Chinese and consequently made subject to Chinese Law.

2.   CONFLICT MANAGEMENT

There are three types of enterprise in China: state-owned, collective-owned firms (similar to those in public sectors in the West) and private firms (similar to those in private sectors in the West). In the building industry, state-owned and collective-owned firms have always dominated the market. Together they have more than 12 million employees and account for more than 70% of construction work (Statistical Yearbook of China, 1994).

Four main players are involved in the industry: the development unit (commonly referred to as Party A in a contractual agreement in China), the construction unit (referred to as Party B), the design institute (normally comprising not only architects but also other consultants) and the People's Construction Bank of China (PCBOC). The development unit is the client and all the investment is audited and controlled by PCBOC through either state or local government investment allocation. Usually a "Preparatory Office" staffed by the development unit will act as the client's representatives on the site. The construction unit is the builder and is responsible for the construction work and sub-contracting. The design institute is the consultant and is normally paid by the developer directly. However, there is no strong binding relationship between the client and the consultant as compared with what exists in the West. This is gradually changing now since the government deregulation which forces design institutes to form strong links with clients in order to secure enough work. In recent years, especially in coastal cities with rapid economic growth, a new player called the Construction Supervision and Management Unit has emerged which acts on the client's behalf to ensure projects conform to the national standard and are completed within the scheduled time, similar to project management firms in the West.

Different interest focuses inevitably result in conflicts and disputes between the parties concerned. However, avoidance of conflict is preferred and carefully managed by most people. Developing and maintaining a friendly relationship with one's business partners is considered ultra important in China. It is rare to find any dispute settled by either arbitration or litigation leaving the involved parties still hand in hand, smiling at each other and claiming "business as usual". The mere diminishing of goodwill that results from the adversarial procedures is a loss to both parties, regardless of who wins the case.

Stronger government intervention may also contribute to the avoidance of disputes in China. Since the government still insists that the market economy is not a purely free economy, it may intervene both directly and indirectly. For this purpose, many negotiations have been urged by the government to avoid further damage to both the parties concerned and the general public and in most cases, if not all, a government organisation may assume the role of a mediator to ease the tension between the parties.

2.1   Non-binding

2.1.1   Dispute Review Boards

Not specifically existing in China.

2.1.2   Dispute Review Advisers

Not specifically existing in China.

2.1.3   Negotiation

This is by far the most used method in dealing with conflicts and disputes. Traditionally, with strong influence from Buddhism and Confucianism, Chinese culture favours collectivism rather than individualism. Social harmony is highly regarded while individuals are obliged to compromise with other members in the society on conflicts and disputes. Using "mutual trust and understanding" in resolving conflicts and disputes is an integrated part of Chinese philosophy.

Negotiation gives people more space to deal with difficult issues while on the other hand, it does not cast any shadow on the business relationship between the parties involved. Negotiation frequently happens in the construction industry and proves to be extremely successful within the business environment in China. The common approach is that both formal and informal meetings and talks are held between the CEOs of the parties whenever there is potential for conflict, sometimes with coercion from a supervising government agency. Usually compromises are made and agreements are reached thus avoiding any unnecessary disputes.

By and large, negotiation is a practical skill learnt through individual experience though formal training is available. Chinese people are culturally conditioned to negotiation. The traditional Chinese culture plays a very important role in gaining the experience, as mentioned above, informal friendly talks sometimes are far more effective than legal procedures.

2.1.4   Quality Matters

Apart from the Total Quality Management (TQM) adopted in the construction industry, there is another mechanism to provide quality assurance. The government has institutionalised a watchdog under the control of the Ministry of Construction: Quality Supervision Authority for Construction Project (QSACP) which exists at all levels of government. Its pure function is to monitor the quality of construction projects. This mechanism is governed by the Regulation of Quality Supervision for Construction Projects (1990).

A project is required to be registered with the relevant authority one month before commencement and all relevant design information should be provided. The QSACP will then assign a supervisor to monitor the quality of the project during the construction period. The authority can intervene in the construction project when there has been severe violation of national standards, it can also certify quality projects and recommend awards.

There are two grades of construction quality in China, satisfactory and excellent. Clients requiring projects to be finished with an excellent quality grade are expected to allow for an increase both in cost and construction time. The projects then will be monitored on the quality grade agreed upon in the contract. Any disputes concerning the quality grade in the actual construction may be resorted to arbitration conducted by local QSACP.

ISO9000 has been adopted in China. However, since much of the demands for ISO9000 certification exist in importing and exporting industries, the application in the construction industry is very limited. The construction industry's main activity is still domestic and it remains a labour intensive and low technology industry. Therefore ISO9000 registration has not been seen as a necessity. However, due to the large quantity of construction work and the common trade-off of quality for speed, quality assurance still needs to be significantly improved. While the government is certainly pushing the certification of ISO9000, it still needs time for the industry to accommodate itself to this international standard.

2.1.5   Partnering

Partnering is a frequently used business strategy in the construction industry in China. It may refer to three different activities in the Chinese context:

Partnering may occur when each participant of a project joins a round table meeting regularly held by the supervising government agency. Usually a division of the local construction commission will chair the meetings and each party may raise issues which need to be addressed and potential conflicts which need to be avoided. The aim of this loose partnering is to enable a smooth relationship among all parties, encouraging good co-ordination and co-operation, which in turn makes the goal agreed upon with the client more achievable. This informal partnering is usually very successful as the parties can openly discuss the issues concerned and the chairing government agency may also assist in conflict management by acting as a mediator.

2.2   Binding

2.2.1   Partnering

Secondly, partnering may refer to the relationship between a main contractor and specialist subcontractor. Usually a rural construction firm will form a partnership with a municipal firm to provide labour and other low technology services. Often an agreement is signed which binds both parties, although this type of partnering is usually short termed. Today in big cities, almost all general labour is contracted through this type of partnership. It serves as a conduit for redundant peasants in rural areas to move to and work in cities and provides cheaper labour for municipal contractors. While these people certainly contribute to the competitiveness of the construction industry, lack of general construction knowledge and technical training has created quite a few problems, including high risk in occupational health and safety (more than 95% of the death incidents in the construction industry happened to workers from rural areas in the past decade).

Late 80s saw another type of partnering formed between municipal construction firms. For instance, Shanghai Construction Group (SCG) consists of more than ten contractors and employs well over 100,000 people. Compared to the above mentioned partnering, this relationship is mainly formed for tendering large projects and maintaining the market share. This is partly in response to the government liberalisation of the construction market which enables contractors from other regions to compete in the local market (formerly only national contractors were allowed to work across regional areas). Unlike labour partnering which always tends to be short-termed, the partnering between municipal firms is often a long term relationship. However, since this type of partnering is always formed among state-owned firms, an inevitable by-product is entrenched inefficiency. Partnering is not a panacea as attested by the fact, for instance, that over 90% of the firms in SCG in 1995 made losses.

3.   DISPUTE RESOLUTION

Four methods of resolving construction disputes are adopted in China; negotiation, mediation, arbitration and litigation. Non-binding procedures like negotiation and mediation are far more popular than binding methods. Apart from such advantages as low cost and speedy settlements, negotiation and mediation are often sought after as a result of encouragement from the government. The law explicitly recognises mediation as an alternative method of dispute resolution. Even in an advanced stage of litigation, the court can revert to mediation or direct the parties to negotiate provided consent is given from both sides. Therefore, it is not hard to understand that even though arbitration and litigation are totally enforceable under the law, people tend to avoid them as much as possible. Furthermore, with a legal system that still needs to be fully modernised, concerned parties are reluctant to let their cases be judged by others with little control over the final settlement.

3.1   Non-binding

3.1.1   Conciliation (tiao jie)

Conciliation is considered the synonym for mediation in China and they are frequently interposed when the term in Chinese is translated into the English language.

3.1.2   Executive Tribunal

Not specifically existing in China.

3.1.3   Mediation (tiao jie)

Mediation is employed quite frequently in the construction industry. In fact it is almost always the first step once conflicts have escalated to disputes. Both Civil Procedure Law and Arbitration Law stipulate that mediation should be given priority in settling disputes and adjudication such as arbitration and judicial proceedings should only be used after mediation has failed to deliver an agreement signed by the parties involved.

When disputes occur, the mediator appointed (usually from a local conciliation centre) is invited to organise both formal and informal meetings or talks between the parties involved. Though strictly speaking, the role of a mediator is to assist the parties to reach an agreement rather than to make a decision, mediators in China usually have a strong influence over the final settlement since conciliation centres are usually attached to government administrative departments which. prefer to wield its administrative power to persuade parties involved to reach an agreement in order to prevent further economic or social damage. Mediation is an integral part of both an arbitration in a tribunal and a judicial procedure in a court. The agreement reached and signed by the parties becomes enforceable under the law. When parties involved fail to reach an agreement in an arbitration tribunal or court directed mediation, a mediator can revert to his or her role either as an arbitrator or a court judge and can make an adjudication instead.

3.1.4   Negotiation (tan pan)

Negotiation is the most commonly used method in contractual disputes in the construction industry. This reflects the attitude held by quite a few people in the industry that a contract is open ended and can be negotiated through different stages of the project. Negotiation can avoid the escalation of disputes and in most cases is an effective tool in resolving the disputes. It is usually conducted between the senior management of the parties involved and trade-offs occur frequently. This avoids the "winner takes all" scenario in an arbitration or litigation and people can still maintain good business relationships.

However, negotiation tends to be abused in the industry especially since the introduction of the market economy where deregulation is encouraged while a healthy legal framework has yet to be fully established. Due to the ineffectiveness of the law enforcement, quite a few people still rely heavily on internal rules and even back door transactions through negotiation. This sometimes harbours corruptive activities in resolving the disputes where bribery is offered in return for favourable terms in negotiation. Not surprisingly, with the constant improvement in the legal system, negotiation is losing its significance and legal binding resolution has started to gain general acceptance.

3.2   Binding

3.2.1   Adjudication (cai jue)

Adjudication in China usually refers to binding dispute resolution such as arbitration and litigation where a binding decision is made by a third neutral party in comparison with non-binding settlement such as negotiation and mediation.

3.2.2   Arbitration (zhong cai)

Arbitration in China is governed by Arbitration Law 1995. It features a double regime system which differentiates domestic disputes from foreign related disputes (Chan, 1997). For the construction industry, a well structured network of national arbitration committees are responsible for settling domestic disputes, while foreign related arbitration is always conducted before the China International Economic and Trade Arbitration Commission (CIETAC) under the China Council for the Promotion of International Trade (CCPIT). The bifurcation of arbitration in China has a significant implication on both Chinese and foreign parties involved. In a case China International Engineering Co. v. Beijing Lido Holiday Inn Hotel (1992), a Beijing court refused to enforce the arbitral award made by CIETAC because the court argued that since the hotel was legally incorporated in the country, it should not be treated as a foreign company and the dispute should have been dealt with in a domestic arbitration tribunal.

When resorting to arbitration, the initiating party is required to specify the nature of the dispute, the amount sought and provide detailed evidence to a relevant arbitration tribunal. The arbitration tribunal will then investigate the case and reply to the initiating party within 7 days whether the case is acceptable or otherwise not acceptable together with the reasons. A respondent must file an answering statement within 15 days after notice for the arbitration tribunal. Failure to file an answering statement will not delay the arbitration.

Arbitrators are full time career civil servants, however, part-time arbitrators may be called in when specific technical issues are involved. Generally, the arbitration tribunal is formed by two arbitrators and an additional principal arbitrator appointed by the local arbitration commission. For some simple disputes, there may be only one arbitrator making the decision. The arbitrators may adopt an inquisitorial process and settle the dispute ex aequo et bono or act as amiable compositeur provided that the parties consent in writing. An arbitration tribunal will usually try mediation first but can not compel the parties into mediation. Mediation may be limited and suspended at any time whenever insincerity is sensed by any party and formal arbitration requested to proceed usually with the same arbitrator/mediator. Unless there are exclusion clauses which preclude any disputes from being resolved by litigation, arbitration will not be conducted if any of the parties files a lawsuit in a court.

The arbitration award is enforceable under the law if there is no appeal within 15 days after the issue of award. However, either party has the right to appeal to a court unless, again, exclusion clauses prevent them from resorting disputes to litigation. If the party concerned has failed to follow the award, the other party can apply to a court to enforce the award under the law.

China is a signatory to the New York Convention on the Recognition and Enforcement of Foreign Arbitral Awards 1958 and provides for reciprocity of arbitral awards by other countries ratified to the convention. Although the Arbitration Law permits the parties to select their preferred arbitration institution, international arbitral awards are not automatically recognised and enforceable in China. The parties concerned have to apply to an appropriate intermediate court for the enforcement of the arbitral award made overseas. Normally this will be subject to Chinese law and the ratification of certain international convention.

3.2.3   Expert Determination

This technique is not used in China.

3.2.4   Litigation (su song)

Lack of legal consciousness is often observed in China and the construction industry is no exception. Historically, Chinese society does not favour litigation. Concurrently, lack of proper procedure of law enforcement and low transparency of government administration has further deterred people from appreciating and utilising judicial procedures. However, the current government has made much effort to encourage legal awareness while also establishing a modern legal framework.

The law recognises the binding clauses in a contract that prevent disputes from court proceedings. However, if no such clauses exist, either party may apply for a court hearing which will take precedence over any other forms of dispute resolution. Usually mediation will be the first option tried by the court and any agreement reached will be legally binding. However, mediation can not be used as a coercive measure and all the parties involved have the right to pursue further litigation.

Adversarial procedures are adopted in China's court system. Legal professionals are widely employed as representatives for the parties involved. Cross examination is always conducted and so is courtroom debate. However, litigants still have a very limited right to request information from the other party. "Without prejudice" voluntary mediation is always encouraged and conducted by a court and statements obtained can not be used as evidence if litigation is finally resorted to. However, it is still doubtful whether a mediator reverting to a court judge later on can easily discard these statements and make an impartial adjudication.

All contracts to be executed in China are subject to Chinese law unless there are applicable ratified international conventions which will take precedence. Foreign parties entering a contractual agreement enjoy the same rights as Chinese parties, however, they are subject to reciprocity agreements between China and the foreign countries concerned. Arbitration awards made by foreign arbitrators are not automatically enforceable within China. The enforcement must be applied through the relevant Intermediate People's Court and is subject to either Chinese law or any international convention ratified.

4.   THE ZEITGEIST

One of the major issues in the reform agenda of the Ministry of Construction is to develop a healthy legal system through legislation and to establish an efficient administrative mechanism. Finding better ways to avoid and resolve disputes has become extremely important as the nation's construction industry is expanding rapidly. It is interesting to note that while in most developed countries, people are seeking mediation and conciliation as preferred ADR, in China, the strengthening of the legal system is actually helping people to fully appreciate the role of arbitration and litigation in resolving disputes, which both are considered as foreign concepts in comparison to mediation and conciliation.

Currently the construction industry is still enjoying its boom and many new projects are being constructed, especially in coastal regions where the influx of foreign investment is helping to reshape the landscapes of old cities into modern wonderlands. Although the industry is still entangled with such entrenched problems as low education levels among employees, low levels of automation in construction and lack of legal consciousness, significant improvement has been observed during the past decade. In fact the government has made the modernisation and internationalisation of the construction industry a first priority and has devised detailed measures to implement its strategies (Mayo and Liu, 1995). With the constant evolution, the construction industry will be able to embrace the international practice in the near future and, more importantly, to compete efficiently in the global market.

Bibliography

Baker & McKenzie (1997a): *Asia Pacific Legal Developments Bulletin.* (March) Vol. 12, No. 1-2.
Baker & McKenzie (1997b): *China Legal Developments Bulletin* (Jan-July) Vol.4, No. 1-6.
Baker & McKenzie (1997c): *Dispute Resolution in Asia.*
Cao, SQ (Ed) (1994): Construction Project Investment Control and Supervision. China WuJia Publishing House (In Chinese).
Cardoso, E; Yosof, S (1994): Red Capitalism: Growth and Inflation in China. Challenge 37(3), 49-56.
Chan, HW (1997): Amicable Dispute Resolution in the People's Republic of China and Its Implications For Foreign-related Construction Disputes. *Construction Management and Economics* Vol. 15, No 6.
Davis Langdon & Seah International (Ed.) (1997): *Sponis Asia Pacific Construction Costs Handbook. E & FN Spon.*
Du, YM (1993): Construction Project Administration and Management. Dizheng Publishing House (In Chinese).
Epling, JA (1987): Resolving International Construction Disputes. Project Management (Nov) Butterworth & Co (Publishing) Ltd. Vol 5, No.4, 217-219.
Han, F; Sha, K; Chen, Q; Xie, B (1994): Chinaís Construction Industry: from Planned Economy to Market Economy. Building Research and Information Vol.22, No.4, 206-208
Lau, S (1994): *Property Investment Guide for Foreign Investors in the Shenzhen Special Economic Zone.* Unpublished project, University of Technology, Sydney.
Li, Shirong (1995): The CPAM System: A New PM System in the Chinese Construction Industry. *PM Network* (April), 7-12.
Lin, KC; Lin, X (1992): Management Myopia: A By-product of the Business Contract System in China. Asia Pacific Journal of Management Vol 11. No. 1, 125-131.
Mayo, RE;. Liu G (1995): Reform Agenda of Chinese Construction Industry. Journal of Construction Engineering and Management (March) Vol.121, No.1, 80-84.

Nixon, MA (1994): Building Grain Storage Facilities in China - International Contract Negotiation. Cost Engineering. (May) 36(5), 7-11.

People's Construction Bank of China (1993): Legal Guideline for Real Estate Development. Economic Management Publishing House (In Chinese).

Qian, KR; Du, X; Ge, JP (Eds.) (1994): *Practical Handbook for Construction Management.* 1st ed. Southeast University Press, Nanjing. 838 pages. (In Chinese).

Ross, MC; Rosen, KT (1992): China's Real Estate Revolution. *China Business Review* 19(6, Nov/Dec), 44-51.

State Statistical Bureau of China (1994): *Statistical Yearbook of China.* China Statistical Publishing House, Beijing. 795 pages.

Walker, A (1991): Land. Property and Construction in the Peopleís Republic of China. Hong Kong University Press.

Wang, YC (1995): Chinaís Economic Reform: The Next Step. Contemporary Economic Policy (Jan) 13(1) 18-27.

Wilkins, B (1994): Factors Affecting the Performance for Foreign Construction Enterprises in China During the First Ten Years of the 'Open Door Policy'. *Australian Institute of Building Papers* (No.5), 45-55.

Xu, DT (1989): *Construction Cost Management.* Tianjing University Press, Tianjing. 526 *pages.* (In Chinese).

1   A Directly Administered Municipality (DAM) enjoys the same status as a province in China. Currently there are four DAMs: Beijing, Tianjin, Shanghai and Chongqing.

ENGLAND AND WALES

Peter Fenn
University of Kentucky
Michael O'Shea and
Edward Davies
Masons Solicitors, Manchester, England

1.   **BACKGROUND**

England and Wales are part of the United Kingdom (UK) of Great Britain and Northern Ireland which itself became a member of the European Union (EU) on the first enlargement of the EU in 1973. Within the UK countries enjoy separate legal systems, notably Scotland, but all take the House of Lords as their supreme court.

The UK is a small densely populated country with a mostly temperate climate.

Construction conflict management and dispute resolution in England and Wales have been dominated in recent years by a joint industry/government review of the construction industries. This review took place between 1993 and 1994 and was led by a former construction minister and Conservative member of parliament Sir Michael Latham, *The Latham Report.*[1] The review had been discussed for many years and follows on similar review processes carried out in the 1950's, 60's and 70's. The recommendations of the Latham Report were widely welcomed and the Conservative Government of 1996 provided legislation via *The Housing Grants, Construction and Regeneration Act 1996 (HGCRA)* to implement some, but not all, of the recommendations. The Conservative Government was replaced in an election during the Spring of 1997 by a Labour Government which had already made plain their commitment to HGCRA; indeed the Labour administration have intimated that they will go further than the previous government and consider legislation to implement the remainder of the recommendations of the Latham Report.

1.1   Economic

In 1979 the Conservative Government set a target to expose the economy to the invigorating influence of market forces and to encourage freer movement of capital and labour. It pursued a policy to reduce the role of the state in the economy, cut public spending, and to revive UK industry. Specific measures taken include tax reform, abolition of capital controls, privatisation of national industries, deregulation of financial services, telecommunications and transportation, and labour law reform. These steps have generated significant structural changes in the economy and increased its efficiency. As a result, England and Wales has entered a period of sustained, if modest, economic expansion with a leaner, more competitive business sector. The Labour government elected in 1997 seems set to continue along a similar path.

The construction industry has historically accounted for 8-12 % of Gross Domestic Product.[2]

1.1.1   Trends

The biggest single factor affecting the economy has been political moves towards a economic and monetary union (EMU) within the EU. The future of not only England and Wales; the United Kingdom but the European Union will be dominated by this in the near future.

On the one hand the economic outlook is depressing, the recession of the late eighties still lingers output is low and construction prices lag behind costs[3]. On the other hand the period of sustained, if modest, economic expansion with a leaner, more competitive business sector will continue.

1.2   Legal

England and Wales are Common Law countries with an adversarial court system. This description (adversarial) is not a deprecatory term; although it is often presented as such. English Law exists within the law of the United Kingdom of Great Britain and Northern Ireland which itself exists within the law of the EU. European law is by no means always stronger than the law of the member states. The rule is not that the legal system within the larger jurisdiction takes precedence over that which covers a smaller geographical area. European law does not always take precedence over German law, for instance, just as the jurisdiction of the United Nations cannot always take precedence over European law. Rather, European law forms a legal system in its own right, which is closely bound up with the law of the member states. European law is created by the law-making activity of European institutions; however, its application in the member states is dependent upon a decision by the latter's parliaments. The principle of subsidiarity says that the European Union should adopt a measure only when it can achieve the desired objective better than the Member States.

1.3   The Official Referee

A specialist tribunal exists in the High Court for construction cases. The Court of the Official Referee. Fay [4] describes the general dissatisfaction with all civil litigation in England and Wales during the mid nineteenth century.

Commerce and industry complained bitterly about the failings of the courts which included:

> The expense of proceedings;
> The delay in hearings;
> The quirks of the outmoded legal systems of law and equity;
> The uncertainties of decisions;
> The concentration of court business in London.

In response to these criticisms, and also the loss of business to arbitrators, the government appointed a Royal Commission (September 1867) to consider the judicature. The report of this Commission led to a major reform sweeping away the old courts of common law and equity to be replaced by one Supreme Court (The Judicature Act 1867). Part of this reform included the appointment of permanent officials "Official Referees" (OR) the business of these officials to include, *inter alia*, disputes arising from civil or mechanical engineering, building and other construction works generally. The court of the official referee has become recognised as the construction industry's court, although other cases are tried there. Unofficial estimates put the work of the OR as 85% construction disputes (Fay 1988 p136).

Despite an acute interest by building professionals in all things legal very little is written about the OR or their courts. Fay (1988) describes the response of the Attorney General to criticisms of delays in trying construction cases.

"The Official Referees are an important part of the administration of justice, although clearly one about which not much is known." [5]

Official Referees' business is defined as:

"including any case or matter which includes a prolonged examination of documents or accounts, or a technical, scientific or local investigation such as could more conveniently be conducted by an Official Referee, or for which trial by an Official Referee is desirable in the interest of one or more of the parties on grounds of expedition, economy or otherwise".[6]

This is expanded upon in The Supreme Court Practice in the notes to Order 36/1 - 9/13 to include:

(i)   Civil or mechanical engineering;
(ii)   Building or other construction work generally;
(iii)   Claims by or against construction professionals;
(iv)   Claims by and against Local Authorities relating to statutory development of land or construction of buildings;
(v)   Claims between neighbours in certain torts;
(vi)   Claims between landlord and tenant for breaches of repairing covenants;
(vii)   Claims relating to quality of goods sold/hired;
(viii)   Claims relating to work done, materials supplied and services rendered;
(ix)   Claims involving the taking of complicated accounts.

There is also a provision for the OR to hear claims arising out of fires or which are concerned with computers or the environment.

OR's business has certain important characteristics of procedure which differentiate from ordinary trials in the Chancery or Queen's Bench Division; these include:

Cases are assigned by rota to a named judge, who hears and decides interlocutory applications;

Interlocutory appeals go straight to the Court of Appeal;

Historically important is the fact that there can be no jury trial, this was relevant when other civil cases allowed trial by jury; but now trial by jury has all but disappeared from civil courts.

The legal profession is separated into Solicitors and Barristers. Until recently only Barristers had rights of audience in higher Courts and carried out the advocacy work. Whilst rights of audience can now be achieved by Solicitors there has not been a great rush amongst them to this; the feeling is that the traditional separation will mostly continue.

A special relationship exists between the clients and professional advisors. In English law a special contractual relationship exists under the broad term of agency, used to describe the relationship between the two parties. The agent acts on behalf of the other party the principal. Historically standard forms of contract were written so that the client's engineer or architect acted as "honest broker" between the two parties. In recent times this agent/principal relationship has become more problematical and contractors have queried how the agent can act in the best interests of his principal and remain an honest broker intermediary.

1.3.1   Construction Law

Construction Law exists as a recognised legal specialism; because most construction law is heard by OR's the specialists have organised themselves around the court (The Official Referees Solicitors Association and The Official Referees Barristers Association ORSA and ORBA). Construction Law is taught as post-graduate courses at universities and organisations have developed to accommodate the lawyers and non-lawyers e.g. The Society of Construction Law; The Society of Construction Arbitrators.

The decisions in construction cases and cases relevant to construction are reported in specialist journals e.g. Building Law Reports; Construction Law Reports; Construction Industry Law Letter.

1.3.2   Construction contracts

A great many standard forms of contract for construction exist; these have traditionally been split, like the industry itself, along the lines of building and civil engineering. The major standard forms for building work have been produced by the Joint Contracts Tribunal (JCT) and the civil engineering forms by the Institution of Civil Engineers (ICE). Both organisations produce forms for minor and full scale projects and both produce forms for alternative procurements systems (see 2.1.4) notably design and build.

One of the conclusions of the Latham Review was that a single standard form of contract should be produced. Whilst this did not find its way into the legislation the ICE produced *The New Engineering and Construction Contract* (NECC) in 1996. NECC includes provisions to deal with the proposals of the Latham Review and its use is expected to increase.

2.   CONFLICT MANAGEMENT

2.1   Non-binding

2.1.1   Dispute Review Boards and Advisors/Advisers

There have been some high-profile projects which have included provisions which might be described in these terms.   The most notable perhaps being the provisions included in the Channel Tunnel Project which were only binding if not subsequently reversed by more formal methods of dispute resolution[7].   Developments associated with the Latham Review have taken this technique into what is described under 3.2.1 Binding Adjudication.   There have been few examples of totally non-binding boards and advisors which produce only a recommendation, the parties being free to accept or reject.

2.1.3   Negotiation

It could be said that negotiation does not come easily to most Englishmen, there is no culture of haggling; prices quoted are not normally negotiable.   The exceptions which come to mind are second-hand car purchases and house buying.

Negotiation is not normally formally taught at university and it is expected that negotiation skills are acquired experientially as a young professional develops.   Many successful courses are offered by professional training organisations; fashionable techniques often follow the lead taken by the USA.   Currently the fashion is for Principled Negotiations as proposed by Fisher and Ury[8].

2.1.4   Quality Matters

Total Quality Assurance (TQA) Total Quality Management (TQM) and derivatives have been applied to construction. Construction companies offer their qualification to BS 5750 and ISO 9000 as evidence of ability, there is no system of national contractor registration or qualification and such TQA/M qualifications help fill that void.

A system Co-ordinated Project Information (CPI) [9] was launched by all the construction professions in the 1980's; the principle, based on research, was that buildings were more likely to be defect free when the various design professions liaised to avoid design conflicts and told contractors clearly what was required. These simple axioms remain, for the most part, ignored. The Latham Review included the proposal that CPI should be adopted on all construction projects; the proposal did not find its way into legislation and there is no evidence that CPI is, or will be, adopted.

In the UK the traditional arrangement for buying buildings and structures was via an Architect, Engineer or Masterbuilder controlling craftsmen, journeymen and labourers. The industrial revolution saw the design professionals take an independent role and the traditional system became one where an architect or engineer produced an outline design and took this forward to full working drawings. These complete and fully detailed designs were sent out to competitive tender where main contractors were responsible for the co-ordination of site activities via sub-contractors.

As construction work became more sophisticated certain aspects of the design work passed from the architect or engineer to special sub-contractors e.g. lifts and heating and ventilating installations. These sub-contractors were selected by the client or his representative and carried out the design and installation of aspects of the project. The ability of the client to nominate certain of the sub-contractors for who the main contractor was responsible brought about their name Nominated Sub-Contractors (NSC). The other sub-contractors became known as Domestic Sub-Contractors. Standard forms of sub-contract are available; the JCT produce NSC forms and DOM forms for domestic sub-contracts are also available; the ICE makes no formalised distinction between the types of sub-contract.

The traditional form of procurement therefore became one where a separate designer employed by the client produced a complete design which was priced by a contractor responsible for the construction. The tendering contractors at first joined forces to pay for a measurement surveyor to produce a survey of quantities of materials and labour to avoid the need for repetition. Later a system developed where the client employed this surveyor to produce a complete set of tender documents for use by all the tendering contractors. The profession of Quantity Surveying grew from this technical background.

The separation of design from production has been roundly criticised for many years.[10]

Many alternative forms of procurement are now offered to clients and much literature exists.

The major systems might be classified as:

Traditional
Lump Sum
Measure and Value
Cost Plus
Design and Build
Management Types
Construction Management
Management Contracting

Within the classification many specialist systems are also offered.

Serious quantitative studies of the usage of the differing systems are rare and many claims are made that a particular system is more prevalent. The received wisdom is that Design and Build has steadily taken a larger market share since the 1980's; and that the management types were very popular in the boom of the late eighties, notably for very large mega-projects but have declined since.

Some data is available to support the received wisdoms from the Royal Institution of Chartered Surveyors; a survey carried out on behalf of the Junior Organisations captures data on new building projects only, civil engineering and repair and maintenance are not covered.   The researchers claim to cover 15% by value of new orders and that therefore the survey may be taken as broadly representative of the industry as a whole. The data on procurement methods use between 1984 and 1995 is shown in Table 1.[11]

Procurement Method	1984 %	1985 %	1987 %	1989 %	1991 %	1993 %	1995 %
Lump Sum	72	70	70	62	55	50	56
Measure and Value	7	5	3	4	3	4	2
Cost Plus	4	3	5	1	0	0	1
Design and Build	5	8	12	11	15	36	30
Management Type	12	14	10	22	27	10	11
Totals	100	100	100	100	100	100	100

Table 1 Trends in Procurement - by percentage value of all contracts

2.2   Binding

2.2.1   Partnering

There has been considerable interest in Partnering in construction; once again this has received extra impetus from the Latham recommendations. In addition policies to bring private finance into public sector schemes developed by the previous government and culminating in the *Private Finance Initiative* have supported moves towards construction partnering.

Many commentators have described construction as a necessarily team business. Parties with diverse experience and skills come together to build something, and the interdependent nature of the work demands a joint approach. The futility of creating Temporary Management Organisations on a project by project basis is obvious; at every level, complaints are voiced about protracted dispute resolution, delays, distrust, dissatisfaction and cost increases. The concept of partnering provides ready solutions to the complaints. Indeed many would argue that this is a return to the old-fashioned way of doing business. Great emphasis is placed on mutual trust, and the goals of the project come before each party's agenda. Above all, partnering is an attitude. It is voluntary. It need not be legally binding. It does not necessarily alter the legal and contractual relationships between the parties to a project, but it can shift the focus of responsibility for problems from the individual to the team.

Concern has been expressed that the timing of entering into the relationship may be vital.

If the partnering agreement is concluded before the contract for the works has been agreed, there is a risk that one of the parties may argue that the partnering agreement constituted a misrepresentation which induced him or her to enter into the contract for the works. But that argument cannot run if the parties sign the works contract before concluding the partnering agreement.

Framework agreements and supply chain partnership are arrangements where partnering is more of a long-term relationship or collaboration between a supplier of services and a purchaser. These arrangements can apply equally to the construction industry, the IT industry or even the legal profession; and there is a fast growing trend among clients in various sectors to outsource their non-core business activities to contractors. These services, which were previously provided by in-house staff, can be provided under a form of long-term partnering arrangement, sometimes known as a "framework agreement". The concept has attracted strong support. In the UK Construction Challenge, BAA/Bovis said that "forming a new team for every project on the basis of competitive selection may reduce project costs by a few per cent, but the research findings show that the potential for realising savings are of an order of magnitude greater than this through a closer integration of the design, procurement and construction processes". [12] Sir Michael Latham in his report Constructing the Team (July 1994) promoted framework agreements between contractors and sub-contractors, saying that "such arrangements should have the principal objective of improving performance and reducing costs for clients", adding that "good relationships based on mutual trust benefit clients". [13]

2.2.2   Public sector partnering

The report of the Levene Scrutiny committee published in November 1995 recommends that government rules should be reviewed so as to allow partnering to develop. It has been estimated that, if the UK could reproduce the success of partnering in the USA, central government could save between £600 million and £1 billion each year on its construction activities alone. However, there are several factors which complicate the issue: government policy, and in some cases the law, requires that all goods and services are procured by competition, the motivations of which are value for money and the competitiveness of the supplier. The government's aim is to secure the best value for money possible for taxpayers and to use its commercial influence to promote the UK in world markets. Some of the principles of long-term partnering run contrary to these aims. For example, a long-term relationship may be seen to create a barrier to new suppliers. It may also give the supplier/partner a commercial advantage to the extent that it can obtain lower finance on the strength of its government connections. EU directives call for competition in the provision of goods and services. Consequently, some framework agreements may possibly breach EU Directives and even Article 85 of the Treaty of Rome as regards the free movement of goods and services.

However, it should, in principle at least, be possible to ensure that a particular partnering arrangement does not infringe EU competition and procurement rules by, for example, restricting the terms and ambit of such agreements. However, partnering is still a developing area and consequently there is little guidance on the way in which it will be treated by the European Union. In addition, when setting up the framework agreement care must be taken to ensure that the relationship reflects clearly the partners' defined needs and objectives over a specified period of time.

Effecting a cultural change may well be the biggest difficulty in the adoption of partnering by construction. Construction procurement is an area where the government is seen at its most bureaucratic. Fear of public accountability and the need to demonstrate value for money to the public accounts committee or National Audit Office may put off the government from partnering in cases where the benefits are not immediately apparent. It is to be hoped that the report of the Levene Scrutiny committee will help change official attitudes here.

In the UK, Bovis Construction and Marks and Spencer, a major retailer, have been described as forerunners in the development of partnering. British Airways Construction Group has partnered with Mace, a professional management company, to achieve improved efficiency. British Airways Property are also developing partnered relationships with Ferguson Bucknall Austin to provide facilities management support and with selected contractors for medium to long-term facilities maintenance contracts. Norwich Union are also interested in partnering. Its objective is to identify and work with approximately four organisations within each category of service which it requires, so as to benefit from a closer working relationship. [14]

Other examples include the energy industry where there has been a tradition of partnering in the North Sea for several years. Brown and Root are seen as leading exponents of this form of procurement, having successfully allied with BP and Total. The Forties Field is held out as a success story for partnering where the benefits are said to include better resource management, improvements in the delivery of materials, the removal of duplication and simpler processes.

Partnership Sourcing Ltd, which is a CBI/DTI sponsored not-for-profit company, has identified projects of high strategic priority and high value spend as prime candidates for partnering.

3.   DISPUTE RESOLUTION

3.1   Non-binding

3.1.1   Conciliation

Conciliation is a well recognised form of dispute resolution in England and Wales. There is however a widespread confusion over the terms. Conciliation is recognised as distinct from mediation (See 3.1.3) and within the ADR community conciliation would be thought as lying at the adjudicative end of the mediation/conciliation continuum. Conciliators, and conciliation, are more likely to produce a recommendation should the parties be unable to reach agreement.

There has been no widespread adoption of the UNCITRAL Conciliation Rules.

The ICE forms of contract generally contain a conciliation procedure. The ICE publish Rules of Conciliation. There is no data on the use of conciliation on construction contracts but the ICE Rules provide that if the parties cannot agree on a conciliator the President of the ICE will appoint one. Such appointments are rare and are shown in Table 2.

Year	1993	1994	1995	1996	1997
Number of Presidential Appointments	1	4	10	6	0

Table 2: Presidential Appointments of Conciliator ICE (Source: the Institution)

3.1.2   Executive Tribunal

The Executive Tribunal (or Mini-Trial) is a more formal type of mediation hearing or meeting. The major feature of the process is to allow senior company executives to hear, and assess, the case or dispute between their companies. Formal presentations are made in the presence of the executives and a neutral chairman in an attempt to provoke settlement. There is no evidence of any adoption of this technique in construction.

3.1.3   Mediation

Conciliation and Mediation lie at each end of a spectrum or continuum of non-binding dispute resolution. Conciliation often provides, or results in a decision, or recommendation, from the conciliator; conciliation lies at the adjudicative end of the continuum. Mediators are reluctant to provide such decisions or recommendations, indeed some mediators will not. Mediation lies at the facilitative end of the spectrum. Any settlement which occurs is the parties' own and the analogy of chemical catalysts is often made.

Med Arb is a process that has been much debated in recent times. However, although the idea is attracting interest there remains some fundamental matters of principle, which may affect the validity of a binding award for such a tribunal. These include the argument that such an award would be against the natural rules of justice in that a party to the process is unaware of what is being said in one of the private caucusing sessions and is therefore unable to reply. There is little evidence of the adoption of this method of dispute resolution but no doubt the debate is likely to continue for some time.

3.2   Binding

The main tribunals for binding dispute resolution are litigation and arbitration. Litigation in the courts is governed by codified rules; in the High Court the rules are set out in an official publication are colloquially known as the White Book[15]. The system of delivering evidence is governed by case law with occasional legislation e.g. the rules on hearsay in civil litigation were recently relaxed by the Civil Evidence Act 1995.

In court witnesses are normally permitted to describe the facts; however it is recognised that the court may require assistance with technical matters and in these cases witnesses can give their opinion. The expert witnesses are given guidance and their duties were reinforced recently by case law[16]. The overall principle that underlies the case law is that the expert is independent of the party which instructs them. Whether this will lead to a change of attitude amongst experts remains to be seen. At present a great deal of time and money is spent in the resolution of technical issues by expert witnesses and dissatisfaction from clients and legal professions is apparent.

Courts and arbitrators apportion legal and expert witness costs on the basis that the loser pays the winner's costs, subject to assessment (called "taxation") by the court or arbitrator and also subject to practices whereby a party can limit recoverable costs by e.g. payments into court and sealed offers in arbitration.

Contingency fees / Conditional fee arrangements

A contingency fee is any sum (whether fixed, or calculated either as a percentage of the proceeds or otherwise howsoever) payable only in the event of success in the prosecution or defence of any action or other contentious proceedings. Contingency fee arrangements are prohibited by the Law Society in England and Wales for contentious business[17]. However, such an arrangement is permitted where an action is being pursued outside England and Wales and a contingency fee would be permitted in that country in respect to the proceedings. This form of arrangement is also available in relation to non - contentious work.

A conditional fee arrangement allows a solicitor to act on a "no win, no fee" basis in a limited number of circumstances[18]. In the event of the successful outcome of the case an "uplift" on the solicitor's normal fees will be payable.

There is at present extensive debate within the profession regarding the extension of such arrangements to other categories of litigation.

3.2.1   Adjudication

Adjudication as dispute resolution presents a problem for the construction industry. The problem lies in adjudication as an established practice, and adjudication as envisaged by The Latham Report and provided for by HGCRA and The Scheme for Construction Contracts.

Adjudication has existed as established construction practice for some time. Standard forms have included express provision for adjudication; although the grounds for such adjudication were normally restricted to payment and *set-off* matters.[19]

Early debate on the subject of adjudication merely treated adjudication from the position of what it is not. Adjudication is not arbitration. Adjudication was classified together with the "other" adjudicative binding processes e.g. reference to an expert; independent valuation and expert determination. [20]

The HGCRA gives little guidance on the intended nature of the adjudication process; although benchmarks for timescales are laid down in The Scheme for Construction Contracts. Adjudication is clearly distinguished from arbitration e.g.

"An exchange of written submissions in adjudication proceedings or in arbitral or legal proceedings" s107 (5)

"The contract shall provide that the decision of the adjudicator is binding until the dispute is finally determined by ... arbitration" s108 (3)

The current debate is how the courts will treat adjudicator's decisions. It is clear that routes to enforcement of contractual provisions and decisions exist, what is less clear is how the courts will intervene in decisions.

Initial drafts of the Scheme for Construction included a list of Nominating Bodies, with the idea that parties would turn to these Nominating Bodies for their adjudicator, in the absence of agreement. The drafters then proposed that no Nominating Bodies would be listed and that the market would decide. The position at the time of writing is that a list of potential Nominating Bodies will be issued by the Government.

3.2.2   Arbitration

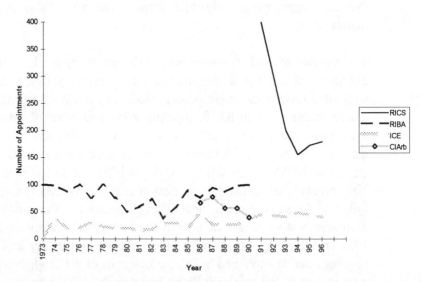

Figure 1 Presidential Arbitration Appointments (Source The Institutions).

All standard forms of contract provide for an arbitrator to be agreed by the parties, if they cannot agree between themselves the form provides that an arbitrator will be appointed by the President of one of the Professional Institutions.   The Institutions most frequently given this task are: Chartered Institute of Arbitrators (CIArb), Institution of Civil Engineers (ICE), Royal Institute of British Architects (RIBA) and The Royal Institution of Chartered Surveyors (RICS).

There are others but it is proposed that these Institutions are the main ones in terms of the numbers of appointments.   All deal with appointment in slightly different manners; but the result is that the President appoints an arbitrator from a list of suitably qualified members.

The number of such appointments is shown in Figure 1.   With the exception of the RICS the figures are relatively stable, and it is suggested lower than might be expected.   The RICS figures demand closer attention.   The RICS appoints a large number of arbitrators each year, these are mostly property related disputes e.g. lease disputes. In the late 1980's the number of such appointments was in excess of 20,000.   This fell to 16,000 in 1991 and to 5,000 in 1996. The number of construction disputes while much smaller has followed the same pattern; unconfirmed figures put the construction arbitrations at in excess of 600 in the late 1980's falling to 350 in 1991 and levelling out at ~170.   The collapse in property arbitration numbers can be explained by the parlous state of the English property markets; rents fell markedly from highs in the 1980's boom.   There were few leases to produce disputes and the practice of upward only rent reviews was forced out by tenants during lease negotiations. The number of construction arbitrations has also reflected the economic climate; construction output has declined markedly since 1988.

There is, once again, very little data on the use of arbitration.   The research which has been done centres on the opinions of arbitrators and the satisfaction of the parties.   There is no data on the frequency of arbitrations; the amount in dispute or the cost of the process.   The discussion of arbitration in the literature is dominated by anecdote and it is suggested that anecdote can prove anything

England and Wales has ratified the New York Convention 1958 and this is included in the Arbitration Act of 1996. The UNCITRAL Model Law for Arbitration has not been adopted. The Department of Trade and Industry's Departmental Advisory Committee on Arbitration Law (the DAC) considered this and concluded that the UNCITRAL model law should not be adopted into English Law [21]. However the DAC recommended that 'there should be a new Arbitration Act setting out in statutory form ... the more important principles of the English law of Arbitration .... in logical order, and expressed in language which is sufficiently clear and free from technicalities to be readily comprehensible to the layman".

The decision not to adopt the UNCITRAL Model law was not a spurning of that law. Commentators have argued that English Law has taken the torch of the UNCITRAL model and further developed international arbitration law [22]. England is a major centre of International Arbitration and the importance of the invisible earnings to UK companies are recognised.

The Arbitration Act 1996 identifies three principles :

1.   Fair, speedy and cost-effective dispute resolution by impartial tribunals.
2.   Autonomy of the parties.
3.   Minimum court intervention in the arbitration process and support for the process.

Equity Clauses; *ex aequeo et bono* and *amiable compositeur*

There has been concern that arbitrators should not decide under unascertainable principles of equity; and that arbitrators should not be allowed to act as mediator or amiable compositeur.

Therefore neither UNCITRAL Model or The Arbitration Act 1996 allows the arbitrator such free reign. However, the tribunal is allowed with the agreement of the parties to decide *in accordance with such other considerations as are agreed* (Section 46(1)).

Immunity of arbitrators and bodies appointing arbitrators

The Arbitration Act 1996 provides that the arbitrator is not liable for anything done or omitted in the function of an arbitrator provided the act or omission is not shown to be in bad faith (Section 29(1)). The position of appointing bodies is also covered; Section 74 provides for the same exclusion of liability.

Separability, Kompetenz-Kompetenz and arbitrability

The problem which arises when the arbitration agreement is contained in an agreement which is challengable as invalid, is dealt with at Section 7 of The Arbitration Act 1996.

When the jurisdiction of the arbitrator is being challenged the arbitrator is given express powers to "rule on its own substantive jurisdiction" at Sections 30, 31 and 32 of The Arbitration Act 1996 although it remains to be seen how this will work in practice if one party asserts that there is no arbitration agreement at all and refuses to participate.

The arbitrator is permitted to use his own knowledge and expertise during the course of the proceedings provided that the parties are made aware of such area of knowledge and expertise and are give the opportunity to address the arbitrator on the matter.

Arbitration awards are not officially published.

3.2.3   Expert Determination

Expert determination is a private means of commercial dispute resolution. The parties to a contract jointly instruct a third party to decide a dispute. The essentials of the role of the expert have been described as:[23]

1.   No requirement to act judicially but may carry out investigation;

2.   No requirement to receive submissions or evidence;

3.   No legislation to cover process;

4.   No statutory right of appeal or review;

5.   No statutory right of registration of and enforcement of decision.

However the expert must act in accordance with the terms of his appointment (which may have implied terms or incorporate parts of the contract) if his award is to be valid.

3.2.4   Litigation

As described in 1.1 above a specialist tribunal exists in the High Court for construction cases (the Court of the Official Referee). Actions can be started elsewhere in the High Court and not all construction litigation takes place here; but it is suggested that the OR's court is representative of the majority of construction litigation.

Activity in the OR's court is recorded by the Lord Chancellor's Department in statistics on the number of trials per year and the number of summonses issued[24]. This data has been collected and is shown in Figure 2.

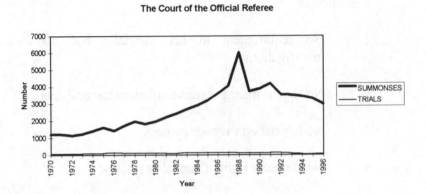

Figure 2 Activity in the Court of the Official Referee

The salient features of the data include the increase in activity when measured by summonses and the relatively stable trial activity. The increase in summonses may be explained by increased litigation in the construction industry matching a general widely accepted increase in litigation. Although this is not proven and there are other variables, which may contribute. For example the OR's court has become more widely known and more actions are now entered there as opposed to elsewhere in the High Court. In addition, it could simply reflect more activity within particular cases.

The stability in the number of trials is also difficult to explain. Despite an increase in the number of full-time OR's in London the number of trials has remained stable and has fallen recently from a peak of 151 in 1987 to 86 in 1996. Explanations offered for this include the propensity of the parties to reach a settlement before trial and the increasing length of trials due to complexity.

Certainly trial dates are currently being set 3 years ahead; there is some indication that the courts are busy with lengthy complex trials.

Avoiding Court Jurisdiction

The parties may avoid the jurisdiction of the court e.g. by an arbitration agreement. The court will recognise such agreements and by virtue of The Arbitration Act 1996 will issue a stay of legal proceeding (Section 9). This has been reinforced by the Courts in the case of Halki Shipping -v- Sopex (1998).

Court's Powers to direct the use of external procedures[25]

The courts will on an individual basis enforce any agreement to use some other dispute resolution than litigation and arbitration; or as a precursor to litigation and arbitration.

There have also been some "pilot" schemes for court annexed ADR and statements of practice on ADR and the courts:

1. 10 December 1993. The judge in charge of the commercial list issued a Practice Statement stating that the judges wished to encourage the parties to use ADR.
2. 24 January 1995. The Lord Chief Justice issued a Practice Direction (Civil Litigation: Case Management).

None of these appear to have produced any volume of ADR and although the English courts do appear to be sympathetic to and enthusiastic for ADR the parties to disputes have not shared this.

Court Appointed Experts

Expert witnesses and their reports have been the subject of criticism by the judiciary for more than 100 years.[26] Judges tired of having to choose between such opposing expert witness reports resorted to a referee out of sheer frustration at being unable to assess the relevant information and then to reach a decision. Systems of court appointed experts developed, where matters of opinion and fact finding were referred to a single expert appointed by the courts. This removed the need for both parties to appoint their own expert and obviated the often wasteful, and sometimes pointless, system of adversarial examination and cross-examination. Since 1932 Order 40 of the Rules of the Supreme Court has allowed the judge to appoint a court expert when requested by the parties, however this rarely happens.[27]

What are the Court's powers to enforce foreign judgments and arbitration awards?

The Foreign Judgments (Reciprocal Enforcement) Act 1933 allows judgments from a foreign country to be recognised by the courts of England and Wales if that foreign country has a reciprocal arrangement with the United Kingdom to recognise United Kingdom judgments.

The judgment creditor may apply within 6 years of the date of the judgment to the High Court to have the judgment registered. Under section 2 (1) of the Act (as amended) once the judgment has been registered it can be treated as if it were a judgment of a Court in England and Wales. Section 2 (6) also allows for interest and the reasonable costs of and incidental to registration, including the costs of obtaining the certified copy of the judgment from the original court.

Order 73 Rule 8 of the Rules of the of the Rules of the Supreme Court states that if an arbitration award has become enforceable in a foreign country then it can be enforced as if it were a Court Judgment for the purposes of the 1993 Act.

The Brussels Convention

The Brussels Convention had the object of determining which Courts have jurisdiction in disputes relating to civil and commercial matters between the contracting states and to facilitate enforcement of judgments. The convention establishes an enforcement procedure which constitutes an autonomous and complete system independent of the legal systems of the contracting states[28]

The Civil Jurisdiction and Judgments Act 1982 implemented the Brussels Convention into England and Wales. A judgment given in a contracting state and enforceable in that state can be enforced in another contracting state when on the application of an interested party it has been declared enforceable within that country.

In England and Wales an application for registration is made to the High Court.

The Lugano Convention

The Lugano Convention was made between the member states of the EU and the European Free Trade Association (EFTA). It is very closely modeled on the Brussels Convention and achieves similar enforceability but it should be noted that whilst the United Kingdom has ratified the Lugano Convention it has not been ratified by all EU and EFTA member states.[29]

Geneva Convention and New York Convention 1958

Sections 99 - 101 of the Arbitration Act 1996 Act preserves the ability to enforce awards under the Geneva Convention and the 1958 New York Convention subject to certain requirements and restrictions.

Common Law

It is also possible to enforce Arbitration Awards under common law. This derives from the contract between the parties giving rise to the arbitration agreement and will depend on applicable laws and ordinary questions of jurisdiction.

Appeals

In most cases an appeal to the Court of Appeal lies from Orders of the High Court (which includes the Official Referees Court).

In circumstances where the decision of the High Court is to determine an appeal from the award of an arbitrator it is necessary to obtain leave of this court or from the Court of Appeal.

In addition no appeal will lie without the certificate of the High Court that the question of law either is one of general public importance or is one which for some other special reason should be considered by the Court of Appeal.

Appeals from decisions of the Court of Appeal are made to the House of Lords. It is necessary to obtain leave to appeal from either the Court of Appeal itself or by petitioning the House of Lords. The criteria upon which leave is granted include:

(a)   the general importance of the case;

(b)   the likelihood of success; and

(c)   the degree of dissension the case may have previously caused i.e. whether the Court of Appeal decide the case by majority or overturned the decision of the lower court.

Appeals from the award of an Arbitrator

It is necessary, due to the vast number of arbitration agreements still subject to the 1979 Arbitration Act, to examine the law in relation to this legislation in addition to the newly enacted Arbitration Act 1996.

Under section 1(2) of the Arbitration Act 1979 it is possible to appeal any question of law arising out of an arbitrator's award with either the consent of all other parties to the reference or with the leave of the court.   A number of standard form contracts used within the construction industry contain clauses expressly agreeing to such an appeal.

The occasions in which leave will be granted have been severely restricted by the courts.   The House of Lords has set down strict guidelines in two judgments "The Nema"[30] and "The Antanios"[31] which must be followed and satisfied if leave is to be granted.

The "Nema" guidelines in general terms restrict the granting of leave to cases in the following categories:

(a)   where the subject matter is a "one-off" point the arbitrator's decision must be "obviously wrong on mere perusal of the reasoned award";

(b)   where "it appears upon perusal of the award either that the arbitrator misdirected himself in law or that his decision was such that no reasonable arbitrator could reach";

(c)   where the subject matter is a more standard and commonly used term the appeal has to be "seen to add clarity and certainty to English Law" and there must be a strong prima facie case that the decision of the arbitrator is wrong.

It is possible to contract out of the right to appeal under Sections 1 and 2 of the 1979 Act.

Sections 68 and 69 of the 1996 Arbitration Act deal with challenges to the award of an arbitrator and appeals on a point of law.

Section 68 is a mandatory provision of the Act which states that a party;

> *"may (upon notice to the other party and the tribunal) apply to the court challenging an award of on the ground of serious irregularity affecting the tribunal, the proceedings or the award."*

Section 68(2) defines serious irregularity and provides a list of grounds upon which an application must be founded. The leave of the court is required for any appeal from a decision of the court under this section.

Section 69 deals with appeals on points of law but the parties are free to exclude such a right of appeal. All other avenues of appeal must be exhausted before an application for leave or appeal to the courts is made (Section 70(2)) and this must be brought within a time limit of 28 days (Section 70(3)).

Section 69(3) states that leave to appeal shall only be given if the court is satisfied:

(a)   that the determination of the question will substantially affect the rights of one or more of the parties,

(b)   that the question is one that the tribunal was asked to determine,

(c)   that, on the basis of the findings of fact in the award -

   (i)   the decision of the tribunal on the question is obviously wrong, or

   (ii)   the question is one of general public importance and the decision of the tribunal is at least open to serious doubt, and

(d)   that, despite the agreement of the parties to resolve the matter by arbitration, it is proper in all the circumstances for the court to determine the question

This provision is a codification of the "Nema" guidelines.

Leave of the court is required for any appeal from the court under this section (Section 69(9))

Without Prejudice

The concept of "without prejudice" exists within the legal system of England and Wales. All letters and oral communications which are written or made for the purpose of settling a dispute which are headed or alternatively proved to be "without prejudice" cannot generally be admissible in evidence at trial and cannot be used to establish an admission or partial admission.

This protection extends not only to the particular letter or conversation itself but to all later parts of the same series of correspondence or discussions between the parties even if they are not each marked "without prejudice" or said to be so. This remains so until there is a clear break in the chain of correspondence or discussions.

The Courts have made it clear that only genuine negotiations with a view to the settlement of a dispute are protected from disclosure. This applies whether or not the term "without prejudice" has been expressly applied to the negotiations[32]. Therefore, this protection may arise expressly or be implied from the parties conduct.

Evidence

In all cases in the High Court the parties are required by RSC Order 38 rule 2A to exchange witness statements. It is usual, if the matter proceeds to trial, for the Court to allow the witness's statement to stand as evidence in chief.

It is commonplace for this procedure to be adopted in arbitration. The witness will then be orally cross-examined and re-examined.

In certain proceedings, for example applications for summary judgment, evidence is provided in written form by a sworn affidavit. RSC Order 38 rule 2 empowers judges to order, at their discretion, that a witness's evidence is provided by sworn affidavit and read out at the trial. The deponent of the affidavit may be ordered to appear to be cross examined on its contents. This is a particularly useful method of adducing evidence which will not be contested by the opposing party.

3.2.6   Ombudsman

English definitions of ombudsman [33] centre on officials to investigate complaints against public bodies. However there is also a development in the UK of Ombudsman operating as consumer protection eg insurance, banking and legal services[34]. There is no system of Ombudsman in construction.

4.   THE ZEITGEIST

In the England and Wales the legal zeitgeist in litigation may be summarised by the aims of the review of the current rules and procedure of the civil courts in England and Wales carried out by Lord Woolf: the aims of the review were:

- to improve access to justice and to reduce the cost of litigation;
- to reduce the complexity of the rules and modernise terminology;
- to remove unnecessary distinctions of practice and procedure.[35]

As noted earlier the construction industry has been dominated by the Latham Review. Construction dispute resolution, because it so often depends on arbitration, has eagerly awaited the Arbitration Act and rather more nervously awaited the Adjudication provisions of HGCRA. These will undoubtedly have a profound effect on construction.

There is a trend to the use of ADR and, in particular adjudication, as a forerunner to litigation or arbitration or both in anticipation of the HGCRA and because parties are beginning to see ADR as "mainstream" rather than "quirky". Mediation is being used occasionally but is still the subject of much caution and reluctance particularly from those who have not experienced it yet.

The Private Finance Initiative is leading to new types of contract structures which have led to the re-appraisal by many contractors of the type and extent of risk which they are willing to take on. Some "re-inventing of the wheel" has taken place leading to lengthy negotiations increasing tendering costs.

1   Latham, M. (1993) Trust and Money, Interim Report of the Joint
         Government/Industry Review of Procurement and Contractual
         Arrangements in the United Kingdom Construction Industries, The
         Department of the Environment, London.
         Latham, M. (1994), Constructing the Team, HMSO ISBN 0 11 75
         2994X, London.

2   Construction Industry Board   (1997) The State of the Construction
         Industry, November - Issue 8, Department of the Environment,
         Transport and the Regions.

3   For an excellent discussion of the economy see: Hutton (1995), The
         State We're In, Cape, London.

4   Fay, E. (1988), Official Referees' Business, Sweet and Maxwell,
         London.

5   Hansard (6th Series) vol 57 col 626.

6   Rules of the Supreme Court Ord 36, r 1.

7   See The Channel Tunnel Group v Balfour Beatty Construction Ltd
         [1993] AC 334.

8   See Fisher and Ury (1997), Getting to Yes, Arrow Books.

9   Building Project Information Committee (1987), Co-ordinated
         Project Information Common arrangement for building work
         sections.

10   In no other industry is the responsibility for design so far removed
         from the responsibility for production The most urgent problem
         which confronts the construction industry is the necessity of thinking

and acting as a whole... it has come to think of itself... as a series of different parts. These attitudes must change.
The Banwell Report 1964.

11   The Royal Institution of Chartered Surveyors (1996), Survey of Contracts In Use.

12   The Lynton Report Lynton plc, Bovis Construction & Davis Langdon & Everest. The UK construction challenge: main report. Lynton plc. April 1993.

13   Latham op cit.

14   http://www.masons.com/library/journals/irwin96/win96-05.hts.

15   The Supreme Court Practice 1997 (the Rules of the Supreme Court) Sweet and Maxwell 1997. Colloquially The White Book.

16   National Justice Compania Naviera SA v Prudential Assurance Company Ltd [1993] 2Lloyds Rep 68 and Cala Homes (South) Ltd and Other v Alfred McAlpine Homes East Ltd (1995) 13-CLD-10-27.

17   Solicitors Practice Rules 1990 rule 8

18   Solicitors Practice Rules 1990 rule 8(1A), Conditional Fee Arrangement Order 1995.

19   Bentley, B. (1992) 'Adjudication - a Temporary Diversion', in Fenn, P and Gameson, R (Eds) (1992), Construction Conflict: Management and Resolution, Chapman Hall, London.

20   See in particular McGaw, M. (1991) 'Adjudicators, Experts and
        Keeping out of Court', Current Developments in Construction Law,
        Fifth annual conference the Centre of Construction Law and
        Management, Kings College, London and Kendall, J. (1992b)
        Dispute Resolution: Expert Determination, Longman, London.

21   Report of June 1989 of the Departmental Advisory Committee on
        Arbitration Law.

22   Lord Hacking, (1997) "Arbitration Law Reform" Arbitration, Vol 63
        No 4 p291-299.

23   Fernyhough, R., (1997) The Power of the Courts to Intervene, The
        Annual Construction Law Conference, Masons.

24   The data has been compiled from Civil Judicial Statistics and
        Parliamentary Command Papers. The data does not include work
        carried out by Official Referees in the provinces.

25   This section relies heavily on Black (1997) ADR of Commercial
        Disputes and the English Courts, Bulletin 1, Swiss Arbitration
        Association p3-12.

26   Kennard v Ashman [1894] 10 TLR 213.

27   Fenn, P. O'Shea, M. and Speck, C. (1995), Scientific Witness of
        Opinion (Expert Witness) in Civil Litigation, Nature, Vol 378 No
        6558.

28   Societe d'Informitique Reaslisation Organisation -v- Ampersand
        Software BV The Times 25 September 1995.

29   Up to date information as to the progress of ratification of other
        member states may be obtained from the Lord Chancellor's

Department, International Division, 28 Old Queen Street, London.
SW1H 9HP.

30   Pioneer Shipping v BTP Tioxide, The Nema [1982] A.C. 724.

31   Antaios Cia Naviera S.A. v Salen Rederierna A B, The Antaios
      [1985] A.C. 191.

32   Rush & Tompkins Limited -v- Greater London Council [1989] 1AC
      1280.

33   See for example The Little Oxford and Collins Concise.

34   See Brown and Marriot (1993), ADR Principles and Practice, Sweet
      and Maxwell, London, pages 279-281.

35   The Woolf Review Access to Justice, Woolf Enquiry Team, Room
      438, Southside, 105 Victoria Street, London.

ESTONIA

Nicholas Gould
Masons Solicitors, Manchester, England

1.   BACKGROUND

1.1   Economic

Estonia took advantage of Russia's internal turmoil in 1991 and promptly declared independence.[1]   Since then Estonia has established a market-based economy and shifted trade from the East towards the West.   This transformation has been brought about by a number of youthful Estonian entrepreneurs.   Indeed, Estonia's commercial foresight is to be admired as is forges ahead taking full advantage of its geographical location.   Superficially it's location does not appear to be of particular advantage, but Estonia does have the deepest ice-free ports in the region, thereby providing an important international trade route between parts of Europe and Russia.   Finnish businessmen dream of making Estonia 'the Hong Kong of the Baltics', a view shared by many aspiring Estonians.[2] This, of course, depends on maintaining a good relationship with Russia.

1.1.1   Geography

Estonia is the most northerly of the Baltic republics and covers an area of 45,100 sq km (117.413 sq miles).   The population is approximately 1.6 million of which a large proportion are ethnic Russians (30-35%).   The density of population is of the order of 35.5 per sq km although one third of the population is concentrated at the capital, Tallinn.   In comparison the UK's population density is in the region of 235 per sq km.

Estonia is highly forested and surprisingly flat.   There are about 1500 lakes and 500 rivers, together with some 1500 coastal islands.[3] Coupled with its large port and proximity to Finland means that economic activity is heavily concentrated in Tallinn. Helsinki is only 78 km from Tallinn, north across the Gulf of Finland.

1.1.2   Culture

Estonians are ethnically related to the Finns and linguistically to
the Hungarians. The Estonian language derives from the Baltic-
Finnish brand of the Finno-Ugric languages.[4]   This may well
account for the close modern-day relationship with Finland. Most
Estonians can not only speak Estonian, but also Russian, as they
had to do so under Soviet rule. In addition, a great many can
speak German, and other European languages (including
English). The Estonian language can only boast approximately
1.0 million fluent native speakers. Further languages are,
therefore, vital for economic export development.

1.1.3   Currency and Economy

Within 12 months of achieving independence, Estonia became the
first state to emerge from the Soviet empire, to abandon the
plummeting rouble and to issue its own legal tender (June 20th,
1992). Estonia redirected most of its trade from the East to the
West following independence and Finland quickly replaced
Russia as the main trading partner. Russia took 95% of Estonian
exports in 1991. This is now less than 30%, whilst Estonia no
longer relies on Russia for its imports, dropping from 85% in
1991 to around 30% in 1994. These remaining imports are
mainly in the form of oil and natural gas.

1.2   Construction

1.2.1   Construction Firms[5]

The latest data suggest that there are 4,263 construction firms in
Estonia.[6]   According to a survey, 60% of the total construction
industry turnover was carried out by privately-owned companies.
The current regulations require all of these firms to employ at
least one person with a Diploma qualification, who then appears
on their letterhead. There are six large companies who dominate
the market, which suggests a market structure comparable to that

of the UK. The turnover of one of the largest construction companies was 110.9 million kroons in 1993 with a gross profit of 9.9 million kroons. (8.93% gross profit margin).[7] It is interesting to note that compared to the previous year, turnover grew by 215%. Taking into account the inflation rate, this appears to be a rather good result, particularly as the number of employees decreased during the same period from 944 to 824.[8]

1.2.2   Planning and Building Control

Estonia operates a directive-issuing, highly centralised investment-based planning system. This can be contrasted with the Finnish system which is not only community consultative based, but also requires a great deal of consultation with local business in order to encourage new investment or support for decisions. Urmo Kala, the Vice Chancellor of the Ministry of the Environment, has stated that he would like to see this type of system developed in Estonia.[9]

1.2.3   Construction Professionals

The two predominant professions in Estonia are the architect and the engineer. Both have separate training routes. Architects training at the academy of arts and according to Roode Liias the architect is the "author" of the building, preparing the design and drawings but not managing the contract in any way.[10] The engineer receives training at the Technical University in Tallinn, the only one in Estonia. The Association of Civil Engineers in Estonia is probably the most important professional body. Working parties set up by the Association are currently considering civil engineering practice and in particular changes to the design standards.[11]

The Estonian Association of Construction Entrepreneurs comprises voluntary membership of building companies with the aim of harmonising and co-ordinating the activities of its members. In addition to this, they recognise that by acting together they will be able to achieve their common objectives more easily.

The Association's main aim is to aid the implementation of an "... *economic mechanism based upon an open market as well as to contribute to the elaboration of the construction policy of the republic of Estonia.*"[12]

1.2.4   The Construction Market[13]

The turnover of the Estonia construction industry in 1993 was 2.497 million kroons which rose to 3,500 million kroons in 1994.[14] The Investments and Construction Statistics Department of the Government produces a quarterly bulletin with some limited industry information, based upon 1725 companies who are listed on the EER (Estonian Enterprises Register) whose main activity is construction work. In the first quarter of 1994 the turnover of the EER was 647.1 million kroons, 85% of which could clearly be attributed to their main activity of construction, with 880 million kroons for the fourth quarter.[15] Compared to the fourth quarter 1993 this represents a decrease of nearly 24% The concept of a single legal entity engaging in the activities of different industries seems an unusual concept (unless of course that company is a parent holding company), but it must be understood that the Estonian swing towards a free market has only just emerged and is undergoing vast development. The "entrepreneurs" who are taking advantage of this situation will engage in any profit making venture and are not willing to restrict their activities. In addition to this 91% of the total turnover of the EER was attributed to work in Estonia, whilst the remaining 9% was due to the efforts of the Estonia enterprises in overseas countries.[16]

Maintenance accounts for a small amount of the total construction workload and it would seem that in the view of the vital repairs needed this proportion of the construction workload should expand.[17] The vital areas of work are generally related to the domestic multi-storey blocks, in particular repairs to roofs, replacement of windows and improvements to thermal insulation being highest on the list of priorities. Urmo Kala has suggested that the maintenance figure should increase to approximately 40% of the total workload. However Allen, points out that this appears to be a hope rather than an actual planned increase.

If additional finance was required for an individual project during the Soviet central planning period, then it would be provided, on certification of the engineer. However the overall budget does not change, the effect is that less of a certain type of construction is procured as a result of the overspending. This system was not conducive to increases in productivity or efficiency. Recent changes in the supply side of the construction industry may lead to a dramatic change in this situation as the increased private sector gears itself to the "West's" free market ideals of profit maximisation.

1.2.5   Market Opportunities

The market opportunities in Estonia can be considered from two perspectives.   First, the domestic opportunities, which the international consumer industry has already started to explore. In addition to this the availability of cheap skilled labour and intellectual ability suggests that competitive advantages could be gained from establishing engineering and manufacturing ventures in Estonia. This opportunity has already been exploited by Finish engineers.

The second limb of opportunity may prove in the long run to be the more attractive. If Estonia is established as the "gateway" to Russia then the opportunity for countries like the UK to benefit from the deep ice free ports, rather than cross Europe are potentially immense. The result would lead to a huge derived demand in the construction industry to supply the necessary infrastructure together with commercial and industrial installations.

In the short term, the most important tasks for the construction industry will be called on to fulfil are in the areas of maintenance and repair, housing and new commercial development. Whilst the latter is being turned into reality with the injection of foreign capital, the most important issue of repair work to existing housing does not appear to finding favour with investors. Even the World Bank which is planning to provide large sums of money for housing will only fund the construction of completely new buildings, rather than carry out the urgently needed repairs and renovations to the existing housing stock.

1.3   Legal System

The legal system is currently undergoing reform and this is resulting in conflicting legislation, particularly in the area of property transactions. Although the Soviet Civil Code is still in place, new legislation has replaced the old property law. In addition to this the old contractual system is still in operation, which leads to uncertainty in property transactions. The new property law is consistent with EC law and was drafted in the light of German property law.[18]

1.3.1   Law

The existing law was imposed by the then USSR, and it is still in use today, for example, in the area of contract law. For this reason much of the civil law is currently under review and in many cases being updated. The law in relation to former state owned or "public" property represents the largest changes, with a move towards the support of private ownership. It is hardly surprising to learn that there is no separate court for construction. In fact "generally, professionals in legal matters know nothing about construction."[19]   Nonetheless, advocacy is a separate profession.

1.3.2   Insurance

The oldest insurance companies in Estonia have only been established for a matter of years. At present there are a total of 14 companies operating there. Hansa insurance was formed in the spring of 1991 as a joint venture limited company. The majority holding is Swedish and Finnish; it would appear that all the other companies have a large stake foreign investor. Insurance is a product of the de-sovietisation and a relatively new issue in Estonia. The need to insure has grown from the evolving free market and in particular the banks insistence as a pre-requisite for lending. It would seem that whilst most contractors realise that they need to insure, they avoid the costs of doing so to remain competitive.

Institutions and laws have been established for consumer protection, and planning laws were established in 1995. As yet, there is no concept of agency.

1.3.3   Construction contracts

There are five contracts forms in use for construction works.

>   client - contractors
>   contractor - sub-contractor
>   client - consultant
>   management contract (client - professional manager)
>   client - professional supervisor

These forms are currently being updated to try and keep pace with the changes to the law. Some of these standard forms make reference to arbitration as the default dispute resolution technique. These clauses are being updated, and will, apparently, reflect a similar approach to the FIDIC form. The FIDIC form was translated into Estonian in order to train Estonian construction professionals in the use of the contract which has been used in Estonia by projects funded from the World Bank.

2.   CONFLICT MANAGEMENT

Advanced conflict management techniques, such as DRBs or DRAs are unknown in Estonia. This is not surprising, the size and nature of the construction industry simple does not demand this level of sophistication. Quality control is still very much in its infancy. However, some of the major companies are now introducing quality assurance schemes. Estonian standards are on a parallel with ISO 9000

Direct negotiation is the key conflict management tool. In general the Estonians are not adversarial in nature, in fact, quite the opposite appears to be the case. Tolerance is the key word for a country as small as Estonia which has had to survive Soviet domination for 50 years. As a result, the construction industry does not suffer from claims to any where near the same extend as the UK. The former Soviet thinking also plays a part in this philosophy, where projects were completed without sophisticated cost control mechanisms.

Formal partnering does not appear to have developed; the Estonians do not appear to need it.

3.   DISPUTE RESOLUTION

Currently, the courts have a large backlog of business with delays of up to three years before a case is heard.[20] Although in the UK construction disputes of any magnitude may take a similar time to reach the Official Referee's Court the size and turnover of the industry in the UK is far greater. In addition to the time period expected for trial, de-sovietisation often results in a change of the law to confuse the issue further. The main cause of the court's backlog is a result of independence itself.

The New York Convention 1958 was incorporated into Estonia law in 1993. Interesting, it would appear that no data on the UNCITRAL model law or model rules of arbitration exist in Estonian. The legal adviser at the Association of Construction Entrepreneurs can only recall one arbitration and one court case relating specifically to construction. An arbitration court is provided by the Council of Commerce which provides a set of rules. Arbitrators are given immunity, but are required to act independently during the arbitration.

In general, there appears to be little evidence of formal dispute resolution. At an informal level, the picture is much the same. Very occasionally professors from the Technical University are consulted over disputes. This often only extends to explaining the matter to the parties - how the parties finally resolve the matter is their business. Perhaps this could be considers mediation, at an informal level; institutionalised formal mediation does not exist.

It is clear that, once again, negotiation plays the central role in the resolution of disputes.

4.   ZEITGEIST

Estonia's move towards independence and the free market has provided a challenge to the Estonian people and their industries. The construction industry has risen to the challenge and adopted western management philosophies. This raises the concluding question; is there an Estonian market for overseas construction related services?

There is little evidence of foreign competition in the Estonian construction industry. The size of the construction industry is such that all of the operators are aware of each other and all contracting organisations must incorporate a qualified Estonian construction professional. A joint venture arrangement with a local company, which would overcome these barriers, could provide opportunities for foreign contracting organisations. Experience has shown that this is often the only way to infiltrate a foreign market; the local company providing local market knowledge whilst the foreign partner often provides management and capital. However, risks are higher than in the home market and this factor needs to be reflected in the profit/risk margin of the project.

Nevertheless, there is a growing market for insurance services (mainly led by Finland) and western plant and machinery; the old Russian made site equipment is dated in inefficient. Overseas suppliers of insurance services and contractors' heavy equipment may initially prove welcome by the Estonians. Further opportunities could include establishing business operations in Estonia, which would benefit from relatively cheap skilled labour and a readily available land resource. Finnish engineers have taken advantage of this situation already, but many other industries could consider this option.

Arbitration, although known in Estonia, is not used. It could be that this will provide a more suitable alternative to the courts in the future. At a recent conference in Tallinn the members of the Society of Entrepreneurs were introduced to the concepts of avoiding disputes through risk management, in addition to the concepts of ADR (alternative dispute resolution).[21]   It is suggested that in such a close knit community such as Estonia (especially Tallinn) ADR may well provide a real alternative to the courts for several reasons.

First, the providers to the industry rely upon maintaining a good relationship with their clients in order to ensure that vital repeat business is secured. The industry is small enough to ensure that everyone is aware of those who are gaining an unacceptable reputation. Secondly, in a culture which is already akin to a non adversarial stance, ADR would seem the most appropriate option, especially in the light of the court backlog and lack of arbitration experience.

It will be interesting to watch the developments of this new independent state. Will the move away from state control and towards a profit driven free market lead to an increase in disputes as contractors compete for work, or will the Estonians maintain a culture of non-adversarialism?

1   For a construction market perspective see Gould, N. (1996) Estonia: An Emerging Opportunity,   CIOB Occasional Paper, No. 58.

2   (1993) The Baltics' would-be Hong Kong: Estonia's new face, *Economist*, 6 November.

3   ADA Publications, *Insight Guide*, Eastern Europe.

4   Ibid.

5   Gould, N. (1994) Interview with Roode Liias, Professor of Construction Economics, Technical   University of Tallinn, 13 October.

6   Data current at 1 April 1995.

7   EMV certified annual report 1993.

8   Ibid.

9   Allen, J. (1994) Interview with Mr Urmo Kala, Vice Chancellor of the Ministry of the Environment and the President of the Association of Civil Engineers in Estonia, May.

10   Ibid.

11   Allen, J. Interview with Mr Urmo Kala, President of the Association of Engineers, Op. cit.

12   Estonian Association of Construction Entrepreneurs (1994) Information pack.

13   See also Siinmaa, U. (1996) Estonian Building Market Handbook, Ulo & Tiit Siinmaa Ltd, Estonia

14   Krabbi, L. (1994) Senior Economist, Investments and Construction Statistics Section, Quarterly   Report, Estonian Association of Construction Entrepreneurs, First Quarter.

15   Ibid.

16   Ibid.

17   Allen, J. Interview with Urmo Kala, Op. cit

18   Allen, J (1994) Interview with Andrus Lauren, Deputy General Manager, Mendelson Real Estate Company, May.

19   Quoted from recent correspondence with Roode Liias, January 30, 1997.

20   Allen, J (1994) Interview with Olev Saadoja, Hansa Insurance, 31 May.

21   Gould, N. (1994) Conflict Avoidance through Risk Management, Risk Ehitustegevususes ja Ehitusriski Kindlustsatmine (Risk Management in Construction Seminar) Tallinn, Estonia. Sponsored jointly by the University of Westminster and the Estonian Association of Construction Entrepreneurs, 27-31 October.

HONG KONG

Colin Wall
Commercial Mediation & Arbitration Services Limited
Timothy Hill
Masons Solicitors, Hong Kong

1.   BACKGROUND

1.1   Economic

The Construction Industry in Hong Kong had a manual worker labour force of approximately 83,500 in the first quarter of 1997. Of these, approximately 46,000 were employed in the public sector and 37,000 in the private sector. When the manual labour force is analysed by type of project the statistics are 58,107 employed in the building industry and 25,393 in the civil engineering industry. The figure for the overall labour force in Hong Kong in 1996 was 3,075,000 and the figure for manual workers in the construction industry amounted to 77,000.

In the first quarter of 1997 the gross value of construction work performed by main contractors at construction sites in Hong Kong amounted to HK$23,710,000 (HK$7.8 =US$1). Of this value, just over half was for public sector projects, the figures are HK$12,041,000 for public sector projects and HK$11,669,000 for private sector. When the total value is analysed in terms of project type, buildings account for HK$16,534,000 and civil engineering projects HK$7,176,000.

The Gross Domestic Product per capita for Hong Kong in 1995 was HK$176,178.00 of which construction accounted for 4.9%.

1.1.1   Trends

In 1991 there were 64,000 manual workers engaged on construction sites. The figures were 66,000 in 1995 and increased to 77,000 in 1996. The total number of manual workers engaged in construction during the first quarter of 1997 was 2.2% higher than the figure for the fourth quarter of 1996. This figure is the highest on record since 1983, and was 9.4% higher than the previous record of 76,340 reached in the fourth quarter of 1988.

Schemes have been introduced into Hong Kong for the importation of foreign labour to work in the construction industry. There is a general scheme and a specific scheme for projects related to the Airport Core Programme (ACP). The Government has launched an ambitious housing programme that will put further demands on the need for construction workers. Moves are under way to improve the method of importing foreign workers for construction projects in Hong Kong. Notwithstanding these importation schemes, the number of foreign workers engaged in construction projects in Hong Kong remains quite small.

Although work on the ACP projects is drawing to a close and the new airport and airport railway will open to the public in July 1998, it is not anticipated that the overall volume of construction in Hong Kong will slow down. This is because any down turn in the amount of civil engineering work is likely to be absorbed with the increase in the building sector, and in particular, in public housing. In addition, there are major infrastructure projects due to commence in 1998. These include, the Kowloon Canton Railway Corporation (KCRC) who are embarking on a major new railway line, known as West Rail and extending the existing line into Tsim Sha Tsui and the Mass Transit Railway Corporation (MTRC) who are extending the existing railway system.

1.2   Legal

The construction law of the Hong Kong Special Administrative Region (which the authors shall refer to as Hong Kong) has its roots in the English common law system. The common law system consists of legislation, in Hong Kong called Ordinances, supplemented by a body of decisions of the Courts known as precedents. Courts are generally bound to follow decisions of superior courts and will generally follow prior decisions of courts of the same level.

The highest court is the Hong Kong Court of Final Appeal. Prior to July 1st, 1997 when sovereignty over Hong Kong reverted to China, the highest court was the Privy Council, which sat in London. Beneath the Court of Final Appeal is the Court of Appeal and beneath that the Court of First Instance (formerly known as the High Court). Whilst not strictly bound to do so the Courts of Hong Kong have historically followed the decisions of the English Courts. Following the change of sovereignty over Hong Kong it remains to be seen whether the appellate courts will be more inclined to follow the jurisprudence of other common law jurisdictions e.g. Australia and Canada in preference to that of England where views differ on specific issues. The Joint Declaration on the Question of Hong Kong, signed between the Peoples' Republic of China (PRC) and the United Kingdom on 19th December 1984, by which sovereignty over Hong Kong was returned to China, expressly preserved the pre-existing system of laws, i.e. the common law for a period of 50 years from 30th June 1997.

Resolution of disputes in the Courts in Hong Kong is by adversarial procedures. Although the Arbitration Ordinance (Cap. 341) makes reference to decisions ex aequo et bono and as amiable compositeur[1], in the context of the international regime, there is no practice of such an approach being adopted and it remains questionable whether such a provision would be enforced.

Hong Kong is fortunate to have within the Court of First Instance a specialist Construction and Arbitration List, in which matters related to the Construction Industry will be heard either upon the request of the parties or direction of the Court.

The legal profession in Hong Kong is split between solicitors and barristers. Barristers tend to be specialist advocates and have exclusive rights of audience at hearings in the Court of First Instance in open court (subject to very limited exceptions) and in the appellate jurisdictions. The barristers' profession is divided into Junior and Senior Counsel (formerly entitled Queen's Counsel).

In arbitration the widest possible rights of audience are permitted. A party may be represented by anyone it chooses[2] and may seek to recover the costs of such representation[3]. The representative of a party has, in the absence of notice, ostensible authority to bind the party by consenting to orders or commercial compromises.

1.2.1   Construction Law

Hong Kong does not have a single codified construction law. Reference must be had to a number of Ordinances to establish the body of construction law. The following Ordinances are of particular relevance to the Construction Industry but are by no means a comprehensive list of all relevant Ordinances, the Arbitration Ordinance, the Buildings Ordinance (Cap. 123), the Employees Compensation Ordinance (Cap. 282), the Sale of Goods Ordinance (Cap. 26) and the Supply of Services (Implied Terms) Ordinance (Cap. 457).

The relationships between parties involved in construction projects will primarily be governed by the terms of the contracts entered into by the parties. The common law principles of negligence applied in Hong Kong may impose duties on parties to take reasonable care to exercise reasonable skill and care when dealing with others. Other tortious duties may also be relevant e.g. nuisance, trespass. In addition, certain statutory obligations are imposed on the parties pursuant to the various Ordinances referred to above.

There is a substantial body of professionals with specialist expertise in construction matters in Hong Kong. It should be noted that pursuant to the Buildings Ordinance (Cap. 123) professionals carrying out certain tasks in connection with construction works must be registered with the Buildings Authority e.g. approvals of drawings and structural plans etc.[4]

1.2.2   Construction Contracts

(a)   Government of Hong Kong - The Government of Hong Kong has drafted a number of contracts for use as standard forms in connection with the procurement of construction works by it. The current editions of the principal forms are listed in Appendix 1. In general terms, several of these standard forms are loosely based on the FIDIC provisions. Copies of these forms can be purchased from the Government Printer, Hong Kong. In addition to the general standard conditions, the Hong Kong Government has drafted specific terms for the ACP infrastructure developments associated with the New Airport at Chek Lap Kok. A common feature of these forms of contract is that they provide for the contractor to build in accordance with the Government's design.

The Hong Kong Government has produced a standard form of design and build contract. It is not uncommon for the forms to be subject to amendment by special conditions to reflect the needs of a particular project. Recently this has been particularly prevalent in respect of the Design and Build form used in building projects where amendments have been made to introduce fitness for purpose type obligations.

The most striking departure from the risk allocation contained in FIDIC that is embodied in these forms is the allocation to the contractor of the entire risk of unforeseen ground conditions.[5]

(b)   Private - Most civil engineering work executed in Hong Kong is executed for the Government. There have however been a number of tunnels, roads and power projects constructed using private finance to which bespoke contracts have applied. This section therefore focuses on building contracts.

The most widely adopted form for private building works is the first Royal Institution of Chartered Surveyors (RICS) (HK Branch) edition 1986, which, because its predecessor was published by the Hong Kong Institute of Architects (HKIA), is widely referred to as the HKIA Form. The authors will adopt this abbreviation in this report. This form of contract is based on the Joint Contracts Tribunal (JCT) Form 1963 Edition. A number of the private developers make standard amendments to the Form.

Minor works are executed under the first RICS (HK Branch) Minor Works Form of Contract, a bilingual document.

(c)   Dispute Resolution - It is a standard feature of the Hong Kong Government contracts listed in Appendix 1, that any dispute to be referred to and settled initially by the Architect or Engineer. (The Supervising Officer in Design and Build Contracts.) Since many disputes will themselves arise from a decision or direction of the Architect or Engineer few disputes are resolved by this decision. In the event that a party is dissatisfied with the decision of the Architect or Engineer it may, within a 28 day period, refer the matter to mediation in accordance with the Hong Kong Government's Mediation Rules. Under such rules the party receiving a request for mediation may refuse to participate in the mediation. In the event of such refusal or the failure of the mediation, a dispute may be referred to arbitration within 90 days of such refusal or failure. In the absence of agreement any such arbitration may, save in limited circumstances, not be pursued until after the works have been completed.

The conditions for the ACP contain a more cumbersome procedure for dispute resolution.[6] A formal Notice of Dispute must be prepared and submitted to the Engineer for a decision. In the event that either party is dissatisfied with the decision it may refer the matter to mediation. Parties are compelled to participate in the mediation, which in the absence of agreement to the contrary must be concluded within 42 days of the appointment of the mediator. In addition to conducting a facilitative mediation within this time period the mediator is expected to also prepare a report containing his recommendations to the parties in the event that the mediation is not otherwise going to succeed. This requirement to deliver a report has been the subject of much debate within the dispute resolution community in Hong Kong[7]. On the one hand it is argued that such a report is contrary to the process of mediation, inhibiting parties from being entirely open with the mediator. On the other hand, in many situations one party to the dispute is Government, represented by officials who may find it difficult to make "commercial" concessions. In such circumstances it is argued that it is the report which has facilitated settlement.

In the event that the mediation does not succeed, either party may then refer a dispute to adjudication, provided that the dispute relates to time or money. In practice most disputes fall into these categories. The decision of the adjudicator is final and binding on the parties unless a contrary settlement agreement is made or the matter has been referred to arbitration in accordance with the conditions. The adjudicator is required to reach his decision within 28 days of termination of the mediation. In practice, this time period has often proved too short. If the matter is not referred to adjudication or the adjudicator has given a decision with which either party is dissatisfied the dispute may be referred to arbitration. In the event that a dispute is referred to arbitration, completion of the works is a condition precedent to any step being taken in the reference.

The HKIA form contains provision for the reference of dispute to arbitration. The RICS (HK Branch) Minor Works Form of Contract contains a three-tier dispute resolution system, comprising negotiation, mediation and arbitration.

All the arbitration provisions referred to provide for references to be to a single arbitrator. With the exception of the 1993 Building Conditions the Hong Kong Government forms all opt into the domestic regime of the Arbitration Ordinance.

2.   CONFLICT MANAGEMENT

In Hong Kong there is little in the way of formal conflict management[8]. There is only one known use of Dispute Review Boards, only one known use of the Dispute Resolution Adviser system in a pre-contract dispute prevention role and there is limited use of partnering.

2.1   Non-binding

2.1.1   Dispute Review Boards

The only example of a Dispute Review Board known to the authors is the one in use at Hong Kong's new airport. The background to the establishment of the Dispute Review Board is set out in another TG 15 paper[9] and is not reproduced here. One of the authors was involved in the decision to establish the Dispute Review Board, the design of the procedures and the selection of the members. This author is aware that from the contractors' viewpoint at least, that the primary intention of the Dispute Review Board, was the avoidance of disputes. Thus far, this objective has, to a large extent, been achieved, as there has only been one formal dispute.

The then Provisional Airport Authority, now the Hong Kong Airport Authority (HKAA), had been in negotiations with the Hong Kong Construction Association (HKCA) in respect of the proposed conditions of contract. The proposed conditions of contract were regarded as onerous and disputes were thought to be inevitable. The contractors on the advice of the HKCA originally boycotted the tenders. Amongst the contractors' demands for improvements was a quicker and more effective dispute resolution system. The proposed conditions of contract only contained arbitration after project completion. The HKCA and HKAA negotiations led to the establishment of the Dispute Review Board. The Dispute Review Board is actually called the Dispute Review Group and comprises seven individuals including a chairman known as the Convenor. The Convenor, a retired High Court judge, is a resident of Hong Kong. The remaining members are from outside of Hong Kong. Two of the other members of the Dispute Review Group are arbitration experts from the PRC. The remainder are from the United Kingdom[10] and comprise two civil engineers, a quantity surveyor and one electrical & mechanical engineer.

The members of the Group visit the airport site once every three months and spend approximately four and a half days on the project. The Group members spend half a day visiting the airport terminal building, which is the single largest and the most complex contract. This gives them the opportunity to view physical progress. They then attend a meeting between the HKAA and the contractor constructing the terminal building. This meeting is known as the Quarterly Review Meeting. The Group members spend half a day visiting the remaining contracts let by the HKAA, including the runway and apron works contract, the Ground Transportation Centre and contracts for ancillary buildings. In the afternoon, immediately following the second site visit, the Group members again sit in on the various

Quarterly Review Meetings. The Group members spend the remaining time attending other meetings.

As noted above, up to 31st December 1997, there had only been one formal dispute that needed to be considered by the Dispute Review Group. Three members of the Group decided this dispute recently.

2.1.2   Dispute Review Advisor

To the authors' knowledge this term is not used in Hong Kong.

2.1.3   Negotiation

In the private sector negotiations are common place and it is usually possible for contracting parties to negotiate terms between them which vary the original terms of the proposed construction contract. As negotiations are conducted in private, the authors have no details of the extent to which negotiations assist in the management of conflict or avoidance of disputes. In the public sector it is not usually possible for individual contractors to negotiate with the Works Bureau Departments inviting tenders. Occasionally matters of general concern on individual contracts are negotiated by the HKCA on behalf of all its members (rather than the tenderers), directly with the relevant Works Bureau Department or the Bureau itself. These negotiations can lead to the issue of tender addenda. It is standard for public sector tender documents to warn tenderers that tender qualifications may lead to the disqualification of the tender. Negotiations do take place with tenderers on Government design & build contracts by way of "clarifications" to the scope of works. In contrast to public sector contracts, negotiations are commonplace between tenderers and the MTRC or HKAA. See also section 2.1.1 above in relation to HKAA and HKCA negotiations. HKCA negotiations with MTRC have also led to improvements in the dispute resolution provisions of MTRC's airport related projects.[11]

2.1.4   Quality Matters

The Hong Kong Housing Authority, the Government body responsible for public housing and Works Bureau, the branch of Government responsible for the general public works programme, have embarked upon using ISO9000 as the standard to be applied in the construction industry for both contractors and consultants.

In 1997 the Government agreed to adopt the World Trade Organisation Government Procurement procedures.   As a consequence, Works Bureau is currently revising their procurement regulations.

2.1.5   Partnering

Informal partnering is used in Hong Kong, particularly by the Hong Kong Housing Authority, as the contractors who undertake the public housing work tend to specialise in this area and thus both parties are familiar with what is expected from them.   The housing element of the public housing projects in Hong Kong are based upon standard and known designs.   There are usually relatively few disputes on these elements of the housing projects. However, disputes are more common place on Hong Kong Housing Authority contracts for site formation, foundations and on the commercial developments associated with the projects. (These commercial developments tend to be customised to the needs of the individual project.)

The Hospital Authority has employed a more formal type of partnering.   An Australian contractor on a particular hospital project originally suggested partnering and this proved successful.   As a consequence, the Hospital Authority has introduced partnering on two further large design and build hospital projects.   The type of partnering adopted involves the use of a third party facilitator and a retreat for the key project participants. Partnering has been used in the private sector but the authors know of only one such example. This again involved the same Australian contractor who operates in Hong Kong.

2.1.6   Dispute Resolution Adviser

The Dispute Resolution Adviser (DRA) system was invented in Hong Kong[12]. It was originally developed for use on one particular Government refurbishment contract but has been used by Works Bureau's Architectural Services Department on a trial basis on several other contracts, including those relating to new construction. The development of the DRA system has been widely written about. [13] [14] When used in its proper form, the DRA system incorporates informal partnering techniques pre- as well as post-contract. Some of the dispute avoidance and resolution techniques contained within the DRA system have been incorporated into private sector contracts.[15] The DRA system itself, or at least components of it, has now been incorporated into private sector contracts, the most notable one being the Hong Kong Convention and Exhibition Centre extension.

2.2   Binding

2.2.1   Partnering

The only form of binding partnering which the authors know of, is the way that Works Bureau adopt lists of approved contractors. Works Bureau maintain three lists of contractors who are permitted to carry out Government construction works, each list enables contractors to tender for work up to a particular specified value but in the case of list C an unlimited amount. Within the list the contractors are approved for certain project types e.g. building. Works Bureau also maintains lists of specialist contractors for carrying out specialist work. In practice these works are often undertaken by the specialist approved contractors acting as sub-contractors for a main contractor. Examples of specialist approved contractors are fire services and electrical works. Works Bureau specifies within the main contract documentation that such specialist works have to be carried out by approved contractors from its lists.

Works Bureau has a detailed contractor reporting system. Reports are submitted quarterly and an adverse report can result in suspension, which will deny a contractor the opportunity to tender for Government works. The Hong Kong Housing Authority has its own list of approved contractors. As a consequence of these list systems, there is a tendency for the various Works Bureau Departments and the Hong Kong Housing Authority to work together with particular contractors over a long period of time.

3.   DISPUTE RESOLUTION

The systems of dispute resolution used in Hong Kong in the construction industry include negotiation, arbitration, mediation, adjudication, dispute review boards, the Dispute Resolution Adviser, partnering and other ADR techniques.[16] As all of these processes are to a lesser or greater extent private, it is difficult to obtain reliable statistics. Little attention is currently given to dispute avoidance. Conflicts in construction contracts are common place. A general overview on construction disputes in Hong Kong was published as a TG15 paper in 1996.[17]

Fortunately, the Hong Kong Government has kept statistics related to the airport infrastructure projects and a summary of how and when disputes were resolved appears in the two tables below.

Table 1 records the number of disputes for each of the years from 1992 to 1997 (August). Under an ACP construction contract, a dispute is a defined term and it is therefore possible to monitor resolution.   The ACP dispute resolution system comprises a three-tier system, excluding the Engineer's decision.[18]

Table 1

Year	1992	1993	1994	1995	1996	1997	Total
Number of Disputes	1	6	21	40	28	26	122

Table 2 records how and when these 122 disputes were resolved.

Table 2

Stage	Number of Disputes
Mediation Stage	
Awaiting mediation request	1
Resolved prior to mediation	36
Resolution being attempted by mediation	7
Resolved during the mediation stage (not necessarily by mediation)	51
On hold during the mediation stage	6
On hold prior to mediation	3
Adjudication Stage	
Resolved during the adjudication stage (not necessarily by adjudication)	6
Adjudications in progress	1
Pending adjudication	1
Arbitration Stage	
Resolved during the arbitration stage (not necessarily by arbitration)	4
Pending arbitration (after completion of contract)	5
Arbitration in progress	1

In summary, negotiation or mediation in the mediation stage settled 71% of all disputes. 5% of all disputes were settled at the adjudication stage, but some of these were settled by negotiation. 3% of all disputes were settled at the arbitration stage but again some of these were resolved by negotiation. Overall as at 1st

August 1997, 79% of all disputes had been resolved and about 90% of these disputes were settled either by negotiation at some stage or by mediation.

An analysis of the component parts of the ACP dispute resolution scheme appears elsewhere in this paper

3.1   Non-binding

3.1.1   Conciliation

The term conciliation is used interchangeably with mediation in Hong Kong.   Since June 1997 Section 3 of the Arbitration (Amendment) Ordinance 1996, has amended Section 2 of the Arbitration Ordinance to provide that the term "conciliation" includes mediation. However, the Arbitration Ordinance has a particular procedure called conciliation.   This is contained in Section 2B. If the parties to a reference consent in writing and no party withdraws that consent, an arbitrator (or umpire) may act as a conciliator.   This enable the arbitrator to communicate separately with the parties and treat in confidence any information that party provides. However, if the conciliation proceedings are terminated without the parties reaching agreement, the arbitrator, before resuming the arbitration process, must disclose to all other parties so much of that confidential information as he thinks is material. Thus the procedure can be likened to ARB:MED:ARB.

The procedure has been criticised and is regarded by many as flawed. This is because it is felt that parties will not confide in an arbitrator acting as a conciliator, knowing that if the conciliation fails, the confidential information, which could include admissions, would be disclosed to the other party or parties. It is also extremely difficult, if not impossible, for an arbitrator to put such confidential information out of his mind. The authors know of only one reported case where this procedure has been used.[19]

3.1.2   Executive Tribunal

The authors know of no formal procedures contained within Hong Kong construction contracts for the use of the Executive Tribunal or "mini trial". See section 3.1.4 for negotiation by executives.

The authors are aware of the use of the Executive Tribunal to resolve construction disputes in only two cases. One was in respect of a Government building contract and the other was in respect of a private sector power supply contract. In both cases the dispute was successfully resolved by the use of the Executive Tribunal. The proposed Administrative Guidelines for the "Government of Hong Kong Special Administrative Region Mediation Rules" which will accompany the proposed Works Bureau Technical Circular, "Mediation in Construction Disputes" encourages the use of other forms of alternative dispute resolution in addition to mediation, its preferred form.

3.1.3   Mediation

Mediation is the preferred method of dispute resolution adopted by the Hong Kong Government for resolving its construction disputes. As noted in the section of this paper on construction contracts, mediation is incorporated into the dispute resolution procedures of Government's construction contracts and, in the case of the ACP contracts, an attempt at mediation is a condition precedent to dispute moving on to the binding forms of dispute resolution.[20] Mediation is used by MTRC in its airport railway contracts.

Mediation is being increasingly used in the private sector but as noted in Section 1.2.2 of this paper, mediation has not yet been adopted in the widely used standard form of contract for private building works (the HKIA Form).

There is a pool of trained construction mediators in Hong Kong. The Hong Kong International Arbitration Centre (HKIAC) maintains three panels of mediators. There are two general panels, one for family mediators and one for others, the majority of who are construction industry professionals or construction lawyers. In addition, the HKIAC maintains a panel of mediators for construction disputes arising out of the ACP contracts. (It also maintains panels of ACP adjudicators and arbitrators). Parties are not bound to choose mediators, adjudicators or arbitrators from these panels. Although there is a pool of trained construction mediators in Hong Kong, those with actual mediation experience are few in number. Parties to mediation prefer experienced mediators and thus actual mediations are concentrated in the hands of only a few people. For large value or complex mediations, overseas mediators have been used; sometimes co-mediating with a Hong Kong based mediator. There would appear to be a tendency for construction mediations to be evaluative in Hong Kong, as opposed to facilitative. Parties tend to choose mediators with subject matter expertise that encourages a more evaluative type of process. The authors consider that the evaluative approach has contributed to the failure of a number of mediations.

Extensive mediator training courses have been held in Hong Kong. Although recent training courses in Hong Kong have concentrated on the growing area of family mediation, there is still a demand for construction mediation training.

Hong Kong has developed mediation rules to be used for the resolution of construction disputes. As noted in the section 1.2.2, these mediation rules are incorporated by reference in to the various construction contracts. If ad hoc mediation takes place, parties tend to adopt the HKIAC mediation rules, which although general in nature, are suitable for construction disputes. It is the authors' experience that the parties to mediation and/or their advisers tend to adapt the mediation rules to suit the particular needs of the dispute. This applies even in the case of Government contracts. The most common area for amending the rules is to extend the overall timeframe for conducting the mediation.

In the authors' experience, mediations are still taking place long after the event that has caused the conflict. This trait even applies to ACP mediations, where the early resolution of disputes is encouraged. To help determine if a dispute really existed, Government Departments engaged in ACP contracts have taken the unusual step of requesting Engineer's decisions on inactive claims. Once this occurs and the Engineer publishes his decision it becomes binding on both parties unless notice to mediate is given within a particular time frame. This has led to the withdrawal of a number of claims and a number of Government instigated mediations. It may also have resulted in formal dispute resolution procedures being used in respect of claims that might otherwise have been resolved by negotiation or abandoned.

The HKIAC formed a Mediation Group in January 1994. It is a broad-based organisation dealing with all forms of mediation. Approximately two years ago the HKIAC Mediation Group restructured itself and formed interest sub-groups in particular areas. Currently there are three such interest sub-groups covering construction, commercial and family. The construction interest sub-group is the largest, and currently the most active. It has an outreach policy and regularly runs mediation seminars to professional organisations engaged in the construction industry. The main HKIAC Mediation Group Committee also runs a regular series of evening meetings, many of which deal with construction matters as well as mediation in general. The HKIAC Mediation Group is also co-operating with the Law Society of Hong Kong in running introductory courses as part of the Society's Continuing Professional Development programme.

Despite extensive attempts to explain the true nature of the mediation process, mediation is not, as yet, well understood in the construction industry. Many construction professionals and some construction lawyers still regard mediation as some form of non-binding arbitration.

The Chief Justice is currently looking at introducing court attached mediation schemes into Hong Kong and has instructed the formation of two committees to look at pilot schemes. These pilot schemes are looking at court attached family mediation and court attached personal injury mediation. There have been some indications that a third pilot scheme might be tried in respect of construction disputes. As noted in Section 3.2.4, parties commencing litigation on the construction and arbitration list will be required to state, at the time of setting down the matter for trial, whether they have considered use of alternative dispute resolution.

3.1.4   Negotiation

As noted above, negotiation is used extensively to resolve construction disputes. It is said that in Hong Kong negotiation is a less confrontational means of resolving disputes and is suited to the indigenous population. The authors' experience of construction industry negotiations is that they tend to be positional rather than principled. There are several organisations that run courses in Hong Kong dealing with negotiation training, both generally and in the construction industry.

The settlement of construction disputes by negotiation with Government officials sometimes proves difficult for contractors. This would appear to be so because of levels of authority, the Government bureaucracy and apparently, a concern regarding possible subsequent investigation of the negotiation by the Independent Commission Against Corruption (ICAC).[21] The Corruption Prevention Department of the ICAC actively works with Government, quasi Government organisations and others in devising rules and procedures with sufficient transparency so as to reduce corruption opportunities. The ICAC are active supporters of both negotiation and mediation and certain officers of this organisation feel that the ICAC is often used as an excuse for civil servants failing to resolve disputes by negotiation.

In the private sector, Clause 9.1 of the RICS Minor Works Form of Contract requires negotiation to be attempted in good faith as the first stage of the dispute resolution process. The Clause states that the "Employer and the Contractor shall each designate a senior person, preferably not involved in the day to day administration of the Contract, with authority to settle disputes". To this extent, these senior persons might be regarded as executives.

3.2   Binding

3.2.1   Adjudication

The Hong Kong Government has adopted mediation as a precursor to arbitration in its standard forms. The Hong Kong Government Mediation Rules pursuant to which such mediations are conducted impose on the mediator a requirement, if requested by either party, to render a report stating his views on the dispute (Rule 15).[22] The report is required to set out the facts as he finds them, his opinion in relation to the matters in dispute and propose terms of settlement. The views expressed in the document are not binding and are intended to assist in facilitating a settlement. This provision has excited considerable controversy[23].

For the first time in the ACP standard form of contract the Hong Kong Government went one step further by introducing a formal adjudication mechanism. Under this procedure a dispute is referred to an adjudicator only after mediation has been unsuccessfully conducted. The dispute in question must relate to time or money issues. If neither party refers the matter to adjudication, the dispute can proceed to arbitration without being considered by an adjudicator.

The ACP Adjudication Rules provide for the appointment of an adjudicator. This is done by adopting a list system which requires each party to advise the HKIAC of at least three people willing and able to act as adjudicator. The HKIAC combines these lists. Each party is then required to place the individuals in an order of preference deleting the names of persons to which it objects. The Rules do not identify the grounds of objection. If an appointment cannot be made in accordance with the procedure, the HKIAC may in its sole discretion appoint an adjudicator of its choice. The adjudicator is charged with the widest discretion permitted by law to determine the procedure of the adjudication and to ensure a just, expeditious and economical determination of the dispute. In particular, the adjudicator is entitled to decide whether to hold a hearing, to receive oral evidence or whether to proceed on a documents only basis. The adjudicator is empowered (i) to conduct his own examination and inspection of properties in the absence of representatives of the parties; (ii) to continue in default of the appearance or participation of the party; (iii) to order the production of documents; (iv) to order the inspection, preservation, storage, or interim custody of property or other things relevant to the dispute; (v) to require the parties to provide written statements of their case. It is considered that the adjudicator is not bound by the rules of evidence. The adjudicator is required to make a decision with 42 days of his appointment or within such other period that the parties may agree or he shall decide. However, in the absence of an agreement from the parties the adjudicator may not extend the time for his decision by more than 28 days. The decision of the adjudicator should identify the dispute referred to him, state reasons for his decision and record any admissions that were made. The adjudicator has discretion to award that the unsuccessful party pays the costs of the successful party. The decision of the adjudicator is binding unless varied by a settlement agreement or decision of an arbitrator.

The Hong Kong Government did not adopt adjudication in its General Conditions published after the ACP Conditions were drafted.

3.2.2   Arbitration

HKIAC is the principal appointing body in Hong Kong. It was established in 1985. The HKIAC is a non-profit making and charitable company limited by guarantee. It was initially provided with seed capital by the Government and is now financially independent of Government. The HKIAC is run by a management committee drawn from a large cross section of the business community of Hong Kong.

The law stated in this paper is stated at 31st July 1997.[24] Amendments were made to the Arbitration Ordinance effective on 27th June 1997 and 6th April 1990 which may affect the position of parties. The amendments of 6th April 1990 are particularly significant. If a dispute relates to an arbitration agreement executed prior to either of these dates, reference should be made to the relevant transitional provisions.

There are two arbitral regimes in force in Hong Kong. The domestic regime is loosely based on the English Arbitration Acts 1950-1979. The international arbitration regime is based upon the UNCITRAL Model Law regime. The Arbitration Ordinance adopts the definitions contained in Article 1(3) of the Model Law to distinguish between the domestic and international regimes. However, parties may elect to opt into either the domestic or international arbitration regimes even if their agreement falls into the other regime. An election may be made in the original arbitration agreement or at any time thereafter. The forms adopted by the Hong Kong Government (save the General Conditions for Building Works), the HKAA and MTRC all opt into the domestic regime. The HKIA Form is silent on the point. The RICS (HK Branch) Minor Works Form of Contract opts for the domestic regime. A construction contract, under the HKIA

Form involving a foreign contractor will therefore fall within the international regime. For the sake of simplicity this paper confines itself to the domestic regime which is generally adopted for construction disputes.

(a)   Arbitration Agreements - The Courts of Hong Kong respect arbitration provisions and will automatically stay Court proceedings in favour of arbitration where parties proceed in the Court in defiance of an arbitration clause. The Arbitration Ordinance adopts Article 7(1) of the Model Law as a definition of an arbitration agreement. This provides: - *(1) "Arbitration Agreement" is an agreement by the parties to submit to arbitration all or certain disputes which have arisen or which may arise between them in respect of a defined legal relationship, whether contractual or not. An arbitration agreement may be in the form of an arbitration clause in the contract or in the form of a separate agreement. (2) The arbitration agreement shall be in writing. An agreement is in writing if it is contained in a document signed by the parties where an exchange of letters, telex, telegrams or other means of telecommunication which provide a record of the agreement, or an exchange of statements of claim and defence in which the existence of an agreement is alleged by one party and not denied by another. The reference in a contract to a document containing an arbitration clause constitutes an arbitration agreement provided that the contract is in writing and the reference is such as to make the clause part of the contract.*

Following the initial enactment of this provision the Courts were faced with a number of applications regarding meaning of the agreement to submit. The Courts adopted a liberal interpretation and decided that a party asserting an arbitration provision need only establish an arguable case that an arbitration agreement existed before the matter would be stayed to arbitration (William Co. v. Chu Kong Agency Co. Ltd. [1995] 2 HKLR 139) - this was on the basis that the arbitrator should initially resolve disputes regarding jurisdiction.

One issue, which troubled the Courts on a number of occasions, is when an agreement was in writing. To overcome these difficulties Section 2AC was introduced in 1997. This amendment adopts a liberal and common sense approach to the question of whether an agreement is in writing, in particular making it clear that as agreement is in writing whether or not the document is signed by the parties.

(b)   Separability of Arbitration Clauses - The Courts in Hong Kong recognise an arbitration clause as being separable from the other provisions of the agreement. Thus it will refer to arbitration disputes where one party asserts that a contract has been discharged by performance, frustrated, fundamentally breached or repudiated. The decision of the English Court of Appeal in Harbour Assurance Co. (UK) Ltd. v. Kansa General Insurance Co Ltd [1993] 3 All ER 897 was adopted in Hong Kong in Fung Sang Trading Ltd v Kai Sun Sea Products [1992] 1 HKLR 40. Any doubt regarding the position was removed by the introduction of Section 13B of the Arbitration Ordinance that applies Article 16 of the Model Law to domestic arbitration. This provides that an arbitrator may rule upon objections to the validity or existence of the arbitration agreement. It also provides that the arbitration clause shall be treated as an agreement independent of the other terms of the contract.

(c)   Consolidation of Arbitration - The Court has power (Section 6B of the Arbitration Ordinance) to direct that where there is some common question of law or fact, or the rights to relief claimed arise out of the same transaction or series of transactions, or for some other reason it is desirable, that two or more arbitrations be consolidated and heard at the same time, or immediately after another, or one of them stayed until after the determination of the other. Although this provision was enacted in 1985 there had been surprisingly few applications under this provision. Those applications that have been made were considered by the Court on the same basis as the Court would consider an application to consolidate Court proceedings.

For the purposes of giving effect to an order made pursuant to this power the Court may appoint an arbitrator and the appointment of any previously appointed arbitrator ceases. (See Shui On Construction Co. Ltd. v. Schindler Lifts (HK) Limited [1986] HKLR 1177, Shui On Construction Co. Ltd. v. Moon Yik Company [1987] 2 HKLR 1224, Harlifax Limited v. Transatlantic Schiffahrtskontr (unreported, 25 October 1988), Ming Kee Shipping Service Co. Ltd. v. Autogain Limited (unreported, 30 April 1992) and Dickson Construction Co. Ltd. v. Schindler Lifts (HK) Ltd. [1993] 1 HKLR 45).

(d)   Domain of Arbitration - Proceedings may not be stayed to arbitration where a dispute has arisen which involves the question of whether a party is guilty of fraud. In these circumstances, the Court of First Instance has power to bring the arbitration to an end, such that the Court (Section 26(2)) determines the issue. However, if the party against whom the fraud is alleged seeks to continue the arbitration proceedings the Court will normally exercise its discretion in favour of arbitration unless good reasons exist to the contrary. (Cunningham-Reid v. Buchanan-Jardine [1988] 2 All ER 438). Equally the Court will probably exercise its discretion in favour of the continuance of the arbitration where the parties entered into the arbitration agreement after the allegation of fraud had been made (<u>Yeu Shing Construction Co. Ltd. v. Attorney General</u> [1988] HKC 710).

(e)   Effect of the Agreement - Prior to 1997 a party was required to establish a dispute before a matter would be stayed to arbitration. Section 6 of the Arbitration Ordinance, enacted in 1997, introduced the compulsory stay provided for in Article 8 of the Model Law. The Court will therefore automatically refer disputes to arbitration. This is subject only to the Court being satisfied that there is sufficient reason why the matter should not be referred in accordance with the agreement. At the date of this paper this qualification has not been tested in the Courts.

In recent years the Hong Kong Courts consistently recognised that a low threshold applied in deciding whether a dispute existed. There have been a number of judicial comments to the effect that it is unwise for parties to incur costs and time in pursuing applications for summary judgment where they are almost certainly bound to fail. The Courts instead urge parties to seek early interim awards from arbitrators where it is contended that there is no dispute. Summary judgement will only be granted in very clear situations.

(f)   Appointment of Arbitrators - It is a common feature that the standard forms of construction contract, with the exception of the HKIA Form, that in default of agreement between the parties as to the identity of an arbitrator the HKIAC be requested to assist in filling the vacancy. This is effected in accordance with the HKIAC's Domestic Arbitration Rules.

The HKIA Form provides for the appointment to be made by the President or Vice-President of the Hong Kong

Institute of Architects co-jointly with the Chairman or Vice-Chairman of the RICS (Hong Kong Branch). In mid 1997 the RICS (Hong Kong Branch) ceased to exist in Hong Kong.

The role of the HKIAC as an appointing body was further enhanced by amendment made to the Arbitration Ordinance. The amended Section 12 of the Arbitration Ordinance provides that in the event of the parties being unable to concur in the appointment of an arbitrator or the appointment mechanism breaking down the appointment should be referred to the HKIAC who will fill the vacancy.

(g) Removal of Arbitrators - The Court retains a number of powers to remove an arbitrator in certain circumstances: -

(i) the Court may remove an arbitrator who fails to use all reasonable dispatch in proceeding with the reference in making an award; (Section 15(3))

(ii) the Court may remove or set aside an award where the arbitrator has misconducted himself (Section 25). Misconduct has been interpreted broadly to cover the appearance of misconduct or bias as well as the fact of misconduct or bias. The Court has identified types of technical misconduct justifying the intervention of the Court including ambiguities or uncertainties, excess of jurisdiction, incompleteness or breach of the natural justice. (Kong Kee Brothers Construction Co. Ltd. v. Attorney General [1986] 2 HKLR 767)

(h) Conduct of the Arbitration - The arbitrator has considerable discretion in the conduct of proceeding under most forms of construction contracts in Hong Kong. As noted above, Government Forms of Contract adopt Domestic Rules of the HKIAC[25]. Where a set of rules is identified, failure to adopt the required set of rules will amount to misconduct and may lead to an arbitrator being removed. However, the primary intent of the HKIAC Domestic Rules is to provide the arbitrator with the widest possible discretion to adopt procedures to cost effectively and speedily resolve the particular dispute[26].

In the absence of an agreed procedure the Arbitration Ordinance now expressly incorporates into all arbitration agreements certain powers (Section 2GB), including powers regarding the collection and preservation of evidence, discussed below, and certain procedural powers. The procedural powers include:-

(i)   the power to require the provision of security for costs;

(ii)   the power to secure the money in dispute, and

(iii)   the power to grant injunctions.

(i)   Commencement of Arbitration - an arbitration commences upon service of a written notice requiring a party to appoint or concur in the appointment of an arbitrator. The effect of such notice is to stop time running for the purpose of Limitation Ordinance (Cap. 347). Particular arbitration rules may impose more onerous requirements. For example under the HKIAC Domestic Rules the arbitration notice is required to include details of the names and addresses of the parties, their representatives, the principal contract documents, a brief statement of the dispute and the name and address of a proposed arbitrator. They also require a copy of the notice to be sent to the Secretary-General of the HKIAC.

(j)   Evidence - The strict rules of evidence do not apply to arbitration. Section 2GA provides that an arbitrator may receive any evidence that he considers relevant and shall not be bound by the rules of evidence.

The Arbitration Ordinance contains various powers to assist in the collection and presentation of evidence in arbitration. The Court may, under Section 2GC of the Arbitration Ordinance, make orders in respect of: -

(i) the preservation, custody, detention or sale of any property which is relevant to the reference;

(ii) the authorising of samples to be taken and observations to be made or experiments to be conducted upon the property;

(iii) require persons to attend before the arbitrator to give evidence, produce documents or other material evidence.

Where both the Court and the arbitrator have jurisdiction to make orders addressing the same subject e.g. inspection, the Court is likely to be reluctant to exercise its discretion to make an order until such time as the powers of the arbitrator have been exhausted. However, the orders of the Court can be enforced by way of contempt of Court proceedings a sanction that is not available to an arbitrator. There are therefore, at least in theory, circumstances where the Court will exercise jurisdiction before the arbitrator has exercised his jurisdiction.

The Arbitration Ordinance (Section 2GB) also gives certain powers to the arbitrator including, directing the inspection, photographing, preservations, custody detention or sale of relevant property and directing samples be taken from, observations be made of or experiments conducted on property.

(k)   Representation and Legal Assistance - The Arbitration Ordinance provides that the parties may represent themselves or be represented by any party of its choice (Section 2F). However, the rights of audience in the Court of First Instance remain restricted.

(l)   Types of Awards - An arbitrator may make one or more awards in relation to a particular dispute. "Interim Awards" resolve something less than the whole dispute. In practice it is very common to find such interim awards dealing either with liability or all issues other than costs. An interim award will be final in respect of the issues with which it deals, leaving only matters which are not considered in that award to be dealt with at a later date. An interim award should be clearly marked as being interim otherwise it may lead to confusion with the final award and arguments as to whether the arbitrator is in a position to render further awards. (Sections 16 and 18).

A final award resolves all the issues submitted to the arbitrator. Once the final award is published, the arbitrator ceases to have jurisdiction over the dispute with the exception of very limited powers to correct slips or omissions (Sections 18 and 19) or as otherwise provided in the relevant set of rules. For example the HKIAC Domestic Rules provide that the arbitrator may make interim awards or separate awards on different issues at different times (Article 16.4). The Rules also provide that the arbitrator may upon the request of a party, made within 14 days of receiving the award, make an additional award in respect of claims presented in the reference but not dealt with in the award (Article 17.3).

(m)   Making the Award - The Arbitration Ordinance does not provide a statutory time limit for the making of awards. However, the Court may, on the application of any party, remove any arbitrator who fails to use all reasonable despatch in making an award (Section 15(3)).   If the parties have agreed on the time limit for making an award, such agreement will be enforced.   The HKIAC Domestic Rules do not provide a time limit for the making of an award.

(n)   Publication of Awards - Arbitration awards are not published in Hong Kong.

(o)   Form of Awards - There is no formal requirements for the format of the award.   However, it is the practice for awards to be in writing and sign by the arbitrator.   The Courts have identified that there are strong policy reasons why awards should be reasoned (see Kong Kee Brothers Construction Co. Ltd. v. Attorney General [1986] 2 HKLR 767).   It is the invariable practice of arbitrators in Hong Kong to provide reasons for their decisions.

If a party has requested a reasoned award, then the arbitrator should provide reasons when the award is made.   If an arbitrator fails, following such request, to give reasons, then the Court will require the arbitrator to give sufficient reasons to allow the Court to consider any question of law arising out of the award.   The Court may exercise this power even if the request has not been made prior to the award being made, however the party applying will have to establish special circumstances as to why the request is not made.   (Sections 23(5) and (6)).

A provision that purports to exclude the parties' right to seek reasons is of no effect unless entered into after commencement of the arbitration.   Such exclusion agreement may apply generally in the reference or to particular issues and may be revoked or reinstated by agreement between the parties from time to time (Section 23B).

The HKIAC Domestic Rules provide that arbitrator's awards should be in writing, signed, dated and provide reasons for the decision (Article 16.1).

(p)   The Award of Interest - Section 2GH of the Arbitration Ordinance provides that the Arbitration Tribunal may award simple or compound interest from such dates at such rates, and with such rests as the Tribunal considers appropriate for any period ending not later than the date of the award. Such power extends to money claimed and outstanding at the commencement of the arbitration, but paid before the award is made. Interest is payable on amounts awarded in the arbitration award at the judgment rate (see Section 2GI of the Arbitration Ordinance). Under the HKIAC Domestic Rules an arbitrator is empowered to award compound interest (Article 20).

(q)   Settlement - If the parties reach a settlement to a dispute, the arbitrator's general jurisdiction comes to an end. Where the parties request the arbitrator to record a settlement in the form of award he may do so. However, in practice this is rarely necessary because Section 2C of the Arbitration Ordinance provides that a written settlement agreement between the parties to an arbitration agreement shall, for the purpose of its enforcement, be treated as an award on an arbitration agreement. This may be enforced by way of judgment. The HKIAC Domestic Rules provide that an arbitrator may issue an order terminating the reference or record the settlement in the form of a consent order (Article 16.5).

(r)   Correction and Interpretation of the Award - Once the arbitrator makes an award, his authority comes to an end and he has no further authority. Section 19 of the Arbitration Ordinance provides that an arbitrator may correct a clerical mistake or error in the award unless a contrary intention is expressed in the arbitration agreement. The Court has given the concept of a clerical error a wide meaning. However it will not allow an arbitrator to reverse a material decision of fact or finding of law. Under the HKIAC Domestic Rules, provided a party applies within 14 days of receipt of the award, an arbitrator may interpret or correct his award (Article 17.3)

(s)   Enforcement of Award - The Court will grant leave to enforce awards pursuant to Section 2GG of the Arbitration Ordinance. An award will be enforced in the same manner as a judgment. The Court has adopted an administrative procedure by which judgment will be granted in the terms of the award unless the respondent party makes an application to set aside the Court order. The wide range of remedies available in respect of any judgment of the Court can be used to enforce the award.

(t)   Challenges to Award - The Arbitration Ordinance seeks to give finality by allowing appeals in only limited circumstances. Section 23 of the Arbitration Ordinance permits appeals upon questions of law arising out of an award, made with the consent of the other parties to the proceedings or leave of the Court. Leave of the Court will only be granted in limited circumstances. Where the question of law involves construction of a one-off clause, leave to appeal should not be granted unless it is apparent to the judge upon the reading of a reasoned award, without the benefit of argument, that the meaning ascribed is obviously wrong. Slightly less strict criteria are applied where the construction of standard forms of contract is concerned. However, even in these cases leave will only be granted where the judge considers that a strong prima facie case has been made out and the arbitrator was wrong in his construction and the events giving rise to the construction of the standard clause are likely to reoccur in similar transactions between other parties engaged in the same trade. In recent years there have only been a very small number of cases where leave to appeal has been granted. Pioneer Shipping Limited v. BTP Tioxide Ltd. (The "Nema") [1982] AC 724; PT Dover Chemical Co. v. Lee Chang Yung Chemical Industry Corporation [1990] 2 HKLR 257.

This limited right of appeal can be excluded by agreement of the parties. Such an exclusion agreement must be made in writing after the reference has been commenced (Section 23B of the Arbitration Ordinance).

(u)   Conventions and Treaties - Hong Kong is bound by the following multi-lateral conventions:-

(i)   Geneva Protocol on Arbitration Clauses, 1923(Arbitration Ordinance Schedule I);

(ii)   Geneva Convention on the Execution of Foreign Arbitral Awards, 1927 (Arbitration Ordinance Schedule II);

(iii)   New York Convention on Recognition of an Enforcement of Foreign Arbitral Awards, 1958 (Arbitration Ordinance Schedule III).

3.2.3   Expert Determination

The Courts of the Hong Kong recognised the validity of expert determination. The role of an expert is distinguishable from that of an arbitrator in a number of principal respects: -

(a)   The expert is required to form an opinion to determine the issue. In so doing he can take into account such materials he considers to be relevant without drawing such materials to the attention of the parties.

(b)   The Arbitration Ordinance does not apply to expert determination, consequently the powers of the Court to assist arbitrators in the collection of evidence, the powers given to arbitrators, the power to fill vacancies and the power to enforce awards do not apply to expert determination.

(c)   Arbitrators are immune from suit for their acts carried out as arbitrators. An expert owes the parties a duty of care in the execution of his responsibilities and will be held liable in negligence if there is a want of care in the way in which they are carried out.

Expert determination has been rarely used in Hong Kong and does not presently form a significant part of the dispute resolution process in the Construction Industry.

3.2.4   Litigation

In the absence of an express arbitration provision or where the parties choose not to enforce an arbitration provision, disputes will fall to be resolved by the Courts of Hong Kong. The Courts of Hong Kong are a three-tier system. The Court of First Instance (formerly the High Court) consists of a single judge. The Court of Appeal consists of three judges. The Final Court of Appeal consists of five judges. The judges sitting in the Final Court of Appeal are mainly drawn from the Hong Kong judiciary but are supplemented by overseas experts drawn from common law jurisdictions.

As noted above, the Courts of Hong Kong support arbitration as a means of resolving construction disputes and will therefore not exercise jurisdiction where there is a valid arbitration provision. Pursuant to the Arbitration Ordinance the Court has a number of powers which can be exercised in support of arbitration. These powers are readily exercised in appropriate cases.

In the Court of First Instance there are a number of specialist lists, including the Construction and Arbitration List. All cases regarding construction matters or arbitration should be referred to the judges of this list. Matters may be transferred into this list upon the application of any of the parties or at the Court's discretion. At any one time there are likely to be one or two judges who deal with the bulk of the business in this list.

Upon the setting down of a dispute for trial in this list the parties are required to state whether they have considered the use of alternative means of dispute resolution. However, at present there is no compulsory procedure that can be used to compel a party to submit a dispute to mediation or any other form of alternative dispute resolution.[27]

The procedures adopted in the Hong Kong Courts are adversarial. Judges are therefore not permitted to carry out their own investigations. Only in cases where a party has failed to attend a pre-arranged hearing will a judge hear evidence in the absence of one of the parties. The Court has power to appoint a technical expert to assist in resolving technical issues in respect of Court proceedings. Historically this power has been rarely exercised; instead the Courts have reached their decision on the basis of expert evidence produced by the opposing parties.

The Courts of Hong Kong will respect any choice of law made by the parties. Thus if a contract expressly provides for the adoption of the law of a foreign jurisdiction, it will adopt those laws in reaching a decision. Issues regarding the law of the foreign jurisdiction will be dealt with as an issue of fact on the basis of expert evidence led by the parties. In the absence of a statement from either party that the law of the foreign jurisdiction differs from that of Hong Kong, the assumption will be that the law of that jurisdiction is the same as the law of Hong Kong. The procedural law adopted will be the law of Hong Kong. The law relating to conflicts of law is a complex area and goes beyond the scope of this paper. In practice, many Hong Kong construction contracts expressly provide for the adoption of Hong Kong law.

In general terms, the Court will exercise jurisdiction over the property of a foreign individual which is located within the jurisdiction of Hong Kong in order to assist in the enforcement of awards and judgments. Such awards will include awards made in countries that are signatories of the New York Convention. It also enforces judgments made in a limited number of foreign jurisdictions.

Like arbitrators, the Court will respect the concept of "without prejudice" correspondence. Correspondence of this sort refers to attempts by parties to resolve the dispute or any part of the dispute. The Courts may look at the substance of the communication to identify whether the communication forms part of a chain of communications to compromise the proceedings or any issue. Whilst the inclusion of the words "without prejudice" in the communication will tend to indicate that the document should be afforded this protection and will not be conclusive.

With a view to saving costs and allowing cases to be resolved in a more timely manner the Courts of the Hong Kong have adopted the practice of requiring increasing amounts of materials to be placed before them in writing. It is therefore common for hearings to be preceded by the presentation of a formal written opening, delivery of written statements and a formal written closing. Witnesses when called will generally be asked to confirm the accuracy of their statements without being examined, by the party calling them, on the content of their statement. They may be asked questions to elaborate or place emphasis on particular parts of their statements, and will then be cross-examined by the opposing party.

3.2.5   Negotiation

Most construction disagreements and disputes are resolved by negotiation. Negotiated settlements are binding as a contract and if the negotiated settlement results from a dispute in arbitration, the settlement agreement, as it is known, is enforceable as a judgment. See Section 3.2.2(q) of this paper.

3.2.6   Ombudsman

Hong Kong has an Ombudsman who has specific responsibilities for administrative complaints against Government officials and departments.   The Ombudsman does not make binding recommendations.   There is not however an Ombudsman for construction matters.   The Ombudsman's office has a mediation service.

4.   THE ZEITGEIST

Following the transition of sovereignty to the PRC, the uncertainties that surrounded the process of transition have been removed.   The new administration has largely continued the economic and social policies that were pursued in the final years of colonial rule.   This has boosted confidence.

The new Chief Executive has announced an ambitious plan to add to Hong Kong's housing stock building over 85,000 flats a year. The MTRC and the KCRC, Hong Kong's two railway companies, have announced plans to expand their networks.   Despite the opportunities offered by these projects the short-term outlook for the construction is uncertain.   A number of smaller companies have run into financial difficulties and a number of liquidations have occurred.   Such problems will be exacerbated by the current regional turmoil in Asian currencies.

The current trend is towards an increase in activity in building work whilst the level of civil work is declining following completion of a number of the projects related to the new airport at Chek Lap Kok.

APPENDIX 1

Government of Hong Kong - Standard Forms

General Conditions of Contract for Building Works 1993 Edition

General Conditions of Contract for Civil Engineering Works 1993 Edition

General Conditions of Contract for Electrical and Mechanical Engineering Works 1994 Edition

Chek Lap Kok Airport Related

General Conditions of Contract for the Airport Core Programme Civil Engineering Works 1992 Edition

1   Article 28(3) of the UNCITRAL Model Law incorporated as the
        Fifth Schedule of the Arbitration Ordinance.
2   Section 2F, Arbitration Ordinance (Cap. 341).
3   Section 2G, Arbitration Ordinance (Cap. 341).

4   For further information see, McInnis, J (1997) Hong Kong
        Construction Law, Butterworths Asia and Halsbury's Laws of
        Hong Kong (Vol.1) Butterworths Asia 1996.
5   See Yogeswaran, Miller, Kumaraswamy (1994) *Allocation of
        Risk in Respect of Ground Conditions in Hong Kong*,
        Construction Conflict Management and Resolution, Fenn. P.,
        (Ed) CIB Publication 171, CIB, The Netherlands, pp141-152.
6   See Lewis, D., (1993) *Dispute Resolution in the New Hong
        Kong International Airport Core Programme Projects*, ICLR
        76-87, for an overview of the procedure.
7   See for example Wall, C., (1994), *ACP Rule 16 Is it Mediation?*,
        HKIAC Mediation Group Newsletter Issue 4, Hong Kong.

8   For an analysis of causes of conflict in the construction industry in
        Hong Kong see Kumaraswamy, M., (1997) *Common Categories
        and Causes of Construction Claims* (1997) 13 Const. LJ p21.
9   Wall, C., (1996) *Hong Kong's Provisional Airport Authority
        Conditions of Contract: The Balance of Risk and Dispute
        Resolution Process,* Heath (Ed) CIB Publication No 196 CIB
        The Netherlands, pp *184-197.*
10   Originally there were eight members of the Board. The eighth
        member was a construction professional from the United States
        of America.
11   See Lewis, D. (1994) *Dispute Resolution in the New Hong Kong
        International Airport Core Programme Projects (Part 2)* ICLR
        25-45, which describes the original MTRC dispute resolution
        system and Lewis, D. (1995) *Dispute Resolution in the New
        Hong Kong International Airport Core Programme Projects
        (Part 3)* ICLR 131-136, which describes the revised system.

12   Wall, C., (1992) *The Dispute Resolution Adviser in the
        Construction Industry*, Construction Conflict Management and
        Resolution, Fenn and Gameson (Eds.), Spon, London, pp328-
        339.

13   See Wall, C., (1994) *The Dispute Resolution Adviser System in
        Practice*, Construction Conflict Management and Resolution,
        Fenn P., (Ed), CIB Publication No 171, CIB, The Netherlands
        pp154-167.

14   See Tsin, H. (1997) *Dispute Resolution Adviser System in Hong
        Kong - Design and Development*, 63, JCI Arb. 2(S), -pp67-78.

15   Wall, C., (1994) *Dispute Prevention and Resolution for Design
        and Build Contracts in Hong Kong*, Rowlinson, (Ed), CIB
        Publication No 175, CIB, The Netherlands pp353-360.

16   See Hill, T., (1994) *Alternative Dispute Resolution in Hong
        Kong* Vol. 3, No. 1 Asian Law Journal, for a general overview.

17   Cheung, S. & Lui, A. (1996) *Resolving Construction Disputes: A
        Hong Kong Perspective*, Heath (Ed) CIB Publication No 196
        CIB The Netherlands, pp 173-183.

18   Wall, C., (1992) *Hong Kong's Airport Project Provides
        Innovative ADR System*, World Arbitration & Mediation Report,
        Vol. 3, No 6, pp. 150-153

19   See Thomas, M. (1992) *Mediation at Work in Hong Kong, 58*,
        JCI Arb. 1, p29.

20   See Lewis, D., (1994) *Dispute Resolution in the New Hong
        Kong International Airport Core Programme Projects (Part 2)
        ICLR 29-33, for details of the rules.*

21   Since 1st July 1997 the ICAC has become known as the
        Commission Against Corruption.

22   See Lewis, D., (1994) *Dispute Resolution in the New Hong
        Kong International Airport Core Programme Projects (Part 2)*
        ICLR 34-38, for details of the rules.

23   See footnote 7. Rule 15 of the Hong Kong Government
        Mediation Rules is similar to ACP Rule 16. In the latest draft of
        the Hong Kong Government Mediation Rules - all references to
        the mediator producing a report have been removed.
24   For a comprehensive guide to arbitration in Hong Kong including
        the effects of the amendments in June 1997, see Morgan, R.,
        (1997) *The Arbitration Ordinance of Hong Kong A
        Commentary*, Butterworths Asia, Hong Kong.
25   Copies of these can be purchased from the Secretary-General of
        the Hong Kong International Arbitration Centre, 38th Floor, Two
        Exchange Square, Hong Kong Fax: +852 2524 2171/Tel: +852
        2525 2381.
26   The HKIAC Domestic Rules were published in 1993 and there is
        now a considerable overlap between the provisions, especially as
        they relate to powers and jurisdiction and the amendments
        introduced into the Arbitration Ordinance in 1997.
27   See the last paragraph of Section 3.1.3

IRAQ

Mohammed Dulaimi
University of the West of England, Bristol, England

1.   BACKGROUND

It has been called Mesopotamia, or the country between the two rivers. The cradle of civilisations has witnessed the great changes and developments of mankind and societies over thousand of years. Great cities and empires were built on this land, from Babel and Sumer to Baghdad and Basrah; history shows the great achievement of this country. It has been recognised that history and engineering were born in what's known today as Iraq.[1] This country has witnessed an unprecedented turmoil from 1980-1988 during the Iran-Iraq war and the 1990 the Gulf war, and is facing great challenges in trying to rebuild in the presence of crippling international sanctions.

Iraq occupies an important geographical position at the west gates of Asia to Africa. The size of the country is 432,162 sq. km, about twice the size of the United Kingdom, with a population of over 22 million. The people of Iraq are predominantly Arabs with a number of other minority groups, the largest group being the Kurds with an estimated population of over three million. The overwhelming majority of the population, 95-97%, are Muslims. Iraq joined the League of Nations as an independent state in 1932. In 1958 a revolution led by the military toppled the pro-British monarchy and established the Republic of Iraq.

In 1968 another revolution brought the present government to power. The supreme ruling body is now the Revolutionary Command Council which is dominated by the Arab Baath Socialist Party. This power structure has dominated and controlled all aspects of life, socially, politically and economically. The declared philosophy driving the government policies was akin to that of the Eastern European countries before the collapse of the Soviet Union. However, the government adopted a more pragmatic approach towards its links with western companies and countries. Faced with the lack of expertise and technological know-how in certain areas of the economy the government did not hesitate in importing expertise and commodities essential to achieve its ambitious development objectives.

Driven by its socialist philosophy the all powerful ruling council assumed the authority to make decisions. The council pursued a central planning ideology and system. After 1968 the government *"expanded the role of the state even more: the base of public ownership was broadened; the government tightened its grip further on economic establishments and reorganised the public sector, controlling the top levels; foreign investment was all but eliminated; the agrarian reform laws were modified in 1970; and planning was promoted 'to a higher level', becoming more comprehensive and sophisticated"*.[2]

The leadership of Iraq's approach to economic development was conditioned by their belief that the solution to Iraqi problems was through the rapid economic development of the country to become a developed country by using its finite oil resources. The political instability of the region, wars, and likely external interference in the national and regional policies has made the issue of economic independence a prerequisite of political independence.

1.1   Economic

The economy has been dominated by the oil sector which accounts for over 95% of the country's foreign exchange earnings. Iraq's crude oil exports were only $1,030 million in 1970. But this figure grew to $26,136 million in 1980. The 1980 figure was more than halved on the start of Iran-Iraq war in 1980 and reduced to $10,388 million in 1981.[3]

The modern economic history of Iraq may be divided into three periods.

1.1.1   Post 1968:

After the 1968 revolution and the establishment of the current government the oil industry was nationalised in 1972. This was followed by the 1973 rise in oil prices which marked the start of an economic boom. Oil wealth enabled Iraq to embark on far ranging and rapid development. However, through its central planning system the government set an accelerated timescale for development outpacing the county's own labour pool and infrastructure. This eventually re-initiated the return and participation of western companies in Iraq's development process.

It is not surprising to find the private sector weakened even further in such a political and economic climate. Extensive regulations coupled with sometimes erratic changes in planning policies and resource allocations have squeezed the private sector, especially in construction, into a very narrow area concentrating their activities in one city or town. In addition to restrictions on imports the government had a monopoly on the workforce. The government introduced free education which has provided the country's economy with a very valuable scientific base. However, this was accompanied with a policy of "central appointment" of university graduates. This policy reserved the right of the government to employ graduates at its establishments for at least twice their higher education period.

The national development plan (1976-1980) was described by the government as the first integral and comprehensive plan in Iraq's history. The aim was to raise the rate of national economic growth and the standard of living of the Iraqi people. Agriculture, irrigation and drainage figure prominently as areas for development. Health and education are also included. New sewerage and sewerage treatment schemes are dotted throughout the country. The work in the 'Medical City', an extensive hospital complex in Baghdad, is one example of an ambitious programme to improve the health service across the country. Airports, bridges, petrochemical plants and major hotels were under construction.

The construction industry enjoyed a significant share of the government investment and development plan. Figures show that construction share of investment grew from an average of 16.8% of total investment during 1976-1980 to 28% in 1981. The period of 1976 to 1978 is considered the boom years for the economy and especially for construction. This is not surprising as oil revenues grew from 562.1 million Iraqi Dinars (ID) in 1973, 1,982 million in 1974, to 7,800 million in 1980.[4] Figure 1[5] shows the level of growth in the Iraqi economy over this period. GDP has more than tripled in five years from ID 4,022 million in 1975 to ID 15,825 million in 1980. It is useful to bear in mind the average annual rate of inflation was estimated for the years 1970-1980 at 14.1%.[6] While the construction sector attracted only 3% of the total workforce in 1970 this number grew to 9.2% in 1979 after peaking at 10.7% in 1977.[7]

This rapid development created severe shortages in construction materials and a skilled workforce. Figure 3 shows Iraq's production of cement which failed to meet demand which was at least double the country's production capacity. This led to foreign contractors importing their own materials and workers. This put an upward pressure on contractors tender prices.[8]   The shortage in cement had a more devastating effect on the struggling private sector especially where the contract was not with the government.

1.1.2   The 1980's:

The start of the Iran-Iraq war in 1980 posed a significant challenge to an economy for which the government was keen to maintain an upward direction. The war was a heavy burden on the Iraqi economy in two main aspects. The first was the need to finance a very expensive war and at the same time continue with development plans. Such plans were significantly expanded to provide the necessary infrastructure and capacity for the military.

The second was that the mobilisation for the war caused a significant reduction in the available workforce. This has led to the government to borrow heavily and to import workers in significant numbers.

After the first 2-3 years of the war with Iran the economy started to grow again (figure 1). This was further helped by some relaxation of trade controls to allow the importation of foreign goods and services by the private sector. In its attempt to encourage greater efficiency in government departments and therefore reduce the heavy burden on the country's finances a framework of an internal market policy was introduced. Government departments and ministries with resources and expertise in construction, building and engineering were structured on a footing similar to that of private enterprises. They could bid for and carry out work and in theory make a profit. One has to bear in mind that the main client was still the government. The allocated investments for the construction sector were increased to 28% of total allocations in 1981 compared to 16.8% in the period from 1976-1980. Shabandar (1987) reported other estimates which show a quantum jump in the government investment in construction (table 1).

Figure 1   GDP in Million ID (at current prices)

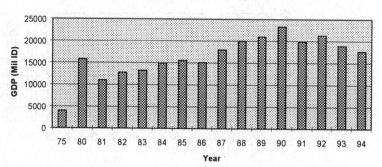

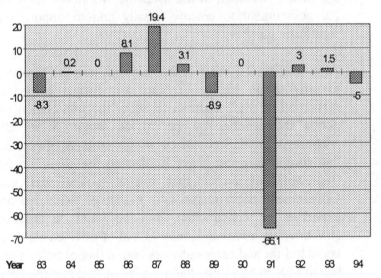

Figure 2 Percentage Growth per Year

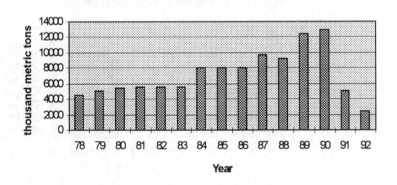

Figure 3 Cement Production

Table 1: Construction Activity between 1979-1985

	1979	1980	1981	1982	1983	1984	1985
No of Employees	208737	209256	240923	303635	283530	221157	176843
Building *	592.3	567.8	1053.5	1130.9	1172.6	1699.0	1673.4
Civil Engineering*	118.6	156.7	135.4	345.9	888.1	1094.0	1091.0
New Building Construction Authorised	509.2	786.6	1778.2	1282.9	912.5	1079.8	1050.1
New Buildings Completed	592.3	567.9	1053.5	1130.9	1172.6	1966.0	1673.4

(Source Construction Statistics Yearbook, United Nations 1985)

In 1982 the government allocated ID 7700 million to construction compare with an average of ID 3038.62 million between 1976-1980. Iraq is estimated to have awarded contracts to the value of $33 billion for the whole of 1981 exceeding that of Saudi Arabia. In his analysis of construction in the Arab world Shahbandar (1987) suggests that 75% of construction expenditure was used to buy the services of foreign contractors, consultants, equipment, materials, and a large labour force.

1.1.3   The 1990's:

The start of the Gulf war and the imposition of the most comprehensive economic embargo on Iraq has brought the whole country to a standstill. Figure 2 shows that the United Nations estimated that the economy shrunk in 1991 by a devastating 66%. The war caused serious damage to Iraq's infrastructure and production capacity. Figure 3 shows an example of how the cement production capacity slumped after the end of the war. This was not only the result of materials and equipment shortages but also because of the physical destruction of such facilities. The construction industry's contribution to the whole economy slumped to only 4.1% in 1991. The priority was to rebuild what was essential to sustain the population. Sewage systems, water treatment plants and electricity power generation stations were given first priority. In addition rebuilding the country's road networks and bridges were achieved in record times. The national currency, the Dinar, collapsed from its 1970's rate of $3.216 to a Dinar to reach an all-time low of ID 3000 to the dollar. Further analysis of current and future planned developments will be discussed in section 4.

1.2   Legal

Constitutionally, Iraq is a republic with a presidential system of government. The president is the head of the government and the armed forces. The government consists of a number of cabinet ministers. The ministries have a cabinet head office with permanent secretaries, under secretaries, various divisions, directorates and other normal civil service organisational departments. The Revolutionary Command Council, considered to be the highest legislative authority in the country, would make decisions which have the force of law. Laws and regulations can only be enforced after publication in the official Gazette.

Many Middle Eastern countries, including Iraq, have declared that they are a Muslim country and that Islamic law (Shari'ah) is the basis of the country's legal system. However, Islamic law forms only a part of the applicable law. These countries have borrowed and adopted laws from other legal systems such as the French Civil Code. Bunni (1993)[9] explains that the sources of law, in these countries, to which the courts would refer would be in the following order of priority:

- legislation and the codified law;
- if no applicable rule is found, then the principles of Shari'ah would apply;
- the prevailing custom and practice unless it is contrary to laws, public policy or public morality;

- if no help is forthcoming from the above sources, then reliance is placed on natural justice, equity and fairness.

Iraq law is an amalgamation of state law and Shari'ah. The 1970 constitution does not proclaim that Shari'a is a source of law. *Where a point law is not covered by statue, the Iraqi Civil Code gives Shari'a a second place after custom (Saleh 1984).*[10] However, as a Muslim country the people's customs and traditions are very strongly linked to their religion.

The legal system is a modern and reasonably well developed. It is a codified system of law based on a combination of Islamic and Roman jurisprudence. There are five main codes:

- The Civil Code;
- Criminal Code;
- Civil Procedures Code;
- Criminal Procedures Code; and
- Commercial Code.

The Iraqi Civil Code contains provisions which would be, in principle, comparable in the English Law to the laws of contract, tort, evidence, misrepresentation, property and to the rules of equity amongst others. In addition, the code contains the provisions relating generally to contract formation, its form and terms, defects, assignment, performance and discharge.[11]

There is an independent judiciary in basically three levels:[12]

- Courts of First Instance;
- Courts of Appeal;
- The Court of Cassation of Iraq.

In 1977 the Administrative court was established to deal with disputes between ministries and state owned organisations.

1.2.1   Construction Law

Construction contracts in Iraq are governed by the Iraqi Civil
Code number 40 for the year 1951 and its subsequent
amendments. This particular code contains several articles which
are regarded as the basis of construction contracts in Iraq.
Articles numbered 864-890 which are contained in chapter three
of the above code provide the framework of the main requirement
for the "Aqd Al-Muqawala" (contract for works) which is used in
construction contracts. Bunni (1993) traces the origin of the
contract for works to the Islamic Law of the contract of
manufacture which is defined as an agreement for delivery of
future, non-existent goods, for a certain price that is not due until
delivery. Article 864 of the Iraqi Civil Code defines "Aqd Al-
Muqawala" as a contract by which one party promise to
manufacture something or undertake work for an "A'ger" (fee)
promised by the other party.

In a Middle Eastern country such as Iraq it is very important for
contractors to understand the difference between public law and
private law. When the contract is with the government, which is
the case in most contracts in Iraq, then this contract will be
governed by the public law. Iraq's Administrative Law court
jurisdiction was extended in 1980 to cover disputes with only one
party being a public sector entity.[13] The significance is that in
case of default of the other party which may entail a penalty
and/or compensation, for example, the government does not have
to prove that damage has been sustained. The assumption is that
damage has occurred when the breach is confirmed.[14]

1.2.2   Construction Contracts

Until 1957 there were two kinds of forms of general conditions of
contract in existence, one for national contractors and the other
for the international contractor. In 1957 the Iraqi Society of Civil
Engineers (SCE) published the first edition of the General
Conditions of Contract for Civil Engineering Works. This was

based on the British Institute of Civil Engineers (ICE) conditions of contract. [15]

The boom in the economy in general, and in construction in particular, in the 70's led to a significant rise in the use of foreign contractors to such an extent that it dominated the construction industry. The government preference for the use of foreign contractors in most, if not all, its construction contracts has squeezed the local contracting firms out of the main contracting business and into subcontracting. Apart from a handful, local contractors were mostly involved in supplying materials, earth moving, and other works as sub-contractors to foreign companies and the state owned contracting establishment. In addition, the size of mnay projects, in terms of value, put them beyond the reach of most local contractors.

There was then a need for the development of a new general conditions of contract that would reflect the needs of Iraq and the international nature of its contracting business. The Ministry of Planning published a new General Conditions of Contract for Civil Engineering Works (GCCCE) in 1972 and the General Conditions for Electrical, Mechanical and Process Works in 1979. The different Ministries and Government Establishment has adapted this form, under different titles, to reflect the nature of their work. Its interesting to point out that the term "Standard" does not appear in name of the published conditions of contract.

The latest version[16] of the GCCCE has 72 clauses drafted following the principles of Federation Internationale des Ingenieurs-Conseils (FIDIC) General Conditions for Civil Engineering Works, third edition, with a number of changes and additions. These changes and amendments are significant as they would give the employer (the government) more power, protection and reduce its share of the risk. It is prudent to pay special attention to differences in language, as words in Arabic or English might have several translations in the other language. In comparison with the third edition of the FIDIC forms of contract a number of important differences were identified. The main changes to FIDIC are:[17]

- Clause 59 item 4 in FIDIC is amended in Clause 60 item 2C to restrict the payment to the contractor in respect of his cost, expenses, profits, supervision and services provided in connection with the work of the Nominated Subcontractor (NSC) to a percentage of the cost of the NSC work;
- The items in Clause 65 which refers to special risks are amended to reduce the employers liability;
- Clause 69 which provides for employers default has been deleted;
- Clause 70 which deals with changes in costs and legislation has been deleted;
- Clause 71 and 72 which deal with currency and rate exchange have been deleted;
- Clause 72, in the GCCCE form, stipulated that contract will be subject to the laws, regulations, and the courts of Iraq.

The parties to a contract may introduce new clauses, amend or delete from the GCCCE. Majid (1988)[18] argues that although the use of the GCCCE is binding upon government establishments their compulsory nature is often ignored by the Iraqi employers, and deviations and changes are in practice acceptable. Al-Kashtaini (1975) reported that the Public Directory of Roads and Bridges amended the national form to fit its own needs.[19]

All disputes between the employer and the contractor should be referred to the Engineer. This is stipulated in Clause 69 of the GCCCE which is in line with FIDIC clause 67. The main differences between the two conditions are:

- The GCCCE does not limit the time by which the Engineer has to give his decision regarding a dispute;

- If one of the parties is not happy with the Engineer's decision then the matter should be referred to arbitration within 30 days of the notification of such decision, in the GCCCE, and within ninety days in the FIDIC conditions;
- In FIDIC conditions if the Engineer fails to make a decision within ninety days of a dispute being referred to him then it goes to arbitration. The GCCCE is silent about this issue;
- The GCCCE confirms that arbitration will follow the Iraqi Law even if the other party is a foreign organisation or national.

2.   CONFLICT MANAGEMENT

There is an argument that there is a significant difference between conflict and dispute. Conflict is seen to exist when there is an incompatibility of interest. When all available methods and techniques for managing this conflict are exhausted and conflict becomes irreconcilable, techniques for the dispute resolution are required.[20] In reviewing the approaches to resolving such differences in Middle Eastern countries, and especially Iraq, such distinction does not seem to exist.

Regulations and laws, whether of the state or of the Islamic Shari'a, do not seem to differentiate between conflict and dispute as far as their approach to their resolution or management. Arbitration seems to be the umbrella that embraces the Shari'a approach to solving differences between individuals. This is deeply rooted in the Islamic teaching which preaches conciliation and harmony. Arbitration would then refer to the intervention of an external in a conflict or dispute. However, practice and the use of available techniques, mainly arbitration, seems to take account of the possibility of resolving and managing differences before moving to a more formal phase. Such techniques will be mainly discussed under dispute resolution, section 3.

The domination of the government has become evident in the construction industry, both as a regulator and client. The ambitious national development programmes during the 1970's and the 1980's focused and directed all developments in construction procurement processes and strategies. This lucrative market has attracted international contractors, consultants, material/ components and construction equipment suppliers from around the world. International firms have experienced great difficulties in trying to establish themselves in the Iraqi market. One of the main factors is the insistence of the Iraqi government in many occasions to link trade to economic and political relationships. However, in this kind of market and environment the contracting parties would be very wise to rely on conflict management and dispute avoidance in order not to upset the client (the government), the only client.

2.1   Non-binding

Conflict management and dispute resolution in the Middle East are dealt with, at large, in the literature, under arbitration (see 3.2.2).

2.1.1   Dispute Review Boards

In Iraq there are arrangements to deal with conflict similar to that of the Dispute Review Boards. If the individuals directly involved in a conflict such as the client's engineer and the contractors representative can not resolve their differences then the matter could be referred to a higher authority. The relevant government establishment may be asked to clarify a technical or contractual issue. In many cases, especially on major contracts, there will be a periodical meetings of senior directors to discuss progress and ways of solving any unresolved conflicts.

2.1.2   Dispute Review Advisers

A practice similar to the Dispute Review Advisers may be used deal with conflict. More information can be found under 3.2.2.

2.1.3   Negotiation

A practice similar to "Negotiation" is available to deal with conflict. More information could be found under 3.2.2. There is no evidence available to suggest that there is a national theory or training programmes to develop such skills.

2.1.4   Quality Matters

The structure of the industry, the political and economical forces, and the demise of the indigenous (non-government) contracting firms impeded the development and improvement of procurement strategies. Zahlan (1991)[21] reported a survey of Arab contractors which shows Iraq have declared to have only 12 contractors, classified as "first category" and none reported at the "second category"[22]. It is not clear if this figure includes the government's own contracting "companies".

To start with most of the ministries in Iraq has their own Direct Labour Organisation (DLO). The government has also established, in the 1970's, the State Company for Construction Contracts the country's largest civil engineering group (SCCC). SCCC was established with the aim to stimulate competition in the construction industry which was dominated by DLOs. The other major player is the Ministry of Housing and Construction with its two powerful establishments, the State Organisation for Roads and Bridges (SORB) and the State Contracting Company for Sewerage and Water Projects (SCCSWP). In addition, the ministry of defence has its own contracting divisions. Another interesting development in 1970's is the establishment of the State Establishment for Prefabricated Construction (SEPC) which has developed the skills, expertise, and infrastructure in this important aspect of construction.

Construction projects may be procured using two main routes. Unfortunately, there are no statistical data available nor a formal classification other than the following:

- Direct Execution.
  Most ministries would have their own Direct Labour Organisation (DLO). DLOs will have varying degrees of sophistication, experience and expertise. Such organisations would have the resources to provide for certain aspects of their ministry's development needs. DLOs may subcontract the works, or part of the works, in packages to local contractors, government contracting establishments, and if appropriate to foreign companies.

- General Contracting.
  General contracting in Iraq takes different forms depending on the particulars and urgency of the project. One of the main declared objectives guiding the government procurement strategies is the acquiring of the technical skills that would allow the country to reduce its reliance on foreign expertise. However, the urgency of some of the government development plans has overridden this objective on certain projects.

One of the main procurement methods in use is the traditional procurement system. This is very similar to that used, for example, in the UK where the designer(s) is engaged to develop the full design prior to the involvement of the main contractor. However, the government pragmatic approach to procurement has produced a number of variants to this method as dictated by the needs of the project as shown by the following illustration. The Ministry of Housing and Construction (MHC) had awarded the contract to build a bridge north of Baghdad to a joint venture, set-up between the government SCCC and a Japanese contractor, beating a rival bid by SORB, which is part of MHC. A British consultant was engaged to develop the design and provide supervision of construction works. Another example is the project to build the prestigious Baghdad University. The client was the MHC who appointed a US consultant to provide advice to its own staff. A British contractor was engaged to provide the management role, working with the SCCC who won the contract.

Package deals, Turnkey and Design and Build contracts are other forms of procurement that have been utilised. These methods have been used in a variety of projects such as industrial buildings and hotels, as too extreme examples. One key characteristic in all the different procurement methods is the involvement of the client, through his representatives, in the design and construction process. Client engineers would be seconded to work with the consultants and contractors. This is seen not only to safeguard the client's interests but also to create greater opportunities for the successful transfer of technology and expertise. One of the great examples of the success of such policies is the ability of Iraq to rebuild most of its infrastructure, such as the roads and bridges, after the Gulf war with no external assistance.

The main professions in the Iraqi construction industry are the architects and civil engineers. Civil engineers in Iraq would perform other roles which in the UK, for example, would be performed by the quantity surveyor. Services engineers, mechanical and electrical, would contribute to the procurement process their design and supervision services. The key institution that represent all these professions is 'Nagabat Al-Mohandessine' (the Engineering Union). This union publishes a set of conditions that regulate the roles and responsibilities of engineers in providing engineering and consultancy services. The union also provides different levels of membership to engineers that would reflect their professional experience and expertise. This classification system is essential when it comes to making appointments and in selecting engineers for particular tasks.[23]

The terms quality control and quality assurance are more familiar when dealing with materials and products quality. The 'Establishment for Specifications and Standards' is a government body which would examine the degree to which products and materials meet the relevant standards. The issue of quality of work on construction sites is the degree to which the executed work meets the agreed design and specifications. Communication is a major issue, especially when one considers the makeup of the staff involved on projects and the workers. In the 1970's and 1980's when there were serious shortages of skilled workers it was not unusual to find at least two different nationalities on site. This scene may well be repeated as and when the current UN sanctions are removed. Although having the client's engineers working closely with the consultants and contractors is a great advantage in improving communications the frequent changing of such personnel may be unsettling. The shortage in skilled staff on many occasions has forced the client to move staff between projects.

2.1.5   Partnering

Not applicable.

2.2   Binding

2.2.1   Partnering

The government objective to become self-sufficient and technologically independent would not lend itself to long term partnering. However, there are examples of short term partnering and alliances where a national organisation/establishment join forces with an international firm. In 2.1.4 the joint venture between a Japanese contractor and the SCCC to tender for contracts to build bridges is an example. The example was their second project. The aim of SCCC was to benefit from the Japanese experience and technology in a the field of bridge building. The Japanese contractor's objective was to break into a difficult but potentially rewarding market.

3.   DISPUTE RESOLUTION

The influence of the Islamic Shari'a is evident in the rules and regulations of the country. It has also a significant influence in shaping the peoples' customs and practices. When dealing with dispute resolution arising from contracts the available conditions of contract and legislation refers only to arbitration. Failing that litigation would be the next course of action. However, there is no evidence that any other formal method of dispute resolution exists. The term "arbitration" on the other hand, as defined in Islamic shari'a, include other forms of dispute resolution such as conciliation.

3.1   Non-Binding

3.1.1   Conciliation

Conciliation or 'Sulh' is closely linked to arbitration in Islamic Shari'a. The "arbitrator" may be given the authority to settle a dispute by conciliation. Saleh (1984) argues that in this case the arbitrators, unlike *amiable compositeurs*, are not completely free from the obligation of applying the shari'a rules. However, the procedure is simplified by *"avoiding the application of Shari'a rules of evidence and even some of its strict substantive laws".* [24]
The term conciliation was also used interchangeably with arbitration in discussing the cases that can be dealt with under arbitration according to article 254 of the CCP. [25]

3.1.2   Executive Tribunal

There is no available evidence of an established organisation or procedure for such tribunal. However, there were attempts by the

Council of Planning in 1965 to set-up a committee to study "requests" made by foreign companies and to provide the council with an impartial opinion about such requests.[26]

3.1.3   Mediation

There is no available evidence of an established organisation or procedure for such method. However, one would argue that the arbitrator may well be able to perform such a role (see section 3.2.2).

3.1.4   Negotiation

In practice on most projects involving a state establishment or government department conflicts and disputes that can not be resolved on site will be subject to negotiations between the contractor and the relevant government department. Such disputes would be referred to the design or contract section of the employer's organisation who may have the authority to negotiate a compromise. If the contract is at a ministry level, such as that with the Eastern Block in the 1970s and 1980s, then if the dispute is not resolved it would be referred to the relevant ministers. The contract maybe amended and clauses added to prescribe the procedure for such negotiations.

3.2   Binding

3.2.1   Adjudication

There is no evidence that there is a third party adjudication under any of the available forms of contract. However, the GCCCE has followed the practice in the FIDIC form that sets the Engineer as adjudicator. The main concern expressed about this form of adjudication [27] is the issue of the impartiality of the Engineer, especially as the Engineer's fees are paid by the Employer. In the case of contracts in Iraq the Employer, in most cases, is a government body, and is the Engineer's only employer (see section 1.1.1). It is not rare for the Engineer, as referred to and named in the contract, to be another government department. For example, a Brazilian contractor signed a contract with the Ministry of Irrigation. The contract has named the Rafidain State Organisation for Irrigation projects as the Employer with the General Establishment for the Main Outfall Drain as the Engineer.[28] The Engineer in this case is a public body with a strong culture biased towards the protection of the interest of the public and the state. This culture has been shaped and influenced by the law and regulation that governs the internal organisation of such bodies. More importantly such bodies may have acted as an employer, representing the state, in a previous contract.

Article 69 requires the Engineer to notify the contractor and employer of his decision. However, the article does not specify a time limit within which the Engineer need to make his decision. The article also requires the contractor to continue his work without any delay even if the contractor disagrees with the Engineer's decision. If either party is unhappy with the Engineer's decision then a they would have to serve notice of arbitration within 30 days of the notification of the Engineer's decision.

3.2.2   Arbitration

The Arbitration agreement in the Iraqi GCCCE has been modelled on that of the FIDIC.

The Iraqi GCCCE stipulates in article 69 that arbitration is the first course of action to resolving disputes between the employer and the contractor in matters relating to the construction contract. Although GCCCE modified and deleted the reference to disputes between the contractor and the engineer as was in the third edition of FIDIC, however, the fourth edition of FIDIC followed the same route.[29] Arbitration is a more advisable course of action for a contractor to resolve disputes with the employer, especially if the employer is a government body more likely to posses a strong legal position with a wide administrative authority.

(a)   The GCCCE does not distinguish between conflict and dispute as far as what resolution methods are to be used. Article 69 has the title Dispute Resolution, 'taswiat al-neza't' - Arbitration, 'al-tahkeem'. The opening sentence of this article starts by saying *"if a dispute or conflict occurs between the employer and the contractor ... then this conflict or dispute shall be referred to the Engineer, ... the Engineers decision ... shall be binding to the Employer and Contractor"*. The roots of such an approach is in the Islamic Shari'a. 'Al-tahkeem' or arbitration has been practised since the early days of the Islamic era. The term arbitration would refer to all available techniques and approaches to sorting out differences between people.[30]

(b)   The Iraqi legislator dealt with rules of arbitration in Code of Civil Procedures (CCP) of 1969, no. 83, articles 251-276. Judicial precedents will be mainly the Iraqi Court of Cassation. The 1973 law no. 43, which created the Iraqi Chambers of Commerce, allows these chambers to act as arbitrators in commercial disputes with the parties' consent. [31] Although the parties to a construction contract under GCCCE would have agreed to refer matters to arbitration, the law would allow arbitration, even in the absence of an arbitration agreement, by the consent of the parties involved. The parties may even decide during litigation to move to arbitration. In this case any party may claim the stay of these proceedings until the award is made (article 252 of CCP). It is important to note that if one party to a dispute brings an action in a court of law, then the other party must object to the litigation at the first hearing. Failure to do so would render the arbitration agreement null and void and the litigation must then continue.[32]

(c)   The arbitration clause in the GCCCE requires each party to appoint an arbitrator. These two arbitrators are then required to appoint a third to chair the committee. An acceptance to be an arbitrator should be in writing. Once the position has been accepted the arbitrator can not resign without an "acceptable justification". There does not seem to be any special requirements as to who can be an arbitrator. However, if the two arbitrators fail to agree on the third appointment within 14 days or if one of the parties fails to appoint an arbitrator then the competent court may assume responsibility for their appointment. The court's appointment is final and may not contested. Although there are no restrictions as to the nationality, religion or sex of the arbitrators, judges need a special authorisation from Majlis Al-Qadaa (the Judiciary Council) to become arbitrators.[33]   There are two categories of arbitration in Iraq: arbitration according to law and arbitration by amiable composition. The CCP provided for the appointment of *amiable compositeurs* who would have greater freedom from any procedure law to reach a decision as long as the chosen procedure is not contrary to public policy.[34] The doctrine has emphasised that the right of defence of the parties may not be disregarded in either category of arbitration.

(d)   The Iraqi legislator has given the arbitrators full power to deal with disputes on equal footing with a court judge. The arbitrators would have to follow the same procedures of the courts, as laid down in the CCP, in terms of setting a timetable, notifications, hearing witnesses, exchange of documents, gathering and requesting from the parties any relevant evidence. However, the parties may agree to change such procedure. Arbitrators are allowed to hear witnesses, appoint experts, and carry out surveys and investigations in the place of dispute. The issue of taking of appropriate measures against a witness who fails to appear or fails to testify are matters exclusively for the courts.[35] If one party fails to submit his statements or documents within the time limit prescribed by the arbitrator then an award based on the available submissions and documents is made. Article 69 gives the arbitrator full authority to change or amend any decision, opinion, order, certificate or evaluation made by the Engineer that is relevant to the dispute.

(e)   Arbitration procedures should not commence until after completion unless both parties agree otherwise. However, article 69 of GCCCE allows the arbitration to start immediately in the absence of such agreement if *"the postponement of arbitration would lead to harm/ damage to any party or may obstruct further progress of the works"*. Amin (1989) argues that disputes and incidents that would be covered by this exemption are very wide ranging and can be used by the contractor to obtain immediate recourse to arbitration.[36]

The arbitrators must reach their decision within the period set in the arbitration agreement or any extension thereof agreed between the parties. If no period is specified then, by default, the decision should be made within six months of the arbitrator agreeing to arbitrate. The arbitrator should stay the arbitration if during the process, for example, a question of law or an issue outside their jurisdiction should arise. The arbitrator, depending on the issue, may either instruct the parties to request the competent court for an order or the arbitrators may request the court to intervene.

(f)   The arbitrators make their award, in writing, in the same way as that of the court. Articles 154-163 of CCP describe the form of an arbitration award. The award should give a summary of the contract, incident/dispute details, date of arbitration, submissions, investigations, inquiries made etc. The award must also describe the arguments, documents, evidence and defences of the parties and the reasons for accepting or rejecting them. The award must also include the actual decision and reasoning on every point of the dispute.[37] The decision can be unanimous or by majority vote. A copy of the award must be given to each party, the original together with the contract of arbitration must be formally deposited, within three days, with the competent court. One of the main shortcomings of the set framework for arbitration is that the award may not be enforced unless granted a confirmation by the competent court at the request of one of the parties. This stipulation would weaken greatly the power of the arbitrators. It would also strip this route of dispute resolution of its main benefit of allowing the parties a speedy and simplified way to settle their differences.

Al-Mukhtar (1985) criticised the reporting system of the courts of not being properly organised. There are three main reporting sources:

- The Publication Department of the Court issues an annual publication containing some, but not all, of the decisions;
- The Ministry of Justice issues a special publication to report on court decisions;
- The Iraqi Bar Association has a quarterly publication reporting cases and issues of interest.

(g)   The applicable law is that of Iraq. In the absence of any statutory provision the arbitrators must make the award in accordance with customs; in the absence of any relevant custom the arbitrators must apply Shari'a[38], failing which they must apply "general principles of justice". The Civil Code also allows judicial precedents, whether Iraqi or foreign, to be taken into account, provided that the laws from which the foreign precedent has been derived are similar to the corresponding Iraqi law.[39] The parties to the contract may agree a different procedure for arbitration which may well be that of another country provided it does not violate public policy and morals of Iraq.[40] The data available from the United Nations office of legal affairs shows that Iraq has not ratified the 1958, New York, Convention on the Recognition and Enforcement of Foreign Arbitral award nor the UNCITRAL Model Law for Arbitration. Iraq, however, has enforced the United Nations Convention on Contracts for International Sale of Goods on April 1 1991. Iraq is a party to the Geneva Protocol of 1923 (Law no. 4 of 1928). It is also a party to the Arab League Convention in the Enforcement of Foreign Judgements and Awards of September 14, 1952.

3.2.3   Expert Determination

The intervention of a third party to advise on a dispute may be introduced by the parties to a construction contract as an amendment. In one particular example a design consultant was contracted to design and supervise the construction of a sewage system. The Engineer, a government department, examined and approved the proposed design. However, during construction the Engineer raised an objection about a certain element of work which the consultant insisted was adequate and conformed to the design as approved by the employer. A third party, another consultant, was engaged by the agreement of both parties. This expert studied the problem and gave his recommendations which were accepted and the work continued.

3.2.4   Litigation

The dominant role of the government as the main client/employer to the construction industry is clear. Any dispute where a state entity is a party may be referred to the Administrative court, but on a non-exclusive basis. The jurisdiction of the court is determined by the CCP 1969, as amended in 1973, which states in article 41 that *"when the defendant has no domicile or residence in Iraq, the legal action will be instituted in the court within his jurisdiction the domicile or residence of the plaintiff is situated"*.[41]

The parties to a dispute can resolve to resolution arbitration with no interference from the courts as long as they agree on the appointment of arbitrators, procedure of arbitration, and accept and implement the award.[42]

(a)   If there is a disagreement about the appointment of arbitrators or one party refuses to appoint his arbitrator then the matter should be referred to the courts. The referral can be made jointly or by one of the parties. The courts decision is final and is not subject to any appeal.

(b)   During arbitration if the arbitrators fail to make their award within the contractual or legal time limit or are prevented from doing so by force majeure then the competent court, at the request of either party, may intervene. The courts may be asked to extend the expired time limit, appoint a new arbitrator(s), or decide the case, as appropriate.[43]

(c)   At a request of one of the parties, or both, the competent court may intervene to confirm the award, invalidate partially the award for defects, or to set the award aside. In the latter two the court may refer the case back to the arbitrators or may rectify or adjudge the case itself. The court judgment, although is not subject to objection, is subject to appeal and other judicial remedies.[44]

(d)   It has been already established that the parties to a contract may set and agree any preferred arbitration procedure as long as does not contradict public morals and policy. It follows from this that they may choose to adopt the International Chamber of Commerce (ICC) arbitration. Al-Mukhtar (1985) argues that the legal system in Iraq will tolerate arbitration conducted in accordance with a foreign law. Article 25, of the Civil Code states that *"contractual obligations are governed by the law of the domicile when such domicile is common to the contracting parties, and in the absence of a common domicile by the law of the place where the contract was concluded. These provisions are applicable unless the parties agree, or the circumstance indicate that the intention was to apply another law".*[45]

(e)   The 1928 Law number 30 dealing with the enforcement of foreign judgments stipulates that only the courts of the state where a foreign judgment is issued are competent to deal with enforcement and in certain cases only.[46] Another restriction is that such law will only apply by virtue of bilateral treaties and Iraqi internal regulations subject to a condition of reciprocity. Al-Sungegali (1983) argues that it is not advisable for arbitration to be held outside Iraq for a contract to be executed inside Iraq as it will be considered as a foreign judgement only enforceable by the above law.

3.2.5   Negotiation

Refer back to section 3.2.2.

3.2.6   Ombudsman

There is no evidence to suggest that such a system exists in Iraq.

4.   THE ZEITGEIST

At present the country is overwhelmed by the international embargo it has been underneath for the last seven years. The great uncertainty hanging over the country's future politically and economically would make it difficult to have clear view of future. What is clear is the need for a major reconstruction programme. It is safe to say that the country will not be able to afford to resource such a programme at the necessary speed without serious external help. Procurement strategies may well involve funding and operation of the facility. This would require changes in the legal and commercial framework to allow, if not to attract, foreign investors. This would require a move toward a more liberal system. There is very little detail published outside the country to give a more detailed account of current developments.[47]

1   Z. Cywinski, (1993), *Structural Preservation of the Architectural Heritage*, International Association for Bridge and Structural Engineering (IABSE) Symposium, Rome.

2   Y. A. Sayigh, (1978), *The Economics of the Arab World*, Croom Helm, London.

3   The Economist, (1987), *The World in Figures*, The Economist Publication Ltd, London.

4   M. Shabandar, (1987) *Clients and Consultants in Less-Industrialised Countries: Architectural and Planning Consultancies in Iraq*, PhD thesis, University of Sussex, UK.

5   The figures were compiled using statistics published by the Department of International Economic and Social Affairs & the Department for Economic and Social Information and Policy Analysis, Statistical Office in the United Nations.

6   The International Bank for Reconstruction and Development (IBRID), (1982*), World Development Report*, 1982, New York: Oxford University Press.

7   P. Marr, (1985), *The modern history of Iraq*, Colorado/London: Westview Press/ Longmans.

8   New Civil Engineer (1979) *Middle East Change*, Special Supplement, July.

9   N. Bunni (1993), *Liability of Contractors for Design and Construction*, International Construction Law Review, pp 441-451.

10   S. Saleh, (1984), *Commercial Arbitration in the Arab Middle East*, Graham & Trotman Ltd, London.

11   For detailed analysis of Iraqi legal system read S. Al-Mukhtar, (1987), *Disputes and Arbitration in Iraqi Construction Contracts*, The International Construction Law Review, pp. 216-238.

12   S. Al-Mukhtar, (1985), *An Introduction to Construction Contracts in Iraq*, The International Construction Law Review, pp. 35-50..

13   For more details about public works contracts and the Iraqi Judiciary read S. Al-Mukhtar, (1985), *Introduction to*

Construction Contracts in Iraq, The International Construction Law Review, pp. 35-50.

14   A. A. Al-Sungegali, (1983), *Contracts for Engineering Consultancies,* (in Arabic), The Law Research Centre, Baghdad.

15   S. N. Al-Kashtaini, (1975), *Study of the Iraqi General Conditions for Civil Engineering Works as a Model Contract,* College of Law and Politics, University of Baghdad, Al-Maaraf, Baghdad.

16   Any recent developments will be acknowledged if the relevant information was accessible internationally. Otherwise the author will rely on personal experience and contacts.

17   Further comparison of the two forms of contract is in S. Al Mukhtar (1985), ibid.

18   S. Majid, (1988), *Contractors Remedies in Case of Delay in Payment in Iraq,* International Construction Law Review, pp. 112-117.

19   S. N. Al-Kashtaini, (1975), ibid.

20   Brown & Marriot, (1993), *ADR Principles and Practice,* Sweet and Maxwell, London.

21   A. B Zahlan, (1991), *Acquiring Technological Capacity: A Study of Arab Consulting and Contracting Firms,* International Labour Organisation, Macmillan, London.

22   Zahlan (1991) explain that he requirement to be included in the first category are a capital of $0.5 Million, a minimum of two professional engineers, a surveyor, a quantity surveyor, a draughtsman and basic management staff. To qualify as a general contractor in the first category a firm has to own equipment worth $2 Million.

23   Read chapter twelve of Ihsan Al-Ataar, (1989), Construction Management and Professional Relationships, (in Arabic), Ministry of High Education and Scientific Research, University of Technology, Baghdad.

24   S. Saleh (1984), ibid.

25   S. Saleh (1984), ibid, read section IV.

26   S. N. Al-Kashtaini (1975), ibid.

27   For more details about the FIDIC form of contract read N. Bunni (1991), *The FIDIC Form of Contract*, BSP Professional Books.

28   S. Amin, (1989), *The Legal System of Iraq*, Royston Publishers, Glasgow, UK.

29   Ibid.

30   Arbitration is described in shari'a texts as the spontaneous, and more or less improvised, move by two or more parties in dispute to submit their case to a third party called a 'hakam' or 'muhakkam' (arbitrator) (Saleh 1984), ibid.

31   The Iraqi publication containing judicial precedents in commercial arbitration is Al-Nashra Al-Qada'iyya.

32   Article 253.2

33   Article 255.

34   Article 265.2

35   Article 269.

36   Ibid. Amin lists argues that if the dispute is over adverse physical conditions, artificial obstruction, or the withholding of a certificate this would be an acceptable reason for immediate arbitration.

37   S. Al-Mukhtar (1987), ibid.

38   The Islamic legal principles require contracting parties to perform their contractual obligations in good faith and under the principal of *"there should no harming nor reciprocating harm" (la tharara wa la therar)*, therefore, all losses- including what results from delay or error- must be compensated. Amin, (1989), ibid.

39   Civil Code , article 1.

40   Civil Code, article 32.

41   As translated by S. Amin ,(1989), ibid.

42   Read S. Saleh, (1984) section IX for a detail analysis of the "Court Intervention", ibid.

43   CCP, Article 263

44   Read S. Saleh, (1984), ibid, for further details.

45   S. Al-Mukhtar, (1985), ibid.

46   S. Saleh, (1984), ibid.

47   This monograph represents the views of the author, based on his research and analysis of publicly available information.

IRELAND

Nael Bunni and Anne Bunni
Bunni & Associates, Dublin, Eire

1.   BACKGROUND

1.1   Economic

The value of the output of the construction sector is estimated at IR£5.7 billion in 1996. After a period of declining output (-7%) between 1991 and 1993, the industry experienced a period of rapid growth between 1994 and 1996, when output is estimated to have increased by a cumulative 40%. The Central Bank forecasts a growth of 8% in 1997. The main positive factors affecting the performance of the industry have been the continuing rapid growth of the economy (average 4.5% per annum since 1991); the large volume of inward industrial investment; an increase in household formation; a significant growth in finance at relatively low interest rates (currently around 7% APR); and the success of the Urban Renewal Scheme underpinned by tax incentives. An all time record of 33,700 houses were built in 1996.

The bringing of Ireland's infrastructure gradually up to European standards has been greatly assisted by grant funding from the European Union. A total of almost IR£5 billion worth of construction investment will be grant aided in Ireland by the European Union over the period 1994 to 1999.

As can be seen from the following table, Ireland has the fastest growth construction sector in the European Union.

Changes in Construction Output 1993 to 1997

EU Member State	1993	1994	1995	1996(F)	1997(F)
Austria	+2%	+4%	-1%	-1%	+0.5%
Belgium	-1.5%	+1%	+1.5%	+2%	+2.5%
Denmark	-5%	+2.5%	+7%	+4.5%	+4%
Finland	-14%	-3%	+2%	+3.5%	+7%
France	-6%	-0.5%	0%	-3%	+0.5%
Germany	+1%	+8%	+1.5%	-3.5%	-1%
Greece	NA	NA	NA	NA	NA
Ireland	**-6.4%**	**+12%**	**+13.5 %**	**+10%**	**+8%**
Italy	-6%	-3.5%	+1%	+2.5%	+1%
Luxembourg	NA	NA	NA	NA	NA
Netherlands	-2.5%	+1.5%	+1%	2.5%	+2.5%
Portugal	+6.5%	+1%	+6%	+4%	+6%
Spain	-7.5%	+1%	+5%	+1%	+3%
Sweden	-8.5%	-3%	+2.5%	+2%	+1%
U.K.	-2%	+3.5%	-0.5%	+0.5%	+2%
EU(13)	**-4%**	**+2%**	**+3%**	**+2%**	**+2.5%**

Source: EUROCONSTRUCT (via DKM), December, 1996.
F denotes forecast.

The prospects for the construction industry between now and the end of the century are good. The positive economic, monetary and demographic factors underlying the recent excellent performance of the industry look set to continue.

The construction industry employs 120,000 persons directly and indirectly (April, 1996) which represents one in eleven of the 1.3 million persons now at work. Direct employment in construction increased from 71,000 in April, 1993 to 82,000 in April, 1995 and to 86,000 in April, 1996. The increase of 11,000 jobs between 1993 and 1995 accounts for over one in three of the 32,000 industrial jobs created during that period, according to the Central Statistics Office Labour Force Survey.

1.2   Legal

Ireland has a Common Law legal system. There are two distinct legal jurisdictions in the island of Ireland. The six counties of Northern Ireland are governed by the laws promulgated by the British Government and administered under the legal common law system of England, Wales and Northern Ireland.

In the Republic of Ireland, Irish law is an adversarial common law legal system. It has its foundation, since 1937, on the Irish Constitution (Bunreacht nah Eireann) which provided that justice was to be administrated "in courts established by law, by judges appointed in the manner provided by this constitution".

The court system established under the Constitution can be summarised as follows:

a)   District Court - This court has jurisdictions to deal with civil cases where the claim does not exceed £5,000.00

b)   Circuit Court - The jurisdiction of this court is confined to actions in contract or in Tort which do not exceed £30,000.00

c)   The High Court - The High Court has jurisdictions over all civil cases irrespective of the amount of the claim.

d)   The Supreme Court - The Supreme court has appellate jurisdiction from all decisions of the High Court and also from other Courts as may be prescribed by law.

Legal Professions: In Ireland, there is a division between solicitors and barristers. Traditionally the barrister is the advocate, but solicitors in Ireland have a right of audience within the court system. However, this right is not widely availed of in higher courts, and in general terms, the solicitor instructs counsel on behalf of his clients for cases being heard before the higher courts.

Relationship between Clients and Professional advisors: The relationship between clients and their technical professional advisers is contractual and there are standard forms of contract for the provision of such services.

1.2.1   Construction Law

In Ireland, as in most other jurisdictions, there is no single body of law that could be termed "construction law". Contract law is of course a fundamental element of construction law, but many other areas of law are also involved e.g. tort, labour, copyright, patent and of course in many instances administrative and procurement law. The principle sources of construction law obviously stem from standard forms of contract in use with construction projects. The two most widely used forms in this jurisdiction are the IEI (Institution of Engineers of Ireland) and the RIAI (Royal Institute of Architects in Ireland) Forms and their derivatives, see section 1.2.2 below.

1.2.2   Construction Contracts

As stated above, the most widely used construction contracts in the domestic field are:

> (a)   IEI Conditions of Contract for Works of Civil Engineering Construction
> (b)   RIAI Articles of Agreement and Schedule of Conditions of Building Contract

Both of these forms stem from the standard English forms of contract, the ICE and the RIBA forms of contract, respectively.

IEI Conditions of Contracts: The recently Revised Fourth Edition of the IEI Conditions of Contract has just been published. It incorporates a Conciliation Clause into the Form of Contract. This clause provides for a mandatory conciliation process to be entered into prior to referring a dispute to arbitration. It is in the form of a sub-clause within the arbitration clause of the Standard Form. The Arbitration Clause itself, Clause 66, follows the two-tier system, i.e. the dispute is in the first instance referred to the Engineer for his decision. In the event of (i) dissatisfaction with the decision or, (ii) failure of the Engineer to give a decision, within specified periods of time, then the dispute would have to be referred to arbitration. Copies of the IEI Conditions of Contract may be obtained from;

The Institution of Engineers of Ireland, 22 Clyde Road, Dublin 4.
Association of Consulting Engineers of Ireland, 51 Northumberland Road, Dublin 4.
Civil Engineers Contractors Association, Federation House, Canal Road, Rathmines, Dublin 6.

It is notable that the Fourth Edition of the FIDIC Form of Contract for Works of Civil Engineering Construction, popularly known as the Red Book, is also widely used in Ireland by owners who have to seek tenders in the international field. The other forms of contract published by FIDIC are also used in Ireland, particularly as there are no equivalent forms published by the IEI. FIDIC's Orange Book is used for design & build contracts and the Yellow Book for Electrical and Mechanical Works.

RIAI Contracts: The latest edition was published in 1996 and is obtainable from the RIAI at 8 Merrion Square, Dublin 2, Ireland. It also incorporates an Arbitration Clause, (Clause 38), which stipulates that in the event of a dispute, either party may refer such dispute to arbitration and similar to the IEI Form incorporates a mandatory conciliation clause.

2.   CONFLICT MANAGEMENT

2.1   Non-binding

2.1.1   Dispute Review Boards: At the present time, the concept of Dispute Review Boards has as yet not been adopted in Ireland.

2.1.2   Dispute Review Advisors: As in 2.1.1 above, this concept has not been adopted in this jurisdiction.

2.1.3   Negotiation

This method of conflict management is widely used in this jurisdiction as a first step in the resolution of disputes. There is no formal training given to negotiators, but see discussion at 3.1.4 below.

2.1.4   Quality Matters

Most reputable companies in Ireland have set up their own in-house management systems and a growing number of these companies have had their systems certified under the ISO 9000. The certifying body here is the National Standards Authority of Ireland.

2.1.5   Partnering

Whilst in Ireland partnering is expected to be a part of all well managed construction contracts, it is not formally acknowledged as a conflict management instrument. This is because there is a significant amount of scepticism as to whether or not it is fundamentally different to the proper norm.

2.2   Binding

2.2.1 Partnering

As above for non-binding procedures, partnering is not formally acknowledged as a conflict management instrument.

3.   DISPUTE RESOLUTION

3.1   Non-Binding

3.1.1 Conciliation

As mentioned earlier, the new Fourth Edition of the IEI form of Contract provides for a mandatory conciliation procedure prior to arbitration. It is recognised as being distinct from Mediation in that a recommendation is required from the conciliator in the event of not reaching a settlement at the end of conciliation process. Similarly, the RIAI has adopted a mandatory conciliation procedure in the 1996 Edition of their standard form.

In Ireland conciliation is understood to be more at the adjudicative end of the spectrum of ADR techniques as it requires the conciliator to provide a recommendation should the parties fail to reach an agreed settlement. The conciliator's Recommendation should state his or her solution to the dispute, which has been referred for conciliation. It is usually based on how the parties can best dispose of the dispute between them and need not necessarily be based on any principles of law or equity. The Recommendation should not disclose any information, which any party has provided in confidence. The effect of the Recommendation is extremely valuable in settling a difficult dispute in that if no party rejects the Recommendation by notice in writing to the conciliator and the other party within a specified time limit, the Recommendation would become final and binding.

The UNCITRAL Rules have not been used or adopted in this jurisdiction up to the time of writing. However, in drafting the conciliation Procedure of the IEI, cognance of the UNCITRAL Conciliation Rules was taken and many aspects were incorporated.

3.1.2   Executive Tribunal

This method has not been used in Ireland.

3.1.3   Mediation

Mediation like conciliation is a well recognised alternative method of dispute resolution. Again like conciliation (except in the IEI and RIAI Forms) it is a voluntary process entered into by the disputing parties the outcome of which is of a non-binding nature. There is no legal provision in this jurisdiction whereby the decision of a mediator can be deemed to be binding on the parties.

3.1.4   Negotiation

This method of dispute resolution would, in general terms, be the starting point of any amicable settlement endeavours by the parties. If negotiations between the disputing parties are unsuccessful then the dispute will be referred to either Mediation, Conciliation or Arbitration. Negotiations however between the disputing parties can be ongoing despite the fact that one of the other methods of dispute resolution is under way.

It is the authors considered opinion that settlement of the majority of construction disputes is effected through negotiation with only the most complex being referred ultimately to Arbitration. Of course the success or otherwise of negotiation depends to a very large extent on the skill, expertise and experience of the chosen negotiation. The ability for negotiation is a skill few are born with and is mostly learned from one's daily inter-acted with those around us.

3.2   Binding

There are within this jurisdiction only two methods whereby a final binding and enforceable decision on any dispute can be obtained, these are Arbitration or Litigation.

3.2.1   Adjudication

This concept has not as yet found its way into dispute resolution in Ireland.

3.2.2   Arbitration

The 1954 and 1980 Arbitration Acts govern arbitration in Ireland. However, in this connection, see below in respect of a new Draft Arbitration Bill for international arbitration. The acts provide, inter alia, that unless a contrary intention is expressed in the agreement, parties must, subject to legal objectives, submit to examination on oath or affirmation, produce documents within their power or possession and do all other things the arbitrator may require.

Arbitrator's Powers: Article 22(2) of the 1954 Arbitration Act allows for the parties to vest in the Arbitrator, if they so wish and agree, the same powers as those vested in the High Court with respect of:

- security for costs;
- discovery and inspection of documents and interrogatories;
- giving evidence by affidavit ;
- examination of any witness on oath, either within or outside the jurisdiction;
- preservation, interim custody or sale of any goods which are the subject matter of the reference;
- securing the amount in dispute in the reference
- detention, preservation or inspection of any property or thing and authorising entry authorising of any samples to be taken,

or observation to be made or experiments to be tried for the
purpose of obtaining full information or evidence
interim injunctions or appoint a receiver.

Unless a contrary intention is expressed in the arbitration
agreement, Article 25 of the 1954 Act empowers the arbitrator to
make interim awards and Article 26 confers upon the arbitrator the
same power as that of the High Court to order specific
performance of the contract, (other than a contract relating to land
or an interest in land).

The costs of the reference and of the award are at the discretion of
the arbitrator. Any provision in an arbitration agreement to the
effect that parties shall pay their own costs (or any part of them) of
the reference or the award is deemed to be void. This does not
affect the validity of the agreement and has the same effect as if
the provision did not exist.

Once appointed, an arbitrator cannot be removed except by order
of the High Court and then only in very limited circumstances, e.g.
misconduct, lack of impartiality, not proceeding with due dispatch.

The Courts in Ireland are, and have been, very supportive of
arbitration and are given the power under Section 5 of the 1980
Act to stay court proceedings where a valid arbitration agreement
is in existence. This power is used extensively by the Courts.

The agreement to arbitrate a dispute which arises under a contract
stands independently and separately from the other contractual
provisions of the contract.

On the question of international arbitration, a Draft Arbitration Bill
was published on 30th September, 1997, which is intended to
enable effect to be given, in the Republic of Ireland, to the
UNCITRAL Model Law on International Commercial Arbitration.
Essentially, the Bill adopts the Model Law in its entirety with a
number of additional provisions, which include the following:

- In its Article 8, the Bill refers to consolidation of arbitral proceedings and concurrent hearings leaving it to the parties to agree on such a procedure.
- In its Article 9, the Bill confers power on the arbitrator to award simple or compound interest.
- In its Article 10, the Bill deals comprehensively with the question of costs of an international commercial arbitration.
- In its Article 11, the Bill restricts the liability of arbitrators to matters of bad faith. The liability of appointing institutions is also restricted in this Article to matters of bad faith. An interesting addition to matters of immunity is in Articles 11(6) and 11(7), which refer to the privileges and immunities of witnesses and lawyers in arbitral proceedings. However, the matter of the duties, which these persons would owe to the arbitrator, is not dealt with.

The draft legislation is also intended to amend the present 1954 Arbitration Act, in regard to the matter of interest in order to make further and better provision for arbitration proceedings in the domestic scene. It is envisaged that when this Bill becomes law in the very near future, Ireland will become an attractive forum for International Arbitration. With this in mind, the Bar Council of Ireland will in February 1998 inaugurate a purpose built International Arbitration Centre.

3.2.3   Equity Clauses

Ex aequo et bono (what is fair and right) – All decisions made by an arbitrator in this jurisdiction must comply with the relevant laws and legal principles together with the provisions of the Irish Constitution (Bunreacht na hEireann) and the principles and rules of Natural Justice.

The concept of an amiable compositeur is not widely used in this jurisdiction however should the parties wish and agree then such an appointment can be made.

Kompetenz – Kompetenz: The 1954 and 1980 Arbitration Acts in this jurisdiction do not confer on an arbitrator the power to determine the question of his/her jurisdiction. However, if and when the Draft Arbitration Bill becomes law, then Article 16 of the Model Law would govern this concept in international arbitration.

Immunity of arbitrators: The 1954 Act is silent on the question of immunity of an arbitrator. However the proposed new Arbitration Act for international arbitration, as discussed above, addresses this question and the Bill provides at Article 11 for such immunity.

Immunity of the various appointing bodies: As for arbitrators, the 1954 Act is silent on this topic, but see above under the new Draft Arbitration Bill.

The use of his own knowledge: the arbitrator is permitted to use his own knowledge and expertise during the course of the proceedings provided always that the parties are made aware of such area of knowledge and expertise and are asked to comment. The case of the *Annie Fox and Others* v. *P. G. Wellfair Ltd. (in liquidation)*, (1981) Lloyd's Law Reports C. A. 514 is persuasive in this jurisdiction.

Privacy of Awards: Because of the private and confidential nature of the Arbitral process in Ireland, arbitration awards are not published.

The 1958 New York Convention: this Convention has been ratified in Ireland and the 1980 Arbitration Act gives effect to it.

Expert Determination: This procedure is not usually used in Ireland except in very specific cases.

3.2.4   Litigation

In the absence of an arbitration agreement in a construction contract the only other means whereby the parties can obtain a final, binding and enforceable decision is through the Court System - Litigation.   If however parties have included a valid arbitration clause in their contract then the courts, in the absence of extenuating circumstances, will refuse jurisdiction, and stay any proceedings commenced by a party to the contract who seeks to ignore the arbitration clause (see section 3.2.2 above).   There are no specialist courts or judges reserved to hear construction disputes in Ireland the procedure is adversarial and court appointed technical experts are seldom, if ever, used.

Within this jurisdiction the Courts would be most reluctant to allow evidence on statements or correspondence made or submitted on a "without prejudice" basis.

Under the adversarial legal system, the evidence before the Courts is most frequently oral, but written sworn testimony can also be given.   Parties are always given the opportunity of cross-examination.   Cross-examination by a court under the adversarial system does not usually take place.

Under the Irish Conflict of Laws the Irish Courts will recognise and enforce foreign judgements on the basis of the "doctrine of obligation".   The basis of this doctrine is that where a court of competent jurisdiction has adjudged a certain sum to be due from one person to another, a legal obligation arises to pay that sum on which an action of debt to enforce the judgement may be maintained.   There are certain conditions, which must be fulfilled before an Irish Court will recognise and enforce a foreign judgment:

- the judgment to be enforced must have been handed down by a court of competent jurisdiction in relation to the particular defendant according to Irish Conflicts of law rules.

- the judgment to be enforced must be final and conclusive.

Where the enforcement of foreign arbitral awards is concerned there are several ways in which these may be enforced. They may be sued upon in an action at common law, or enforced under the 1954, or the 1980 Arbitration Acts.

Enforcement at Common Law – In order to have an arbitral award enforced at Common Law the following requirements must be complied with:

(a)   the defendant must have validly submitted to Arbitration;

(b)   the arbitration must have been conducted in accordance with the submission;

(c)   the award must be valid by the law governing the arbitration proceedings; and

(d)   the award must be final and conclusive under the law governing the arbitration proceedings.

From relevant case law in this area it appears that neither a foreign judgement nor an arbitration award would be enforced in this jurisdiction if:

·   the arbitrator (or court) had not the jurisdiction to make it;
·   it was obtained by fraud;
·   its recognition or enforcement would be against public policy; and
·   the proceedings were contrary to natural justice.

Under the 1954 Arbitration Act, the Courts are, in limited circumstances, empowered to review an arbitrator's award in a domestic arbitration. A foreign judgement however cannot be examined on its merits by the courts here, the reasons for this are based on the maxims *interest rei publecae ut sit finis litium* and *nemo debit bis vexari pro eadem causa,* it is "contrary to principle and expediency for the same question to be again submitted to a jury in this country".

3.2.5   Negotiation

See Sections 2.1.3 and 3.1.4 above.

3.2.6   Ombudsman

The concept of Ombudsman – the representative of the people is relatively new in Ireland.   In 1980 legislation was passed to establish the Ombudsman's Office and was brought into effect in 1983.   The Ombudsman has power to investigate complaints against Government Departments and Offices, Local Authorities, Health Boards etc.   This power does not extend to the Private Sector.   Two Institutions have established an Ombudsman office to deal with their own particular sectors, these are the insurance and lending agencies (Banks and Building Societies).   Neither one of these three Ombudsman offices in Ireland have authority to investigate construction matters.

4.   THE ZEITGEIST

The legal Zeitgeist in Ireland as we move towards the new millennium is most encouraging.   Following the publication of the New Arbitration Bill in September, 1997, a new Court Service Bill was published in October, 1997.   The publication of the Court Service Bill was the result of the recommendation of the Working Group on a Courts Commission which was set up with a view to reviewing and reforming the present court system.   The Working Group looked at systems in other countries but the framework that it set out for the proposed new system is uniquely Irish.   The present system the Group said was "cumbersome unwieldy and outmoded" and had produced crippling inefficiencies.   The new system will lead to a more accessible and community friendly service which will enable decisions to be made speedily.

The economic zeitgeist in Ireland is buoyant, the recession of the late eighties and early nineties has had no lasting effect and the country is economically very well positioned. The construction industry is booming, this, to some extent, must be attributed to EU funding for major infrastructure projects. One of the most important contribution to the present economic climate is however the enormous expansion here of the hi-tech computer industries.

ITALY

Alberto Feliciani and Carlo Massini
Pavia e Ansaldo, Rome, Italy

1.   BACKGROUND

1.1   Economic[1]

1.1.1   General Trends

The Italian construction industry has experienced a serious downturn over the past few years, and has started recovering only very recently. The building industry has been faced with the consequences of four years of economic recession that has involved all areas of Italian national economy. From 1991 to 1996 the drop in investments has been close to 12.4%, with effective losses in terms of production of about 20,000 billion Lira.

The consequences on employment and productive resources have been substantial: 150,000 employees lost their position (i.e. 42% of the overall losses in Italian industry in general, whereas only 20% of the industry occupation is covered by the construction industry).

Statistic data on the volume of investments show an improvement in the second semester of 1995: after an overall decline of 5.9% in real terms for 1994, and a slight decline during the first semester of 1995, the second half of the year has brought about an increase of 2.9%. The overall figure for 1995, accordingly has been an increase of 0.7%.

As to specific data for 1996, however, investments in construction industry have started increasing, mainly as a reflection of the material expansion in investments for non-residential buildings.

The first semester of 1996 has shown an increase in construction investments of 1.9%, while performance in the second six months has been slightly recessive: the overall figure for 1996 is an increase of 1.1%. National accounting data published in the

General Report on the Economic Situation of the Country estimates the overall amount of investments in construction is 154,697 billion Lira, which reflects an increase of 3.6% of the value of investments which equates to 1.1% after deduction of inflation quotas.

As to public works investments, the reduction in investments for 1996 was equal to 2.1%. According to the Annual Report of the Bank of Italy, investments in public works only amount to 1.2% of the GDP (Gross Domestic Product) in 1996.

As to residential housing, 84.697 billion Lira were invested in 1996, which represents an increase of 1.0% in the value of investments, whereas the overall number of construction projects decreased by 2.4%. Investments in new residential buildings dropped by 3.8%, while restructuring and reorganisation operations for existing premises were stationary.

As to the non-residential construction industry, the trend was positive in 1996 as well as in the preceding year. The investments in this specific sector amounted to 43,587 billion Lira in 1996, which represents an increase of 13.6% in value and of 10.9% in quantity. The overall increases for 1995 have been estimated to be around the 7.0% mark. The named positive figures have been brought about by specific tax-exemption regulations that involved non residential investments over the past two years.

The latest data on tax-exemption for investments in construction industry relate to a 41% income-tax exemption bill for restructuring and restoration of residential buildings, that was enacted by the Government in September. 1997, and is currently being negotiated. A substantial amount of investment in public works is planned as a consequence of the earthquake that hit central Italy in Autumn of 1997.

1.1.2   Residential construction industry

Market research underlines the increasing percentage (from 11.7 in 1995 to 21.1 in 1996) of individuals who deem the fiscal pressure on housing investments to be excessive, as a major ground for the reduction in the volume of operations. 236,696 sale transactions have been executed in the first six months of 1996, corresponding to a -3.7% decrease as compared to the first semester of 1995.

A substantial reason for the decline is the increase of the offer of public-owned real property on the market. The overall negative trend of market prices is another factor for the loss of interest by investors in the real property market. Nowadays, 36.7% of the Italian population deems an investment in real property as profitable only in the event of personal use of the property.

In general, investing in residential property has lost 17% in the scale of most favoured ventures (from 48.4% in 1995 to 31.4 in 1996).

On the contrary, the rental market has recorded an increase in number of contracts, since the number of transactions for the same period has grown by 8.8%, totalling 436,941 transactions. According to the NOMISMA market research institute, the total number of rental transactions in 1997 should reach the 950,000 mark, i.e. 200,000 more than 1992. Rental prices are decreasing: average figures have descended by 4.1% in large urban areas in 1996.

The average value of the family-owned residence is 223.9 million Lira, while the average value of the rented house is 144.7 million Lira (the difference is grounded on the smaller average size of rented property). The annual average rent in Italy is 4.1 million Lira (plus 17.4% as compared with 1993).

The data concerning construction licenses for residential premises in the third quarter of 1996 reflects a reduction of 5.8% in the size of the houses and of 5.0% in the number of buildings as compared to the same period in the preceding year. The average size of a new apartment has passed from 90.1 to 88.6 square metres, and the average number of rooms for each unit has decreased from 4.10 to 3.97. The overall trend has been negative since 1992.

As to composition materials for residential buildings, the period between 1991 and 1994 shows an increase in the use of bricks and stone. In terms of volume of the buildings, the share of construction having a frame made out of bricks or stone has passed from 18.6% (1991) to 24.4% (1994). The current figure for 1995 is 24.2%. The share of buildings made out of concrete has dropped from 75.1% (1991) to 69.5% (1994).

As to the territorial apportionment of construction ventures, percentage data of the overall number of new ventures show a progressive reduction of the share for southern Italy (from 41.7% in 1987 to 29.5% in 1995) and an increase for northern regions (from 43.2% to 57%).

1.1.3   Non-residential construction industry

As a consequence of the incentives introduced by recent tax regulations with reference to the purchase of new instrumental premises, investments in non-residential construction industry have shown a positive trend over 1995 and 1996.

In 1995 investments increased by 6.3%, while resources allocated for investments in this area for 1996 amount to 43,587 billion Lira (+13.6% in value and +10.9% in volume as compared to 1995). With reference to licences for new construction, after a five-year period of progressive reduction, the amount of operations started growing substantially over the named period.

In 1995, the number of licences issued by the municipalities increased by 7.1% as compared to 1994, while the increase in volume of the new buildings grew by 38.1%. During the same period, developments and improvements of existing premises increased by 42% with reference to the size of the buildings involved. The aggregate figure is 39.3%.

The overall number of licences issued in 1995 with reference to non-residential buildings was 21,906, for a total volume of 126.120 million square metres. The average size of new buildings in 1995 is 4,100 cubic metres.

As to the intended function of the buildings, 65.5% (57.6% in 1993) of the volume of licensed constructions in Italy is destined for industry and handicraft, while 14% is intended for agriculture, 13.2% is for trade and hotel industry and only 1.3% for communication, transportation and insurance businesses.

As to the composition material for non-residential premises, more than 85% of the buildings are made out of concrete, while the share of bricks and stone is only 6%, and the share of steel is 3.4%.

As to the territorial apportionment of constructions, the 39% figure for the overall increase in licences has to be apportioned as follows: 58% in northern Italy, 44.1% in central regions and only 3.4% in the South.

1.1.4   Public construction works

The estimated amount of investment in public works for 1996 is 26,413 billion Lira, a decrease of 2.1% when compared to the preceding year. The figure confirms the negative trend that was experienced over the last years (-6.5% in 1992, -14.4% in 1993, -10.4% in 1994 and -0.2 in 1995)., which resulted in an overall decline for the 1990-1996 period of 32%. However, tendering activities and negotiation of new contracts by public bodies are experiencing a substantial growth. The improvements that have been brought about by recent changes in the regulation of public works, in order to expedite award procedures and negotiation with public bodies, may bring about a material increase of investment in the public sector.

As to the general expenses of central administrative authorities, the decline in 1996 has been around 4.4%, while the ANAS (Public Body for Streets and Highways) has invested 2.9% less than in 1995. Local administrative bodies (Regions, Districts and Municipalities) and health institutions have increased their operations in 1996 by about 9 to 16 %.

Statistic data on existing construction ventures are only available until 1995. The indicated figures refer to the agreed price for the works, and do not include VAT and expenses for designing, inspection and testing.

As to public construction work that were initiated in 1995, the overall value amounts to 13,578 billion Lira. The most active public bodies were the Municipalities, that entered 43.9% of the works. The categories that were most interested in the new ventures are plumbing activities (20%, 2,716 billion Lira), health and medical structures as well as street and airport construction (both sectors 17.5%) and social-educational building industry (13.3%, 1,811 billion Lira). As to territorial apportionment, 55.6% (in value) of the works is to be performed in northern Italy, 17.6% in central regions and 26.8% in southern Italy.

As to public works that were completed in 1995, the overall value amounts to 13,928 billion Lira: 20.8% for plumbing and electrical facilities, 16% for health and medical structures and 15.9% for street and airport construction.

In comparison, the public need for building ventures is increasing, as shown by the specific data on public tendering for works contracts. With reference to invitations for tender for public works construction contracts, the overall value of the offer in 1996 increased by 11% as compared with the preceding year.

The overall amount of works tendered was 25,694 billion Lira, as compared to 23,077 billion in 1995. The average value of tenders in 1996 (1.25 billion Lira) is lower than that of the preceding year (1.281 billion Lira). Small and medium value works have increased in number, while projects between 18 and 30 billion Lira in value have decreased by 20% in overall value, and works for more than 30 billion Lira have decreased by 3% as compared with 1995.

There have been 314 tenders for construction ventures of more than 5 million ECU (i.e. the bottom figure for application of EU regulations) in value, for a total value of 8,817 billion Lira. The share of such tenders in the overall figures is 1,5% in number and 34.3% in value.

A substantial part of the increase in public offers relates to maintenance and repair activities on existing premises, that increased by 20% in value as compared to 1995, while construction of new premises increased by 9% in value.
About 22% of the invitations for tenders (5,500 billion Lira) relate to operations on existing public owned property: there have been 6,475 projects with an average value of about 860 million Lira. Works for new ventures provide for 78.4% of the value of the tenders (20,145 billion Lira), with an average value per initiative of 1.431 billion Lira. In 1996, 59% of the tenders have been issued by local administrative authorities, that offered works for 15,000 billion Lira (28% more than 1995). Local health institutions have tendered works for about 2,000 billion Lira (79% more than 1995), and universities have issued tenders for 463 billion Lira (64% more than 1995).

The analysis of territorial apportionment of the tendering activity shows a decline (-8%) in north-western regions, a slight increase in the south, and increases around 25% for north-east, central Italy and in the islands.

With reference to awarding of construction contracts with public bodies in 1996, 4,879 agreements were entered into, for an overall tender value of 9,696 billion Lira. Information regarding the tendered amount and the final contract price show an average markdown of 17.1% for awarding.

The average difference between the amount of the tender and the award price is lower than the average markdown for the preceding three years. As a matter of fact, the average markdown has been of 25 % in 1995, 27% in 1994 and 27.5% in 1993. The reduction in 1996 has been more substantial for new construction (17.3%) than for works on existing premises (16.1%).

A major opportunity for the development of construction enterprises in Italy is represented by the funds provided by the European Union. The effective support granted by such resources will depend on the ability of the public bodies in meeting the standards set by the specific programs.

1.1.5   Costs

The 1996 general national index of the cost for construction of a residential building show an increase of 1.8% when compared with 1995. The same index shows an increase of 1.9% for labour costs, while materials and transportation costs have increased by 1.4% and 2.4% respectively.

The average annual index of the costs for construction of industrial premises in 1996 shows an increase of 2% when compared to the preceding year. The average increase of cost for transportation has been of 3.2%, while materials and labour have increased by 2.3% and 1.5% respectively.

As to the construction of streets, highways and galleries, average cost for construction in 1996 have increased by 2.5% (2.4% for streets with galleries). Transportation costs have increased by 3.4%, labour by 2.1% and material by 1.4%.

1.2   Legal

Italy has one overall civil law system. Administrative powers are managed by the Government and by the local administrative bodies, i.e. Regions, Provinces and Municipalities.

There are no special courts having subject matter jurisdiction over construction disputes. A special regulation is provided for labour disputes, that are submitted to the jurisdiction of specific sections of the Low Court and of the Tribunal.

Advocacy is a distinct legal profession, separate from magistrature and notarial professions. Lawyers are admitted to practice upon graduating in Law School and passing a specific exam.

There is no discrimination or bias by the Courts or by legal professionals towards corporate concerns as compared to individuals.

The regulation of construction is contained in the Italian Civil Code and in a number of additional statutes. There are two main areas of Italian construction law, concerning public works on one side, and private works, on the other.

As to consumer protection, a new section (articles 1469-bis to 1469-sexies) was added to the Civil Code by law n. 52 of February 6, 1996, implementing the EEC directive n. 93/13. The section introduces a broad system of protection of consumers, in particular by limiting the validity and enforceability of oppressive and unfair contract clauses.

"Public works" are works carried out by the State or any other public body for the benefit of the public or for a public interest, or that are financed with public money. In the absence of one of the named prerequisites, the construction contract belongs to the category of "Private works".

Public works are governed by specific regulations that oftentimes depart from the general civil rules for private works construction contracts contained in the Italian Civil Code, articles 1655 to 1677. However, for matters that are not expressly or implicitly regulated by specific statutes, the rules of the Civil Code apply to public works construction contracts as well (this principle has a narrow scope, since, generally, statutory regulation of public works construction contracts is much more detailed than the general rules for private agreements). The regulation of public construction has been recently amended by Law n. 109 of February 11, 1994, and subsequent statutes which varied the same. Additional amendments are currently in process.

1.2.1. Public Works Construction Law

A. Main Features

The major requirements for public works are that the employer be a public or state body and that the development be intended for the use or enjoyment of the public or for serving a public interest, or that it is financed with public funds. Another important requirement is that the construction involves the development of real estate.

The main sources of rules regarding public works construction contracts are the General Conditions and the Special Conditions, the former being the set of rules generally applicable to any kind of work for which tenders are offered by the State, and the latter being regulations providing terms and conditions applicable to the specific agreements entered into by the public body or administration.

Special Conditions have always been considered as discretionary, i.e. freely negotiable by the parties, while there is uncertainty as to whether the rules under the General Conditions are mandatory or discretionary, since they apply to any contract entered into by a public body. It has been previously held that the General Conditions are mandatory rules, since they necessarily apply, regardless of whether they had been negotiated by the parties. On the contrary, the currently prevailing view is that the General Conditions are merely binding on the officers of the public body, who are prevented from entering an agreement with private contractors unless the latter accept the provisions of the General Conditions. Accordingly, private contractors are not bound by the General Conditions until they accept them by entering into the contract.

As a general rule, consent to a public works construction agreement is freely given, on an equality basis, by the public as well as the private party to the contract. Despite the involvement of a public interest, and despite the fact that the private contractor is actually facing a public authority, there is no general subjection of the private contractor to the public administration, or predominance of the public body over the non-public party. Although there are, of course, some important privileges granted to the public body, such as the *jus variandi* (see under 1.2.1 C), the basic scheme remains one of free acceptance, with no forced imposition on the private contractor.

B.   Formation of the Construction Contract

The General Conditions generally provide for an administrative process consisting in a series of acts (resolutions, decisions, acts of inspection, acts concerning the public tender) regulating the various procedural steps. This segmentation allows the public body to firmly control the contract and its performance, and to repeatedly examine the coincidence of the development with the public interest. The main phases of this process are:

i)   Resolution phase, where the State or public body enacts a series of administrative acts in order to prepare the guidelines for negotiating the construction contract: the most important among these acts is the Resolution to enter into a contract, followed by a temptative agreement. Under art. 322, of Enclosure F of law n. 2248 of March 20, 1865 and under art. 1 of the Royal Decree n. 422 of February 8, 1923, public works have to be performed according to regularly prepared and approved designs. The "temptative agreement" is supplemented by a "design of the works", which reports the technical data and which is accurately regulated by legal provisions in the event that the State is the employer of the construction contract.

Designs are generally classified into preliminary designs, merely containing the information necessary for an approximate evaluation of the works, detailed designs, providing specific data for the execution of the works, as well as the specific indications for the purpose of obtaining all required approvals and permissions, and final designs, containing all calculations, indications of form, price, type, quality and size of all elements, in order to provide a detailed description of the works and a close estimate of their costs. Further, there are maintenance designs (concerning the preservation of existing works), additional designs (concerning variations to a design during its execution) and temptative designs (for works that allow for various technical solutions).

The guidelines to be followed in the development of a design are: a) quality of the works and satisfaction of the specific purposes; b) conformity to environmental and town planning regulations; c) meeting of essential prerequisites, set by State or EC regulations.

Designs are generally carried out by the engineering bureaus of the public administration developing the works, and can be delegated only in case of absolute necessity. Where the design activity to be delegated has a value of 200.000 ECU or more, the selection of the professionals has to comply with the strict prerequisites and rules set by the EEC directive n. 92/50 of June 18, 1992. In other cases, the choice of the expert is generally free, although the selection has to be adopted by a resolution of the appropriate office of the public administration. The public body can also assign the design through *bidding by ideas*, where the public body gives only general indications of the project and chooses the most appropriate draft among several designs submitted by the competitors.

ii)   Award phase, where the private party to perform the public works is selected. This phase is accurately regulated by the statutes, and it is governed by the principles of competition between entrepreneurs (in order to obtain the best terms for the public administration), lowest price as guideline for selection, and equal treatment between competitors. Under art. 30 of Law n. 109/1994, applicants who take part in the awarding phase have to provide the public body with a bid bond in the amount of 2% of the value of the works, in order to restore damages incurred by the employer in the event of a contractor's wrongful refusal to enter into the agreement.

The main methods for selection of the contractor are:

a)   *Public Tender*: under art. 3 of the D.P.R. n. 627 of June 30, 1972, this should be the method generally used by public administrations for awarding construction contracts. The public tender starts with a call for bids by the public administration, and it is open to all contractors having the prerequisites reported in the bid. The general guideline for the selection of the contractor is the lowest price.

b)   *Private bidding*: this is the most widely used method of selection, despite the preference accorded by the law to public tenders (these latter prevail for supply agreements or for those agreements producing income for the State or the public body). Under art. 4 of R.D. n. 422/1923, private bidding is admitted when the public administration deems it as appropriate to confine the competition to certain businesses having specific features, due to the special type or technical demands of the works. Private bidding takes place through an invitation by the State or the public body to take part in the competition. The competitors evaluate the design and submit their prices, and the employer adopts the lowest or most convenient bid according to the procedures under art. 1 of law n.14 of February 2, 1973. In the event of private bidding for construction work higher than 7.5 million ECU in value, the public body has to invite the bidders to a "preliminary conference", in order to allow the competitors to request information and obtain clarifications in connection with the works to be performed.

c)   *Competition by solution*: where the State or public body has not prepared a specific design or technique to be used for the construction, it may invite experts to submit their solution, provided that the works require peculiar expertise or technical skills. Under art. 40 of R.D. n. 827 of May 23, 1924, it is allowed for public works construction contracts, where the administration deem it as convenient to take advantage of technical, artistic or scientific designs and ideas from private businesses. The procedure is limited to contractors who are invited by the employer. The main difference with private bidding, where the employer entirely prepares the object of the competition and the bidders merely submit their prices, is that in a competition by solution contractors draft a design for the works, and submit the price and terms for its execution. As a consequence public administrations have the discretionary power to actually choose the work they deem as most appropriate, while the selection of the best offer in a private bidding is merely a mechanical comparison between different prices. The choice is made by a special commission, and is binding upon the State or public body: the employer may dissent only for just reasons. No compensation is granted to the competitors for the designs. Moreover, the administration may choose one of the solutions, upon indication by the awarding commission, and execute it. In that case, the competitor will receive a compensation to be determined between a maximum and a minimum amount previously notified to the participants in the invitation letter.

d)   *Private negotiation*: when other methods for selecting the contractor cannot be used, the employer enters into this extraordinary bargaining system with private businesses. The main feature of private negotiation is the absence of formalities: the structure and contents of the relationship with the State or public body are entirely negotiated between the parties. This model is very similar to bargaining between private parties. As a consequence, access to private negotiation for public administrations is substantially limited by the statutes. Private negotiation proceedings are only admitted under special and exceptional circumstances, that will have to be described in the decree of approval of the contract, provided that it is impossible to perform public competition procedures. Further, art. 41 of R.D. n. 827/1924, peremptorily limits the use of private negotiations to the following situations: 1) where tenders and private biddings were deserted; 2) contracts of supply of goods that are covered by industrial property rights or that cannot be the object of a competition; 3) contracts for the purchase of delicate machinery or instruments, only manufactured by one business according to the required standards of perfection; 4) lease of premises for governmental services; 5) works, purchases, transports or supplies to be performed under unique time constraints, such as not to allow the delays required by auctions or private biddings. However, under art. 12 of Law n. 1 of January 3, 1978, where certain requirements are met, subsequent public works on construction lots may be awarded through private negotiation to the contractor that was previously granted the lot. Under art. 24 of Law n. 109/1994, access to private negotiation is limited, for designs totalling more than 150,000 ECU in value, to cases of extreme urgency resulting out of unforeseeable events such as natural disasters; restoration and maintenance designs for artistically relevant areas may reach the value of 300,000 ECU. Moreover, under the named provision, at least 15 competitors have to be involved in the negotiation.

The Legislative Decree n. 406 of December 19, 1991, implementing the EEC directive n. 89/440, provides for a special regulation (concerning publicity, participation, selection etc.) for public works construction contracts totalling 5 million ECU or more (excluding VAT) in value.

Under art. 23 of the above mentioned Decree, contractors may submit a joint offer for the execution of the works.

Each contractor will manage the part of the works he assumed under the joint venture, and there is no central direction of the works by one of the businesses. However, the companies are jointly liable for the performance of the entire contract from the moment they submit the joint offer. Moreover, under art. 26, businesses forming a joint venture pursuant to art. 23 may create a corporation for the execution of all or part of the works.

At the outcome of the above mentioned selection procedures, the State or public body formally adjudicates the contract to the private party. The awarding phase ends with the signing of the agreement by the parties. The contract has to be in writing.

iii)   <u>Endorsement phase</u>, where a separate supervising authority examines the legality of the procedure, the compliance with the law and the resolutions adopted by the public body, the economic soundness of the contract and, eventually, declares its approval, which is a legal prerequisite for the validity of the agreement. Prior to the approval by the appropriate authority, the contract has no validity. Under art. 107 of R.D. n. 827/1924, the control authority examines the legality of the awarding procedure, and the compliance of the agreement with the applicable General and Special Conditions and with all applicable clauses and prerequisites. Art. 113 of the same statute enables the supervising body to deny the approval for serious grounds of public interest: this extends the scope of the control as to include the analysis of the convenience and economic soundness of the agreement. As a general rule, the contracts have to be approved by the appropriate Ministry, or by its delegated agents, who have to confer with the Minister prior to denying the approval. The major agreements entered into by local authorities are subject to approval by the Regional Control Committee.

C.   Performance of the Works

Performance under the contract starts with the delivery of the site to the contractor. The agreed construction period begins from this moment, when the contractor gets possession of the area to be developed, and is responsible for its custody. The delivery date is generally not essential to the contract, and its expiration does not authorise the State or public body to terminate the contract. Where the contractor fails to receive delivery of the site on the due date, the public administration may set a peremptory date, upon expiry of which it may withdraw from the agreement and/or have the works carried out otherwise. If the State or public body does not comply with the delivery date, the contractor may request the termination of the agreement: he will be entitled to withdraw from the contract or, where the request is denied, to claim damages for late delivery. A contractor who fails to request the termination of the contract is not entitled to damages for late delivery.

The public authority directs the execution of the works through active technical intervention, in order to ascertain and encourage the regular and accurate performance of the works. This principle is expressed in the establishment of a "supervising body", designated by the public authority and formed by several technical experts, usually officers of the public administration, sometimes private professionals. The public authority, when dealing with the contractor, is represented by the "site manager", who issues orders and instructions in writing, that have to be followed by the contractor. The powers of the site manager are limited by the following principles: the instructions cannot introduce amendments to the design or to the contract, and they must be technically correct, otherwise the contractor may refuse to comply.

Under art. 31 of Law n. 109/1994, construction works have to be performed in compliance with the safety regulations enacted by the Government pursuant to EEC directives n. 89/391 of June 12, 1989 and n. 92/57 of June 24, 1992, and to the relevant implementing statutes. The site manager is responsible for supervising compliance by the contractor with the relevant safety plans. In the event of serious violations or departures from the safety regulations, the public body may terminate the contract.

The *jus variandi*, i.e. the right to depart from the agreed design, used to be regulated by art. 342 of Enclosure F of law n. 2248/1865, which stated that the contractor is not allowed to introduce "variations" to the works, in the absence of a written order or instruction by the site manager, mentioning the approval by the superior offices of the public administration. Under the named regulation, the contractor who departs from the design is liable for damages and has no right to compensation for the variation. On the contrary, the site manager may request variations to the design, provided these were previously approved by the employer.

Art. 25 of Law n. 109/1994 has restricted both parties' right to depart from the design to the following cases: a) necessity due to changes in applicable statutes or regulations; b) unforeseeable events or improvements in materials or techniques that permit a more efficient completion of the works, provided that the general outline of the design is not altered; c) increased costs for materials or labour, resulting in an overall escalation of the expenses for the project higher than 10% (reference is made to art. 1664 of the Civil Code, see under 2.2.2, below); d) errors or omissions in the performance of the works that could cause a seriously prejudice the completion of the works or to the utilisation of the construction.

Further, under art. 20 of R.D. n. 350 of May 25, 1895, in case of necessity or urgency, the chief engineer (who is a public officer) may request slight additions or variations within given cost limits, provided that they do not change the contractual terms or the substance of the project. In case of extreme urgency, the contractor will modify the design if requested by the site manager: the public administration has to be informed immediately: it may stop the performance by the contractor, and pay a just compensation for the urgent work performed.

In addition to the above regulation of variations, there are several limitations to the power of the public administration to amend the agreed upon design. As a general rule, contractors cannot be forced into variations that fall outside the scope of the original contract (i.e. they cannot be required to perform something completely unrelated to the works they started). Further, under art. 344 of Enclosure F of law n. 2248/1865, the contractor has a duty to perform an increase or decrease of the works only up to one fifth of the contract price. If the required variation is greater than that, the contractor has a right to withdraw from the contract.

Under art. 1664 of the Civil Code (applicable to public works construction contracts), in the event of unforeseen difficulties in the execution of the works, due to geological or structural complications, that result in a substantial burden, the contractor is entitled to just compensation. Acts of State or of public authorities and acts of third parties fall outside the scope of the above rule.

As to the termination of the agreement, the general rules (contained in the Civil Code) for private works construction contracts apply. Moreover, under art. 345 of Enclosure F of law. n. 2248/1865, the public body may withdraw from the contract at any time, by paying the performed activities and the price for the materials remaining on the site, added by 10% of the value of the part of the works still to be performed. Under art. 340 of the same statute, the public administration is also entitled to terminate the contract in the event of fraud or gross negligence, or where the contractor fails to comply with the duties under the contract or directions by the employer.

D.   Relationship between Contractor and Employer

Under art. 67 of R.D. n. 827/1924, contractors have to establish their technical ability prior to performing public works. Contractors had to apply for a certificate of ability to be issued by the appropriate public officers. Law n. 511 of March 30, 1942 instituted the National Board of Contractors. Under art. 2 of Law n. 57, of February 10, 1962, effective until December 31, 1999, there is a mandatory registration to the Board for contractors performing public works for more than 75 million Lira. Additional requirements to the admission to public works construction contracts have been introduced by law n. 109/1994. In particular, contractors entering agreements totalling more than 150.000 ECU in value will have to certify the conformity to the European regulations series UNI EN 29000 and UNI EN 29004; additionally, contractors will have to establish their technical-administrative capacity as well as their professional, financial and moral fitness, on the ground of their education, works performed in the last years and size/structure of their businesses. Effective from January

1, 2000, the named regulations will apply to all public works construction contracts, regardless of their value.

However, under art. 19 of the Legislative Decree n. 406 of December 19, 1991, EU contractors are exempted from the filings and registrations with domestic authorities.

Law n. 109/1994, has introduced some major changes in the regulation of the relationship between public body and contractor. Art. 7 has instituted the "Authority for the Supervision over Public Works Contracts", which is formed by five members to be appointed by the Presidents of both Houses of the Parliament, and is independent from any administrative public body. Among the duties of the named Authority are the scrutiny of the economic efficiency of the contracts entered into by the public bodies, and the supervision of compliance with the regulations for public works contracts (with particular reference to contract awarding procedures). Under art. 7 of the named statute, representatives of the relevant public body appoint a "sole co-ordinator", who will be responsible for the preparation of a program of the works to be performed in the following three years. Moreover, the public bodies designate a "head of the execution proceeding", who will be in charge for all necessary acts in relation to the phases of design, awarding and completion of the works.

Public works construction contracts are grounded on the individual talent of the contractor. This principle is reflected in several rules, such as art. 6 of R.D.L. n. 1396 of August 28, 1924, which prohibits the State from entering into public works construction agreements with persons to be later determined or corporations to be later created. Further, art. 339 of enclosure F of law n. 2248/1865 bans all assignments of claims and powers of attorney under the contract, unless previously authorised by the State or public body.

With reference to the financial regime of public works construction contracts, art. 26 of Law n. 109/1994, states that the employer advances to the contractor 10% of the agreed upon price within 15 days from the beginning of the works, against an equivalent advance payment bond. The following payments take place upon the submission, by the contractor, of a Monthly Progress Certificate (SAL) containing a description of the progress in the works (see under 2.1.5 below).

Under art. 30 of Law n. 109/1994, the contractor has to provide the public body with a performance bond in the amount of 10% of the value of the contract. Further, the contractor has to provide the employer with insurance coverage for all risks arising out of completion of the works, mistakes in construction or designing, third party action and force majeure events.

Under art. 35 of Law n. 109/1994, in the event of disposals of assets, mergers or other major changes in the structure of the contractor, the public body has to be given notice. The public body has to verify that the contractor still meets the statutory prerequisites for entering the contract and, in the event that the conditions are no longer present, the contract is immediately terminated.

1.2.2. Private Works Construction Law

Art. 1655 of the Civil Code describes works contracts in general as agreements where "*a party undertakes the execution of works or services, and the organisation of the necessary means and management at its own risk, for money consideration*". Independent organisation and own-risk management are vital features of a works contract: there is no works contract in the event that the principal directs the execution of the works and refunds the costs incurred by the contractor for materials and labour.

Unlike public works, where contracts are regulated by the General and Special Conditions, there is no generally used standard form for private construction contracts. FIDIC-type forms are not adopted in Italy. The grounds of its limited popularity lie in the emphasis that it places on the engineering and designing phase of construction contracts. There are some forms prepared by category associations such as the ANCE, but the only forms that are widely incorporated by reference in private construction contracts are the General Conditions for contracts entered into by the Ministry of Public Works.

A.   The Parties to the Contract

The contractor has to be an entrepreneur.

The parties to a works contract may be individuals or legal entities. The execution of the contract by a construction company is deemed to be an important matter, which calls for appropriate powers. With regard to the employer, entering into a construction contract may be deemed as an ordinary or as an extraordinary matter, depending on the purpose of the works (i.e. whether they are arranged for the maintenance or modification of existing properties, or for building new premises).

A construction contract is based on the personal attributes of the contractor, and on the qualities of its business. However, under art. 1674 of the Civil Code, the employer may not terminate the agreement as a consequence of the death or of the bankruptcy of the contractor, unless the person of the contractor is the essential reason for having entered into the contract.

As a general rule, the rights and obligations under the contract may not be transferred in the absence of the other party's consent. Similarly, the contractor has to obtain the approval by the employer prior to hiring a sub-contractor for the performance of all or part of the works.

As in public works construction contracts (see under 1.2.1.B), private businesses may join in a temporary association of enterprises. Each of the companies manages its part of the project, but they are in principle all jointly liable for the performance under the contract (see under 2.1.7).

B.   The Contract

According to the general definition provided by art. 1655 of the Civil Code, works construction contracts are "onerous" agreements, i.e. the legal scheme provides for a consideration to be given to the party who performs the works. Moreover, under art. 1655, the contractor has to be paid in money. However, failure to determine the amount of the compensation to be paid to the contractor does not invalidate the contract: in such cases, the consideration is calculated through tariffs, customs or ascertained by the courts. As to the determination of the works to be performed, an indication of the fundamental elements of the design is deemed to be sufficient.

The agreement may provide for a lump sum as compensation for the works. Theoretically, the price may be determined at the moment of the execution of the contract, or later, even after completion of the works. Where the parties agree upon a lump sum payment, the contract must describe the works very accurately.

Alternatively, the contract may provide for payment by measurement, i.e. the parties agree upon a given price for each unit, and the final amount to be paid by the employer is calculated upon completion of the works, irrespective of progress payments, which are usually made.

The contractor's obligation to execute the works is a so-called "obligation of result", i.e. the contractor is liable for the regular completion of the works, not merely to perform his obligation with diligence and skill. However, under art. 1664 of the Civil Code, the contractor may require an adjustment of the compensation in the event of unforeseen difficulties, or upon an increase of the costs of labour and/or materials exceeding 10% of the total value of the contract. Correspondingly, the employer may request a reduction of the price in the event of a decrease in the costs for completing the works.

There is no general formal requirement for construction contracts: they do not have to be in writing. However, under the specific regulation provided by the statutes, construction contracts have to be in writing when regulating the development of real property furnished by the contractor, or the building of ships. Of course, the parties can agree to have their relationship governed by a written contract, or refer to the regulations of public works construction contracts, and this is the customary approach.

As a general rule, the contractor's obligation is for the whole, and cannot be divided: the employer's obligation to pay derives from the delivery of the entire works. However, the parties usually agree to have the works periodically inspected, and consequently paid separately upon acceptance of each portion by the employer. Further, in the event of termination of the contract for unexpected impossibility to perform, or for the death or bankruptcy of the contractor, the employer is only liable to the contractor up to the value of the works that were performed and expenses met until termination, to the extent that they are useful to the employer. Moreover, in the event of partial performance of the works, the employer may choose to accept the part and pay the corresponding value.

Law n. 1369 of October 23, 1960 introduces a general prohibition for intermediation and intervention in labour relationships. This regulation is relevant for construction contracts as far as it is intended to prevent arrangements where the contractor uses materials and goods provided by the employer and merely procures the labourers. Under art. 1 of the mentioned statute, such arrangements are treated as direct employment relations between the labourers and the employer. On the contrary, where a genuine works construction relationship is in place, the employees refer to the contractor exclusively. However, under art. 1676 of the Civil Code, labourers employed by the contractor have a direct cause of action against the employer in the event of the contractor's failure to pay their wages, up to the amount of the employer's debts to the contractor. The importance of these rules lies in the peculiarity of the regulation of labour disputes in Italy, that are governed by a specific procedure. Moreover, Courts tend to be much more biased when deciding labour conflicts than in normal civil litigation.

C.   Performance of the Works

The contractor has an obligation of custody in any goods and/or property supplied by the employer.

The execution of the works must comply with the technical specifications agreed by the parties, and, in general, with the "rules of the art", i.e. with the standards and guidelines generally adopted in the construction of equivalent works, with regards to the safety and use of the works, as well as to their shape and appearance. The mentioned standards and guidelines have to be adjusted with reference to the peculiarities and specific requirements of the works, as well as to their function.

The factual impossibility of pre-determining all details of the works is the reason for the contractor's licence to modify minor aspects of the works when necessary. There is a different regulation for agreed variations, necessary variations and for variations requested by the employer.

Under art. 1659 of the Civil Code, *agreed variations* are changes in the original design that are introduced by the contractor after receiving the employer's consent. The contractor has no right to depart from the project in the absence of the employer's prior consent. In the event of a dispute, evidence in writing is needed to establish the employer's consent. As a general rule, the contractor has no right to a compensation for the variation where the agreed price for the works was predetermined as a lump sum, except as otherwise agreed by the parties.

Art. 1660 regulates *necessary variations*. In the event that variations to the original project are necessary to the correct completion of the works, the compensation will be determined by the parties. If the parties fail to reach an agreement on the price for the variations, the Court will determine the variations to be performed and the relevant prices. If the value of the variations exceeds one sixth of the total agreed price, the contractor has the right to withdraw from the contract. The employer has a right to withdraw from the contract if the variations are substantial. In any event, upon withdrawal by any party, the contractor is entitled to equitable compensation under the circumstances.

On the contrary, under art. 1661, the contractor always has a right to compensation for *variations ordered by the employer*, even if the agreed price was predetermined as a lump sum. There is no requirement of a written order by the employer, since the contractor may use any proof in order to show that the change was actually requested. Moreover, the contractor can refuse to perform the variations when their value is greater than one sixth of the price of the works, or where the changes would cause a substantial transformation of the works or of the contractor's obligations. Additionally, any dates and penalties for late delivery are invalid in the event that a substantial change is requested by the employer. Of course, the parties may agree upon a new delivery date.

The employer has an obligation of ensuring the legal feasibility of the works, i.e. he is liable for any authorisation, permit, licence or concession that may be required for the execution of the works. Accordingly, the employer is liable to the contractor for damages arising out of any defect or irregularity in the mentioned permits.

Under art 1662 of the Civil Code, the employer has a right to inspect the works during the execution, and to contest any deviation from the agreed plan or from the "rules of the art". In such cases, the employer may set a deadline for removing the irregularities, and terminate the agreement upon expiration, if the contractor fails to comply.

Upon completion, the contractor delivers the works to the employer, who has a duty to inspect them as soon as possible (art. 1665 of the Civil Code).

In the event of late delivery, the employer is entitled to damages. Moreover, it is customary to have a contractual provision whereby the employer may hire a third party to complete the works, and request the contractor to bear the costs of the completion. Penalty clauses are commonly used for liquidating damages arising out of late delivery.

The contractor has a right to retain the works until payment, unless they were performed on the employer's land.

D.   Defects and inadequacy of the works

Under art. 1667 of the Civil Code, the contractor is liable for any defects and imperfection of the works. The guarantee does not apply to the defects or imperfection that were known or perceivable (unless the contractor wrongfully failed to give notice thereof) by the employer, provided that the employer accepted delivery of the works. In order to take advantage of the statutory guarantee, the employer must notify the contractor within sixty days from the discovery of the flaw. There is no need for such notice in the event that the contractor admitted the irregularity or tried to conceal it.

Under art. 1668, the employer is granted two separate causes of action against the contractor for the enforcement of the mentioned guarantee. It may request the defects to be eliminated at the contractor's expense or a reduction in the price, in addition to the right to reimbursement of damages.

There is a two-years statute of limitations for these claims, running from the delivery date. As mentioned above, the claims are subject to the precondition of notification within sixty days of the discovery of the defects.

Art. 1669 of the Civil Code provides for an additional guarantee which is particularly relevant for construction contracts: it is applicable to buildings and real estate developments intended for long-term use. The contractor is liable to the employer for major defects or for the breakdown, or the danger of breakdown, of the building due to defects in the construction or in the ground. The relevant statute of limitations bars the action after ten years from   delivery. Notice must be given to the contractor within one year from the discovery of the major defect, collapse or danger of collapse, and the employer must initiate the judicial proceedings within one year from the notice to the contractor.

E.   Termination Events

The relationship between employer and contractor ends usually with the completion of performance of the works.

Upon invitation by the contractor, the employer carries out an inspection of the works. In the event of an unreasonable delay or refusal to inspect the works, and where the employer fails to notify the result of the inspection to the contractor, the works are deemed as approved, and the risk of loss or destruction of the works is transferred to the employer, subject always to the rule of art. 1669 of the Civil Code. After the inspection, the employer issues the approval, which terminates the contractor's liability for visible defects in the works and enables the contractor to request payment under the contract.

The same effects result out of an express acceptance of the works by the employer upon delivery without inspection, which is deemed as a waiver of the employer's right to inspect the goods.

Under the contractual obligations general rules (art. 1372 of the Civil Code), the relationship between employer and contractor may also be terminated by mutual agreement. In the absence of a different arrangement between the parties, the employer holds title to the works if they were executed on real property; if the works were performed on movable property, title is vested in the contractor.

Both parties can terminate the contract upon default by the other party. Under art. 1662, the employer may request the agreement to be terminated even prior to the completion of the works, in the event that, upon inspection of the site (see under 2.2.3) any irregularities or deviations from the agreed plans or from the "rules of the art" are found.

Moreover, under art. 1671, the employer is also entitled to withdraw from the contract at any time, provided that he has to indemnify the contractor for all expenses, works performed and loss of profit.

2.   CONFLICT MANAGEMENT

2.1   Non-binding

2.1.1   Dispute Review Boards

The only dispute review procedure that is formally regulated by Italian law is applicable to conflicts arising out of public works. However, the procedure that is described by art. 31-bis of law n. 109/1994 (see 2.1.5) is not effectively comparable to settlement through a Dispute Review Board, since the former is initiated and managed by the head officer of the public body in charge for the completion of the works, whereas an actual Dispute Review Board is supposed to be a neutral third party.

2.1.2   Dispute Review Advisers

See under 2.1.1

2.1.3   Negotiation

Negotiation is not formally taught in Law School and, to our knowledge, there is no significant level of negotiation training on the education market. Negotiation is a basic skill for doing business in Italy, in construction industry as well as in any other economic sector. It is also very useful in everyday life. Accordingly, we would quite recognise that negotiation is a practical skill that is learnt pragmatically by personal experience.

2.1.4   Quality Matters

As a general rule, Italy has adopted the ISO 9000 quality certification standards.

Certification of the quality of materials and systems used by the contractor is increasingly requested in private construction contracts.

With regard to public works construction contract, both law n. 109/1994 and the current bills for amendment thereof, require quality certification under to the ISO 9000 rules, to be issued by private businesses licensed pursuant to EU regulations of the series UNI CEI EN 45000. The Authority for Supervision over Public Works Contracts (see under 1.2.1.D, above) will be in charge of authorising private businesses to issue quality certificates in relation to public works construction contracts. However, compliance with the named quality standards has not yet been effectively tested in public works construction contracts, since the required implementing provisions for public works construction contracts are still in the process of preparation.

2.1.5   Conflicts Arising out of Public Construction Contracts

As a general rule, the regulation of public works construction contracts calls for certain steps that must be followed prior to recourse to either judicial proceedings or arbitration. Specific conflict management mechanisms are prescribed in the General Conditions (see under 1.2.1 above). The main set of rules are contained in the General Conditions for contracts entered into by the Ministry of Public Works, which are also applicable to contracts entered into by different public bodies using State funds.

The conditions provide for a particular system of conflict management during performance of the works and before recourse to arbitration can be allowed.

The General Conditions provide a specific regulation for claims submitted by the contractor to the public body. In conrtrast claims by the public body against the contractor are not subject to any particular regulation.

As a general rule, claims by the contractor have to be registered in certain accounting documents that are prepared and kept during the course of the performance of the works.

In public works contracts, amounts owing to the contractor are paid during the construction phase on the basis of the progress of the construction, verified by ongoing site inspections. The results of the inspections carried out by the employer with the co-operation of the site manager are recorded in a monthly progress certificate (SAL) and submitted to the contractor for his signature. Should the contractor have any claim regarding the part of the construction that was performed within the reference period of the SAL, or regarding special events or the calculation of the outstanding amounts, he must indicate such claims in the SAL. In the event that the contractor cannot quantify or specify its claims at the moment of the execution of the SAL, he may sign the SAL while reserving a right to submit the details at a later date. In such cases, the contractor must specify his claims in the SAL within 15 days from the date of the execution of the SAL. Failure to do so will result in the expiry of the contractor's rights in respect of such claim. The determination of the claim by the contractor must specify the amount of additional compensation he is requesting as well as the methods of its calculation. The determination must be very exact in order to give the public body an opportunity to predetermine the costs of that particular portion of the works.

In practice, the reservation allows the contractor to be paid while he has an outstanding claim, which might otherwise delay the completion of the SAL and payments in respect thereof.

Any claim to be submitted by the contractor, with or without reservations, must be entered in the Accounting Registers provided for by the General Conditions, where all accounting data of the development are recorded. Even though under certain circumstances the Conditions may allow the contractor to make his claim in other accounting records, such claim must be registered in the Accounting Register as well.

In the event that a claim by the contractor remains outstanding for more than one SAL period, the contractor must repeat his reservation when signing any subsequent SAL. The reason for this mechanism is that the ongoing nature of the relationship between employer and contractor relieves the latter only from the duty of quantifying the claim at the moment of the execution of the SAL: the claim itself must still be entered on every SAL in order to acknowledge the fact that it is ongoing and that it is affecting each portion of the construction work that is completed.

The above scheme refers to construction contracts falling within the jurisdiction of the Ministry of Public Works, but also applies to contracts under the jurisdiction of other public bodies, including the Ministry of Defence, with some minor variations. With reference to such contracts, the equivalent of the Accounting Register is the Measurement Booklet ("libretto delle misure") in which the contractor must enter his reservations within 10 days from the date of the execution of the Measurement Booklet. Except for the deadline for executing the different registers, the effects of signing the SAL and the Booklet are the same under any public works system.

Further, the contractor can state on the accounting records and books any "observations" he may have as to the execution of the works or the performance of the obligations of the public body, even without making a claim or reserve. These observations are often the premise of a future reserve or claim by the contractor. No particular formalities are required.

In order to determine the entire compensation for the works, the parties consolidate the figures of the Accounting Register and prepare the Final Account. The outstanding claims and reservations by the contractor will now have to be repeated and recorded in the Final Account according to the same figures and amounts that were expressed in the Accounting Register.

The contractor may also have specific claims relating to the activities that are expressed in the Final Account or in the Certificate of site inspection. Such claims and reservations will have to be entered in the relevant document, in order to be actionable towards the public body.

As to the "general reservations", i.e., reservations or claims relating to the works in general and that cannot be entered with reference to a specific portion thereof (and, accordingly, do not have to be registered in relation to a particular SAL), the contractor will have to express them in the Final Account.

In the event that a contractor expresses a reservation or a claim, the public body commences its internal administrative proceedings in order to verify such claim or reservation. The main feature of such administrative procedure is that it is carried out in the absence of any significant participation by the contractor, who can challenge its results, but not take part in the determination. The first step in the proceedings is the opinion of the site manager regarding the claims or reservations submitted by the contractor in a SAL. The next step is the preparation of a report by the site manager himself to be released simultaneously with the final accounts. The report must include a reasonable proposal for the resolution of all disputes arising under the contract during the completion of the works. The report by the site manager is followed by another report prepared by his superior, the chief engineer, which includes an opinion as to the possible settlement of the claim. A third report is later issued by the site inspector, which has to be "separate and secret", and carries particular weight as the site inspector has not participated in the completion of the works.

The contractor is not heard in these administrative proceedings, but he is entitled to submit a written statement setting out the relevant points of fact and law upon which he has based his claim. Submission of such statement is not compulsory and is not subject to any particular formal or substantial requirements or deadlines. It is within the discretion of the contractor to lodge the statement when he feels it would best serve his case.

At the outcome of the named administrative proceedings, the public body will determine the contractor's claims. Typically, the public body notifies the contractor on an informal basis of its decisions regarding the latter's claims prior to service of any formal notice, so as to settle the dispute as amicably as possible. In the event that the contractor wishes to challenge the decision of the public body, after receipt of the named informal notice, the public body formally notifies the contractor of the decision, and the contractor is entitled to commence proceedings. The applicable regulations require that the decision by the public body be made and notified simultaneously with the completion of the final inspection of the works. However, in practice the inspection and approval of the works is carried out before any decision on the claims is made, so that the contractor can receive payment of the first instalment of the outstanding balance, as well as the retention money and the release of the performance bond.

The decision by the public body on the contractor's claims is not a public act in the sense that the public body uses its public authority: the public body operates as a private party in this context, making its decision as a party to a contract. Accordingly, litigation arising out of the decision should be referred to the ordinary courts (or arbitration, as the case may be).

Since the above mentioned proceedings are mandatory for the contractor prior to commencing any legal action against the public body, the latter could theoretically stall the proceedings indefinitely simply by not making any decision on the contractor's claims. Until 1981, when statutes were enacted to rectify this situation, courts and commentators set guidelines to ensure that the contractors obtained a prompt decision from the public body. These methods were superseded by Law n. 741/1981, which fixed a deadline for the completion of the site inspection (commencing from the date of the completion of the works) and a subsequent period of two months for the public body to approve the inspection. In the event that the inspection is not approved within the second deadline, the contractor is entitled to take his claim directly to the court or to arbitration, without the need to obtain the decision of the public body regarding his claims.

A substantial change in the above mentioned scenario has been introduced by art. 31-bis of law n. 109/1994. In order to ensure a regular performance of the works and to prevent recourse to judicial proceedings, the new regulation provides for a conflict management mechanism to be initiated by the head officer in charge of the execution proceedings. The mechanism is mandatory when the claims expressed by the contractor in his reservations would bring about an increase in the overall costs for the works larger than 10% of the contractual value. In the event that the contractor expresses such a reservation, the officer immediately examines the reports by the site inspector and hears the contractor. Within 90 days from the formulation of the claim by the contractor, the officer submits a proposal for amicable settlement of the claim.

The main guideline for the formulation of the proposal by the head of the execution proceedings is the evaluation of the convenience of reaching an amicable settlement of the claim and the appraisal of the factual and legal grounds for the contractor's claim. However, expiry of the 90 days period does not enable the contractor to initiate litigation or arbitration proceedings. Upon submission of the proposal, the public body has 60 days to make a decision regarding the proposal. In the event that the decision by the public body is not issued within the 60 days period, the contractor is allowed to take his claims to the court or the board of arbitrators.

2.1.6 Conflicts arising out of Private Construction Contracts

No mandatory conflict management mechanism is statutorily available for the parties to a private works construction contract, who are generally free to initiate arbitration or judicial proceedings (as the contract indicates) unless such preliminary steps are incorporated by reference.

Disputes relating to private construction contracts are subject to less compulsory bureaucratic rules. The parties are in fact free to decide whether or not there will be time limits for the contractor's claims, whether litigation will be settled by courts or by arbitration and whether litigation is subject to any particular time frame. Practical experience indicates that the stronger party, i.e. the employer, tends to impose upon the contractor limitations similar to those contained in the public works regulations. It is also considered from experience that arbitration is presently the preferred route for settlement of disputes due to its speed, even though the high costs it involves are increasingly criticised.

2.1.7   Partnering

Under Italian law, partnering by contractors is expressly allowed for the execution of private as well as public construction contracts. Law n. 584 of August 8, 1977 (as amended by law n. 687 of October 8, 1984), implementing the EEC Directive n. 305 of July 26, 1971, has introduced and thoroughly regulated Temporary Associations of Enterprises (ATI). Some major problems for such associations to execute public construction contracts have been removed. Accordingly, ATIs are allowed to submit joint offers for the award of public works, and are capable of being jointly registered in the National Board of Constructors.

There are two main kinds of ATI. In the so-called *horizontal ATI*, all members are responsible for the completion of all works, and are distributed between the members pro-rata. Alternatively, where the contractors choose a *vertical ATI*, each of the associated businesses is responsible for a specific portion of the works, which is assigned with reference to the particular expertise of each individual firm.

As to the liability of the members towards the employer and third parties, all members are jointly and severally liable for the entire works.

In the event that the contractors plan a long term commitment, an ATI is probably not the most appropriate structure. The regulation of ATI is very narrow and focused on the activity of the association when dealing with third parties. ATI are Joint Ventures, intended for a single event of collaboration.

There are no statutory guidelines as to the resolution of disputes between the members of an ATI. In general, the rules for litigation and arbitration apply. However, an ATI as such is not entitled to be a party in an arbitration or litigation proceeding, since it has no independent legal status. Accordingly, it will have to be represented by one or more of its members when entering litigation or alternative dispute resolution proceedings.

The Civil Code contains a specific regulation (articles 2602 to 2620) for long term commitments between entrepreneurs. A *consortium* is created with the main purpose of regulating the relationship between its members, in order to establish a common organisation for the performance of certain activities of their businesses. Except as otherwise agreed upon by the parties, the duration of a consortium is ten years. Where the contract regulates the apportionment of the business between the members, it must specify the share for each member, or set the criteria for its determination. Under art. 2603 of the Civil Code, if the apportionment of the business is deferred to one or more persons, the decision of the relevant person or board can be challenged by recourse to the ordinary courts, within thirty days, only in he event that it is clearly erroneous or unfair. Resolutions are met by the majority of the members, and may be challenged to the ordinary courts within thirty days.

As to conflict management and dispute resolution mechanisms, the rules for litigation and arbitration apply.

As a member of the EU, Italy is subject to the application of the EEC Regulation n. 2137 of July 25, 1987 (implemented by law n. 240 of July 23, 1991), that created the European Group of Economic Interest.

2.2   Binding

2.2.1 Partnering

See under 2.1.7.

3.   DISPUTE RESOLUTION

Disputes arising out of construction contracts may be settled either through court proceedings or by arbitration. Due to the delays involved in the court process, it is more desirable to settle construction disputes by arbitration. However, a significant part of the court procedure is also used in arbitration proceedings. Although it is not the main avenue of relief for the parties to a construction contract, construction disputes are still settled through court litigation, which is less expensive, since court costs are very low in Italy.

Arbitrators cannot issue orders of injunctive relief, or provisional remedies, such as attachments or garnishments. Accordingly, parties to arbitration clauses are free to apply to the Court for provisional remedies. The same is true with reference to injunctions to pay due sums, except that, while in principle the Court will issue such orders regardless of the agreement to arbitrate, the debtor can successfully challenge the injunction for grounds of lack of jurisdiction if there is a valid arbitration clause between the parties.

Law 2248/1865 removed the jurisdiction of administrative courts over construction contract disputes, and stated that General Conditions could provide for disputes between the employer and the contractor to be resolved by arbitration. Accordingly, almost all public bodies amended their own General Conditions for construction contracts in order to introduce the use of arbitration proceedings for dispute resolution.

There is no statutory regulation of Alternative Dispute Resolution mechanisms other than arbitration proceedings and the mechanisms provided for under the rules for public works. Out of court settlement is certainly the most widely used system for composition of disputes in Italy. In general, public bodies as well as private institutions have so far refrained from regulating or instituting appropriate proceedings of Alternative Dispute

Resolution. One exception could be brought about by the Rome Council of Attorneys, which is currently instituting a Chamber of Conciliation.

Negotiation and Mediation institutions, whether public or private, are unlikely to be successful in Italy, since judicial proceedings are generally perceived as the last resort, and negotiation is customarily a basic skill for any market operator (in construction industry as well as in other economic sectors).

3.1   Non-binding

3.1.1   Conciliation

There are no major institutions or regulations for conciliation in Italy. Although some initiatives have been started (such as the Chamber of Conciliation mentioned above), there are no significant records to be reported.

It may be of significance that, under the Code of Civil Proceedings, the judge seeks to reach a conciliation between the parties during the first hearing (see under 3.2.4.A, below). The judicial conciliation, if successful, is binding on the parties.

3.1.2   Executive Tribunals

This process is generally not adopted Italy. Although a substantial amount of conflicts between entrepreneurs is resolved through meetings between executives having the authority to settle the dispute, "mini trials" are uncommon for the settlement of disputes.

3.1.3   Mediation

Mediation is not officially recognised as a dispute resolution mechanism. However, the duties and powers of a mediator can be assigned to a third party arbitrator within a contractual arbitration procedure (see under 3.2.2.C, below). The arbitral award in contractual arbitration is binding on the parties.

3.1.4   Negotiation

See under 2.1.3.

3.2   Binding

3.2.1   Adjudication

This process is generally not used in Italy.

3.2.2   Arbitration

A.   Arbitration in General

In the Italian system, arbitration is regulated by Articles 806 to 840 of the Code of Civil Proceedings. The parties to an arbitral procedure are free to appoint the members of the arbitration panel, in compliance with the rules set forth in the arbitration clause (or in the arbitration agreement). In the absence of an agreement by the parties, the members of the panel will be appointed pursuant to Articles 810 and 811 of the Code of Civil Proceedings. Arbitrators are appointed by serving a deed on the other party, indicating the name of the arbitrator and requesting the other party to appoint its arbitrator. In the event that the other party fails to appoint its arbitrator, or the arbitrators appointed by the parties fail to agree upon the name of the third arbitrators, the second or the third arbitrator are appointed by the Court Chairman of the arbitration venue.

Italy has ratified the New York Convention of 1958 and implemented it by law n. 62 of January 19, 1968. The

UNCITRAL Model Law for Arbitration has not been adopted.

Various institutions have been established in Italy for the purpose of organising and administering arbitration proceedings (Associazione Italiana per l'Arbitrato, Camera Arbitrale Nazionale ed Internazionale di Milano, etc.). Such institutions set down their own arbitration regulations which are enforceable between the parties only if adopted by reference in the arbitration clause (or the arbitration agreement). There is currently no specific institution in charge of resolving disputes arising out of construction contracts, although the Parliament is now examining a bill for the institution of a Chamber of Arbitration for Public Works, to be created within the "Authority for the Supervision over Public Works Contracts" (see under 1.2.1.D, above).

The Italian system provides for two different kinds of arbitration:

a)   quasi-judicial arbitration ("*arbitrato rituale*");

b)   contractual arbitration ("*arbitrato irrituale*").

B.   Quasi-Judicial Arbitration

Quasi-judicial arbitration can be recognised as arbitration in most parts of the world, i.e. proceedings in which there are no requirements for the judges to be qualified as such and are appointed by the parties and not by the Department of Justice as under the Court system. It is regulated by a section of the Code of Civil Proceedings which was amended in 1983. Further amendments have been introduced by Law n. 25 of January 5, 1994, to make arbitration procedures more consistent with the requirements of current national as well as international commercial disputes.

In order to be valid the arbitration agreement has to be in writing, and state the matter of the dispute (in arbitration clauses, this is customarily done by reference to the contract).

In the event that one of the parties to an arbitration agreement submits the dispute to the ordinary courts, the judge will uphold the arbitration agreement, and refuse to examine the case, for lack of jurisdiction. However, the existence of the arbitration agreement will have to be expressly declared by the opponent.

The dispute can be decided by one or more arbitrators, provided that they have to be odd in number. The arbitration agreement or clause must contain the appointment of the arbitrators or the indication of the procedure for their appointment. If one party fails to appoint its arbitrator, the other party may request an appointment to be made by the President of the Tribunal.

Upon acceptance of the appointment, the arbitrators assume a contractual obligation to issue an award within the deadline provided for by the law (180 days from the acceptance of the arbitrators, to be increased by a maximum of additional 180 days where discovery is needed) or by the parties (no limits, provided that the deadline is determined by the parties in writing).

The Code does not regulate the arbitral procedure very accurately. The discovery phase is directed by the arbitrators and by the guidelines indicated by the parties in the arbitration agreement. Under art. 816 of the Code, the arbitrators may set deadlines for the parties to submit evidence or documents. However, they cannot coerce witnesses to appear before the panel, or order the submission of sworn affidavits. All issues that arise prior to the award are decided by the arbitrators through revocable orders. There is no formal requirement for appearance of the parties before the arbitration panel, and third party intervention is not admitted. Parties may appear personally, although they generally prefer to be represented by attorneys.

Under article 822 of the Code, the arbitrators decide the issue by application of legal principles, unless the parties have authorised them to decide the matter according to equitable principles. There is no formal requirement for equity clauses. A widely used wording for equity clauses is to empower the arbitrators to decide the case "*as amiable compositeurs*".

The dispute is decided by the arbitrators through majority vote.

Awards issued by arbitration boards under the quasi-judicial arbitration system have the effectiveness of a court judgment, although they cannot be enforced automatically.

In order to enforce an award, it must be submitted to the Low Court (Pretore), that has jurisdiction over the enforcement of arbitration awards, and that will issue, after checking the formal requirements, an execution order ("exequatur") to enable the award to be enforced. There is no time limitation for the submission of the award to the Low Court.

Arbitral awards can be set aside by the Court of Appeals of the venue of the arbitration, within 90 days from the date of service of the award on the appellant. The Code of Civil Proceedings contains some important limitations to the right of appeal against an arbitral award. As a matter of fact, the parties can request a review of the award only for matters relating to the formation of the award, such as:

- invalidity of the arbitration agreement;

- failure in following the steps and formalities provided for under the Code for issuing the award or appointing the arbitrators;

- award on matters not mentioned in the arbitration clause;

- award issued after expiry of the deadline provided for in the Code (subject to timely claim during the arbitral procedure);

- award inconsistent with a prior arbitral award or judgment having force of res judicata between the parties, provided that such contrast was argued by one of the parties during the arbitral proceedings;

- any failure in following the formalities to allow due defence.

The award may also be set aside on certain grounds relating to the merits of the case, such as:

- evidence being found to be false after the award has been issued;

- discovery of crucial documents after the award which were not or could not have been filed during the arbitration due to force majeure or to misconduct on the side of the counterparty;

- deceit by the successful party towards the other party;

- wilful misconduct by one or more of the arbitrators being established by a final judgment.

The award must be issued within 180 days from the appointment of the arbitrators. The deadline can only be extended for additional 180 days, in the event that evidence has to be collected by the arbitrators.

Under art. 818 of the Code of Civil Proceedings, arbitrators cannot issue provisional remedies, such as attachments and seizures, or injunctive decrees for payments of due sums. The parties can apply for such remedies to the ordinary courts, regardless of the arbitration clause. Injunctions to pay due sums, however, may be challenged by the debtor for lack of jurisdiction of the Court, as a consequence of the arbitration agreement.

As to international arbitration, law n. 257/94 has introduced certain guidelines in the Code of Civil Proceedings. Under art. 832 of the Code, the general regulation for quasi-judicial arbitration applies to arbitration proceedings involving foreign parties (or where a material part of the contract is to be performed abroad) as well, without prejudice to the rules contained in international treaties or conventions. The parties may agree on the law applicable to the case; in the event that the parties fail to agree on the governing law, the arbitrators will refer to the law of the country that has the closest link to the dispute. The language of the arbitration, when not agreed upon by the parties, is chosen by the arbitrators. It is worth mentioning that Law n. 25/1994 provides (as an exception to the general rules set forth by articles 1341 and 1342 of the Civil Code) that arbitration clauses contained in general business forms prepared by one of the parties, and not negotiated between the parties, do not have to be specifically approved by the other party.

The arbitrators' meetings can take place through a "video-telephone" conference, even between arbitrators residing in different nations.

Under art. 838 of the Code of Civil Proceedings, the right to appeal an international arbitration award is more restricted than in domestic arbitration.

With regards to the enforcement of a foreign arbitration award, the Code introduces a special procedure. The award has to be submitted to the President of the Court of Appeals of the place where the other party resides or, in the event that the other party resides outside of Italy, to the Chairman of the Court of Appeals of Rome. An "exequatur" is granted once the Court has verified that (i) the dispute could be arbitrated under Italian law and (ii) the award is not in contrast with Italian public policy principles.

The exequatur can be challenged before the Court of Appeals within 30 days from the date it was served on the other party. The Court will set aside the exequatur in the event that evidence is given in the appeal that:

-    the dispute could not have been submitted to arbitration or the award is contrary to public policy principles, as mentioned above;

-    the arbitration clause is invalid under the applicable law or, in the absence of any indication on the applicable law, under the law of the country where the award was issued;

- the defendant was not properly informed on the arbitral proceedings, or its right to defend was infringed;

- the arbitral award relates to matters which were not submitted to arbitration by the parties;

- the arbitrators were appointed in breach of the procedure or formalities agreed upon by the parties;

- the award is not final or enforceable under the law of the country where it was issued.

C.   Contractual Arbitration

The nature of contractual arbitration has been the subject of a wide discussion by both courts and commentators over the past years. In particular, case law has for a long time been inconsistent on the issue. However, the Cassation Court has settled the fundamental structure of the system over the last few years.

According to the system established by the Cassation Court, parties wishing to settle their disputes out of court (arbitral as well as public courts) may agree in the contract to jointly appoint third parties, called "arbitrators", to negotiate a dispute settlement. The arbitrators are appointed by the parties from a board of arbitration. By signing the contractual arbitration clause, the parties undertake to be bound by the settlement terms contained in the award and to give effect to the execution of the award as if they had been parties to the settlement negotiations. Further, as a consequence of executing the contractual arbitration clause, the parties are deemed to have waived their respective rights to bring a judicial action relating to the issue submitted to arbitration.

The rights of the parties under the contract are superseded by the terms of the award. The nature of the award issued in such arbitration proceeding is purely contractual. Accordingly, the award cannot be appealed through judicial proceedings, and must be challenged, just as any other contract, on grounds relating to its negotiation, execution or lack of formal requirements, such as capacity of the parties, invalidity, duress, deceit, mistake or ultra vires acts by the arbitrators. An appeal on the grounds of mistake will only be available for facts erroneously displayed to the arbitrators, not for any mistake by the arbitrators in applying or construing the law. Any challenge must be brought in compliance with the rules for litigation set out under 3.2.4 below.

Proceedings under contractual arbitration are not regulated by the Code of Civil Proceedings. Accordingly, the parties are free to specify in the arbitration clause whether the arbitrators are to follow the regulations prepared by a particular Chamber of Commerce or Chamber of Arbitration or even to follow no set procedures at all. If the choice for a specific regulation is expressed by the parties, the arbitration proceedings will follow the rule prepared by the relevant arbitration body. If no choice is expressed in the arbitration clause, the arbitrators will be free to determine themselves the steps to be followed.

Usually, the procedure in a contractual arbitration will not deviate greatly from that followed in quasi-judicial arbitration. Of course, some of the rules set down for judicial arbitration, such as Articles 829 (setting out the possible grounds for nullity of the award) and 831 (regulating the grounds upon which an award may be set aside) of the Code of Civil Proceedings, will not be applicable to contractual arbitration proceedings.

D.   Arbitration for the Resolution of Construction Disputes

The General Conditions of the Ministry of Public Works, while in principle upholding the jurisdiction of a board of arbitration, provides that a public body can exclude recourse to arbitration only by inserting an appropriate clause in the tender. As a consequence of this rule, the arbitration proceedings provided for under the General Conditions for public works construction contracts have been construed as not compulsory for the public body. The scenario has been brought about by an amendment made in 1981 to the General Conditions, which previously granted both parties the right to exclude the arbitration jurisdiction by simply suing the counter party in court or, if the defendant wished to proceed to litigation rather than arbitration, by notice in writing to the other party within 30 days from the appointment of the first arbitrator by the plaintiff.

Under art. 32 of law 109/1994, as amended by law n. 216 of June 2, 1995, disputes that could not be settled according to the procedure mentioned in art. 31-bis of law 109/1994 are submitted to arbitration under the rules of the Code of Civil Proceedings, regardless of the amount in dispute. The reference to the rules of the Code clarifies that the existence of an arbitration clause in the contract or an arbitration agreement between the parties is a necessary prerequisite for arbitration proceedings. Accordingly, arbitration requires a prior agreement of both parties.

The arbitration clause in a public works construction contract will have the same effect and nature as a similar arbitration clause in a private contract, where the terms of the clause are negotiated between the parties. The main effects of the clause are derogation of court jurisdiction in favour of the quasi-judicial arbitration system and submission of the parties to the arbitral award. The arbitration clause, although it is a part of the overall construction contract, really takes effect as a separate contract. As a consequence, the invalidity of the construction contract will not necessarily affect the validity of the arbitration clause.

As to the scope of the arbitration clause, the General Conditions of the Ministry of Public Works contain a very general reference to "any disputes between the administration and the contractor, both during the performance of the contract and after completion, regardless of their nature, be they technical, administrative or legal". Although the intention of the legislators was to avoid any confusion arising over the application of the clause, there has been considerable discussion as to whether the arbitration clause will apply to disputes regarding claims under general law, i.e. which are not provided for within the express terms of the contract. It is now generally accepted that the arbitration clause will also apply to such claims, because they are closely connected with matters covered by the contract.

Art. 45 of the General Conditions of the Ministry of Public Works provided for a board of arbitration constituted by five members, including one administrative judge and one ordinary magistrate. The Conditions of the Ministry of Defence provide that, upon agreement by the parties, disputes can be submitted to a sole arbitrator, otherwise to a board of three arbitrators.

The professional qualifications of board members are regulated by the Conditions: for Public Works Contracts, the board must include a magistrate of the State Council, a magistrate of the Court of Appeals of Rome, an officer of the Ministry of Public Works and a professional appointed by the contractor. For railway contracts, the board is made out of a magistrate of the State Council, a magistrate of the Court of Appeals of Rome and an engineer employed in the Railway Company, the latter to be appointed by the Board of directors of the Railway Company. For military contracts, the board is made up from a magistrate of the State Council, a magistrate of the Court of Appeals which would have been competent to hear the case in ordinary judicial proceedings, and a member to be selected among the generals of the particular arm of the forces interested in the contract. Each of the members of the boards of arbitration must serve in the relevant office at the time of his appointment to the board; his continuing participation in the board, however, is not affected by any change in status occurring after the constitution of the board.

There could be problems in co-ordinating the mentioned regulation with art. 32 of Law n. 109/94, that states that matters that could not be resolved amicably pursuant to art. 31-bis, be submitted to arbitration in compliance with the rules of the Code of Civil Proceedings. Under the Code, boards of arbitrators are formed by three members, in the absence of a different agreement by the parties. No magistrate is required to take part in the boards. Since the Parliament is currently examining a bill introducing a preclusion for ordinary, administrative and accounting magistrates from taking part in arbitration boards, the Government is inviting the public bodies to refrain from appointing magistrates as arbitrators. Before the final set-up is indicated by the legislators, the solution appears either entering a different express agreement with the public body, provided that no magistrate should be included in the panel, or in the event of impossibility to reach an agreement, to refer the appointment of the arbitrators to the President of the Tribunal pursuant to art. 810 of the Code of Civil Proceedings.

As a general rule, arbitration cannot be commenced until the works are completed. However, the Conditions call for two exceptions to this rule: i) if the parties so agree and ii) if the subject matter of the dispute is urgent. There are some obvious problems in the determination of the urgency of the matter, since there is a large degree of subjectivity involved (particularly from the contractor's point of view). It has been finally settled that the urgency of the case is to be determined objectively. There is objective urgency, for instance, in the event that it would be difficult to provide evidence to support the claim after the site inspection, either because of its wasting nature or for any other objective reason.

The General Conditions for public works introduce a further exception to the general rule: litigation can be commenced prior to the completion of the site inspection in the event that the value of the claim is significant with respect to the total value of the contract, so as to considerably prejudice the carrying out of the works. Moreover, the general rule will not apply when the subject matter of the claim relates to performance of the works in general (e.g. claims for termination of the contract upon a breach by the public body; claims referring to non-payment of instalments due and payable during the execution of the works; claim of invalidity of the construction contract). As mentioned above, law n. 741/1981 has established the contractor's right to initiate arbitration or judicial proceedings in the event that the public body delays the site inspection procedure for more than two months.

The party who intends to commence an arbitration procedure must submit a request for arbitration to the counter party, specifying the details of the subject matter of the claim. The request must be signed by the party and served through a clerk of the court.

One of the main differences between the arbitration system provided for under the General Conditions and the arbitration system for claims arising under private contracts is that, under the former, there is a time limit for the commencement of the action. In the arbitration system provided for under private contracts, the right to initiate arbitration procedures exists as long as the claim is not barred by statutory prescription or limitation. The deadline set by the Conditions for the commencement of arbitration is 60 days (public works contracts) and 30 days (defence contracts) from the date of formal notice of the decision of the relevant body on the claims lodged by the contractor.

Following service of the request for arbitration, the party commencing the proceedings or, if the parties so agree, both parties, must file an application for the constitution of the board of arbitrators. The named application is not subject to any time limitation, with the exception of contracts governed by the General Conditions of the Ministry of Defence, that provide for a 180 days limitation.

The arbitration system provided for under the General Conditions is quasi-judicial in nature. Accordingly, the parties may challenge the appointment of the arbitrators on the grounds set out in Art. 815 of the Code of Civil Proceedings as well as on the other grounds indicated in the General Conditions. An arbitrator may be disqualified in the event that he has been involved in the works or has acted as a consultant to the works or as a site inspector or has given any advice whatsoever in connection with the claim in dispute. The venue of arbitration is Rome for claims under the General Conditions for Public Works contracts; arbitration under contracts involving the Ministry of Defence will be heard where the claim arises.

The form of the proceedings will be governed in general by the rules under the Code of Civil Proceedings even though certain aspects are regulated by the Conditions. Evidence can either be requested by the parties or called by the board of arbitration. Discovery as well as admission of evidence is not subject to particular formalities.

The award is to be issued within 90 days under the General Conditions of the Ministry of Public Works, while the Conditions of the Ministry of Defence provide for a time limit of 120 days. The award is to be submitted to the low Court for enforcement, as indicated under 3.2.2.A. Any challenge to the award will be governed by the general rules for quasi-judicial arbitration.

3.2.3   Expert Determination

Under art. 1349 of the Civil Code, the parties to a contract may agree on the submission of a particular matter, that is not expressly stated in the contract (e.g. the determination of the price, or quantity of supply) to a third party. The third party, who is commonly referred to as "arbitrateur", determines the matter according to equitable principles. However, he may be vested the authority to determine the matter according to his discretion. In that case, recourse to the ordinary courts is limited to events of clearly erroneous or unfair determinations.

The arbitrateur must not be an expert in the specific field of the matter that is submitted to his decision.

The procedure under art. 1349 cannot be considered as a method for resolving disputes, or managing conflicts, since at the time of the submission of the matter to the third party (i.e. the execution of the contract) there is no actual disagreement between the parties. Expert determination under art. 1349 is merely a peculiar method for fulfilling the parties' consent and the subject of the agreement.

3.2.4   Litigation

The judicial system for disputes involving the State or another public body can be defined as a system of "double jurisdiction". In fact, both "Ordinary Courts" (i.e. courts having jurisdiction over civil law disputes) and "Administrative Courts" (i.e. specialised courts having jurisdiction over administrative matters only) can hear claims involving a public body.

In the event that a public body is involved in a dispute, there are several methods for ascertaining whether the claim falls within the jurisdiction of the Ordinary Courts or in that of the Administrative Courts. As a general rule, Ordinary Courts have jurisdiction over cases where the violation of an absolute right of the private party is claimed, while Administrative Courts have jurisdiction to hear claims of violations of interests of the private individual, which are subordinated to the public interest.

Ordinary Courts may not invalidate the act of the public body, they can only order it not to be applied to the particular case that was brought before them. Administrative Courts can declare the act of the public body as null, and in some particular cases, they can even modify the named acts, being deemed to act on behalf of the public body.

In any event, Ordinary Courts have jurisdiction where the public body acts as a private party (e.g. where a civil law contract is entered into by a public body and a private party). In contrast, where the public body makes use of its public authority, Administrative Courts will have jurisdiction to hear claims brought by private individuals in order to challenge the ways and means of administration and management of the statutory authority of the public body. A great number of construction disputes fall within the jurisdiction of Ordinary Courts, since the contract entered into by the public body for the completion of the works is a civil law contract. However, the rules for litigation in Administrative Courts are relevant for resolution of disputes arising prior to the execution of the agreement (e.g. challenge of acts adopted by the public body in the Resolution phase, or in the Award phase, see under 1.2.1 B, above).

A.   Litigation in Civil Courts

Court proceedings are governed by the rules set out in the Code of Civil Proceedings.

In general, jurisdiction to hear disputes is determined by reference to the subject matter and to the value of the particular claim. There is no subject matter jurisdiction of a particular court over construction contracts; accordingly, jurisdiction with respect to construction contract disputes can only be determined by reference to the value of the relevant claim. Disputes having a value in excess of Lit. 50,000,000 come within the jurisdiction of the Tribunal. Disputes having a value between Lit. 5,000,000 and Lit. 50,000,000 come within the jurisdiction of the Low Court. Where the value of the claim does not exceed Lit. 5,000,000 the Peace Judge will have jurisdiction.

The major disadvantage in bringing an action before an Italian court is the lengthy delay involved in obtaining a final judgment: the latest reports of the Cassation Court contain final judgments of cases commenced in the early eighties. In order to speed up judicial proceedings, Italian legislators have enacted Law n. 353 of November 26, 1990, introducing substantial amendments to the Code of Civil Proceedings.

In order to institute a lawsuit, the plaintiff issues a writ of summons that can only be served on the defendant by a clerk of the Court. The writ must contain particulars of the claim, including the relief sought, the grounds upon which the claim is based and evidence in support of the claim.

The defendant must file a defence no later than 20 days prior to the first date set by the judge for the hearing. The plaintiff and the defendant may amend their claim, and counterclaim, during the initial hearings, after listening to the other party's case. Strict time limits are also provided for the purpose of avoiding lengthy evidentiary procedures: the parties can modify their evidentiary requests indicated in the writ and in the brief of defence only until the second hearing, or within the deadline set by the judge at such hearing.

During the first hearing, the judge tries to convince the parties into an amicable settlement of the dispute.

Discovery is led by the parties. As a general rule, the Court decides the case according to evidence produced by the parties, and the scope of the proceeding is limited by the parties' allegations and requests. However, under art. 115 of the Code, the Judge can ground his decision *on factual notions of common experience*. Moreover, under art. 213 of the Code, the Court can request written information to public bodies relating to documents of the public administration that the Court deems as necessary for deciding the case. Additionally, under articles 117 and 118 of the Code, the Court may order the inspection of persons or things, and the informal examination of the parties.

Evidence to be produced in court by the parties, other than documentary evidence, must have the prior approval of the judge hearing the matter. The Code of Civil Proceedings determines what types of evidence can be admitted. In construction law disputes, there is generally a heavy reliance on documentary evidence as well as expert evidence. Witnesses can also be called by the judge. The Court has the power to appoint a technical expert. The parties may submit affidavits by technical experts, but declarations by expert witnesses at trial are not admitted.

During the course of the proceedings, the claimant can request an immediate order against the defendant to pay any amount which is not challenged by the defendant, or which is clearly shown to be due by written evidence.

When the parties have completed the evidence in support of the case, or when the judge deems that he has sufficient evidence before him to decide the matter, he convenes the parties to finally express their respective claims. The matter will then be referred to the decision of the same judge who heard the evidentiary procedure. Final briefs are filed and exchanged between the parties 60 days after the case was referred to the decision of the judge. The parties have an additional period of 20 days to file and exchange replies to their final briefs. A term of 60 days from the date of filing of the final defence is set for the judge to hand down his decision. The term, however, is not mandatory. Under law n. 353 of November 26, 1990, any first degree judgment is immediately enforceable: this amendment has been introduced to discourage ungrounded appeals aimed at delaying enforcement of the decision.

The costs of litigation, including attorney's fees, are generally awarded to the prevailing party. The judge, however, has the discretion of apportioning the costs between the parties. Court costs are irrelevant in the Italian system.

The successful party will serve the judgment on the other party, who has a right to appeal for reversal or modification of the decision within 30 days from the date of service. Any Appeal is filed with the Court of Appeals or before the Tribunal, in the event that the case was heard by the Low Court. The Court will re-examine the merits of the case. If necessary, the parties may request that further evidence be admitted, but the application for orders sought may not be amended. The appeal proceedings are initiated by serving a writ of summons containing the same elements as the original writ of summons.

The decisions of the Court of Appeals may be challenged in the Cassation Court, but only for legal issues: the merits will not be reviewed. Application for appeal must be filed within 60 days from the date of service of the judgment on the appeal by the successful party.

The regulation of private international law has been thoroughly amended by law n. 218 of May 31, 1995. As to the choice of law for the regulation of contractual obligations, law n. 218/1995 contains a general reference to the rules set out in the Rome Convention of June 19, 1980 and in subsequent international conventions. The main guidelines for the selection of the law applicable to the case are the parties' will (expressed in the contract or implied) and, alternatively, the degree of connection of the law to the contract.

The parties' freedom to choose the law regulating their agreement is very broad: they can select the statutes of a foreign country even if the latter is not a member to the Rome Convention, they can subsequently (even if performance has started already) agree on the laws of a different country to regulate the contract; they can even agree to have different parts of the same contract regulated by the laws of different countries. The only limit to the parties' freedom involves the necessity of applying the *imperative* rules (i.e. the rules parties cannot deviate from) of the country having the closest link to the contract (the degree of connection is evaluated upon the place of execution of the contract, place of performance, nationality of the parties, etc.).

As to the validity of foreign judgments, the new regulation eliminates the need for a formal recognition by Italian judicial authorities. Art. 64 of law n. 218/1995 indicates the prerequisites for the automatic recognition of foreign judgments:

- the judge issuing the decision would have had jurisdiction according to Italian principles

- process was duly served upon the defendant and the fundamental rights of defence were granted

- the decision is final and enforceable under the laws of the country were it was issued

- the decision is not inconsistent with a final judgment by an Italian court

- no process is held in Italy on the same matter and between the same parties, having initiated prior to the foreign process

- the decision is not contrary to Italian public policy principles.

For the enforcement of a foreign judgment in Italy, the party having interest in the enforcement has to apply to the Court of Appeals for an authorisation to enforce the decision. The authorisation is issued by the Court after verifying the existence of the prerequisites for recognition under art. 64 of law n. 218/1995.

B.   Litigation in Administrative Courts

The rules determining jurisdiction of administrative courts are set out in a series of statutes. The main administrative courts are the Regional Administrative Tribunals (TAR), created by Law n. 1034 of December 6, 1971, and the State Council.

The TARs are the courts of first instance, and are formed by magistrates enlisted in a special register, separate from the register of ordinary judges of the civil courts. The State Council is the court of appeals for matters referred to the jurisdiction of the TARs. The State Council also has first instance jurisdiction over a limited number of specific matters. It is a court based in Rome, which also carries out advisory (non-judicial) activity in the interest of the State.

There is no special administrative court having subject matter jurisdiction over construction disputes. However, it may be of interest that under law n. 10 of January 29, 1977, the Regional Administrative Tribunals have exclusive jurisdiction over housing and town planning disputes.

Venue is generally determined by reference to the place where the act of the public body that is challenged by the private party produces its effects.

A petition to the TAR may be based on one of the following grounds: i) the public body was not competent to take the action (i.e. the act was statutorily referred to a different authority); ii) the public body has violated the law in taking the action, or iii) the public body has exceeded its authority.

A petition to the TAR must be entered within sixty days from the date of service of the act on the petitioner (or from when the petitioner acquired full knowledge of the act), provided that the petitioner is directly involved in the act, or from the date of publication of the act in all other cases. The petition has to be in writing and must be served through a clerk of the (ordinary) court upon the relevant public body and upon the private parties (regardless of whether they are mentioned in the challenged act) whose legitimate interest could be infringed by the annulment of the act.

Within thirty days from the last service of process, the original petition must be filed with the TAR. The resisting parties (i.e. the public body and any private parties who are interested in the validity of the act) may appear by filing a brief and documents, within twenty days from the date set for the filing of the petition.

As a general rule, petitions for invalidation of an act of a public body do not suspend the validity of the act. However, the petitioner may request a suspension of the validity of the act, if he proves that his case is prima facie admissible and grounded on the law, and that the performance of the act would cause serious and irreparable damages. The TAR decides the matter quickly, after hearing representatives of the parties. Under art. 5 of law n. 1 of January 3, 1978, in construction disputes the administrative judge cannot decide the issue of suspension of the act prior to determining the date of the hearing on the merits, and the same must take place within four months from the date of the order of suspension. Moreover, under the named provision, the suspension cannot last for more than six months, within which the judgment on the merits has to be published. In practice, however, the proceedings generally take more than six months, and the judge usually orders a new suspension of the validity of the act at the end of the six months period.

Under art. 31-bis, para. 2, of law n. 109/94, the merits of petitions relating to the exclusion of competitors from awarding procedures that were suspended upon application by the excluded contractor, have to be discussed within 90 days from the order of suspension.

Suspension orders may be appealed by petition to the State Council. However, law n. 1/78 has excluded the right to challenge suspension orders relating to works construction disputes.

Under art. 31-bis, para. 3 of law n. 109/94, in the event of a request for a suspension order relating to construction ventures, the public body or other opponents may apply for an immediate decision on the merits. The President of the Tribunal sets the date for the discussion of the case within 90 days from the date of the application. In the event that the application is filed during the hearing on the order of suspension, the hearing on the merits has to take place within 60 days from the date of application. In any event, a request for examination of the merits does not suspend or delay the suspension proceeding.

Discovery in administrative proceedings is quite narrow. The parties may file documentary evidence up to twenty days prior to the hearing, and briefs up to ten days prior to the hearing. At the hearing, after a brief introduction by the reporting judge, the parties' attorneys are admitted to discuss their cases. The President leads the hearing, and may limit the parties' arguments to the fundamental points.

The TAR decides the case by judgment, having immediate enforceability. The judgement also contains a definition of the party's obligations as to the costs for litigation. Administrative judges have broad discretion in deciding which party has to bear the costs of the proceedings and to what extent.

The judgment may be appealed with the State Council within sixty days from service of the judgment of the TAR. An appeal does not automatically suspend the enforceability of the first degree judgment, but the State Council judges may order the suspension upon application of the appellant, provided that the enforcement is likely to bring about severe and irreparable damages to the appellant

The decision of the State Council may only be challenged on grounds of lack of jurisdiction, by petition to the Cassation Court, within sixty days from the service of the decision. In the absence of service of the decision, the petition must be filed within one year from the date of the judgment.

3.2.5   Negotiation

See under 2.1.3.

3.2.6   Ombudsman

There is no system of Ombudsman in Italian construction law.

4.   THE ZEITGEIST

Italian construction industry has experienced serious difficulties over the last years. However, there are increasing symptoms of a rise in the years to come, as a consequence of several initiatives by legislators and by the ongoing pressure by private businesses for the improvement of investment conditions in our country.

The revision of the existing regulation of construction activities that is currently in process has been directed towards certain specific and effective goals.

One of the main intervention areas relates to private investments: efforts by private businesses are increasingly directed towards the current requirements of quality, flexibility of utilities and urban reorganisation, in a general context of closer adherence to the market. Accordingly, the current governmental policies of reducing the pressure of taxation on real property, encouraging enterprises and expanding the rental market, seem appropriate. With reference to maintenance and restoration activities, the current needs for reduction of VAT rates and introduction of a conflict of interests between the Principal and the contractor, in order to reduce tax evasion and black market activities, which are causing serious damages to the building industry, have been considered in the latest bills for enactment of tax regulations.

There is still a need for protection of economically weaker social categories, through the liberalisation of the market of rental housing on one side, and through encouraging public intervention and completion of housing projects on the other.

Improvements are also expected as to simplification and improvement of the procedures for urban development ventures. Associations of contractors, such as ANCE, emphasise the need for effective credit mechanisms in order to encourage investment tendencies of the large middle class that does not fit the parameters of public building industry and is currently excluded from the construction market. The efforts of legislators and public authorities should be directed towards stimulating the interests of private persons through the typical methods of market economy.

Finally, further investment possibilities are brought about by the low degree of expertise by the public bodies in designing activities. Accordingly, the provisions of the Law n. 109/94 that encourage the assignment of the named activities to private professionals or engineering companies should be used as broadly as possible.

1   The kind supply of data by the Italian Contractors Association (ANCE) is acknowledged with thanks.

JAPAN

Fumio Matsushita
Nikken Sekkei, Tokyo, Japan

1.   BACKGROUND

1.1   Economic

According John Bennett et al., Japan produces 10% of the world's total annual products, with 3% of the world's population, on 0.3% of the world's land area.[1]   Though construction investment in Japan reached approximately JPY87 trillion[2] in 1991, larger than any other country in the world, productivity in the construction industry has chronically looked less than promising.   Japan's construction industry employs approximately six million workers or roughly 10% of Japan's total work force.   While the volume of construction has continued to grow, the industry remained plagued by a number of problems, including a chronic labour shortage, ageing of the work force, low productivity, slow improvements in mechanisation and labour saving devices, and perhaps most important, harsh working conditions. In a domestic aspect, the increase in productivity in the construction area has lagged behind the domestic average.   This appears to be a consequence of the low intensity of capital funds.   In other words, large contractors are extremely limited in number, all the remaining ones being small in capital.   The large contractors have grown more and more by improving their technology and mechanisation as well as finance and other general business capabilities, operating their own engineering laboratories for technical advancement.   The medium and small contractors still operate at small capital funding levels.

Table 1a   International Comparison of Construction Data for 1995

	Japan	England	France	Germany	U.S.A.
Construction investment in trillion JPY	79.80 (100)	4.07 (12.3 %)	9.39 (26.0 %)	19.48 (19.8 %)	51.53 (64.6%)
Construction investment compared with GDP in %	16.3%	4.1%	6.5%	8.6%	7.6%
Numbers of construction contractors (in '000)	552 (100)	210 (38.0%) in 1990	307 (55.6 %) in 1993	170 (30.8%) in 1993	578 (104.7%) in 1990
Numbers of construction industry workers (in '000)	6,630 (100)	1,805 (27.2 %) in 1989	1,443 (21.8%) in 1994	2,101 (31.7%) in 1994	7,493 (113.0%) in 1994

Sources: Japan Federation of Construction Contractors, *Nikkenren Handbook '97*, p. 31, as based on the data supplied by the Research Institute of Construction and Economy, Tokyo.

1.1.1   Trends

The ratios of construction investment as compared with the gross domestic product were consistently of the order of 20% from 1975 to 1980, but thereafter gradually declined to a low of 15.6% in 1985.   The trend turned up again since 1986 and reached 18% in 1989 supported by a thrust of increased non-government investment. However, the booming economy called "Bubble Economy" roughly from 1986 to 1992 collapsed and the construction investment in private sector fell to a decreasing trend. In the past, insolvencies were mostly centred on small contractors and the number of such cases were on a decreasing trend; however, ever since 1993, total liability amounts of insolvent contractors has been on a substantial increasing trend as larger contractors were involved.[3]

Table 1b   GDP-Construction Investment and Government-Private Investment

Fiscal Year	1987	1988	1989	1990
GDP (A) in trillion JPY	355.5	379.7	406.5	438.9
Total Construction Investment in trillion JPY (Government and Private Sectors) (B)	61.5	66.6	73.1	81.4
Investment Ratios (B) / (A)	17.3%	17.5%	18.0%	18.5%
Government Investment to Total Investment (B)	36.7%	35.0%	33.2%	31.6%
Private Sector Investment to Total Investment (B)	63.3%	65.0%	66.8%	68.4%

1991	1992	1993	1994	1995
463.9	472.6	476.5	478.6	482.9
82.4	84.0	81.7	78.0	80.4
17.8%	17.8%	17.1%	16.3%	16.6%
34.8%	38.6%	41.7%	41.7%	46.3%
65.2%	61.4%	58.3%	58.3%	53.7%

Table 1c Composition of Construction Investment in 1995

	Private Investment	Public Investment
Housing	30.3%	2.3%
Non-housing buildings	14.0%	6.2%
Civil Engineering	9.4%	37.7%
Total	53.8%	46.2%

Sources to Tables 1b and 1c: Japan Federation of Construction Contractors (1997), *Construction in Japan 1997*, pp. 2-4

1.2   Legal

Overall

As in other modern countries, the State of Japan is governed, under the present Constitution of Japan, 1946, by the Diet, the Cabinet and the judiciary, each having apparent co-equal power and existing independent of each other. Ever since the opening of Japan's doors to the world in 1868, when the feudalistic shognate regime collapsed, the State has modernised the society by importing aspects of western culture, including the legal system. The Constitution of Japan was initially modelled after the German (Prussia) Constitution and, after the defeat in World War II, was drafted by GHQ staff. Japan is a civil-code country and organised as one jurisdiction. Japanese people use a term "Major Six Laws," which consist of the Constitution (though taken, in other countries, as supreme law distinctly separate from the other branches of law), the Civil Code, the Commercial Code, the Penal Code, the Code of Civil Procedure and the Code of Criminal Procedure. The latter five were drafted under strong influence of German and partly French law and are still under the same influence in their interpretation. However, after the defeat of the war, the influence of American Law has been prevailing in other branches of law, such as anti-monopoly law, corporation law (contained in the Commercial Code) and more recently the court proceeding rules in the revised Code of Civil Procedure.[4]

Court System

The judicial system is organised into three tiers. These tiers are the District Courts, the High Courts (or Intermediate Appellate Courts), and the Supreme Court. For detail, see 3.2.4. The Constitution of Japan does not allow for any other special court. All court proceedings are open to any audience, except cases where too many spectators apply in which case audiences are selected by lottery. There is no administrative court which exclusively deals with complaints against the Government or other administrative agencies. Any such complaints must be filed with the courts of general jurisdiction described above. Specifically, before filing an administrative suit, a party can file a complaint with the competent agency or Ministry under the Administrative Complaint Investigation Act, (*Gyosei Fufuku Shinsa Ho*)[5] 1962 No. 160. In Japan, no court uses a jury system. The reason Japanese are reluctant to use the jury system may be traced to the traditional thinking that the law should not be administered by laymen.

Attorneys

Japanese "attorneys" (*bengoshi*) were not accepted as a vocation of high esteem until 1896 at which time the Attorneys Act was enacted. Traditionally in the feudalistic Japanese society, the judicial system was not independent from the administration under the shogunates. Such a function was an ill-favoured one in those days. In the early Meiji Era around 1876, an attorney system was officially authorised; however, those attorneys were utterly subordinated to judges and public prosecutors as public officials. As the society has modernised in a Western way, through great efforts of some attorneys, their status has gradually advanced to its present high esteem position.[6] The Attorneys Act requires candidates to pass the national judicial examination and to complete a two year practice training program at the Legal Research and Training Institute. The judicial examination is to select the candidates for the judges, public prosecutors and attorneys and the successful candidates can select one of these three as their career. The number of registered attorneys is 16,328 as of October 1997. The fact that the number of attorneys is very small as compared with the population (125,569,000 as of 1995) can be traced to the facts that the judicial examination is the most difficult one among all the national qualification systems, i.e. those who pass the examination account for two to three percent of all the applicants, and that private citizens are not apt to call for legal assistance from them in daily life unless they are involved in a desperate dispute.

Quasi-Lawyers

In addition to the attorneys, there are several types of law-related vocations, each being regulated by the respective laws. The "judicial scribes" (shiho-shoshi), whose functions are controlled by a licensing system separate from the attorney's, specialise in document preparation and processing mainly for commercial (corporate entities and real estates) registration purposes. There were 17,020 judicial scribes registered as of October 1997. They cannot represent a client in legal actions, which is the sole domain of the attorney. Also, in respect of the documents filing with the administrative authorities, there are the "administrative scribes" (gyosei-shoshi) as a separate profession. Also, the "notaries public" (koshonin) make up an independent profession separate from the attorneys and the scribes. They belong to the Legal Affairs Bureau, the Ministry of Justice. Their function is strictly limited to notarising documentary records and reviewing the registration documents of corporations to be newly established. There were 550 notaries public registered as of the end of 1997. As for patents, "utility models" (jitsuyo shinan) and trademarks, there is a separate profession called "patent attorneys" (*benri-shi*) whose practice is controlled by a licensing system, and who are authorised to act on these issues on behalf of clients.   The licensed attorneys (at law) are allowed to do scribes' and patent attorneys' business without any special qualification.

Consumer Protection

The key statute for this purpose is the "Consumer Protection Basic Act" (*Shohisha Hogo Kihon Ho*) 1968 No. 76; however, this law declares that the State, administrative authorities, suppliers, and also consumers must make efforts to promote the purpose, but does not set forth more than that, leaving the practical protection methods to each specific law or regulation. Though there are a certain number of statutes, they are for the most part less relevant to construction at large.   For sale of housing, dealings for sales of homes on a commercial basis are controlled by the Housing Premises Deal Act, 1952 No. 176. This act is applicable to such deals as made through professional real-estate agents or brokers who must be licensed under this act.   For the consumer protection, there are various organs or systems to deal with consumer's complaints by way of consultation, mediation and conciliation at the administrative committees, trade organisation's voluntary organs and non-trade institutions, such as a bar association and a consumer protection association.

Corporation

The basic principles for legal entities are set forth in the Civil Code, Book 1 (General Provisions) as to natural and juristic persons.   Out of juristic persons, profit-making business corporations are defined in the Commercial Code and, for smaller ones, the Limited Liability Corporation Act.   Under Japanese Law, practice for profit-making business may be pursued through the following seven types of business vehicles: (1) *kojin* [sole proprietorship], (2) *kumiai* [quasi-partnership],

(3) *tokumei kumiai* [undisclosed quasi-partnership] in which "*tokumei*" means that the name of a fund-investor is undisclosed, (4) *gomei kaisha* [partnership corporation], (5) *goshi kaisha* [limited partnership corporation], (6) *kabushiki kaisha* [stock corporation], and (7) *yugen kaisha* [limited liability corporation]. Of the foregoing seven types, (2) and (3) are, like Anglo-American partnerships, not legal entities, and the remaining (4) through (7) are legal ones.[7] The most preferred type of such business vehicles are the stock corporation which account for more than half of the total juristic legal entities. Though the foregoing (2) and (3) resemble to Anglo-American partnership, they are not preferred in Japan.

1.2.1   Construction Law

The source of construction law is mostly found in the Civil Code and fractionally the Commercial Code. The licensing and practice aspects for architects and construction contractors are respectively governed by the Architects Act and the Construction Business Act. For the procurement aspect of the government contracts, there are the General Accounting Act, (*Kaikei Ho*), 1947 No. 35 for the ministry projects and the Local Government Autonomy Act, (*Chiho Jichi Ho*), 1947 No. 67 for the prefectural local governments. Project planning and engineering aspects are mainly governed by the City Planning Act, (*Toshi Keikaku Ho*), 1968 No. 100; the Building Standard Act, (*Kenchiku Kijun Ho*), 1950 No. 201; and the Fire Prevention Act, (*Shobo Ho*),1948 No. 186. Also, there are special acts, such as the Act concerning Advance Payment Guarantors for Public Works (for the government projects); the Act concerning Delayed Payment to Subcontractors, (*Shitauke Daikin Shiharai Chien tou Boshi Ho*), 1956 No. 120 for A/E and non-construction related business[8]; and for construction operations, the Noise Control Act, (*So'on Kisei Ho*), 1968 No. 98; the Public Nuisance Disputes Resolution Act, (*Kogai Funso Shori Ho*), 1970 No. 108; the Vibration Control Act, (*Shindo Kisei Ho*), 1976 No. 64. In the following, the Architects Act and the Construction Business Act will be briefed as licensing acts and then types of contracts briefed in respect of the interpretation of contracts under the Civil Code.

Architects Act (*Kenchiku-shi Ho*), 1950 No. 202

In Japan, to practice as architects, they must be qualified under the Architects Act which divides the licensed architects into three classes: 1st Class Architects, 2nd Class Architects, and Wood Construction Architects.[9] These classes are delineated on the basis of the types of buildings that the architect is competent to design. The 2nd Class Architects and Wood Construction Architect's domain is to design and supervise small sizes of buildings; thus, ordinary buildings must be designed and supervised by the 1st Class Architects. To be qualified in any of these three classes, an architectural candidate must pass the examination for that particular class. Those who have passed the examination must be registered at the Ministry of Construction as 1st Class Architects and at the Prefectural Governor for both 2nd Class Architects and Wood Construction Architects. To practice as a licensed architect, the registered architect must further register his office as an "Architect's Office" to meet the statutory requirements of the Architects Act. Wherever he completes the registration, he can engage in architectural services even at a construction company. Thus, design-build services provided by a single contractor are very commonplace in Japan. The registered numbers of the Architects were as follows as of 1995: 1st Class Architects - 264,398 persons (84,107 registered firms), 2nd Class Architects - 566,791 (45,762), and Wood Construction Architect - 11,386 (1,332). In the sector of "building service systems" which include mechanical, electrical and plumbing works, there is no substantial compulsory practising act.[10]

Construction Business Act (for Contractors)   (*Kensetsu-gyo Ho*), 1949 No. 100

Construction contractors must be licensed under this Act.[11] The Act aims at improving capabilities and integrity of construction contractors, who undertake any construction as business, and improving practice of construction contracts. The Act was first promulgated in 1949 as a registration Act, but amended to a considerable extent in 1971 as a licensing Act. If they operate their business through a head office or any branch office in two or more prefectures, they must obtain a licence from the Minister of Construction. If they operate in a single prefecture, they must obtain a licence from the Prefectural Governor. The total numbers of licensed contractors in Japan were 350,817 (as of 1975); 518,964 (1985); 508,874 (1990); and 551,661 (1995). According to the 1995 data, the contractors licensed by the Minister accounted for less than one percent of the total number of contractors. Of this total number of contractors, those who have capital exceeding JPY100 million account for less than one percent when based on the data between 1975–1995. The remaining are very small-sized contractors. The licensing requirements are applicable to the following 28 categories of licenses, of which major ones are listed below: 1) Civil Construction Trade; 2) Building Construction Trade; 3) Carpentry Trade; 4) Plastering Trade; 5) Steel Frame-erection and Earthwork Trade; 6) Stone Work Trade; 7) Roofing Trade; 8) Electrical Work Trade; 9) Piping Work Trade; 10) Tiling, Brick and Concrete Block Trade; and 11) Structural Steel Trade. The remaining 12) through 28) are omitted.

To obtain each licence, one must meet the qualification requirements covering: years of experience, academic career and authorised technical competence for key persons, integrity to properly perform construction contracts, sound financial resources, and monetary credibility. As for the foregoing 28 categories, there are further separations. First, there are two types of licenses for each category, i.e., "General License"

(*Ippan Kyoka*) and "Specific Licence" (*Tokutei Kyoka*).[12] The Specific License is required for one who lets a subcontract in an amount of more than JPY20 million (JPY30 million in case of building construction) in a single category out of a contract with an owner. The Specific Licence is more stringent than the General Licence in the requirements. Also, there are other stricter requirements for the major trades which are called "Designated Trades" which include the Civil Construction Trade, Building Construction Trade, Piping Trade, Structural Steel Trade and Pavement Trade. A unique aspect of the act is that the act establishes the normative rules of various elements of the contracts. Though it seems peculiar that the licensing act provides such contractual detail, it was necessary in order to rationalise the business of construction contracting between owners and contractors, and between main contractors and subcontractors. These two relations were in the past one-sided to the advantage of owners and in turn main contractors due to the overwhelming strength of their bargaining powers in contract negotiation. These articles are, for the most part, directory provisions and cannot preempt the basic rules of the Civil Code.

Art. 19 (Contents of Construction Contracts) mandates that the parties to a construction contract shall write down the following matters in a contract and put their signatures, or put their names and seals.

i. Description of construction
ii. Price of construction undertaken
iii. Times of construction start and completion
iv. When and how advance payment or progress payments are made.
v. Changes in construction schedule or contract amount, or sharing and evaluation of loss where construction is changed in design or postponed or cancelled.
vi. Sharing and evaluation of loss in case of Acts of God or other force majeure events

vii.   Changes in contract amount or construction scope due to changes in materials or service

vii-2.   Sharing of liability for damage to third parties

vii-3.   Description and delivery method of owner-supplied materials or owner-rent tools or equipment, if any

viii.   Times and methods of inspection of wholly or partially completed construction, and time of delivery of completed construction

ix.   Time and method of payment upon the completion of construction

x.   Interest, penalty and other damages in case of delay in performance of contractual obligations and other liabilities

xi.   Method of dispute resolution

These are typical elements of a construction contract. However, in the past, failure to include them in the written contract has led to unnecessary disputes. To improve such situations, the foregoing directory provisions have been provided. Under the Civil Code, a meeting of minds of the parties is enough to create a valid construction contract, no more formality required. If any promise is missing from a written contract, the contract itself is valid. In addition, the act provides for the unfair dealing being banned; minimum bidding periods according to the sizes of projects; guarantee for performance of construction contracts; ban on whole subletting; and disputes resolution on construction contracts (see 3.2.3).

1.2.2   Construction Contracts

The subject will be discussed from major two angles, firstly from the Civil Code which governs formation, validity and interpretation of construction contracts and secondly the standardised contracts.

Contracts under the Civil Code

The Civil Code consists of five books. Book 3 deals with the chose in action and contains, among others, contracts, unjust enrichment and tort. On the contracts, the book sets forth the general rules for 13 typical types of contracts.[13] Contracts for design, construction supervision, construction management and construction of buildings and civil engineered works can be usually classified into three types out of the 13 types: the "Employment" (*koyo*), the "Independent Undertaking" (*ukeoi*), or the "Mandate" (*inin*) or alternatively a combination of the last two. In addition to these three, the "Sales" (*baibai*) may be involved for such contracts as plant manufacturing and assembling contracts and sale contracts for a built-for-sale house; however, here the first three only will be briefed. Almost all construction contracts fall under the Independent Undertaking either for civil engineering or building construction. Design and supervision contracts are either the Independent Undertaking or the Mandate or combination of both (i.e. the Independent Undertaking for design and the Mandate for supervision), though authorities are not uniform in the interpretation. All these three contracts become valid when an offer is accepted or by a meeting of the party's minds, and do not require any written instrument or any other formality. Also, reciprocity or consideration is not required to validate a contract.

Employment Contract

This contract (Civil Code, Book 3, §8, Arts. 623 through 631) is a contract whereby one party (an employee) agrees to work for or provide services to another party (an employer) according to the latter's directions and the latter agrees to pay a fee for the services. The services can include a person's labour and professional services. This may be compared to the Anglo-American term "master-servant contract," thus the ordinary construction contracts or design contracts do not fall under this category.

Independent Undertaking Contract

This contract (*ibid*. §9, Arts. 632 through 642) is a contract whereby one party (an independent undertaker) undertakes to complete a scope of services or work at his own risk and another party agrees to pay a fee for completion of the services. This contract is similar to contracts governed by the independent contractor rule in Anglo-American Law. In this category of contracts, the major consideration is warranty that final products, building or completed design documents will not be defective, will comply with the contract, or contain no imperfection in workmanship which decreases the value normally expected of the end product. The accepted authorities consider that the warranty is applicable whether an independent undertaker is negligent or not. If an independent undertaker breaches the warranty, he must be liable usually in three ways, 1) to rectify defects, 2) to be subject to termination of the contract by the owner, and 3) pay damages. Also, no matter how large a defect is, an owner may not terminate a contract for construction attached to land.

Mandate Contract

This contract (*ibid*. §10, Arts. 643 through 656) is a contract whereby one party (mandatory) agrees to perform certain legal acts, business or other services, at his discretion unless otherwise directed by another party (mandator), and the latter agrees to pay a fee for the services performed. Unlike the Independent Undertaking, the Mandate does not impose the warranty for products or services, but demands due care of a mandatory. Separate from the Mandate, there is the rule of "Agency" which has a similar concept to the Mandate, but which is distinctly separate from it as in German law. In the Anglo-American law, the agent is separate from the independent contractor; however, in Japanese law, the Mandate is opposed to the Independent Undertaking. Japanese agency can occur either in the Mandate or the Independent Undertaking. The agency is defined as such manifestation of intention by an agent against a third party as made on behalf of the principal and take a legal effect on the principal. In short, the manifestation must be such as to cause legal effect between the principal and a third party; therefore, the agency cannot occur in respect of non-legally binding process of business or tort.

Japanese Contracts

An intrinsic feature of typical Japanese contracts is their simplicity and vagueness.[14] Japanese do not like to detail and eventually fix the rights and obligations of parties in a contract. To the eyes of the Western legal community, this would seemingly lead to disputes over obligations. On the contrary, the Japanese do not think so and expect without certainty to settle any conflict through negotiation when such necessity arises. They consider the detailed description of rights and obligations as inflexible and causing anxiety. This fact is vividly illustrated by a typical contract clause which is utilised by Japanese. A typical clause contained in many contracts states that if there is any dispute or uncertainty in the performance of a contract, it shall be amicably settled with the sincerity and in good faith of both parties. It is called "a good-faith negotiation clause" or "amicable settlement clause." It is apparent that this clause will not help either party in exercise of his rights, since both parties have to discuss any disputable or uncertain matter with the other party, if it arises.

Similar vague or unclear negotiation provisions are frequently found in design or construction contracts also. For example, it is common that there is a clause to the effect that any change to the construction requirements, time and cost implication shall be decided through negotiations between an owner and a contractor, and, depending on a contract, a construction supervisor. This clause does not clarify when and to what extent an owner may order changes or a contractor may claim any adjustment in time and cost. Usually, they expect they will arrive at a certain settlement through the negotiations (see 2.1.3). In interpretation, such simple, vague Japanese contracts usually must be supplemented by applicable meanings of the code provisions. In this respect, though the interpretation of contract itself consists of the literal one and constructive one, there is no such detail developed interpretation rules as in the Anglo-American law.

Standardised Contracts

The construction contracts now commonly used are two standardised ones, one each for the private construction project issued by long-exited "*Shikai Rengo*" (Four Associations) which now actually consist of eight organisations related to design and construction (except one academic organisation), and for the public construction projects issued by the Central Construction Committee. In September 1997, the associations revised the former form and its title to "Private Construction (Former Four Associations') Form" (hereafter called "Private Contract Form"). The first edition of this form dates back to 1923 and has been traditionally used for commercial building construction projects of ordinary and large sizes. This form is now the best selling form of construction contract.[15]   The Central Construction Committee is empowered by the Construction Business Act to prepare and advise on the use of standardised contract conditions to governmental or other public agencies, private corporations as owners of construction projects and construction contractors.   The committee now issues one public contract form "Standard Public Construction Contract Conditions" ("Public Contract Form") and others. These are issued as advisory or recommendations and are not compulsory. The Public Contract Form was issued in 1950 and revised several times to the current 1995 edition. In the actual use, these two forms have been frequently changed on the owner's side.

The Private Contract Form[16] is, in an overall contractual approach, similar to the Public Contract Form, except that the latter is devised to meet the statutory requirements for government procurement.   Traditionally, most construction contracts were unilaterally advantageous to owners.   For instance, it was rare in the old days that contractor's rights to time extensions or additional payments were expressed in those contracts. Further, even if they had been legally entitled to a valid claim, they did not pursue their rights by bringing a lawsuit. To rely on one's rights, however just, met with social sanctions and the loss of the future contracts from others besides than the very specific owner. It may be said that the superiority in bargaining power governed the rights and obligations in a very feudalistic way. However, the Private Contract Form now can be seen as supplier-oriented one in respect of, among others, defect warranty, delay penalty and suspension. This seems to have been due to successful tactics of the drafting staff representing the contractors by taking advantage of their superior knowledge on contracts.

Another similar feature is the numerous negotiation clauses calling for "agreement through negotiation between the owner, the contractor and the construction supervisor," though such clauses in both forms have been recently replaced by clearer specific stipulations to a considerable extent. It is considered outrageous in Japan for a person to resort to his own rights without prior consultation with the other party, however legally valid the claim.   Even when a contractor consults with the owner, he has to be humble in bringing his claim and explain how he is in a serious situation to obtain the owner's sympathy. Otherwise, his claim might meet with the owner's flat refusal.

Construction Supervisor

The Building Standard Act requires the owner of a building construction project to designate a construction supervisor for obtaining the building permit. The construction supervisor must be a person who is qualified to provide the architectural services for the types and sizes of buildings. In Japan, the construction supervisors have not been recognised as more than technical experts probably due to the fact that A/E's status has not taken as real professional like the attorneys. The A/E's role in construction supervision has been limited to assuring that the building will be completed to the contract documents. Any quasi-arbitrator position has not be given to him and he does not have the authority to decide time extension or additional payment to a contractor; to interpret legal subject of construction contract; and to give first-instance decision to a dispute between the owner and the contractor. All these subjects are negotiated between the owner, the contractor and the construction supervisor in the Private Contract Form.

Guarantee

In Japan, an owner's security for proper performance by a contractor is usually a guarantee other than bond. Usual guarantees include: i) a surety liable for either pecuniary loss or performance of the contract, ii) security deposit, iii) performance bond insurance, iv) advance payment refund bond and insurance, and v) bank guarantee. The most prevailing guarantee was sureties both on commercial projects and government projects; however, the Public Contract Form now takes, in the latest edition, an approach not to use a surety. The surety system is still favoured in the Private Contract Form because of its ready availability.

Completion

The Japanese construction contracts do not contain any provision regarding substantial or practical completion of the works. In Japanese practice, the completion is not two-staged. Usually, when construction comes close to the completion stage, a supervisor issues a punch list and the contractor rectifies defective or incomplete parts of the work so that all the works will be completed by the agreed completion date. If any latent defect appears after the completion of works and owner's acceptance, it is made good during a "warranty period" which corresponds to a maintenance period or defect liability period used in the common-law countries.

Delay Penalty

The Private Contract Form provides for a penalty for delayed performance of construction and for delayed payment to a contractor. Unlike penalties in the common-law countries, such penalty is lawful under Japanese Law. The Civil Code, Art. 420, (3) describes that a penalty is presumed as presumed damages. The usual penalty is effective however large or small the amount and cannot be modified after agreed upon even if a smaller or larger actual amount is testified to by proper evidence at a later stage. The Private Contract Form, Art. 30 sets forth that in case of delay, the Owner may demand from the Contractor penalty at the rate of 1/1000 of the reduced sum of the contract price minus the contract price (portion) proportionate to completed parts of the work and materials which have passed inspection or test and that the Contractor may demand of the Owner penalty at the rate of 1/1000 of the amount of the unpaid portion of contract price for each calendar day. Assume that the total construction is delayed in an incomplete remaining part, say a 10% portion of the entire contract sum. Usually, an owner cannot accept the construction as a whole unless the completed part is separable. Nevertheless, delay penalty are fractional since it is calculated at 10% x 1/1000 despite the fact that the entire construction is unacceptable. These are indicative of the intrinsic features of the Private Contract Form.

Defect Liability

The Private Contract Form sets forth in Art. 27 a contractor's liability for defects in construction. Under this article, a contractor remains liable for one year for wood-framed buildings and two years for rigid (steel or concrete) buildings; however, these periods are extended by five times to maximum limitation (extinctive prescription) periods specified in the Civil Code in case of gross negligence or wilful misconduct of a contractor. These warranty periods are comparable to the correction period, maintenance period or defect liability period utilised in common-law countries, but they are short when compared with the warranty period in other civil-law countries, such as France, Germany and Switzerland. In case of the common-law countries, contractors are liable for breach of contract even after the maintenance period until a period of limitation expires. In Japan, after an owner accepts completed construction, he cannot resort to any remedy other than those under the warranty. Therefore, a breach-of-contract cause of action is not available to him. Then, the warranty period has significant meaning to owners. These shortened warranty periods have been criticised by many academic opinions as being of too short a duration and the court decisions are reluctant to uphold the shortened warranty period.

Contractual Disputes

The Public Contract Form provides for two types of resolution approaches: Approach A is that an owner (public agency) and a contractor agree to solve any dispute through an intermediary designated in a contract by way of mediation or conciliation (see 3.1.1 and 3.1.3) and, if not resolved, finally at the Construction Disputes Resolution Committee (see 3.2.3), and Approach B is to settle such disputes from the outset at the Construction Disputes Resolution Committee. The Private Contract Form adopts similar approaches of the intermediary or the Construction Disputes Resolution Committee. For the latter case, parties to the contract can agree to the arbitration at the committee in a separate designated form. In either form, one party may not demand arbitration pending mediation or conciliation unless the other party so agrees or one party decides that mediation or conciliation failed to reach any solution.

2. CONFLICT MANAGEMENT

Usual conflicts may occur, as in other countries, between project owners and construction contractors. Sometimes, an architect/engineer (A/E) may be involved. There is another type of conflicts and disputes on nuisance which are caused by a building or construction operation. Such conflict usually occur among a building owner, a construction contractor, neighbouring inhabitants affected by construction operation or a building, administration bodies which include the prefectural and municipal authorities, and occasionally A/E. In Japan, there is a commonly used word "*Kenchiku Kogai*" (Construction Nuisance) which covers various public or private nuisances caused by a completed building or construction operation.[17] The Construction Nuisance can be usually a source of the conflict in built-up area in urban district, and has a serious impact the project scheme itself. The nuisance can be classified in a broad sense into the following three groups. 1) noise, vibration, ground settlement, and blocking of pedestrian circulation by construction vehicles, all caused by construction operation. 2) blocked sunlight, blocked outdoor air ventilation, obstruction of landscape view, invasion of privacy, strong wind force (wind gust), and interference of radio waives, all caused by a completed building. 3) non-transitory nature of noise, vibration, danger of explosion and others, and demoralisation by operation of business and manufacturing at a completed building, such as a factory, a gas station, a hotel used for immoral affairs, and a game centre used as a gathering place by "unsound" young generation. The most typical nuisance seems to be type 2); however, nuisance 1) or 3) can cause more serious effects depending on the construction or building area.

2.1   Non-Binding

2.1.1   Dispute Resolution Board

On the construction contract dispute, the Construction Dispute Resolution Committee established under the Construction Business Act provide, among others, consultation at the outset and then mediation. On the Construction Nuisance, conflict management system exists at various levels of the administration body.

Construction Nuisance

In old days, the government was not involved in such troubles as the Construction Nuisance. Ever since 1970 as a turning point (called the "year of public nuisance"), economic benefit as priority in construction projects has been gradually changed into environment priority.   Inhabitants' demand for better living conditions has become noticeable. The basic power clause to regulate conflict on the Construction Nuisance is found in the Local Government Autonomy Act, Art. 2, 3-1 which reads "the local government shall maintain the public order in the local community and uphold the safety, health and social welfare of the inhabitants;" and in Public Nuisance Dispute Resolution Act (*Kogai Funso Torah*), 1970 No. 108, Art. 49 which reads "the local government shall make efforts to dispose of complaints by inhabitants on public nuisance"; and other complaint management procedures in the relevant acts and regulations.

Sanction by Administrative Body and Dispute Resolution Ordinance

The disposition or conflict management of the Construction Nuisance has been provided at the local (prefectural)

governments and more frequently at the level of municipality. Their approach is very Japanese and to demand an amicable settlement of neighbours' complaint by way of what is called the "administrative guidance" (*gyosei shido*) to be described below. In past cases, cities did not issue the building permit or refused supply of city water as sanction for the disobedience.   Some court decisions vacated such city's sanctions as illegal and such a direct control without due recourse to any statute had became obsolete year by year.   In a normal case of the Construction Nuisance, a group of neighbours to the construction site raise a complaint to the owner and sometimes the contractor or directly to the city authority, whether there is any established complaint procedures.   The city will move to settle the complaint with or without any authority to do so according the above-explained approach.   In this process, the administrative guidance has attained a considerable success in the past.

As the Construction Nuisance centred on blocking of sunlight has gained considerable popularity among the citizens, the State revised the Building Standard Act in 1977 and officially allowed the municipalities to regulate the blocking of sunlight through their ordinances.   In line with this movement, various municipalities enacted the dispute resolution ordinance for building construction and provided for the mediation and the conciliation, both not binding the parties unless agreed by them. However, since the ordinances do not have any enforcement power and are not more than the guidance toward a compromise between the disputants, if a building owner or construction contractor refuses to enter the mediation or conciliation, they are legally powerless; however, most ordinances have various dispute-preventive measures or non-legal sanction. They can demand the owner to set up a signboard to advertise building construction in outline and to explain the details of building and construction to neighbours in advance of construction operation. Based on the advertised information, neighbours can demand the owner through construction nuisance office in charge to modify the project in order to alleviate the Construction

Nuisance. However, the owner can still refuses the mediation and the conciliation. In this case, the municipality usually publicises the names of such owner and contractor. This publicising can serve a more effective measure to them than any fine or penalty.

Administrative Guidance

The administrative guidance[18] for the Construction Nuisance is one way of various guidance which are taken by various levels of the administration bodies. The administrative guidance exists in various levels of the administration, such as the administration by the Ministries and down to the administration by a town or village. Though Japanese Government declared to abandon or eliminate it, it will be very difficult to do so, since it is very effective tactics in the administration. The term "*gyosei shido*" is not used in any statute, but is a term which has been used by administration-related people. Since the terms does not have any clearly settled definition, its definition slightly varies from commentator to commentator. According to the dominant definitions, it may be summarised as follows. It is the discretional acts which administrative agencies take toward opponents by means of suggestions, directions, requests, recommendations, advice, encouragement and other similar acts in reliance of voluntary consent from the opponents. Here, the opponents refer to public corporations, local governments (prefectural governments and municipalities), trade associations, business corporations and citizens at large. It is frequently taken without any recourse to a statute, though some have. The administrative guidance may be characterised as voluntary non-juristic acts in non-authoritative nature and sometimes lacks in a valid legal effect in the true legal context.

It is an established rule that any action of an administrative body must have recourse to a statute promulgated in the Diet, the highest organ of state power (Constitution of Japan, Art. 41). It is essential that administrative bodies, central or local, will not exercise their public right to take actions in a way which will bind the opponents without due recourse to any statute or other law. A problem is that administrative agencies take action frequently without due recourse to or in excess of the limit allowed in the law. It has been thought that having no due recourse, itself, is not necessarily illegal if the guidance is followed, as frequently done, by the opponents without any coercive force. The opponent's actions in following the guidance is usually interpreted as a voluntary consent to follow it. Their obedience in following the guidance seems to come from the fact that people, at large, are obedient to the superior power now still under the vestige of the past feudalism, more particularly "to respect the (public) officials and downgrade the people" and are very susceptible to the influence of administrative bodies which have powers to grant permits or approvals on same or other subject matters even if they are reluctant to follow the directions of the body. This aspect is well explained by Prof. Yoriaki Narita.[19] "Although actions of this sort may be called non-authoritative and voluntary, nevertheless, because administrative organs with administrative authority frequently hold their public authority behind their backs, as it were, they thus exert upon parties essentially the same psychological pressure which would be caused by application of public authority. Thus it is not difficult to imagine that parties sometimes unwillingly comply with administrative guidance."

2.1.2   Dispute Review Advisors

See 3.1.3.

2.1.3   Negotiation

In Japan, tactics for contract bargaining or negotiation seem to be not academically or professionally trained, but gained through personal practical or vocational experience. The most important thing underlying the negotiation is to keep the traditional "Confucian virtue of harmony" (*wa*) which has been traditionally fostered in the Japanese mind. The negotiation itself is usually not a battle of rights and obligations, but a place for concession by the participants to reach a compromise. A claimant who is likely to suffer a loss or other disadvantage usually asks for the other party's favour who in turn makes efforts to afford such a favour to a practical extent, without simply refusing it on the ground of non-obligation owned by him. In passing, one typical way for compromising is for claimants and respondents to add up amounts of losses in contention and divide the sum by two, for a two-way split. The negotiation for compromise is more important than pursuing rights and obligations to the last. This aspect is well reflected by the suit-reluctance of Japanese people.

<u>Suit-Reluctance of Japanese People</u>

The following table vividly reflects the suit-reluctance in the Japanese mind.[20] The Confucian virtue of harmony acts as a constraint to bringing a suit.

Table 2a Comparison of Civil Actions in Numbers at First Instance Courts

State and Year	Actual numbers of suits as filed with the first instance courts for one year	Figures adjusted to reflect equalised population in each state by taking the Japanese figures at 100 (Symbol *x* indicates a multiple based on the Japanese figure)
Japan, 1970	492,198	492,198 (x 1)
England & Wales, 1969	2,504,075	5,247,499 (x 10.66)
California, 69-70	1,044,930	5,431,763 (x 11.04)
Massachusetts, 69-70	420,269	7,662,151 (x 15.57)

Source: Tanaka, Hideo, *The Role of Law and Lawyers in Japanese Society*, in Hideo Tanaka (Ed.) (1976), Japanese Legal System, University of Tokyo Press, Tokyo, pp. 255-257

Though conflicts and disputes have existed in the Japanese mind, they have been resolved to a considerable extent by the negotiation in the spirit of Confucian virtue of harmony. Prof. Hideo Tanaka[21] well commented on this issue as follows: "When a dispute arises between two parties, not very many Japanese view the dispute in terms of rights and obligations.... Instead, the traditional value of harmony or Confucian virtue of harmony prevails upon them. To their minds, settlement of disputes without arguing their points of view in a reasoned way and without fighting out their cases to the finish in court, is of supreme virtue." In other words, a conciliatory attitude is more highly valued than an unyielding reasoning. Such attitudes of Japanese are well illustrated in the following survey. This survey result was published in 1973 and confirmed in 1982.

Survey question: If your rights are violated, will you try to bring a lawsuit to court?

1)   I will do so at once   22.8%
2)   Not at once, but I will consider it, depending upon a circumstance.   24.0%
3)   I don't know.   3.3%
4)   I will not consider it unless there is any absolute reason.   49.9%

Quoted from Kato, Masanobu, *Nihon-jin no Hoishiki* [Legal Consciousness in the Japanese Mind], Jurist, No. 1007, p. 19, which is based on the survey results shown in *Nihon Bunka Kaigi* (Ed.) (1973), *Nihon-jin no Ho-ishiki* [Legal Consciousness in the Japanese Mind], p. 90 and (1982) same title, p. 104.

One still unfading trend of Japanese, however less hesitant about lawsuits they may become, is that they do not necessarily follow their reasoned decisions to the last, but they are apt to reach a compromise even if they are likely to win their cases in litigation. With respect to plaintiff's case-winning ratios and compromise ratios at courts, plaintiffs had considerably high ratios of case-winning, but made compromise at higher ratios (more than 50%) in those claims in which a winning prospect may be easily had than others.

Dispute between Neighbours

There is a very interesting case to illustrate the suit reluctance of the people. In Japan, there is the term "*rinjin sosho*," or a "dispute between neighbours." This term has the special connotation that disputes between neighbours are rare. Here, the neighbours include immediate neighbours, friends, or persons in special reliance or confidence relationship. In 1977, there was an accident in which a three-year old-baby drowned and died while it was baby-sat by a couple who were close friends of the child's mother. She brought a damage claim against the couple and won the case at the first instance court.[22] Then, an avalanche of protests rushed to the plaintiff from various parts of the country, saying that the plaintiff's act of bringing a suit against the couple who were such kind-hearted baby-sitters was utterly against humanity. The plaintiff's family were subjected to public censure in their daily life and finally withdrew the suit.

On this case, an editorial in the *Asahi Shimbun*, a leading newspaper, commented as follows: "It is very understandable that people have an antagonistic feeling against the plaintiff. Japanese tend not to like the law to intervene in a family or neighbourhood trouble, since there is a traditional morality in community which regulates any conflicts between the people. An increase in lawsuits among neighbours is indicative of the fact that a certain dispute-control mechanism of the community is becoming less workable." The defendant's attorney also expressed the same view, saying that such a suit would destroy the mutual-help relationship among neighbours and would become one which should not be brought if the plaintiff wanted to be human. Most commentators seem to more or less agree to this view. Despite the fact that dispute between neighbours are rare, why do the Construction Nuisance disputes arise? The answer may be that the nuisance dispute is usually from citizens against private corporations as building owners, thus causing less constraint to the citizens and that the nuisance has a direct irreparable impact to their daily life.

2.1.4   Quality Matters

Total Quality Management

Quality management has been traditionally popular in manufacturing industry, but in construction industry cannot be said to have been prevailing, except for some leading contractors who were awarded the Deming Prize which is the most famous prize for quality control in Japan. Traditionally, skimped construction works have been a serious concern to the owners, especially in the case of small-sized contractors. ISO 9000 and ISO 14000 have become recent topics, and some leading contractors and A/E firms are trying to be qualified; however, such qualification is not demanded for any government procurement system.

Procurement System

In the private or commercial construction industry, no precise data is available on the use of procurement system. Major procurement approach for commercial projects may be first nominated (selective) bidding and next negotiated contracts. As in other countries, all contracts with the government, either central or local, are governed by special procurement laws.[23] Contracts with the Ministries and their agencies ("Government") are regulated by the General Accounting Act which demands auditing by the General Accounting Office. The procurement activities are performed by each ministry or its agencies. There is no such body as General Services Administration of U.S. Federal Government which centralises the procurement activities on behalf of the Government. Each ministry has contracting officers who do not have any adjudicating powers. Contracts with the local governments, which include the prefectural governments, cities, town and villages are controlled by the Local Government Autonomy Act. The latter act is not for procurement itself, but contains certain rules applicable to procurement in them. Under either act, the government is to execute and perform contracts with private entities in an equal standing. When contracting, the government is regarded as a private entity; therefore, it is subject to the rules of the Civil Code and other related business laws.

All the government contracts for design and construction contracts are subject to the foregoing accounting rules and procedures. Under the General Accounting Act and Local Government Autonomy Act, the procurement must be made, as a rule, through "competitive bidding" or "nominated bidding" based on public bid notice. For the nominated bidding, all registered contractors are rated according to the sizes of projects, rendering only those rated contractors eligible to tender bids. Rating systems are widely used in the Government and the local government levels. Under the current trends in the Government and the local governments, the nominated bidding overrides open bidding which allows for entry of any contractor, usually subject to some limitations. A contract must be awarded to the lowest bidder out of those who tendered the prices within the budget estimate determined by the government or agencies. This competitive bidding rule is fully followed in case of construction contracts, but not uniformly followed in case of design and consulting services contracts. Both acts allow negotiated contracts if bidding is improper. This depends upon the nature and purpose of the contracts. According to a survey made by the Japan Institute of Architects[24] nearly 50% of the local governments and municipalities throughout the country relied on full price bidding for architectural services contracts and about 20% mainly used negotiated type of contracts for 1982 and 1983.

Quality Assurance

In normal building construction situation, there are quality assurance systems operated by certain builders; however, it does not come within the purview of this report, since detail is unknown. Certain large builders whose business is devoted to detached houses use their own quality assurance systems, either incorporated in a construction contract or a separate warranty document.   In the case of a wood-framed detached house construction, there is a ten-year housing warranty system called "*Jutaku Seino Hosho Seido*." The system is voluntary in nature and the participants are very limited in number. The system resembles the U.S. warranty system under the Liability Risk Retention Act, 1981. While the U.S. targets the product liability, the Japanese system is focused on quality assurance itself. Under this system, builders for built-for-sales houses have to pass the pre-qualification criteria in advance and quality check during construction by a special supervisory organisation established for this purpose. The system is sponsored by a group of insurance companies. To a purchaser of a house, the builder issues a warranty certificate, assuring that the house is free from defects for two years on every component and for five or ten years, depending on the builder's choice, on structural components.   If defects occur, repair costs or alternative monetary compensation will be paid through an insurance which is bought by the organisation in behalf of the builder. The scheme would be ideal for the protection of owners or consumers. But the problem is that the insurance is far from prevalent when compared with the total number of detached houses constructed in Japan.

2.2   Binding

2.2.1   Partnerning

The true sense of partnering does not seem to exist in Japan; however, similar function may be seen in grouping or voluntary association of companies. The most typical one is the grouping seen in the conglomerates which include the traditional ones and newly developed ones. For example, the traditional ones were originated more than 100 years ago and typically include, in a group, various branches of business, such as bank, trading, insurance, manufacturing and construction. The major ones are Sumitomo, Mitsubishi, Mitsui and Yasuda. Newly developed ones vary from group to group in its formation; however, they have usually a construction company. Within a group or association, conflict may exist; however, it will not develop to any dispute, since to keep a co-operative relationship or more specifically Confucian virtue of harmony is absolute essential to member companies. If there is a contractor in a same group, shopping for another outside contractor is not common, though there may be an exceptional case for the reason of specific type of project. In an actual circumstance, one company can make no profit or even suffer a loss due to the other member company's additional request for something not stipulated in a contract. In such a case, the other member company is expected to make up for the loss under the existing contract or, if it cannot, then on a future occasion. Here there is no room for the operation of law and this is based on the Confucian sense of moral obligations called *"kari."* Apart from such conglomerates, in the case of leading and semi-leading general contractors, they are usually tightly or sometimes loosely associated with their sub-contractors who usually keep an association of sub-contractors for a specific parent general contractor. No court battle is conceivable for conflict among them, since staying as a party to such an association has a stronger effect than theoretical legal effect arising out of a sub-contract. Though infra Table 3a shows a certain number of disputes cases between a main contractor and a sub-contractor, it seems that they are mostly disputes between a first-tier sub-contractor and a second-tier sub-contractor or infrequent cases in which a leading or non-leading contractor sub-let part of his work to a sub-contractor who did not belong to such an association.

3.   **DISPUTE RESOLUTION**

Compromise before Court

In Japan too, litigation is the most typical method of dispute resolution, but with a strong connotation that it is the last resort. Before reaching litigation, there are largely three ways of the alternative dispute resolution: mediation, conciliation and arbitration.[25] In the proceedings of mediation and conciliation, disputants are not bound by a mediation or conciliation commissioner's advice or resolution plan. If they accept such advice or plan, the acceptance is a compromise under the Civil Code. The "compromise" (*wakai*) is one of the 13 typical contracts under the Civil Code and defined as the cessation of a dispute by making mutual concession, whereby parties' rights are changed. Thus, the compromise as one of the typical contracts must be made by mutual concession. The Code of Civil Procedure was substantially amended in its whole extent in 1996, effective in 1998, except for arbitration. The arbitration articles in the previous edition of the code are still effective, but will be amended in the near future, though the time and extent of amendment are uncertain. In the following, article numbers are all those of the amended Code of Civil Procedure, and "the Former Arts. 786-805" refer to the non-amended existing ones on arbitration.

Typical compromises can be largely classified into two categories. The first one is an "out-of-court compromise" (*saiban gai no wakai*, or *jidan*) which is valid, as an agreement, and which remains to be affirmed through a court judgment for its execution. The second is those to be made at a court. They include the following four types. In an "Instantaneous Court Compromise" (Art. 275) (*sokketsu wakai*), disputants agree to settle a dispute by compromise and by which they appear in a summary court to have the compromise reduced in a court record. Next, an "In-Proceeding Compromise" ("*sosho-jo no wakai*), like the instantaneous compromise, is made in a court, but it is made during the course of proceeding while disputants (plaintiff and defendant) attend a court. In conjunction with this In-Proceeding Compromise, the amended code has newly established another two types of compromises: "Acceptance of Compromise Clauses" (Art. 264) (*wakai judaku*) in which disputants agree in writing to accept a compromise proposed by a court even if all disputants do not appear at the court, and "Award of Compromise Clauses" (Art. 265) (*saitei wakai joko*) in which disputants agree in a prior writing to accept compromise clauses to be later awarded by a court. Though the last one seemingly resembles arbitration, but it differs in that before the award, a written agreement can be cancelled unilaterally and a court can seek acceptable clauses in consultation with disputants.

Compromises of the second category are called "compromise before the court" (*saiban-jo no wakai*) and have the same validity as for a peremptory judgment, becoming executive only by obtaining an execution clause of a court. The Instantaneous Court Compromise can be made readily and inexpensively; therefore, the compromise of the first category can be easily documented, by bringing it to a court, as the Instantaneous Court Compromise to give it due executive power. In connection with the In-Proceeding Compromise, it is not rare that during a proceeding, a court judge recommends parties to a lawsuit to try to settle it by compromise before judgment.[26] According to the statistics introduced by Takayuki Yamashita, Esq.,[27] out of 93,502 ordinary civil cases processed in 1979 by all of the district courts as the first instance court, 38,618 cases

(41.3%) terminated with judgment, 30,428 cases (32.5%) terminated with compromise, and 20,610 cases (22.0%) terminated by withdrawal of suit.

3.1   Non-Binding

3.1.1   Conciliation

Conciliation (*chotei*) is the most favoured type of dispute-resolution system on the basis of an intermediary's resolution plan, since people at large generally tend to avoid an all-out struggle, but are hesitant to reconcile a dispute at one's own initiative.   Simply stated, as the most dominant feature of conciliation, conciliation commissioners (in a panel) or intermediaries proposes a resolution plan or advice which is not necessarily required to be based on legal right, but usually in part based on humanity, common sense, rationalism or Confucian morality which will be more adaptable to people. Yet, a resolution plan accepted by parties and put on the court record has an executive power like the compromise before the court.   Today, the number of conciliation cases filed with the courts constitutes one-third the number of lawsuits filed with the courts of first instance.   Out of these conciliation cases, 60% are resolved by the acceptance of the parties.

Conciliation is governed by the Civil Conciliation Act, 1951 No. 222.   This act purports to settle disputes according to the Natural Reason, consistent with the actual circumstances and through mutual concession (Art. 1).   Except for the special types of conciliation, the ordinary civil conciliation is filed with a competent summary court and is processed by a "conciliation panel" (*chotei iinkai*) which consists of three conciliation commissioners, one of which is appointed from among judges by a district court.   The other conciliation commissioners are appointed from among persons who are selected by the Supreme Court out of the general public provided that they must be basically licensed attorneys or equivalent while having expert knowledge on civil disputes or abundant knowledge on social life. Also, they must be between 40 and 70 years old.

More importantly, a party who is summoned by a court must present himself before a conciliation panel. However, he can be represented by his attorney or agent for any unavoidable reason. When parties accept a resolution plan by the panel, they are regarded as having reached a compromise. The result is recorded in a court document which in turn has the same effect as compromise before court and eventually a final court judgment. Where a conciliation panel finds a pending dispute not likely to reach any resolution, it may terminate the conciliation. The parties may then continue their argument in a competent court. In this instance, the panel can send its finding as to the cause and issues to the judge so as to facilitate the judge's evaluation at a new court. In addition to the civil conciliation, there are special conciliations in various branches of industry and daily life, such as sales of premises and houses, agriculture, commerce, traffic accidents, public nuisance and construction disputes (see 3.2.3). Conciliation will be also utilised in the future.

3.1.2   Executive Tribunal

In the Construction Business Act, mediation or conciliation begins when one or both parties to a dispute lodge a petition with the committee or by "mere motion" (*shokken*) of the committee. The latter case is authorised in Art. 25-11, Para. ii) which allows the committee to take mere motion through its resolution for disputes on construction projects bearing public interest, such as for i) railways, bridges, waterworks, ii) public buildings, and iii) power and gas supply. The committee may refuse to accept a petition or abandon with due reason the procedures once started according to the petition. For a construction context, see 3.2.3.

3.1.3 Mediation

Japanese mediation "*assen*" can be said to be a negotiation involving an intermediary or pre-arrangement for the conciliation. Generally, a mediation commissioner assists the disputants to reach a compromise. Conciliation is very similar to mediation, but is based on a commissioner's resolution plan proposed by him. In the case of mediation, commissioner's advice is usually not a resolution plan and he tries to facilitate a compromise by advising concession to disputants. In the case of either the mediation or the conciliation, disputants have their own choice to decide at their own discretion. The mediation is provided in various statutes and voluntary arrangements for dispute resolutions by trade or consumer associations, but is not officially incorporated into the judicial court system, though a judicial recommendation for compromise may be taken as informal mediation. Thus, a compromise reached through a mediation panel, if recorded at the panel, does not have any executive power, such as the conciliation statement recorded in court.

3.2 Binding

3.2.1 Adjudication

If actions by architects or contractors cause the public to complain of a nuisance, there may be a chance that the dispute is put in mediation, conciliation or arbitration, all being processed by the committees established under the Public Nuisance Dispute Resolution Act. In addition, the act provides for the Public Nuisance Adjustment Committee which is empowered to render a quasi-judicial award called "*saitei*." The award procedures may be initiated upon an application of any party who claims compensation for damage or loss. The award binds parties as to identification of the causes and the judgment on liability of the parties unless the award is appealed to a judicial court within 30 days after the award rendered. The committee is an administrative organ of the Government and has an inquisitorial power to look into evidence and facts. Similar committees are also provided at the local governments, but are not capacitated to give such quasi-judicial awards. However, such a quasi-judicial award has not been utilised in full, probably due to fact that these problems are very delicate.

3.2.2   Arbitration

Arbitration[28] may be resorted to instead of litigation. The Code of Civil Procedure sets forth arbitration rules in the Former Arts. 786-805. The articles were enacted by almost literal translation of the old German code, as old as 1890 and have not been amended, though the latter has been frequently amended. However, in keeping pace with the revision of the civil proceeding, these arbitration articles will be revised or at least updated in the immediate future. Under the code, all disputes may be settled through arbitration if they are of such nature where disputants may make autonomous disposal. An "agreement of arbitration" (*chusai keiyaku*) may be made, on either a current dispute or future one. An arbitration award may be vacated when it is not supported by reason or when there was a defect in the procedures (Art. 801, (1)). The reason need not be so thorough as in a court decision. It will not be examined, however rational or irrational it may be, unless it is illegal. It is sufficient if it explains why and how a party owes an obligation or liability. Also parties may dispense with the reason if they so wish by their agreement. If an award meets these requirements, judicial courts will issue an "execution judgment" (*shikko hanketsu*) after the procedural confirmation. Arbitration and litigation are exclusive of each other. Any agreement of arbitration serves as due ground to refuse a court proceeding.

In Japan, the Japan Commercial Arbitration Association (*Kokusai Shoji Chusai Kyokai*) is most commonly used for commercial arbitration and is now entrusted with the international dispute-resolution function of the Japan Chamber of Commerce and Industry as Japanese counterpart of ICC. The Association is now engaged in mediation, conciliation and arbitration, and is now capacitated to employ UNCITRAL arbitration rules for international arbitration. The Association dealt with between 200-300 mediation cases, one conciliation case and less than ten arbitration cases from 1985 to 1989, and thereafter from 1991 to 1995, mediation cases being on a decrease, no conciliation case and 27 arbitration cases, all being newly filed. It appears that these numbers do not include any noticeable number of construction disputes, since a similar resolution system is available at the following Construction Disputes Resolution Committees.

3.2.3   Expert Determination

Construction disputes can be settled by the Construction Disputes Resolution Committees established under the Construction Business Act. Since it is not mandatory on participants in a construction project, a dispute can be brought to a judicial court or an another arbitration panel described. For architects or engineers, there is no such special resolution system, and yet they cannot be parties to a mediation, conciliation or arbitration under the Act. An architect-owner dispute cannot be joined or consolidated with a contractor-owner dispute under the resolution system of the Act. If there are multiple disputes, such as between an architect and an owner, and between an owner and a contractor, there may be a chance in which decisions of these multiple relation-disputes become inconsistent.

Resolution committees include the Central Construction Dispute Resolution Committee at the Ministry of Construction and the Prefectural Construction Dispute Resolution Committees in each prefecture. Also, it is interpreted in practice that filing of petitions must meet two major requirements: first, a dispute must be one involved in construction and second, it must be under an Independent Undertaking Contract, whatever titled, as opposed to the Mandate Contract (see 1.2.2). Here, the construction means the 28 categories described in 1.2.1. Thus, a sales contract between a contractor and a material supplier, a construction contract of the Mandate Contract type and tort claims are not included within the jurisdiction of the committees. A claim by a main contractor against a subcontractor and a claim by subcontractor against a sub-subcontractor are included; however, a claim by a sub-subcontractor against a main contractor is not included, since there is no Independent Undertaking Contract relationship between them. The central committee consists of 15 regular committee members and 130 special members. One third of these members are lawyers. Also, the Ministry of Construction and the Prefectural Governments provide consultation services upon disputants' inquiries. In the year of 1989, the total number of consultation cases reached as many as 5,990 cases.[29]

The resolution processes consist of mediation, conciliation and arbitration. Mediation or conciliation can be started upon one party's petition or by mere motion of the committee; however, unless both or all parties accept the advice or plan of a conciliation panel of the committee, the parties are not bound to obey it. Yet under the current statistics, conciliation cases exceed arbitration. The mediation process is governed by one committee member, while the conciliation and arbitration processes are each headed by a panel consisting of three members. Resolution through mediation or conciliation at these committees is an out-of-court compromise; therefore, unlike the conciliation compromise recorded in court, the compromise of mediation or conciliation does not have any executory power. If a party tries to enforce a compromise, he must file a suit with a court to obtain a court judgment. The claims filed with these committees are as shown in the following tables.

Table 3a Types of Disputes Lodged with the Committees, 1996: (1) Parties

Parties	Central Committee		Pref. Committees		Total	
	Cases		Cases		Cases	
Owner against Main Contractor	30	54%	142	56%	172	55%
Main Contractor against Owner	13	23%	77	30%	90	29%
Sub-Contractor against Main Contractor	13	23%	30	12%	43	14%
Main Contractor against Sub-Contractor	0	0	5	2%	5	2%
Others	0	0	1	0	1	0
Total	56	100%	255	100%	311	100%

Disputes lodged by [Sub-Contractor against Main Contractor] and those lodged by [Main contractor against Sub-Contractor] include respectively those lodged by a second-tier subcontractor against a first-tier subcontractor and by a first-tier subcontractor against a second-tier subcontractor.

Table 3b As above: (2) Issues

Issues	Central Committee		Pref Commitee		Total	
	Cases		Cases		Cases	
Defects in Construction	16	29%	113	44%	129	42%
Delay in Construction	2	3%	8	3%	10	3%
Contractual Payment	15	27%	79	31%	94	30%
Termination of Contract	10	18%	22	9%	32	10%
Payment to Subcontractor	13	23%	29	11%	42	14%
Others	0	0	4	2%	4	1%
Total	56	100%	255	100%	311	100%

Table 3c Numbers of Dispute Cases Lodged with
the Construction Disputes Resolution Committees

Year	Proceeding	Central Committee		Pref Committee		Total	
		Newly lodged	Pending	Newly lodged	Pending	Newly lodged	Pending
1989	Mediation	0	0	5	5	5	5
	Conciliation	16	35	47	87	63	122
	Arbitration	8	19	22	40	30	59
	Total	**24**	**54**	**74**	**132**	**98**	**186**
1990	Mediation	6	8	16	28	22	36
	Conciliation	32	53	148	229	180	282
	Arbitration	16	46	36	105	52	151
	Total	**54**	**107**	**200**	**362**	**254**	**469**
1991	Mediation	2	2	10	16	12	18
	Conciliation	35	54	141	235	176	289
	Arbitration	5	42	41	130	46	172
	Total	**42**	**98**	**192**	**381**	**234**	**479**
1992	Mediation	6	6	14	20	20	26
	Conciliation	21	58	156	252	177	310
	Arbitration	4	26	48	138	52	164
	Total	**31**	**90**	**218**	**410**	**249**	**500**

	Mediation	11	13	21	27	32	40
1993	Conciliation	23	53	154	265	177	318
	Arbitration	6	17	42	138	48	155
	Total	**40**	**83**	**217**	**430**	**257**	**513**
	Mediation	4	6	11	21	15	27
1994	Conciliation	34	56	155	267	189	323
	Arbitration	6	17	51	140	57	157
	Total	**44**	**79**	**217**	**428**	**261**	**507**
	Mediation	13	15	14	21	27	36
1995	Conciliation	33	63	155	268	188	331
	Arbitration	10	23	44	142	54	165
	Total	**56**	**101**	**213**	**431**	**269**	**532**
	Mediation	13	20	17	23	30	43
1996	Conciliation	30	61	187	292	217	353
	Arbitration	13	28	51	154	64	182
	Total	**56**	**109**	**255**	**469**	**311**	**578**

Sources: All data for tables 3a to 3c are kindly provided by the Ministry of Construction.

3.2.4   Litigation

The Japanese litigation system is three-tiered. In an ordinary case, a lawsuit proceeds from a district court to a high court and then to the Supreme Court. As court of first instance, in addition to the District Court there are the Family Courts for the domestic relations and juvenile delinquency cases and the Summary Courts for small-claim-amount cases in which a claimed amount is less than JPY900,000 and for minor crimes and offences. In this instance, an appeal can be brought to a district court as the second instance court and further appeal to a high court as the third instance court. An appeal to a third instance court is limited to certain reasons for appeal; therefore, all claims cannot go to the Supreme Court or a high court as the third instance court.

Some features of Japanese court procedures are well elucidated by Professor Yasuhei Taniguchi[30] as follows: "There is no clear distinction between the pleading and the trial stages; rather these stages are combined. Evidence is introduced and allegations are presented, both in a piecemeal fashion. Japanese procedure adopts an adversary system. The court must decide solely on the basis of the allegations and evidence introduced by the parties. The court has no freedom to find a fact not relied on by either party nor to examine a witness not produced by either party. A judgment on the merits must correspond in nature to the plaintiff's demand and may not exceed it." Also, in Japan, there is no such specialist judge for construction or other technical cases as English Official Referees.

Court judgment is given only by judges, usually one in lower courts and without any jury or assessors. A "peremptory judgment" (*kakutei hanketsu*) is effective for 10 years from its date and may be enforced as an "execution title" (*saimu meigi*) by obtaining endorsement by way of an execution clause at a competent court. Other types of execution titles include "judicial records of compromise" (*wakai chosho*) and, as far as monetary obligations are concerned, "notarised deeds" (*kosei shosho*) which can be easily prepared at a notary public's office (see 1.2) and are commonly used to secure pecuniary rights. Also, a record of arbitration award is given the same effect as a court judgment and may be executive after the review by a court for its procedures in the form of "execution judgment" (*shikko hanketsu*).

The amended Code of Civil Procedure provides, in Art. 118 (Validity of Foreign Judgment), for certain reservations in the execution of a foreign court decision, such as that it must not be against Japanese public policy and good morality and that Japan is assured to enjoy a mutual reciprocal arrangement with the foreign country on execution of court decisions. Here, a difficulty lies in meeting the requirement for the mutual reciprocal arrangement and public policy. If such mutual reciprocal requirement is not met, a winning party of a foreign court decision has to in substance apply for another proceeding in a competent Japanese court for execution. Anglo-American punitive damages and validity in chose in action of longer than 10 years are likely to be taken as against the public policy, since the former is in excess of Japanese compensatory damages and the latter exceeds Japanese ten years extinctive prescription for chose in action. Unlike a court decision, a foreign arbitration award can be readily enforced in the same way as a domestic award if the foreign country is a party to the New York Convention on the Recognition and Enforcement of Foreign Arbitral Awards, 1958, as it meets the mutual reciprocal requirement, and unless it is against the public policy and certain procedural requirements. If parties to a contract agree to any foreign law to be applied, that law may be relied on provided that a plaintiff must certify or substantiate the law, for instance, by having a competent lawyer in that country prepare a proper report on the status of law. If a defendant resides in any foreign country, the complaint will be delivered through a Japanese embassy or similar administrative detachments located in that country.

3.2.6 Ombudsman

The ombudsman system is at a premature stage. No official system exists in the State level, though such necessity has been discussed. In the local government level, certain prefectures and cities have now established their own systems. Their major purposes are directed to the administration business and social welfare; however, irregularity in bidding for construction projects as found from the publicised records are sometimes accused as collusive bidding.

4.0 THE ZEITGEIST

In Japan also, the economic zeitgeist is seriously depressing. After the "Bubble Economy" collapsed, the past trend of increasing private sector investment has stopped. To worsen the situation, a certain number of construction contractors are now in a hopeless struggle. They were forced, as a condition of undertaking the construction, to guarantee banks that they would pay back construction loans the banks made to owners, when the owners failed to pay back the loans. In addition, contractors sought larger profits during the Bubble Economy, investing considerable loan funds in building premises for project development and furthering their own leisure-oriented projects. In consequence, several contractors, not all small concerns, became insolvent in 1997.

The bidding system poses an insurmountable barrier, both in the government procurement and on the side of bidders. To date, contractors have been ranked to participate in certain sizes of projects for the government procurement and cannot enjoy a chance to enter open bidding. The collusive bidding is heavily tied to the government bidding. The Government has started to improve the situation by modifying the system toward open bidding.

Administrative guidance has permeated persistently through every level of public administration, though the Government has sought new plans which would establish clarified standards for administrative procedures at large. However, since improvement involves the relevant existing acts totalling as many as 361 in number, all having bearing on such administrative procedures, it is not expected that all the past confusion will suddenly dissolve in the immediate future.

Product Liability Act, 1994 No. 85 is not applicable to real property so design and construction practitioners are less concerned with it, though product manufacturers are on keen alert. However, no noticeable number of PL liability suits has been brought to date.

The author's cordial thanks are due to Mr. John Dickison, Schal Bovis, Inc., Japan for his advice given in the course of preparation of this monograph.

1   Bennett, J., Flanagan, R., and Norman, G. (1987), *Capital & Countries Report - Japanese Construction Industry*, Centre for Strategic Studies in Construction, Univ. of Reading, p. 7.

2   i.e. 1,000,000 million

3   Relevant data obtainable from: Japan Federation of Construction Contractors (1997), *Construction in Japan* (annual), JFCC at 5-1 Hacchobori 2-chome, Chuo-ku, Tokyo 104; Japan Institute of International Affairs (Ed.), *White Paper of Japan: Annual Abstract of Official Reports and Statistics of Japanese Government*, JIIA at 19th Mori Bldg., Toranomon 1-2-2, Minato-ku, Tokyo

4   For Japanese law generally, see, *e.g.* Tanaka, Hideo (Ed.) (1976), *The Japanese Legal System*, University of Tokyo Press, Tokyo; Kitagawa, Zentaro (Ed.), *Doing Business in Japan*, Matthew Bender & Co., Inc., N.Y., 1982-1992 (consisting of six volumes, comprehensive introduction to Japanese Law, and compiled in loose-leaf supplement system). For a construction aspect, see Matsushita, Fumio (1994), *Design and Construction Practice in Japan: A Practical Guide*, Kaibunsha Ltd. (26 Sakamachi, Shinjuku-ku, Tokyo 160), pp. 26-30.

5   Through this monograph, *italic letters* refer to Japanese words and are provided in order to avoid any confusion, since English translations may vary with Japanese commentators; however, long vowel marks (macrons) are omitted for word processing reason.

     For major statutes, English translations are available from Eibun-Horei-Sha, Inc. at Kiyose Kaikan, 2-4-7, Hirakawa-cho, Chiyoda-ku, Tokyo 102.

6   For Japanese lawyers, see, *e.g.* Kato, Masanobu (1987), *The Role of Law and Lawyers in Japan and the United States*,

Brigham Young Univ. Law Review, p. 627 *et seq*; Meyerson, Adam (1981), *Why There Are So Few Lawyers in Japan*, the Wall Street Journal, Feb. 9, p16.

7   On the Japanese corporation law, there are some English textbooks written by Japanese commentators. The English translation of these legal entities varies with the commentators, some using English usage and other using U.S. usage. For corporation law, see, *e.g.* Kawamura, Akira (Ed.) (1982), *Law and Business in Japan*, The Japan Institute of International Business Centre, Tokyo; Matsueda and Ihara (1991), Vol. 4, Part VII, Chapter 1 *Company Law in General*, in Kitagawa, *supra* note 5; Matsushita, *supra* note 5, pp. 272-297.

8   Payment to construction subcontractors are regulated by the Construction Business Act.

9   See Matsushita, *supra* note 5, pp. 62-73.

10   See Matsushita, *supra* note 5, pp. 20-23.

11   See Matsushita, *supra* note 5, pp. 98-116.

12   In the normal sense of terms, "Special License" may be a more appropriate term than Specific License; however, the act uses a Japanese word "*Tokutei*" which is equivalent to Specific.

13   For the Civil Code and related business law, see, *e.g.* DeBecker, Joseph (1979), *The Principles and Practice of the Civil Code of Japan*, University Publications of America, Washington, D.C.; Hahn, Elliott J. (1984), *Japanese Business Law and the Legal System*, Quorum Books (Greenwood Press), Westport, Connecticut; Gosling, Barker & Matsushita, Mitsuo (Ed.) (1988), *Japan Business Law Guide* (two volumes), CCH international, Sydney.   For a construction aspect, see Matsushita, *supra* note 5, pp. 31-49.

14   For Japanese contracts and negotiation approach, see , *e.g.*
      Hahn, Elliott (1982), *Essay: Negotiating Contracts with the*
      Japanese, Case Western Reserve Journal of International Law,
      Vol. 14, Spring, pp. 377-385; Hayakawa, Takeo (Nov. 1987),
      Understanding Japanese Business and the Japanese Legal
      System, International Computer Law Adviser, pp. 8-13; Hall,
      John Carey (1979), *Japanese Feudal Law*, University
      Publications of America, Washington, D.C.; Kawashima,
      Takeyoshi (1979), *Japanese Way of Legal Thinking*,
      International Journal of Law Libraries, Vol. 7, July, pp. 127-
      132.

15   Copies selling in the order of 260,000 p.a. Its English version
      is available from Management Research Society for
      Construction Industry, 2-5-1 Hatchobori, Chuo-ku, Tokyo 104.

16   Matsushita, *supra* note 5, pp. 120-148 & 323-343 show and
      comment on the Private Contract Form in the previous edition
      which is not substantially different from the revised one as for
      the essential features.

17   For Construction Nuisance, see Matsushita, *supra* note 5, pp.
      187-211.

18   See, e.g. Lury, R.R. (1976), *Japanese Administrative Practice:*
      The Discretionary Role of the Japanese Government Official,
      Business Law, Vol. 31, pp. 2109-2121; Narita, Yoriaki
      (Anderson, James L. Trans.), *Administrative Guidance*, in
      Tanaka, *supra* note 5, pp. 353-388.

19   Narita, *ibid.*, p. 356.

20   On the suit allergy or reluctance, there has been emerging an
      another view based on a rational approach. According to this,
      the reluctance has been caused by the alternative dispute-
      resolution systems at large and by shortcomings in the judicial
      system. One typical paper on this approach is Professor John

O. Haley's *The Myth of the Reluctant Litigant*, 4-2 Journal of Japanese Study 359, 1978.

See also, e.g. Haley, John O (1984), *Introduction: Legal vs. Social Controls*, Law in Japan, Vol. 17, pp. 1-6; Mayer, Cynthia (July/August 1984), *Japan: Behind the Myth of Japanese Justice*, American Lawyer Vol. 6, pp. 113 *et seq*; Miyazawa, Setsuo (1987), *Taking Kawashima Seriously: A Review of Japanese Research on Japanese Legal Consciousness and Disputing Behaviour*, Law & Society Review, Vol. 21, pp. 219-241.

21   Tanaka, Hideo (1976), *The Role of Law and Lawyers in Japanese Society*, in Hideo Tanaka, *supra* note 5, pp. 261.

22   *Yamanaka v. Kondo*, 1083 Hanrei Jiho 125 (Tsu District Court February 25, 1983 decision)

23   For further detail, see Matsushita, *supra* note 5, pp. 27-55.

24   JIA (Japan Institute of Architects) News, No. 538 (July 15, 1985), p. 3.

25   For dispute resolution generally, see, *e.g.* Hattori, Takaaki & Henderson, Dan Fenno (1983), *Civil Procedure in Japan*, Matthew Bender & Co., Inc., New York (two volumes in loose-leaf); Fujita, Y. (1978), *Procedural Fairness to Foreign Litigants as Stressed by Japanese Courts*, International Lawyer, Vol. 12, pp. 795-811; Haley, John O. (1984), *Introduction: Legal vs. Social Controls*, Law in Japan, Vol. 17, pp. 1-6; Kojima, Takeshi and Taniguchi, Yasuhei (1978), *Access to Justice in Japan*, in Cappelletti, Mauro, and Garth, Bryant (Eds.), Access to Justice: A World Survey, Vol. I, Book II, Sijthoff and Noordhoff, Milan; Obuchi, Tetsuya (1987), *Role of the Court in the Process of Informal Dispute Resolution in Japan - Traditional and Modern Aspects, with Special Emphasis on In-Court Compromise*, Law in Japan, Vol. 20, pp.

74-101. For other articles, see Matsushita, *supra* note 5, pp. 270-271.

26   See, e.g. Yoshikawa, Seiichi (1978), *The Judge's Power to Propose Terms for Settlement: the S.M.O.N. case (the Proposals for Settlement and Accompanying Opinions presented to the Parties by Judge Tsuneo Kabe in the 34th Civil Division of the Tokyo District Court on January 17, 1977 and May 23, 1977)*, Law in Japan, Vol. 11, pp. 76-90.

27   Yamashita, Takayuki (1991), Chapter 2 *Compromise*, in Kitagawa, *supra* note 5.

28   For a further discussion, see Hayakawa, Takeo (1987), *Arbitration Law in Japan*, in Simmonds, K.R. & Hill, B.H.W. (Ed.), Commercial Arbitration Law in Asia and the Pacific, ICC Publishing S.A. Paris & Oceana Publications, Inc., pp. 84-104.

29   Kensetsu Koji Funso Kenkyu-kai (1981), *Kensetsu Koji no Funso Shori* [Dispute Resolution on Construction Projects], Seibun-sha, Tokyo, pp. 11-12.

30   Taniguchi,Yasuhei, Vol. 7, Part XIV, pp. XIV 1-11/12, in Kitagawa, *supra* note 5.

MALAYSIA

Inpamathi Natkunasingam and
Satkunabalan K Sabaratnam
Rashid and Lee, Kuala Lumpur, Malaysia

1.   BACKGROUND

1.1   Economic

The GDP in Malaysia expanded in real terms by 8.2% in 1996 as compared to 9.5% in 1995 and is expected to moderate to 8% in 1997. [1] The GDP in 1996 amounted to RM 130,628 million of which the construction sector contributed 4.7%.

The Malaysian economy has generally been transformed from being an agricultural/ commodity based economy in the early 80's to a manufacturing based economy in the 90's. Agriculture, forestry and fishing which contributed to 23 % of GDP in 1984 accounted for 12.2% in 1996, while manufacturing which contributed 18% in 1984, accounted for about 34.2% of GDP in 1996.

The transformation was a result of Government led initiatives and policies to pursue growth and strengthen the manufacturing base. Underpinning economic growth throughout the period was the natural resources that Malaysia has been blessed with namely, oil and gas and timber. As a result of strong growth policies advocated by the Government and the substantial infrastructure requirements of the nation, the construction industry has experienced double digit growth through the nineties. In the past ten years, major infrastructure development has been privatised to ease the Government's financial burden and ensure that greater emphasis is placed on the financial/economic merits of a project.

However, since June 1997, countries in South East Asia have been hit by a currency crisis which has resulted in a substantial devaluation of local currencies and outflow of foreign capital. Malaysia has not been spared and the ringgit depreciated from 2.49 in May 1997 to 3.5 in October 1997 against the US dollar. Consequently the per capita income is expected to increase at a slower rate of 7.7% to RM12,102 (as compared to 11.7% in 1996 to RM11,234). Domestic demand is also expected to grow at a slower pace in 1998.

The construction industry is especially expected to be affected as the Government has deferred several "mega" projects to placate the capital/currency market amid concerns on the country's ability to financially shoulder these projects. Six of these projects, worth RM65.5 billion are the Bakun Hydro-Electric Dam Project, the Putrajaya Administrative Centre Phase II, the Northern Regional International Airport, the Kuala Lumpur Linear City Project, the Cameron Highlands-Fraser Hill-Genting Highlands Road Project and the Malaysia-Indonesia Bridge. In addition, slowdown in construction is compounded by the oversupply of office space and the cautious stance of developers in view of the anticipated slower growth in demand. [2]

The civil engineering sub-sector which expanded by 14.4% in 1996 (1995:18%) is therefore expected to be affected after years of continued growth. The rapid growth was largely a result of considerable strain on the country's infrastructure facilities over the years and a growing demand for more efficient modes of transportation. Hence construction projects in road and rail transport, airport and port facilities as well as power plants constituted the bulk of activities in the civil engineering sub-sector. The hosting of the Commonwealth Games in Kuala Lumpur in September 1998 has resulted in several projects such as the LRT System Two for Kuala Lumpur, the Bukit Jalil Stadium and the Kuala Lumpur International Airport ("KLIA") at Sepang, being constructed on a fast track schedule in time for the opening of the Commonwealth Games. [3]

While it is difficult to estimate at this stage, it is almost certain that the economy will experience slower growth in 1998, with some industry analyst forecasts estimating growth of about 6%. Despite this, the Government appears to remain confident that the construction industry will not be greatly affected, with a 9% predicted growth for this industry due to the continued demand for construction of landed residential properties and lower end apartments and factories as well as the construction activities related to infrastructure projects such as the rail link between the KLIA and the city centre, the Multimedia Super Corridor and the East Coast Highway.

1.2 Legal

"The Malaysian Constitution is not the product of an overnight thought but represents the end result of a century of British colonial administration which transformed the country from being a number of separate Malay States and colonies of Britain into a single Federation with a modern constitution". [4]

Malaysia is a Federation of thirteen States, eleven in Peninsular Malaysia and the other two in East Malaysia. Each of the thirteen States either has a Legislative Assembly with a Sultan as its Head (in the case of nine States[5]) or Governors (in the case of the remaining four[6]). Thus there is a Federal Parliament and the thirteen states have their respective State Legislative Assemblies. The constitutional monarch is His Majesty the Yang Dipertuan Agong, who is chosen on a rotation basis, from among the Sultans of the nine states, only for a term of five years. The head of the Government is the Prime Minister, assisted by his Cabinet. Members of Parliament are elected every five years and the Cabinet Ministers are appointed by the Prime Minister.

The doctrine of separation of powers is evident as the powers of government are separated into the legislative, executive and judiciary. State laws exist side by side with Federal laws but when such laws are found to be in conflict with each other, Federal law prevails.[7] The division of responsibility between the State and Federal governments are found in the Federal Constitution. The powers of the Federal Parliament and the State Legislative Assemblies are not supreme in that they have to enact laws within the limits allowed by the Federal and State Constitution. The administration of justice is enshrined in the

Federal Constitution, as a federal matter so that only the Federal Parliament may legislate on it. Hence apart from Muslim religious ("Syariah") courts and native courts, all the courts are Federal Courts.[8]

The judicial system is an adversarial as opposed to an inquisitorial system. The sources of law include both written and unwritten law. The written law comprises the Federal Constitution, legislation enacted by Parliament and the State Legislative Assemblies and delegated or subsidiary legislation made by persons or bodies under powers conferred by the Acts of Parliament or Enactments or State Legislative Assemblies.[9] The unwritten law is comprised of principles of English law (in the form of the common law and the rules of equity) and judicial decisions. The doctrine of judicial precedent is adhered to.

The personal law of the Muslims in Malaysia is only applicable to followers of Islam and it administration may vary from state to state as it is a matter within the sphere of the States' jurisdiction.

The court system

Article 121(1) of the Federal Constitution establishes two High Courts of co-ordinate jurisdiction and status namely, the High Court of Malaya and the High Court of Sabah and Sarawak. The Superior courts comprise the Federal Court, Court of Appeal and the High Court in order of their hierarchy. The Subordinate Courts consist of the Penghulu (or native courts), Juveniles, Magistrates and Sessions Courts.

The original civil jurisdiction of the courts in Malaysia is conferred by article 121 of the Federal Constitution and the Subordinate Courts Act 1948. There are three tiers of first instance courts dealing with civil matters in Malaysia: the Magistrates, Sessions and High Courts. The jurisdiction and power of the courts is based on the type of action, value of the claim and the geographical area in which the court operates.[10]

The Courts of Judicature Act 1964 governs the jurisdiction of the High Court to hear civil appeals from Subordinate Courts. Where the amount in dispute or the value of the subject matter is RM10,000.00 or less, no appeal lies in the High Court except on a question of law. The High Court has like powers and jurisdiction on the hearing of appeals from the Subordinate Courts, as the Court of Appeal has on the hearing of appeals from the High Court.

Under Article 121 (1B) of the Federal Constitution, the Court of Appeal, comprised of three judges sitting together, has jurisdiction to determine appeals from decisions of the High Court or a Judge thereof. However, no appeal lies before the Court of Appeal where the amount or value of the subject matter of the claim is less than RM250,000.00 except with the leave of the Court of Appeal.[11] No appeal lies from the Court of Appeal to the Federal Court (comprised of three judges or such greater uneven number of judges as the Chief Justice may prescribe) without leave if the matter in dispute in the appeal is less than RM250,000.00. [12] The Federal Court may summarily refuse leave if it appears that the intended appeal basically involves a factual dispute.

There are two levels of first instance criminal courts, the Subordinate Courts and the High Court. The High Court's criminal jurisdiction is contained in s. 22 of the Courts of Judicature Act and it also has the appellate criminal jurisdiction to hear appeals from subordinate courts within the territorial jurisdiction of the High Court. The Court of Appeal similarly has the jurisdiction to hear any appeals against decisions of the High Court in the exercise of its local jurisdiction, appellate jurisdiction in respect of a Sessions Court decision and in the case of the Magistrates Court, on questions of law. The Federal Court has the jurisdiction to hear any appeal from the Court of Appeal in its appellate capacity in respect of any criminal matter decided by the High Court.

There are no specialist courts in Malaysia. The closest thing to a specialist court in Malaysia is the Industrial Court set up under the Industrial Relations Act, 1967 and which exercises its jurisdiction on matters relating to trade disputes and dismissal of workmen who are not union members.[13]

The legal profession

The legal profession in Malaysia is a fused profession. There is no division of solicitors and barristers and a practising lawyer acts in both capacities and is known as an advocate and solicitor of the High Court. The governing Act for the legal profession in Malaysia is the Legal Profession Act 1976 ("the LPA").[14] An advocate and solicitor who is a "qualified person" under the LPA[15] is given the exclusive right by law to appear and plead in all Courts of Justice in Malaya, so long as the advocate and solicitor is on the roll and has a valid practising certificate and is not an "unauthorised person" under the Act.[16] Advocates and solicitors have the exclusive right of audience without differentiation. The recent proposals by the Chief Justice of Malaysia to limit the rights of audience of advocates and solicitors in the Superior Courts in relation to their years of experience was met with strong objection from the Malaysian Bar.

The Attorney General has the power to issue a special certificate for admission as an advocate and solicitor for a person in possession of a qualification which renders him eligible to practice by whatever name called in any country or place or territory outside Malaysia and who has been employed in that capacity for over seven years.[17]

The Bar in East Malaysia is separate from the Bar in Peninsula Malaysia.[18] Therefore admission as an advocate and solicitor in the High Court of Malaya does not grant such person a right of audience in the High Court of Sabah and Sarawak except by admission on an ad hoc basis for specific cases. Calls for the review of the provisions of what appear to be outdated ordinances in Sabah and Sarawak have been met with resistance from the East Malaysian Bar.

Consumer protection

A Malaysian consumer's legal rights are derived primarily from the common law, particularly the law of negligence and from the Contracts Act, 1950.   The overall legal system is not self-implementing meaning that the Malaysian consumer must take the initiative to enforce his legal rights.

There is no single comprehensive consumer protection law in Malaysia as there exists in some other countries.   There is various piecemeal legislation in this country which attempts to cover consumer protection such as the Sale of Goods Act 1957, the Trade Descriptions Act 1972, the Price Control Act 1946 and the Hire Purchase Act, 1967.[19] Other recent consumer protection legislation is the Fair Competition Bill which is still on the drafting board, the Direct Sales Act 1993 to encourage the growth of ethical direct selling and to eradicate pyramid schemes and the Industrial Design Act [20] aimed at regulating transfer of technology. These statutes provide no redress mechanism specifically for consumer disputes. However the Consumer Protection Bill which has been under discussion for 4 years, is in the final stages of drafting and is expected to be tabled in Parliament at any time.

The Housing Developers Control and Licensing Act 1966 ("the Housing Developers Act") and the Housing Developers (Control and Licensing) Regulations 1982 ("the 1982 Regulations") is particularly worth a mention as it was enacted principally for the regulation of housing developers. Purchasers monies are protected in the event of a developer's insolvency through a "housing developers account" and prescribed sale and purchase agreements have to be adopted failing which the transaction would be rendered void ab initio. The defects liability period in respect of such houses has been extended from six to eighteen months.

There is ample authority in Malaysia to the effect that any attempt to contract out of the Housing Developers Act and the 1982 Regulations will not be countenanced by the Courts. The court in the decision of SEA Housing Corpn Sdn Bhd v Lee Poh Choo[21] observed that "it would appear that only "contracting out" in favour of a weaker party i.e. the purchaser might be countenanced by the courts". It therefore appears that if it is the purchaser who commits a breach of the Act, that the only sanction will be the criminal penalty under the Act and in the case of a housing developer, the courts would render the contract null and void.[22] However the recent Federal Court decision in Insun Development Sdn Bhd v Azali bin. Bakar[23] that held that the purchaser's claim for late delivery was time barred, is perhaps an indication that the courts may be moving away from the principle of the zealous protection of purchasers.

It is also interesting to note however that the courts in Malaysia have moved away from the position in England that restricts recovery for economic loss, holding that such a restriction would "leave the entire group of subsequent purchasers in this country without relief against errant builders, architects, engineers and related personnel who are found to have erred."[24] This recent decision will have far reaching ramifications for developers, contractors and professionals in the construction industry.

1.2.1   Construction Law

Sources of construction law

Construction law in Malaysia would involve consideration of the English law of contract and tort as received locally, local legislation and case law and the terms of the particular contract. It is therefore in this context fair to state that no obvious specialised area of construction law exists within the Malaysian legal system.

The major statutory source for Malaysian construction law would be the Contracts Act, 1950 (Rev 1974), which is a codification of old case law and modelled on the Contracts Act of India, and the Civil Law Act 1956 (Revised 1972). However, the Contracts Act has to a certain extent contributed to the slow pace of statutory development and limited the continued reception of UK statutes by local enactment.

The Malaysian construction industry also relies heavily on judicial precedents as sources of the law of contract and tort.[25] By reason of the close relationship between English law and local practice, and the greater incidence of construction litigation in England and related jurisdictions, local cases frequently cite precedents from English and other common law jurisdictions such as Canada, Australia, New Zealand and Hong Kong. Where the statutes are pari materia with statutes of other Commonwealth countries, for instance decisions on interpretation of the Contracts Act in India, such foreign decisions are persuasive.

Specialist reports

Although there is specialism in construction law in this country, there are no specialist construction law reports published or brought out in Malaysia. This could be attributable to the fact that construction law is still to a large extent regarded as an extension of contract law and the reliance on UK and other common law jurisdictions for precedents. As such Malaysian construction law cases are usually reported in the major law journals of this country, namely the Malayan Law Journal, the Current Law Journal and the All Malaysia Reports. Practitioners rely heavily on foreign construction journals such as the Construction Law Reports, the Building Law Reports and the Asia Building Reports.

1.2.2.   Construction Contracts

Standard form contracts in Malaysia

The most commonly used construction contracts in Malaysia are
the local standard form contracts, the choice of which is mainly
between the PWD 203 and PWD 203A issued by the Public
Works Department, commonly used for public sector projects and
the PAM69 and PAM69NQ (issued by the "Pertubuhan Akitek
Malaysia"), widely used by the private sector. The other standard
forms, not so commonly used, are the Institute of Engineers
Malaysia's standard form contracts for civil engineering works
and mechanical and engineering works.

Given the rapid expansion of the construction industry in
Malaysia over the past ten years and the large number of major
infrastructure projects, it is not surprising that there has been an
increased use of international standard forms like the FIDIC
contracts, ICE Contracts and the JCT contracts with necessary
modifications to suit local arrangements. This recourse to UK
standard forms is largely due to the lack of development to the
local standard forms and the absence of local provisions suited for
specific procurement systems such as "cost reimbursement"
contracts and turnkey packages. Another factor is the increase in
foreign participation and expertise in the construction industry
and their lack of familiarity with local standard forms.

PAM 69 (Without Quantities) and 69(With Quantities)   Private
Edition

The PAM 69 Private Edition (With Quantities) and the PAM69
Private Edition (Without Quantities) was issued under the
sanction of the Pertubuhan Akitek Malaysia[26] (PAM) and the
Institution of Surveyors Malaysia (ISM) and first printed in 1969
("the PAM Forms"). [27] It is modelled on the form published by
the Joint Contracts Tribunal in the UK 1963 Edition which has
been subjected to extensive criticism. [28] This is, sad to say, the

form that has been inherited by the PAM Forms with all the legal and procedural defects of its English parent. The only amendments made to the PAM Forms are mainly those necessary to bring the document into line with the law of Malaysia.

The difference between the two forms is that one is a lump sum contract with Specifications and the other, a re-measured contract with a Bill of Quantities. The Architect[29] features prominently in the administration and supervision of the works under the contract and has wide but strictly defined powers to issue instructions on behalf of the Employer. He also acts as the independent certifier of payments, quality of work and the contractor's performance. The Quantity Surveyor is also featured, his role limited to measuring and valuing work executed under the Contract, variations and provisional sums included in the Contract Bills.

The provisions for dispute resolution are contained in Clause 34 of the PAM Forms and provides for settlement of disputes under the contract by arbitration. Any dispute or difference which arises during the progress or after the completion or abandonment of the Works "on any matter or thing of whatsoever nature arising thereunder or in connection therewith" is referred to arbitration.

Unless both parties consent in writing, no reference to arbitration is permitted until after practical completion, termination or abandonment of the Works. However, there are circumstances where immediate arbitration is available for matters that need to be settled immediately such as the appointment of another architect or quantity surveyor or the alleged improper withholding of any certificate.

The arbitrator is conferred the power to "open up, review and revise any certificate, opinion, decision, requirement or notice" [30] and is thus entitled to substitute his own opinions and decisions for those of the architect or quantity surveyor. The arbitrator cannot however go behind the final certificate which is made final and conclusive or review the validity of an instruction acted upon by the contractor in reliance on an architect's direction which he is empowered to issue under the Contract.

The proceedings for arbitration are commenced on the service of a written notice by either party to the effect that the dispute or difference be referred to a mutually agreed arbitrator. Failing agreement, or in the absence of a reply after fourteen days from the date of the notice to concur in the appointment of an arbitrator, the person seeking arbitration can apply to the President of PAM to appoint an arbitrator. The award of the arbitrator is final and binding.[31]

The PAM Forms continue to be the most widely used form of contract in the private sector, not because of an unwillingness to recognise its shortcomings but more from a reluctance to do away with a document that most of the present managers and consultants are familiar with. This often results in the standard form being substantially amended by those preparing the contract documentation, often with the amendments running to dozens of revisions and additions. A result of this is uncertainty in the interpretation of these additional clauses both in the administration of the contracts as well as in the settlement of disputes.

Recognising the grave flaws in the current PAM contracts and the uncertainties resulting from contractual amendments to this standard form, PAM has substantially amended the previous PAM 69 forms and drawn up two new standard forms, with and without quantities, due to be issued in early 1998.[32] It is proposed by PAM that the PAM 69 Forms will no longer be printed in the hope that the new forms will be embraced immediately by the industry.

Rather than taking its cue from the evolvement of the JCT 1963 form, the amended form does not in any way resemble the JCT 1980 form and purports instead to be a document that is user-friendly yet appropriate for larger scale projects. In line with this philosophy, the order of the clauses are preserved so that the transition from the PAM 69 forms to the new forms will be painless for those accustomed to the old forms. Whilst the minor amendments relate to the insertion of a table of contents and the addition of definitions, the other changes relate to the introduction of sectional completion, the introduction of a Works Schedule, an ability for the Architect to accept errors in setting out, expanded testing and inspection provisions, the introduction of a Resident Architect and Resident Engineer, expanded provisions on partial possession by the Employer, more detailed procedures for applications for extension of time, and an extended list of relevant events. The role of the Quantity Surveyor has also been limited as he can now only assist the Architect with the valuation and measurement of the work under the instructions of the Architect.

However the more significant amendments are in relation to the deletion of the Employer's rights of set-off and the removal of the clause providing for the conclusive nature of the Architect's final certificate. The arbitration provisions have been amended to make reference to the Arbitration Act "or any statutory modification or re-enactment thereof" in anticipation of further amendments to the Arbitration Act and to provide for the arbitrator to award interest on the whole or part of the amount awarded for any period up to the date of the award and to award interest from the date of the award at such rate as he thinks fit. There is an additional clause which enables the parties to opt for mediation in respect of any matter arising out of or in connection with the carrying out of the Works whether in contract or tort, or as to any direction or instruction or certificate of the Architect or as to contents of or granting or refusal of or reasons for any such direction, instruction or certificate. This is to be conducted under the Mediation Rules of PAM (which are still being drawn up) and before a mediator that will be appointed by the President or Deputy President of PAM.

A form of nominated sub-contract will also be released with the new standard forms. A minor works standard form is being drafted by PAM for use on projects of a smaller scale.

PWD 203 and 203A

The PWD 203 and PWD 203A forms, Rev 10/83 ("the PWD Forms") are the "government" forms of contract, without and with bills of quantities. These forms are issued by the Public Works Department for use on public sector projects. This document, as well as the Singapore PWD form, traces its origins to the RIBA/JCT family although Singapore has now come out with the CIDB 1995 (with or without quantities) Pubic Sector Standard Conditions.

Under the PWD Forms there is a Superintending Officer ("the S.O.") instead of an architect or engineer, who acts as the Government's representative and who is responsible for the overall supervision and direction of the Works. The PWD Forms contain detailed provisions preferring to keep the obligations and duties of the Contractor express rather than implied as in the PAM Forms. The settlement of disputes under these contracts are also by way of arbitration. Any dispute or difference is first referred to the S.O. prior to a reference to arbitration. It is only on failure of the S.O. to give a decision within 45 days of being requested or on dissatisfaction with the S.O.'s decision that the arbitration proceedings kick in. The arbitration is to be held at the Kuala Lumpur Regional Centre for Arbitration. As with the PAM Forms, such reference may not be commenced until after the completion of the Works unless agreed otherwise by the parties concerned. The arbitrator has the jurisdiction to review and revise any certificate, opinion, decision, requisition or notice and to determine all matters in dispute. Any reference is deemed to be a reference to arbitration within the Arbitration Act 1952 and the award of the arbitrator is also final and binding on the parties.

Institute of Engineers Malaysia standard forms

The Institute of Engineers, Malaysia (IEM) has issued two standard form construction contracts, one for civil engineering construction works and one for mechanical and electrical works[33]. The civil engineering form is very similar to the PWD forms except that the Engineer is the consultant responsible for the overall supervision and direction of the Works. Disputes are settled by arbitration but like the PWD Forms, is first referred to the Engineer for a decision. The decision of the Engineer is binding until the completion of the Works. The arbitration is conducted under the IEM's own Rules for Arbitration. As with the PAM and the PWD forms, the arbitrator can review and revise any certificate, opinion, decision, requisition or notice of the Engineer and determine all matters in dispute. The arbitrator's award is of course final and binding.

The IEM mechanical and engineering form is radically different from the civil engineering form, containing design provisions and more detailed provisions on the duties and responsibilities of the Contractor. The Engineer is to exercise his discretion impartially within the terms of the contract. The arbitration provisions are the same as in the civil engineering form. There is also a form of sub-contract for use in conjunction with the conditions of contract for civil engineering works.

2.   CONFLICT MANAGEMENT

The impetus to develop conflict management systems in most jurisdictions have been attributed to cost and delays in litigation and arbitration proceedings. Parties are left dissatisfied with the outcome of the adversarial technique of dispute resolution which does not consider the future commercial relations of the parties. The losing party emerges disillusioned with the findings. Despite this, conflict management techniques have not received the attention they deserve in Malaysia.   However, the present economic crisis ailing the region will probably provide the necessary impetus to changes in this attitude. Furthermore recent contractual disputes in relation to major infrastructure contracts between the employer and contractor will probably act as a catalyst for contracting parties to consider proactive conflict resolution.

2.1   Non Binding

2.1.1   Dispute Review Boards

Dispute Review Boards ("DRB") are yet to be utilised in the construction industry in Malaysia. It requires DRB members who are selected for their knowledge and technical expertise in the type of project to be constructed and for the owner and contractor to have complete confidence in the impartiality of the DRB. As the construction industry in Malaysia is still largely conservative, the managing of disputes is still the domain of the persons administering the contracts such as the project manager, architect or engineer rather than independent outsiders. Other contributing factors are the shortage of professionals in the construction industry due to rapid growth in the economy over the last seven years and scepticism on the impartiality of the DRB members. Pioneering work on the introduction of this concept into the local construction industry it is hoped will be undertaken by the professional organisations themselves or by the formation of a national committee under the auspices of the Construction Industry Development Board.[34]

2.1.2   Dispute Review Advisers

This concept is also not utilised for the same reasons mentioned in 2.1.1 above.

2.1.3   Negotiation

Malaysia is a heterogeneous society. The three major ethnic groups are Malays, Chinese and Indians. However the dominant races involved in commerce are the Malays and the Chinese.

The blending of the two cultures has developed a unique negotiation style where great stress is given to the observance of courtesy, and the saving of face.[35]The preservation of commercial relations are emphasised in any negotiation with the losing party being allowed to emerge having gained something and without a loss of face. There is a heavy reliance on reassurances and trust as opposed to legalistic safeguards[36].

However, as in many other countries, there is a lack of research in Malaysia into negotiation styles and no formal or structured training is provided. It is a skill learnt pragmatically by personal experience and influenced by the blending of Islamic values and Asiatic culture. Nonetheless negotiation is very much a facet of the Malaysian construction industry and an effective means of disputes avoidance.

2.1.4   Quality Matters

The Standards and Industrial Research Institute of Malaysia (SIRIM) launched a scheme for the certification of Quality Systems to provide Certification of Quality Systems to the ISO 9000 series in 1987.

The following excerpt is self-explanatory of the current state of quality matters in Malaysia.[37]

"Up to the end of October 1995, the number of applications for certification has reached nine hundred and sixty five (965)".

The number of certificates awarded to successful applicants totalled six hundred and eighty (680) as at the end of October

1995. The slow growth in the number of certificates awarded could be attributed to a number of reasons. Many Malaysian companies particularly the small and medium sized companies are still practising the traditional concept of quality control, i.e. quality by inspection. Very few are aware or understand the terms "Quality Assurance", "Quality Systems", and "Total Quality Management". It is often found that the managers of these organisations practice the "inspect in quality" concept by carrying out inspection-oriented quality control on incoming materials and components, in process intermediate products and final products.

Other reasons that can be attributed to the failure to achieve the requirements of the standards are as follows :

(i)   Lack of infrastructure necessary to establish and implement the system;

(ii)   Lack of the clear directions, i.e. absence of a quality policy and quality objectives;

(iii)   Lack of the necessary documentation such as procedures, work instructions and records;

(iv)   Lack of clear lines of authority and responsibility; and

(v)   Lack of suitably trained personnel."

Less than 10 of the 40,000 contractors registered with the Construction Industry Development Board have achieved Certification of Quality System to the ISO 9000 series.

In order to encourage contractors to achieve certification, there should be a mandatory requirement for all contractors tendering for government projects to obtain certification, as required in Singapore. This will certainly provide contractors with the impetus towards striving for quality control and reversing the lack of emphasis on quality and safety on construction projects.

A major step in Malaysia towards improvement of the overall quality and regulation of the construction industry is the enactment of the Malaysian Construction Industry Development Board Act 1994 ("the CIDB Act").[38]The principle objectives of the CIDB Act are to regulate the construction industry and to lay the groundwork for participation of the Malaysian construction industry in the international construction market. The CIDB Act established a Board, entrusted with the task of promoting and stimulating the development, improvement and expansion of the construction industry, to undertake research and to advise and make recommendations to the Federal and State Governments.[39] The Board is to also provide consultancy services and promote quality assurance such as encouraging the standardisation and improvement of construction techniques and materials and to upgrade safety standards.

One of the main functions of the Board is to accredit and register contractors, construction workers and construction site supervisors.[40] The CIDB Act makes it mandatory for all contractors whether local or foreign to register with the CIDB before they undertake to execute and complete any construction work in Malaysia. Failure to comply with the registration provisions would subject the Contractor to a fine not exceeding RM50,000.00 or to issuance of a stop work order on his construction activities. The CIDB Act has established the Registration and Levy Unit to register contractors and a pre-requisite for registration is the demonstration of relevant experience, financial, technical and management capability. [41] The Board obtains its financing from the collection of levies from contractors. This levy is imposed upon contractors before the commencement of any construction works failure of which will subject the contractor to a fine.

Another important piece of legislation passed in recent years is the Occupational Safety and Health Act 1994 ("the OSHA")[42] amidst growing concern over the number of deaths and injuries at work sites. The OSHA comes under the regulation of the Department of Occupational Safety and Health, Ministry of Human Resources, with the primary purpose of securing the safety, health and welfare of persons at work. The OSHA provides for the setting up of regulatory bodies, imposes general duties to be observed by specified persons and empowers the Minister to make new sets of regulations and industry codes of practice. Most importantly, the OSHA contains enforcement and investigative procedures which enables officers appointed by the Minister to enter upon premises, conduct inspections, orally examine witnesses and issue improvement or prohibitive notices. A myriad of duties are placed on the employer in respect of his employees and even to persons other than his employees who are exposed to risks to health and safety.

Despite the existence of the OSHA, the number of death related accidents has continued to increase causing the Department of Occupational Safety and Health to step up enforcement measures and to issue stop work orders on project sites. [43]

2.1.5   Partnering

The concept of partnering where two organisations with shared interests come together to combine their resources is, in the context of Malaysia, usually facilitated by the formation of joint venture partnerships. Such collaborations are usually formed for the purpose of providing a service or resources to meet the needs of a third or external party.

The partnering ideal that is advocated in countries like the United States, which requires two organisations (who in some cases worked in the past at arm's length or had a adversarial relationship) embarking on a collaborative process that focuses on co-operative problem solving, avoiding or managing disputes, is not a feature in the Malaysian construction industry.

The Malaysian construction industry has not as yet acknowledged dispute avoidance mechanisms such as partnering as a means of doing business. At the most, parties would convene regular project meetings and ensure that on site relationships are maintained. The main difficulty with introducing such a concept is that it will only work if there is a commitment to place the best interests of the project or the common goal, ahead of the organisation's individual interests. However, the attitude of loyalty and wholehearted devotion cultivated in the corporate culture in Malaysia is a major impediment to the viability of partnering in this country. A further barrier is the recent presence of claims consultants employed by local participants to safeguard their interests on projects by the use of "letter wars". Therefore rather than moving towards the avoidance or minimising of disputes, the industry appears to be moving in the opposite direction.

2.2   Binding

2.2.1   Partnering

For the same reasons outlined above, partnering that is contractual and binding on parties will be difficult to implement or introduce in Malaysia.

3.   DISPUTE RESOLUTION

3.1   Non- Binding

3.1.1   Conciliation

As the use of conciliation and mediation is at its infancy in this country, the terms are often used interchangeably to simply mean a non binding process by which a neutral person facilitates and finds a solution acceptable to both parties. This is consonant with the attitude in most Asian societies where the term conciliation, arbitration and mediation would be considered synonymous with the rendering of assistance by a neutral party to compromise the differences of disputing parties. The term conciliation is however more familiar in Malaysia because it was originally utilised in the context of statute based systems of dispute settlement such as in family law and industrial disputes.[44] Provisions for conciliation are found in both the Law Reform (Marriage and Divorce) Act 1976 and the Industrial Relations Act 1967 which lays stress on direct negotiation and conciliation as modes of settling disputes. It is therefore difficult to discuss the concepts of mediation and conciliation in isolation as the term mediation is often in reality used to mean a conciliation process where the mediator performs an evaluative function and actually imposes a decision on the parties as would a conciliator.[45]

Conciliation is nonetheless being actively promoted in this country by organisations such as the Chartered Institute of Arbitrators, the Malaysian Institute of Arbitrators and the Kuala Lumpur Regional Centre for Arbitration ("the KLRCA").[46]

One of the functions of the KLRCA is to provide other options for settlement of disputes such as negotiation, mediation and conciliation. Parties may refer matters for conciliation by a written request to the Director of the KLRCA and the Rules for Conciliation held under the auspices of the Centre are the UNCITRAL Conciliation Rules. The KLRCA has recently drafted a modified form of the UNCITRAL Conciliation Rules for use at the KLRCA which will be published in January 1998. A standard clause for referring disputes for conciliation under the Rules of the KLRCA is suggested and the modified Rules for Conciliation are intended to provide parties with more flexibility so that the procedures may be adapted according to the form of dispute resolution preferred by the parties i.e. mediation or conciliation.

However the KLRCA has not been frequently used for conciliation and the preference is still to use the KLRCA for arbitration.

3.1.2   Executive Tribunal

Although this form of dispute resolution is not commonly used in Malaysia, it is likely to be the most popular method as the parties involved, are limited to a senior executive from each disputing party and a neutral third party, appointed under terms of the contract. Malaysian parties would be more receptive to this form of dispute resolution as it would in effect be merely elevating the disputes from lower or middle management, for consideration and discussion by senior management. It is also more consensual in nature and may focus on the on-going or future relationships. The neutral facilitator who would probably be named at the outset during negotiation of the contract, would probably be linked or connected in some way to both parties. The senior executives concerned will also have the authority to negotiate to see whether they can resolve the dispute.

This dispute resolution process is already a feature in some major infrastructure contracts in this country and is often limited to resolving specific issues such as extensions of time and valuation of variations. The disadvantages that have been apparent in this process is the fact that senior executives are often out of touch with site administration issues and do not necessarily possess the expertise to appraise and make a determination of the issue at hand. Further their decisions will not necessarily have the element of independent appraisal essential to the fair settlement of the issue.

3.1.3   Mediation

Whilst the courts remains the main forum for resolving disputes, mediation is gaining popularity in Malaysia as an alternative form of dispute resolution. There is no statutory legislation providing for mediation as a method of resolving commercial disputes. However, as mediation is consensual, the lack of legislation or structure on mediation has not impeded those advocating mediation as a means of dispute resolution.

The promotion of mediation within other industries is much more advanced. The Insurance Mediation Bureau ("the IMB") is the first formal alternative dispute resolution centre, initiated by an industry in the country. It was established in 1992 to receive references within certain perimeters in relation to complaints, disputes and claims in relation to policies of insurance.

The Banking Mediation Bureau ("the BMB") was established by the commercial banks, merchant banks and finance companies along the lines of the IMB in early 1997, to principally provide dispute-resolving services for the benefit of their customers. It only began to operate in June 1997.

However, both the IMB and the BMB have been modelled along the Ombudsman schemes that are in operation in the UK and are therefore not strictly a mediatory forum as the Mediator will actually evaluate the dispute and make a binding decision. The workings of both bureaus will thus be discussed in detail in 3.2.6. below.

Many construction contracts are now drafted to compel parties to attempt to settle disputes amicably by mediation with a set of defined rules prior to a referral to arbitration. Although the parties may not be able to resolve their differences at the point of time when the mediation process is being undertaken, they leave the mediation process having explored the disputed issues, needs and settlement options. It serves to narrow the issues to be arbitrated or litigated. There is a growing lobby among professional advisors in Malaysia for reform in legislation to promote the use of mediation procedures. In a paper presented at the 1995 Pacific Rim Advisory Council Conference entitled "Mediation for Dispute Resolution", it was recommended that an effective mediation would be one that imposed mediation as a condition precedent to the commencement of Court or arbitration proceedings and laid down with sufficient certainty the mediation procedure that parties would follow.[47]

PAM's proposed new standard construction forms will contain an option for parties to refer the matter to mediation prior to arbitration in accordance with defined Mediation Rules drafted by PAM. The new PAM Contract therefore does distinguish between the evaluation process involved in conciliation and arbitration with the merely facilitative process involved in mediation. As mediation is purely voluntary and not a condition precedent for reference to arbitration, the effectiveness of such a clause is left to be seen since parties may well opt to proceed straight to arbitration.

However, the future of mediation in this country still lies in the promotion of mediation and the availability of structured mediation training to ensure that mediation skills are acquired. The KLRCA has played a role in this regard by conducting various advance training courses for both conciliation and mediation at the KLRCA. Other organisations actively promoting mediation are the Chartered Institute of Arbitrators and the Malaysian Institute of Arbitrators.

However, there is perhaps a need for a Malaysian Commercial Dispute Centre which will influence the incorporation of non-adversarial facilitative dispute resolution techniques into contracts in Malaysia. Organisations like the KLRCA with a regional focus, could therefore be able to concentrate on evaluative dispute resolution methods.

3.2   Binding

3.2.1   Adjudication

Adjudication in Malaysia is perceived as the function of administrative statutory bodies outside the judicial system, which may sometimes appear to be indistinguishable from courts but which come within the executive branch of the government. Unlike the courts, these adjudication bodies are more informal, not being required to rigidly observe the doctrine of precedent or rules of evidence. However, there is no statutory adjudicatory body regulating the construction industry in Malaysia.

A good example of an existing adjudicatory body in other sectors is in industrial relations. The Industrial Relations Act 1967 set up an Industrial Court to deal with trade disputes. [48] Parties are initially required to attempt mediation and conciliation and it is only upon the failure of these means of settlement that the Industrial Court is resorted to. The regulation of the proceedings is at the discretion of the President and is therefore free from the more restrictive and cumbersome formal procedure of the courts.[49] Awards of the Industrial Court are conclusive and final but questions of law may be referred to the High Court.[50]

On the other hand, the use of adjudication in the Malaysian construction industry whereby either party invokes an adjudication provision to refer a dispute to an independent person for decision, is rare if not non-existent. Whilst there are obvious advantages to the industry in making contractual provision for binding interim decisions by an independent third party, one of the reasons for the lack of progress in this dispute resolution process is the common misconception that the engineer, architect or superintending officer as manager of a project on behalf of the employer is already carrying out such an adjudicatory function. The preliminary reference to the architect or engineer is often made binding on the parties until the matter is referred to arbitration, to commence only on completion of the works concerned.[51]

Whilst this position is less than satisfactory, given that the engineer or architect may himself have been the cause of the dispute, parties often prefer to refer any dispute on such decisions straight to arbitration or litigation rather than to make use of a separate and independent expert adjudicator. Even if an adjudicator or independent third party is used by the parties to resolve disputes on site, it is done informally, there being no contractual provisions regulating the adjudication mechanism or making the decision of the adjudicator binding on the parties. Further whilst the courts in Malaysia are generally respectful of a parties decision to arbitrate on a dispute, it is more likely than not that the courts would hold an adjudicator's decision as having "an ephemeral and subordinate character" [52] and would therefore not treat such a decision on the same footing as an award made under an arbitration agreement despite its binding nature under the contract between the parties.

3.2.2 Arbitration

The preferred mode of alternative dispute resolution in Malaysia is arbitration. It is a well established part of the Malaysian construction industry and most construction agreements will contain an arbitration clause. It is still lauded by the industry as being speedy as compared to the litigation process. There are also other crucial factors in Malaysia which have advanced the cause for arbitration, namely the government's policy to shift the language of courts from English to the National Language, which is Bahasa Malaysia, the fact that foreign lawyers are permitted to represent parties before arbitrations in Malaysia [53] and the non applicability of the Malaysian Evidence Act 1956, enabling arbitrators to deal with evidence in a practical and less legalistic manner. [54] Whilst it was the case that there was no alternative but to have recourse to the private arbitral institutions in the West,[55] the setting up of the KLRCA has advanced the cause for arbitration, both domestic and international, in Malaysia and the region.

Parties in Malaysia are free to resolve by arbitration whatever kinds of disputes they wish and may choose whatever procedure they think most suitable. In most cases, parties agree to incorporate in the arbitration agreement a submission to a set of institutional rules, common examples of which are the UNCITRAL Rules, the ICC Rules, the PAM Rules of Arbitration or the IEM Rules of Arbitration.

The law

The law relating to arbitration in Malaysia is contained in the Arbitration Act 1952 ("the Arbitration Act") which is similar to the English 1950 Arbitration Act. Therefore English case law based on the UK 1950 Arbitration Act is relied on by Malaysian courts. An arbitration agreement is defined under the Arbitration Act to mean " a written agreement to submit present or future differences to arbitration, whether an arbitrator is named therein or not".[56] There is no law or rule prescribing a specific form for an arbitration agreement, the only requirement being that the agreement should be in writing.

Under the Arbitration Act, the High Court is given a wide range of supervisory powers, making no distinction between local and international arbitrations with the exception of arbitration falling within s. 34.[57] The High Court's supervisory jurisdiction covers different aspects of the arbitrations ranging from the power to stay court proceedings,[58] to appoint or set aside the appointment of an arbitrator, umpire or third arbitrator in defined circumstances, to issue subpoena ad testificandum or duces tecum to compel attendance before an arbitrator or umpire, to order security for costs, for taxation in the High Court of the costs of a reference and the umpire's or arbitrator's fees, discovery of documents and interrogatories, giving of evidence by affidavit, securing the amount in dispute, interim injunctions or the appointment of a receiver, the inspection of property, the taking of samples and the investigation of any matter necessary or expedient for the purpose of obtaining full information or evidence, to extend any time limited for commencement of arbitration and to revoke the authority of an arbitrator or umpire and declare an arbitration to cease to have effect where allegations of fraud are raised by one party to the agreement against the other. [59] There are also important provisions which provides the High Court with jurisdiction to deal with questions of law arising in the course of the reference and lays down the right for a case to be stated, for the remission of matters for the reconsideration of the arbitrator or umpire for the setting aside of award and removal of an arbitrator or umpire.[60]

The Arbitration Act also contains several deeming provisions to increase the power of the arbitrator not conferred on the arbitrator by the arbitration agreement. For example, the arbitration agreement is "deemed to contain a provision that the arbitrator or umpire shall have the same power as the High Court to order specific performance of any contract and that the award shall be final and binding upon the parties.[61]

s. 27 of the Arbitration Act provides for arbitration awards to be enforced in the same manner as a judgment or order of the High Court. Where leave is so given, judgment may be entered in terms of the award.

The Malaysian Limitation Act 1953 which prescribes periods of limitation from the date of accrual of the cause of action, applies to arbitrations. Unless specific rules provide otherwise, an arbitration is usually commenced when notice requiring the appointment of an arbitrator is given. A limitation period of six years is prescribed for the enforcement of an award.

Procedures for arbitration

There are no set procedures for the conduct of arbitrations and pre-hearing procedures are at the discretion of the arbitrator unless otherwise provided. Subject to specific rules applicable to the arbitration, the procedure for the hearing is determined at the first or any subsequent preliminary meeting. The fee is sometimes fixed in accordance with the rules under which the arbitration is conducted for instance the PAM Rules for Arbitration or sometimes by reference to fees fixed by organisations such as the Chartered Institute of Arbitrators. However in the absence of a specific agreement to the contrary, an arbitrator is entitled to fix his own fee provided it is reasonable.

Discovery and inspection of documents by list or affidavit and interrogatories are normally part of the process. All documents including internal confidential memoranda are expected to be disclosed with privileged documents being confined to solicitor-client communications and documents created for the purpose of the arbitration. The proceedings are more often than not adversarial in nature with the claimant first presenting its case and the defendant responding with its case after the conclusion of the claimant's case. Witnesses are also examined, cross-examined and re-examined orally but there is a growing practice towards obtaining written depositions to replace or reduce the length of an examination-in chief. As stated previously, the Evidence Act 1950 states that it shall not apply "to proceedings before an arbitrator".[62] However, the arbitrator will rarely completely disregard the rules of evidence and will be guided by the Evidence Act in the conduct of proceedings.

Unless otherwise stated in the arbitration agreement, an arbitrator or umpire may make an award at any time except in the case of a remitted award which must be made within three months after the order (unless the order otherwise states). The High Court or a judge may enlarge any time limit for making the award.[63] In the absence of an expressed contrary intention, an arbitration agreement is deemed to include a provision that the costs of the reference and award will be in the discretion of the arbitrator or umpire. The costs are taxable in the High Court. Arbitral awards are not published as they are kept confidential between the parties.

The Kuala Lumpur Regional Centre for Arbitration

The KLRCA was established by the Asian-African Legal Consultative Committee (AALCC) [64] in 1978 to provide a venue for arbitration, both domestic and international, within the Asian region. The KLRCA is a non-profit and independent international institution. This has been recognised by the Government of Malaysia by gazetting it as an international organisation under the International Organisations (Privileges and Immunities) Act 1992. The KLRCA may also hold proceedings under the Convention on the Settlement of Investment Disputes, 1965 in Kuala Lumpur as a result of an agreement concluded between the Centre for the Settlement of Investment Disputes ("the ICSID"), the AALCC and the KLRCA on February 5, 1979.

One of the principal functions of the KLRCA is the provision of facilities for arbitration under the KLRCA Rules which are the UNCITRAL Rules 1976 as modified by the AALCC.[65] These facilities may be used by parties who may request them, whether governments, individuals or bodies corporate, provided the dispute is of an international character, i.e., the parties belong to or are resident in two different jurisdictions or the dispute involves international commercial interests.[66] Although set up to deal with international arbitrations as opposed to domestic arbitrations, the KLRCA makes itself available for domestic arbitrations to be conducted under its own Rules. The KLRCA may also hold proceedings under the ICSID Rules which has made it possible for parties to an investment dispute from the Asian- African region to settle their disputes in Cairo or Kuala Lumpur instead of going to Washington. The KLRCA has also concluded an agreement with the Tokyo Maritime Arbitration

Commission, the only specialised institution in the Asian- African region in shipping matters[67] and mutual co-operation agreements with a number of national arbitral institutions for the provision of facilities for the conduct of arbitral proceedings at the seat of one institution but under the auspices of another, particularly for assistance in the enforcement of awards.

The number of international arbitrations held under the KLRCA's Rules average around seven to ten a year. In addition, the Director of the KLRCA has been designated as the appointing authority by the Permanent Court of Arbitration at the Hague in some international arbitration conducted under the UNCITRAL Rules. Six international arbitrations conducted under the rules of other institutions have been held at the Centre this year including two ICC and one ICSID arbitration. This year, seven new international arbitrations have been referred to the KLRCA thus far. The number of domestic arbitrations average ten a year and are increasing with nine new referrals so far this year. The KLRCA has confirmed that the arbitrations have been largely been in relation to construction disputes.[68]

The initiation of the proceedings at the KLRCA is commenced by a written request to the Director of the KLRCA, stating that the parties have agreed to refer their disputes and differences for settlement by arbitration under the auspices and rules of the KLRCA. Unless the parties have agreed otherwise or the appointing authority refuses to act or fails to appoint an arbitrator, the KLRCA is the appointing authority for the purpose of the UNCITRAL Arbitration Rules.[69] The arbitration may be held either at the seat of the KLRCA at Kuala Lumpur or at any place chosen by the parties. Where parties have agreed to arbitrate under the KLRCA's Rules, administrative services are provided and the Director of the KLRCA arranges for facilities and assistance in the conduct of the proceedings including suitable accommodation for sittings of the arbitral tribunal, secretarial assistance and interpretation facilities. If arbitrations are not held under the KLRCA's Rules, the KLRCA will also provide administrative support to such arbitrations, extending to international arbitral institutions with which the KLRCA has concluded co-operation agreements.

A recent amendment to the KLRCA Rules is a provision imposing a time limit of six months after close of pleadings on arbitrators for rendering their final award (r.6(1)) with a provision for extension of time. The Director of the KLRCA will prepare an estimate of the costs of arbitration. Deposits by the parties are used towards the disbursements for the costs of the arbitration and any unexpended balance is returned to the parties.[70]

The determination of the arbitrators' fees is left to the arbitral tribunal's discretion, with the Director of the KLRCA fulfilling the role of a co-ordinator.[71] The fees are settled in accordance with a newly introduced Fee Schedule. It is not fully binding on the arbitrators but serves to be persuasive in an attempt to keep the fees manageable.[72] The KLRCA has just drafted some amendments to the KLRCA's Rules. Most of the amendments are procedural and minor but the significant amendments relate to provisions on confidentiality and an exclusion of the liability of arbitrators similar to that enacted under the UK Arbitration Act 1996.[73]

S.34 of the Arbitration Act- ouster of court's jurisdiction

With the intention of conferring a degree of independence on international arbitrations, the Arbitration Act was amended in 1980 to provide that arbitrations held under ICSID or under the Rules of the KLRCA are outside the ambit of the Arbitration Act.[74]

Section 34 is a unique provision as it goes further than Article 5 of the UNCITRAL Model Law and has the effect of excluding the jurisdiction of the High Court under the Arbitration Act or other written laws, except for the purpose of enforcing an award, in relation to arbitrations conducted in Malaysia under the

Convention on the Settlement of Investment Disputes between States and Nationals of Other States or under the UNCITRAL Arbitration Rules 1976 and the Rules of the Regional Centre for Arbitration at Kuala Lumpur. It is often erroneously thought that s.34 also applies to a third category of arbitration covering the UNCITRAL Arbitration Rules. This would extend the immunity to ad hoc arbitrations under the UNCITRAL Rules and therefore to domestic arbitrations which was never the intention of s.34. Domestic arbitrations would not be converted into an international arbitration that enjoyed the immunities of s 34 by virtue of being conducted under the KLRCA Rules. [75]

The courts have upheld the spirit of this amendment in endorsing the exclusion of the supervisory jurisdiction of the High Court in relation to international arbitrations. [76] It has also been held that the exclusion of other written law must remove the court's jurisdiction completely unless the court has a residual jurisdiction inherent in its character as a court of justice, not conferred by any written law. [77]

The amendment to Arbitration Act therefore confers true and complete finality upon an award made under the KLRCA Rules and eradicates the use of the case stated procedure or of attempts to refer questions of law to the High Court. It would thus appear that a claimant who has not succeeded because the arbitrator has made a grievous and manifestly indisputable error of substance would have no recourse or remedy.

The enforcement of KLRCA awards are undertaken under s.34(2) which allows for the enforcement of such awards in the High Court in accordance with the provisions of the New York Convention. It is only those provisions which relate to the enforcement of awards in the New York Convention that are incorporated into s.34(2).

The Attorney General's chambers are currently drafting amendments to the Arbitration Act in isolation from the views of people in the industry. It is uncertain what direction the amendments will take but it is hoped by many that it would be towards adopting the UNCITRAL Model Law.

3.2.3   Expert Determination

For the same reasons outlined in 3.2.1 above, the use of expert determination to resolve certain disputes under a contract which already includes an arbitration clause is not widely used. The certifiers under building and engineering contract i.e. the architects and engineers are considered the "expert decision makers" who are entrusted with managerial decision making powers, to assess the issues concerned and resolve the dispute.

3.2.4   Litigation

The formal court process remains the main forum for resolving disputes in civil and commercial matters in Malaysia. This would apply even to the construction industry despite the frequent use of arbitration clauses in construction agreements in the past ten years. Parties to a dispute still prefer to take the more adversarial position, afforded by the litigation process despite the obvious advantages of speed and efficiency in arbitration in comparison to the court process in Malaysia.

The litigation of building and construction disputes is treated similarly to other civil litigation. Unlike the UK which has a separate division of the High Court i.e. the Official Referees Court, there is no such specialist court or judges for the hearing of construction cases. Actions in construction disputes are usually brought in the High Court because of the value of the claims and inherent complexity of most construction disputes. A construction dispute in the High Court will be heard in the civil courts and if conducted in Kuala Lumpur, in the Civil Division.[78]

However, there is provision in the Courts of Judicature Act 1964, recently invoked by some judges, which enables the judge, to refer any question other than criminal proceedings by the Public Prosecutor, for inquiry or report to a special referee or arbitrator who shall be deemed to be an officer of the High Court. This is perhaps a substitute for the Official Referees in the UK as the same considerations for commencing a matter in that court applies in the decision to refer the matter to a special referee or arbitrator i.e. that all the parties are not under disability consent, that the cause or matter requires prolonged examination of documents or any scientific or local investigation which cannot in the opinion of the High Court, conveniently be conducted by the Court through its ordinary offices or if the question in dispute consists wholly or in part of matters of account.

As there are no detailed rules guiding this reference, there is some uncertainly as to how such references are to be conducted. The authority of the arbitrator or special referee, the rules of conduct of the reference and the remuneration of the special referee or arbitrator is determined by the High Court. These provisions in effect enable the High Court to impose arbitration proceedings on parties that never contracted to refer the dispute or difference to arbitration with the same powers conferred on the High Court by the Arbitration Act. The award of any special referee or arbitrator in any such reference can therefore be set aside by the High Court in the same way as under the Arbitration Act.

Ousting of the jurisdiction of courts

Parties in Malaysia may not oust the jurisdiction of the courts either contractually or otherwise so as to prevent another party from invoking the jurisdiction of the court. A recent Court of Appeal decision in Malaysia[79] explained that a prior agreement between contracting parties to refer their disputes to arbitration did not operate to bar either of them from instituting proceedings in the ordinary courts or preclude the court from entertaining a suit filed in breach of the contract to arbitrate. Further subject to the provisions of the Arbitration Act, an arbitration clause which completely ousts the jurisdiction of the court is bad in law and will not be enforced.[80]

However Scott v Avery clauses which make an arbitration award a condition precedent to liability under a contract or to any right of action are not considered ouster clauses by the courts in Malaysia.[81] However despite the presence of such a clause, s.26(4) of Arbitration Act gives the High Court a discretionary power to order that such a provision, or even the whole agreement, shall not have effect as regards any particular dispute. This power does not enure to statutory arbitrations.

s. 6 of the Arbitration Act 1952, enables the High Court to grant a stay of all proceedings before it in any action, initiated by a party seeking to have a matter resolved by arbitration. The general tendency is for the Malaysian courts to uphold arbitration agreements,[82] the approach being that those who make a contract to arbitrate their disputes should be held to their bargain. There must be an agreement containing an arbitration clause, the applicant must not have taken any other steps in the proceedings, there must be sufficient reason why the matter should not be referred in accordance with the arbitration agreement and the applicant is still ready and willing to do all things necessary to the proper conduct of the arbitration. The onus of showing why the dispute should not be referred to arbitration is on the party opposing the stay application.[83] The courts will not meticulously examine the case on the merits or go into the question of the bona fides of disputes,[84] the only consideration being whether the dispute or difference comes within the submission to arbitration.

By s. 25(2) of the Arbitration Act, the court will refuse a stay of proceedings where the matter in dispute involves a charge of fraud against one of the parties to the arbitration (subject to there being a bona fide allegation of fraud).[85] In a finding of fraud, the court may order that the arbitration agreement ceases to have effect or to give leave to revoke the authority of any arbitrator or umpire.[86] Further the courts will also refuse to grant a stay if the dispute is outside the scope of the arbitration agreement[87] and if the claim under which arbitration is being is admitted is not in dispute. [88] The courts will not however uphold the use of this statutory provision if it is being used as "an engine of delay or to frustrate the resolution of the dispute".[89]

Section 6 specifically refers to the High Court as having the jurisdiction and power to consider applications for stay of the court proceedings. Order 69 Rule 3 of the Rules of the High Court 1980 ("the RHC 1980") empowers the Registrar to exercise the jurisdiction under the Arbitration Act, to be exercised concurrently with the High Court. An appeal against the order of the Senior Assistant Registrar lies to the Judge-in Chambers under Order 56 of the RHC 1980 which will virtually be a rehearing of the application for stay despite the fact that the Senior Assistant Registrar has as much discretion as the High Court itself.[90] However, when the matter goes to further appeal to the Court of Appeal, the function of the Court of Appeal is only one of review. It must therefore be demonstrated that the High Court has committed one or more of those errors which would entitle appellate interference i.e. if it is clear that the court has misdirected itself in law or where the order is not justifiable on the facts of the particular case.

Choice of law

Malaysian courts would uphold the choice of law upon which the parties have contracted on. An agreement to submit disputes to arbitration or to courts of a particular country will not however constitute an express choice of law but may be a relevant factor in determining what the proper law is. In Globus Shipping and Trading Co (Pte) Ltd v Taiping Textiles Bhd,[91] there was a contract to ship cotton from Karachi to Penang under a bill of lading which provided that the dispute would be decided in the country where the carrier has his principal place of business and the law of that country would apply. The appellants-defendants place of business was in Singapore and they objected to the jurisdiction of the court of Penang. The Federal Court chose to treat the term as a jurisdictional clause rather than the proper law of contract. It stated that where the cause of action arises within the court's jurisdiction, the court has a discretion whether or not to adjudicate on the claim even in circumstances where the parties have agreed to refer such dispute to a foreign court. [92]

It is therefore necessary for parties to use the clear required terminology which would suggest that the contract is governed by the chosen legal system.

Procedure and practice in the courts

An action is commenced in the High Court by writ, or if a particular form of relief is being sought, by originating summons[93] (or summons in a subordinate court action). In the case where one of the parties is not resident in Malaysia, the service of process is governed by O.11 of the RHC 1980 which following English practice, and gives authority to the court to assume jurisdiction over overseas defendants in limited circumstances and defines the principal cases in which service of notice of a writ out of jurisdiction is permissible. [94] A writ which is to be issued out of the jurisdiction must be issued with the leave of the court.[95] Such an application is made by way of ex parte summons supported by an affidavit setting out the grounds of application, the deponent's belief that the plaintiff has a good cause of action, the country in which the defendant is to be found and that the Malaysian court is the most convenient forum for the action to proceed.[96]

Pleadings in construction litigation are no different from those in other types of cases. The plaintiff commences his action by serving a Statement of Claim and the defendant files a Defence to which may be added a counterclaim. The plaintiff may serve a Reply and must serve a Defence to Counterclaim if a counterclaim has been made which he disputes. The Scott Schedule is widely used in construction arbitration proceedings in this country and may also be used in Court, pursuant to Order 18 r.12, RHC 1980, as a form of provision of particulars to pleadings. There is no set form of Scott Schedule to be used and the party raising the items will usually be responsible for preparing a final completed revision of the Scott Schedule for use during the hearing of the action.

Many construction disputes are disposed of at an interlocutory stage by way of a summary judgment application under Order 14 of the RHC, 1980. The application is made by summons in chambers and supported by an affidavit stating that there is no bona fide defence or issue to be tried.

If the matter proceeds to trial, the plaintiff must, within one month after pleadings are deemed to be closed take out a summons so that the judge may give directions to secure the just, expeditious and economical disposal of the case.[97] The judge will give various directions with regard to the state of pleadings, the giving of evidence and directions for discovery and inspection of documents. The judge will fix a period within which the plaintiff is to set the matter down for trial, an estimate of the length of the trial and the number of witnesses. No notice of trial is required but the party setting the action down for trial must notify the other parties that he has done so within 24 hours of setting it down. A set of documents consisting of the writ, the pleadings and all orders on the summon for directions must be lodged with the Registrar.

At the setting down for trial, the matter will be assigned to a fixing list to enable the courts to fix a date for the trial of the action.[98] Many of the judges have recently adopted the practice of "case management" as a means of ensuring that the parties have done the necessary preparations for trial. The Judge will ensure that the parties have prepared bundles of documents, bundles of agreed and non-agreed documents and statements of agreed and non-agreed facts. No hearing date will be fixed until the judge is satisfied that the parties are fully ready to proceed with the trial.

Some judges have even gone further in narrowing the issues to be disputed and on occasions, even as far as persuading the parties to settle the matter.

Evidence

The general principle in the Malaysian courts is that evidence of witnesses must be proven by the examination of witnesses in open court so as not to deny the right of any party to cross-examine the opposite party upon his evidence. This is subject to any other provisions of the RHC 1980 and to any other written law. [99]

The procedures for the provision of evidence is governed principally by Order 38 of the RHC 1980 and the Evidence Act 1950 which applies to all judicial proceedings in or before any court but not to affidavits presented to any court or officer. Examinations in chief and cross examinations are therefore conducted orally and the Evidence Act 1950 governs the law on admissibility of oral and documentary evidence in court. In line with recent proposals for revamping of the rules of the Subordinate Court, High Court and Appellate Courts as well as amendments to the Evidence Act 1950, the courts have already begun to direct that signed statements of witnesses be drawn up to stand as evidence in chief. These will not be filed but will be read out by the witness in court instead of an oral examination in chief. The witness will then be cross examined by opposing counsel.

The provisions on examination of witnesses is contained in the Evidence Act (ss 135-166) and provides that in the absence of any procedural law regulating the order in which witnesses are produced and examined, that the court will have a discretion in the matter. Under s.165 of the Evidence Act, apart from the right to put any questions to or order production of documents by any witness called by the parties, the judge may call any other witness for such purposes but neither party can cross examine the witness except with the leave of the court.

An exception to the rule that evidence in trial must be oral and direct, is that the court can order that the affidavit of any witnesses be read at the trial if in the circumstances it thinks it reasonable to do so such as when the witness is abroad and therefore not available for the trial or where the evidence is uncontested.[100] However, subject to any terms or subsequent order of the Court, the deponent is not subject to cross examination but the opposite party may make use of any admissions in it against the deponent. However, if the matter is by originating summons, origination motion or petition, evidence may also be given by affidavit unless there are any provisions to the contrary and the court may on the application of any party, order the attendance for cross examination of the person making any such affidavit. In the case of absence of the deponent, the affidavit may not be used as evidence without the court's leave.

The court also has the power to order, usually at the summons for direction stage, that evidence of any particular fact shall be given at the trial in any manner specified by the judge for instance by statement on oath of information or belief or by the production of documents or entries in books.[101] The court may also either at or before the trial of any action order that the number of medical or other expert witnesses be limited.

Without prejudice negotiations

The well established rule that statements made by opposing parties (or through their solicitors or agents) to each other, in the course of settlement of negotiations on the express or implied understanding that they are not to be disclosed i.e. "without prejudice", are privileged, is given expression in Malaysia in the form of s.23 of the Evidence Act 1950. Section 23 provides that "in civil cases no admission is relevant if it is made either upon an express condition that evidence of it is not given, or under circumstances from which the court can infer that the parties agreed together that evidence of it should not be given."

As the Malaysian position is codified, the common law rules in respect of "without prejudice" documents are applicable only to the extent that they are not inconsistent with the provisions of the Evidence Act. [102] In the Yeo Hiap Seng v Australian Food Corp Pte. Ltd 7 Anor[103] the court considered if a second defendant, not a party to negotiations between the plaintiff and first defendant, could rely on privilege. It held that the privilege could not be claimed by a party who takes no part in the negotiations, either personally or through an agent. Therefore it appears that the rule of privilege which applies to communications in the course of settlements is not as absolute as a strict literal reading of s.23 of the Evidence Act would suggest.

Further, the privilege may be waived with the consent of both parties or waived by the person entitled to the privilege either expressly or by allowing evidence to be given of matters in respect of which privilege is claimed. A party cannot unilaterally waive the privilege only to claim it later when his opponent makes reference to these communications.

Enforcement of judgments or awards

Foreign judgments or awards may be enforced either at common law or by statute in Malaysia. Statutory enforcement of a foreign judgment is provided under the Reciprocal Enforcement of Judgments Act 1958, which is modelled upon the United Kingdom Foreign Judgments (Reciprocal Enforcement) Act, 1933. It provides for the enforcement of judgments of superior courts of countries which have reciprocal provisions.[104] The definition of "judgment" in s.2 includes Commonwealth arbitral awards. Foreign judgments may only be registered if it is from a superior court, is for a fixed sum or money, is final and conclusive,[105] and was obtained after the country or territory had been added to the schedule.[106]

The Convention on the Recognition and Enforcement of Foreign Arbitral Awards Act 1985 ("the New York Convention Act") was brought into effect to implement the New York Convention as part of the law of Malaysia. This New York Convention Act follows closely the provisions of the New York Convention and specifically provides in s.3 that Convention awards shall be enforceable in the same manner as the award of an arbitrator is enforceable by virtue of s.27 of the Arbitration Act.[107] Objection to enforcement of New York Conventions may be taken on the grounds set out in s. 5 of the New York Convention Act (or art 5 of the New York Convention). A Convention award which satisfies the foregoing criteria for enforceability is treated as binding on the parties to the award, so that they may rely on it in legal proceedings in Malaysia, whether by way of defence, set-off or otherwise.

The Settlement of Investment Disputes Act 1966 Act which enacts the 1965 Washington Convention on the Settlement of Investment Disputes between States and Nationals of Other States provides that an award made under the Convention is enforceable as if it is a decree judgment or order of the Courts. Where the provisions of these Acts are inapplicable, the successful party who has a foreign arbitral award in its favour may file a suit on the award or on a judgment based upon the award.

A judgment creditor seeking to enforce a foreign judgment in Malaysia at common law cannot do so by direct execution of the judgment. An action must be brought on the foreign judgment. However, a Malaysian court will enforce a foreign judgment only if the foreign court had jurisdiction by Malaysian conflict of law rules[108] and the judgment is final and conclusive and if in personam, is for a definite sum of money that is not a tax, fine or penalty.

Enforcement of foreign judgments may be dealt with expeditiously by an action for summary judgment under Order 14 unless it can be established that the foreign court had no jurisdiction, the judgment was obtained by fraud or is contrary to public policy or the proceedings in which the judgment was obtained was contrary to natural justice. [109]

Court's review of awards

The ability of the courts to review the arbitration awards is contained within the provisions of s. 23(1) and 24(2) of the Arbitration Act. The High Court is given a wide discretion to either remit a matter for the reconsideration of the arbitrator or wholly or in part, set aside the award. The courts in choosing to exercise this discretion will often prefer to remit the award to the arbitration for reconsideration on specific issues rather than wholly throwing away an order reached through lengthy and often expensive arbitration. [110] An application to remit an award or set it aside is made by originating motion to a single judge. [111] No order for remission of the award of the arbitrator will be made if the balance of convenience is against the court giving such order. [112] Examples of the grounds that would apply for remitting an award are that the award is bad on the face of it, [113] misconduct on the part of the arbitrator, the existence of an admitted mistake and the discovery of additional evidence after the award is made.

The court may set aside an award where an arbitrator or umpire has misconducted himself or an arbitration award has been improperly procured. The court will not conduct a rehearing when considering whether an award should be set aside and will not readily set aside awards unless it perceives that there has been "something radically wrong and vacuous in the proceedings". [114] Malaysian courts have for instance held misconduct to be failure to perform essential duties cast on an arbitrator not consonant with the general principles of equity as misconduct, [115] failure to hear the evidence of one party in the absence of the other [116] and failure to state a case at the request of the parties. [117] Apart from the express provisions of the Arbitration Act, the court also has a common law jurisdiction to set aside an arbitrator's award for an error of law appearing on the fact of it. However, the court is not entitled to any inference as to the finding of fact by the arbitrator and must take the award at its face value. Awards have also been set aside for its uncertainty, in that the arbitrator has failed to decide on all matters referred to him, or where he has taken a fundamentally wrong view of his function.

A case stated procedure to the High Court is available under s.22 of the Arbitration Act, in respect of any question of law arising during the course of the reference or an award or any part of it. This extends to an interim award or a question of law arising during the course of the proceedings notwithstanding that the proceedings are still pending. There are three conditions that have to be fulfilled; the point of law must be real and substantial, the point of law should be capable of being accurately stated as a point of law and the point of law should be of such importance that the resolution of it is necessary for the proper determination of the case. [118] An arbitrator may with the consent of the parties state a case to the High Court on questions of law even before proceeding with the arbitration.

The decision of the High Court on a case stated issue will be deemed to be a judgment of the High Court within s.67 of the Courts of Judicature Act 1964. No appeal lies from the decision of the High Court without the leave of the Court of Appeal or Federal Court.

3.2.5   Negotiation

Negotiation is still very much a consensual process and parties in Malaysia will not willingly relinquish the conduct of negotiations to a structured or independent assisted negotiation whereby a neutral third party chairs the negotiation and assists where there is a break down of negotiations. This is principally because the "Asian" approach to doing business is governed by an overriding preference for keeping negotiations "in-house". Often when disputes are bogged down at the operative level, it is merely taken to a higher level of management for resolution. The ability to bind parties to negotiate prior to reference to arbitration would at the most be a contractual provision which provides for a mandatory attempt to resolve the matter amicably prior to commencement to arbitration.

3.2.6   Ombudsman

There is no investigative complaints bureau in Malaysia in respect of the construction industry. The Construction Industry Development Board is the only body at present that is attempting to regulate the construction industry and to promote standards within the industry. However, their role does not extend to investigating complaints from the public and other members of the industry and is simply an arm of the Government, empowered to carry out its activities by the Minister of Works and is therefore not an independent forum.

However, as stated in 3.1.3 above, other industries namely the banking and insurance have established a forum for the handling of complaints. Both the Insurance and Banking Mediation Bureaus were set up on the basis of the Ombudsman schemes in the United Kingdom. The Insurance Mediation Bureau whose main objective is to offer consumer protection with regard to fair dealing with policy holders, is authorised to receive references in relation to complaints, disputes and claims not exceeding RM100,000.00. Such an award will be binding on the member company (but not the claimant) except for an award in respect of a claim exceeding RM100,000.00 in which case such an award will only constitute a recommendation as an equitable solution to the dispute. The scheme does not however extend to cover any reference in relation to third party claims. The jurisdiction of the Bureau was recently extended to claims relating to life insurance.

The Mediator is appointed by the Council of the Bureau[119] and although the designation of the adjudicator is termed a Mediator, the position of the Mediator is in reality akin to an Ombudsman who conducts an independent fact finding investigation with the goal of correcting abuses. The Mediator is required to act in conformity with any applicable rule of law or judicial authority with general principles of good insurance, investment or marketing practice. His decision would be reached on the basis of the facts submitted, the company's and policy holder's views together with the evidence obtained from personal interviews and expert opinion.

The Mediator cannot however entertain any reference unless it has been considered by the senior management of the member companies or if the complainant has instituted proceedings in a court or made a reference to arbitration. There is no appeal procedure within the Bureau but the policy holder may institute court proceedings against the company or refer the claim to arbitration.

The 1996 Annual Report of the IMB observes that the rate of increase in the number of references to the IMB has not been encouraging largely due to the lack of public awareness of the role and function of the IMB. The Central Bank has however issued a directive to all insurance companies to adhere to the guidelines issued in 1995 which requires insurance companies to inform claimants in writing of the procedures for reference of disputes to the IMB. For the year 1996, the total number of references and complaints received was 228 of which 76 references were outside the terms of reference of the IMB. Of the 117 cases resolved by the IMB in 1996, the Mediator confirmed 63 insurance companies decisions and revised the other 54 decisions in favour of the policyholders.[120]

The Banking Mediation Bureau. ("the BMB").

The BMB provides services with regard to disputes with banks and finance companies over a claim involving monetary loss arising out of banking services provided by the bank or finance company, namely, the charging of excessive fees, interest and penalties, misleading advertisement, Automatic Teller Machine withdrawals, unauthorised use of credit cards and unfair practice of pursuing actions against a guarantor. It works along the same lines as the IMB except that the Mediator can only make an award up to RM25,000.00. The award is binding on the bank/finance company but the customer is entitled to take the matter further.

4.   THE ZEITGEIST

Proposals for reform of the administration of justice in Malaysia in the past two years have been focused primarily on expediting the litigation process. Hence, the judiciary has directed all its efforts towards ensuring that the backlog of civil cases are reduced, [121] concentrating mainly on minimising the postponement of cases. The Chief Justice of Malaysia implemented a Quality Policy, Slogan and Client's Charter in April 1997 which commits the members of the judiciary to administer justice efficiently and to conduct cases judiciously within a reasonable time frame. A punch-in card system for Judges was even introduced in February 1996 requiring Judges to be "punctual" on pain of disciplinary action. The emphasis on speedy litigation has not always manifested itself in the best light as there are increasing complaints of cases being struck out for minor procedural irregularities.

The High Court Rules Committee has also been working on the amendments to the High Court and Supreme Court Rules. The amendments include the introduction of American style pre-trial conferences which is intended to enable the lawyers of both parties to exchange documents which otherwise would not be allowed to be produced during the trial. As discussed earlier, this is already being practised by some judges in the form of "case management". Evidence may be put in affidavit form and witnesses need only be present in court for cross-examination. There is also proposals to introduce hearing fees as in Singapore where only the first day of trial is free, a fee being charged for subsequent days to prevent prolonged cases. Another proposal is to reduce the life of the writ of summons from 12 to 6 months and to restrict renewal of the writ to two or three times.

However, there is at present no move towards the formal introduction of alternative dispute resolution ("ADR") within the court process or the deregulation of the legal system. Although judges are able to initiate ADR programs by the selection of arbitrators, this role should perhaps be extended to mediators and other third party neutrals. Judges should be free to decide whether ADR is appropriate and to order parties to engage in ADR. It is also possible for judges to use mediation during the course of litigation and to use ADR processes within case management time frames. Education therefore plays an important role to enable judges to have a comprehensive understanding of ADR, its strengths and weaknesses.

A step in the right direction has however been taken by the judiciary in its recognition of the advantages of case management at the summons for directions stage, to maintain control over the litigation process. However, the concept of case management needs to be expanded further to capture the whole litigation process i.e. from the time of commencement of the suit until the resolution of the dispute. It is only in this situation that the aims of case management as expounded by Lord Woolf in his interim report to the Lord Chancellor on the civil justice system in the UK, can be achieved. These are the early settlement of cases, diversion of cases to alternative methods for resolution, encouraging the spirit of co-operation between parties, the identification and reduction of issues as the basis for case preparation and in the case of non-settlement, progressing cases to trial as speedily and at as little cost as is appropriate.

As arbitration will probably remain the most frequently used form of ADR, a large part of the construction industry's efforts should be concentrated on improving the existing arbitration process. Concepts introduced in other countries, such as "fast track" arbitrations for smaller claims and "lean arbitrations" aimed at reducing the cost of arbitration by the identification of the issues to be arbitrated, could be introduced in Malaysia.

1   Economic Report 1997/98 dated 17 October 1997 issued by the Ministry of Finance.

2   About 792,853 square metre of office space is expected to be added in 1997 to the existing supply of 3.4 million sq.m in 1996 in the Klang Valley with close to 60% of the new supply in the Golden Triangle and Central Business District in Kuala Lumpur.

3   As a result of these developments, the civil engineering sub-sector accounted for the largest share (46%) of value added in the construction sector in 1997.

4   F.A.Trindale & H.P.Lee, *The Constitution of Malaysia , Further Perspectives and Developments, Essays in Honour of Tun Mohamed Suffian*, p.1.

5   These are Johore, Negeri Sembilan, Selangor, Perak, Kedah, Perlis, Trengganu, Kelantan and Pahang.

6   These being Malacca, Penang, Sabah and Sarawak.

7   Art. 75 of the Federal Constitution.

8   However the Syariah Courts of the Federal Territory of Kuala Lumpur and Labuan fall within federal jurisdiction.

9   See S.3 of the Interpretation Act 1967.

10   Sections 90 and 65, Subordinate Courts Act 1948 and s. 23 of the Courts of Judicature Act, 1964. Refer to s. 24 of the Courts of Judicature Act for the specific civil jurisdiction of the High Court and the Schedule to s. 25(2) for the additional powers of the High Court.

11   Section 68(1) Courts of Judicature Act 1964.

12   See the new Part IV to the Courts of Judicature Act, 1964.

13   Please refer to 3.2.1 for further discussion.

14   However this Act only applies to Peninsula Malaysia as the Government has not gazetted the provisions to apply to Sabah and Sarawak.

15   The requirements are prescribed under s.11 of the LPA.

16   There are close to 8000 lawyers admitted as advocates and solicitors of the High Court of Malaya as of 1997.

17   S.28A LPA 1976.

18   In Sabah,   the present governing statute is the Advocates
      Ordinance, Cap 2 - Reprinted 1966 and in Sarawak , the
      Advocates Ordinance 1953, Cap 113 - Reprint 1966.

19   The Hire Purchase Act 1967 specifies by legislation the strict
      required terms and content of sale and purchase agreements,
      contravention of which will result in the hire purchase agreement
      being declared void and the owner guilty of an offence.

20   Gazette Notification was on the 26th September 1996 but the Act
      is as yet not in force.

21   (1982) 2 Malayan Law Journal 31.

22   See MK Retnam Holdings Sdn Bhd v Bhagat Singh (1985) 2
      Malayan Law Journal 212; Daiman Development Sdn Bhd v
      Matthew Lui Chin Teck (1978) 2 MLJ 239; Khau Daw Yau v
      Kin Nam Realty Development Sdn Bhd (1983) 1 Malayan Law
      Journal 335 and Beca (Malaysia) Sdn Bhd v Tan Choon Kuang
      (1986) 1 Malayan Law Journal 390.

23   (1996) 2 Malayan Law Journal p. 188

24   Per James Foong J. in Abdul Hamid Abdul Rashid, Dr & Anor v
      Jurusan Malaysia Consultants (sued as a firm) & 4 Ors (1997) 1
      All Malaysia Reports at p. 659. However note that s.95 of the
      Malaysian Street, Drainage and Building Act 1977 prohibits local
      authorities from being sued for negligence in granting approvals
      or inspecting building works.

25   s.3 of the Malaysian Civil Law Act 1956 (as revised) provides for
      the reception of English law where the written law is silent on an
      issue and provides for certain cut-off dates for application of
      English common law and rules of equity in West Malaysia, Sabah
      and Sarawak.

26   Malaysian Association of Architects.

27   The copyright of this document is with PAM and is available for
      sale at the Pertubuhan Akitek Malaysia located at Nos 4 & 6,
      Jalan Tangsi, Kuala Lumpur. The PAM SC, 1970 Edition is the
      sub-contract for use where the sub-contractor is nominated under
      the 1966 or 1969 editions of the PAM Forms..

28   This document was described by Edmund Davis LJ in English
      Industrial Estates Corporation v George Wimpey & Co Ltd

(1973) 1 Lloyds Report 118 as a "farrago of obscurities". See also Vincent Powell Smith *Introduction in the Malaysia Standard Form of Building Contract* , (1990) Malayan Law Journal.

29   It is mandatory that the Architect be registered under the Malaysia Architects Act 1967.

30   This power would not extend to the courts – see Northern Regional Health Authority v Derek Crouch Construction Ltd (1986) 26 Building Law Reports 1 and Central Provident Fund Board v Ho Bock Kee (1981) 17 Building Law Reports 21.

31   This would in effect be the case under the provisions of the Arbitration Act 1952.

32   The copyright in this document is with PAM and it has been endorsed by the Association of Consulting Engineers, Malaysia.

33   First Edition May 1989, Second Reprint September 1994. These forms can be obtained from the Institution of Engineers Malaysia, Bangunan Ingenieur, 60/62, Jalan 52/4, P.O. Box 2223, 46720 Petaling Jaya, Selangor.

34   Please refer to 2.1.4 below for a full discussion on the Construction Industry Development Board.

35   See Joseph M Patti, *Management in the Asian Context.*

36   See Schubert P, *The Making of Managers in Asian Countries: the Experience in Malaysia, Proceedings of Asian Association of Managers Organisation,* October – November 1977, pp. 39-45.

37   Mr. Yeoh Sek Chew and Mr. Lee Ng Chai *"ISO 9002" in the Malaysian Construction Industry : "Guide and Implementation",* McGrawHill Book Co..

38   This Act came into force on the 1 December 1994.

39   s.4(1) of the CIDB Act.

40   Part IV of the CIDB Act.

41   The registration requisites are found in the Registration of Contractors (Construction Industry) Regulations 1995.

42   It is modelled on the UK Health and Safety At Work Etc Act 1974.

43   75 deaths have been reported in worksites within the Klang Valley up to October 1997 – in a statement by Human Resources

Minister , Datuk Lim Ah Lek - The News Straits Times, 7 November 1997.

44   See Sharifah Zubaidah Syed Abdul Kader, *Mediation of Legal Disputes: Whither in Malaysia* ? , Insaf, Journal of the Malaysian Bar, July 1996, p.31.

45   Please refer to discussion in 3.1.3 below.

46   The functions and role of the KLRCA is outlined in 3.2.2 below.

47   K.L.Chen and Vinayak P. Pradhan, *How to Introduce and Establish Mediation Practice in Malaysia.*

48   Other adjudication bodies are the Labour Court, Special Commissioners of Income Tax, set up under the provisions of the Income Tax Act 1967, the Rent Tribunal set up under the Control of Rent Act, 1967 the Disciplinary Board set up under the Legal Profession Act, 1976.

49   See 29 of the Industrial Relations Act, 1967.

50   An Industrial Appellate Court, modelled on the Australian Industrial Appellate Court, is being proposed to hear appeals against decisions of the Industrial Court whose decision is to be final and biniding.

51   This is the case in the Malaysian standard forms.

52   The Court of Appeal in A. Cameron Limited v John Mowlem and Company (1990), 52 Building Law Reports, 24.

53   Zublin Muhibbah Joint Venture v Government of Malaysia (1990) Malayan Law Journal 125 where the plaintiff successfully applied to the High Court and obtained a declaration effectively allowing an American attorney the right to participate in proceedings.

54   For further details refer to Vinayak P Pradhan, *Dispute Resolution and Arbitration in Malaysia* (1992) 2 Malayan Law Journal clxxii.

55   For instance in the area of investment disputes, the Convention on the Settlement of Investment Disputes between States and Nationals of other States, 1965, to which Malaysia became a party in 1966 states the place of proceedings under the Convention is the seat of the Centre for the Settlement of Investment Disputes (I.C.S.I.D) at Washington.

56   See s.2 of the Arbitration Act .

57   Refer to discussion below on the exclusion of the courts jurisdiction under s. 34 of the Act.

58   See s. 6 of the Arbitration Act and the later discussion in 3.2.4.

59   ss. 9, 10,12,13, 19, 20, 25 and 26 of the Arbitration Act.

60   Please refer to discussion in 3.2.4.

61   Sections 16 and 17 of the Arbitration Act 1952.

62   See ss 2 and 3 of the Evidence Act 1950.

63   See 14 of the Arbitration Act.

64   The AALCC is an inter-governmental organisation with 41 members from Asia and Africa, one associate member (Botswana) and two permanent observers (Australia and New Zealand). The AALCC set up two regional centres for arbitration, one in Kuala Lumpur for Asia and the other in Cairo for Africa. The agreement between the Government of Malaysia and the AALCC relating to the KLRCA was renewed on 29 February 1996.

65   See Arbitration under the Auspices of the Kuala Lumpur Centre (Arbitration Rules) published by the KLRCA (October 1991), Revised 1 October 1996. The last amendments which came into force in October 1991, were introduced on the initiative of the Director concentrating mainly on amendments to avoid protracted hearings due to inaction or delay and in relation to the arbitrator's fees.

66   Paragraph 4 of the Introduction to the Rules of the KLRCA.

67   However because its facilities in respect of maritime disputes is under-utilised, the KLRCA has entered into an arrangement with the Tokyo Commission under which maritime disputes referred to the KLRCA are handled by the Tokyo Commission but applying the UNCITRAL Arbitration Rules

68   Statistics provided courtesy of the Kuala Lumpur Regional Centre for Arbitration as at 24/11/97.

69   See r.3 of the KLRCA Rules. The Director is the appointing authority unless the parties have provided for a different appointing authority.

70   r.8(1) of the KLRCA Rules.

71   r.7(1) of the KLRCA Rules.

72   The fee is based on a percentage of the amount in dispute, on a lump sum basis rather than on an hourly or daily rated scale. The maximum scale provided in the Schedule can be exceeded when the circumstances of the case warrant it.

73   The amendments to the KLRCA Rules will be published early next year.

74   Please refer to Homayoon Arfazadeh, *Settlement of International Trade Disputes in South East Asia: The Experience of the Kuala Lumpur Regional Centre for Arbitration* (1992)1 Malayan Law Journal cxxii. for an interesting discussion of the exclusion of domestic law and the jurisdiction of courts with respect to arbitrations conducted at the KLRCA and the shortcomings associated with the independence of the KLRCA awards.

75   Syarikat Yean Tat (M) Sdn Bhd v Ahli Bina Pamong Sari Sdn Bhd (1996) 5 Malayan Law Journal 469. For a discussion on the intended ambit of s.34 see Dato' P.G. Lim, Director , *Practice and Procedure under the Rules of the Kuala Lumpur Regional Centre for Arbitration* (1997) 2 Malayan Law Journal lxxiii.

76   Klockner Industries -Anlagen GmbH v Kien Tat Sdn Bhd & Anor (1990) 3 Malayan Law Journal p. 183.

77   Soilchem Sdn Bhd v Standard-Elektrik Lorenz AG (1993) 3 Malayan Law Journal p. 68.   In rejecting an application for certiorari which was not expressly excluded by s.34, the court relied on the definition of "written law" in s. 3 of the Interpretation Act 1967. The amendments to Art 121 of Malaysia's constitution, brought into effect on the 10 June 1988, removed the reservoir of judicial power which had existed in the courts and circumscribed the court's jurisdiction and powers only to those conferred by or under federal law.

78   The High Court at Kuala Lumpur has been organised into separate divisions, each presided over by a Judge of the High Court; the Appellate Division, the Commercial Division, the Civil Division and the Criminal Division. Building and engineering contracts are to be heard in the Civil Division -Practice Direction No. 7 of 1996.

79   Tan Kok Cheng & Sons Realty Co Sdn Bhd v Lim Ah Pat (t/a Juta Bena) (1995) 3 Malayan Law Journal 273. See also Inter Maritime Management Sdn Bhd v Kai Tai Timber Co. Ltd, Hong Kong (1995) Malayan Law Journal 322.

80   See the case of Perbadanan Kemajuan Negeri Perak v Asean Security Paper Mill Sdn Bhd (1991) 3 Current Law Journal p. 2400. Refer to the discussion on s. 34 of the Act in 3.2.2.

81   See Scott v Avery Co English Reports 1121; Khoo Boo Gay v The Home Insurance Co Ltd (1946) Malayan Law Journal 239; Pembenaan Keng Ting (Sabah) Sdn Bhd v Seloga Jaya Sdn Bhd (1994) 1 Malayan Law Journal 422.

82   The case of Tan Kok Cheng (Ibid) at footnote 80 and Seloga Jaya Sdn Bhd v Pembenaan Keng Ting (Sabah) Sdn Bhd serve as good illustrations of the Malaysian court's approach to applications under s.6.

83   See Lan You Timber Co v United General Insurance Co. Ltd (1968) 1 Malayan Law Journal 181; D & C Finance Bhd,v Overseas Assurance Corpn Ltd (1989) 3 Malayan Law Journal 240; Teknik Cekap Sdn Bhd v Nirwana Indah Sdn Bhd (1996) 4 Malayan Law Journal 154.

84   In the Tan Kok Cheng case (Ibid) the court stated that "It is only in plain and obvious cases that, where a reasonable tribunal, without undertaking a meticulous examination of the case is bound to hold that the issues raised by a defendant are frivolous or vexatious, that a court may be justified in refusing a stay."

85   Malaysia Government Officers Co-operative Housing Society Ltd v United Asia Investment Ltd & Ors (1972) 1 Malayan Law Journal 113.

86   S. 25(3) Arbitration Act, 1952.

87   See both Alagappa Chettiar v Palanivelpillai & Ors (1967) 1 Malayan Law Journal 208 and Hashim bin Majid v Param Cumaraswamy (1993) 2 Malayan Law Journal 20.

88   KSM Insuran Bhd v. Ong Ah Bah & Anor (1986) 1 Malayan Law Journal 237.

89   The case of Dooley v London Assurance (1872) ILTR 22 was cited in the Seloga Jaya case and also the case of Perbadanan

Kemajuan Negeri Perak v Asean Security Paper Mill Sdn Bhd (1991) 2 Malayan Law Journal 309.

90   On appeal in the Seloga Jaya case (Ibid) the Court, the then Supreme Court stated that in an appeal from a discretionary order of the senior assistant registrar, the judge-in–chambers exercises his own discretion as though the matter comes before him for the first time and should accordingly give the decision of the SAR the weight it deserves.

91   (1976) 2 Malayan Law Journal 154.

92   See also Elf Petroleum SE Asia Pte Ltd v Winelf Petroleum Sdn Bhd (1986) 1 Malayan Law Journal 177 where the High Court held that although the parties had agreed that Singapore law would govern any dispute, it did not oust the jurisdiction of the Malaysian court to try the action.

93   Order 5 r 3 of the Rules of the High Court sets out proceedings which have to be made by originating summons.

94   "Permissible" means with the leave of the court and the granting or withholding of that leave is a matter entirely within the discretion of the court, the application being made ex parte with only one side of the matter under review.

95   Order 6 r.6 RHC, 1980.

96   Bank Bumiputra Malaysia Bhd v The International Tin Council & Anor (1989) 3 Malayan Law Journal 286.

97   Order 25 RHC, 1980.

98   Order 34 of the RHC, 1980.

99   Order 38 r.1 ,RHC, 1980; Gomez v Gomez (1969) 1 Malayan Law Journal 228.

100   UMBC Finance Ltd v Woon Kim Yan Orbin (1990) 3 Malayan Law Journal 360.

101   Order 38 r 3, RHC, 1980

102   The Court of Appeal in A.B. Chew Investments Pte Ltd v Lim Tjoen Kong (1991) 3 Malayan Law Journal.

103   (1991) 3 Malayan Law Journal 144.

104   See Order 69 r. 6 read with Order 67 of the RHC, 1980 - for procedures for enforcement of a foreign award to which the Reciprocal Enforcement of Judgments Act, 1958 applies. The

application for registration of the award must be supported by an affidavit. Notice of the registration of the award must be served on the judgement debtor and an application to set aside the registration of an award must be made by summons.

105   S.3(4) explains that a judgment is deemed final and conclusive even if an appeal is pending against it.

106   s.4(1) imposes two negative requirements to the effect that a judgment must not be registered if it has been wholly satisfied or if it could not be enforced in the country of origin and a time limit for registration of six years from the date of the foreign judgment.

107   KLRCA awards are enforced in the same way by virtue of the provisions of s. 34(2) of the Arbitration Act discussed in 3.2.1 above.

108   This is broadly based on the principle of physical presence or implied or voluntary submission to that jurisdiction. See Emanuel v Symon (1908) 1 KB 302 CA; Henry v Groposco Internation Limited (1976) 1 QB 726 CA.

109   Hua Daily News Sdn Bhd v Tan Thien Chin & Ors (1986) 2 Malayan Law Journal 107.

110   Pegang Prospecting Co Ltd v Chan Phooi Hoong & Anor (1957) Malayan Law Journal 231.

111   Order 69 Rule 2, RHC, 1980. Order 69 rule 4(1) of the RHC 1980 sets a period of six weeks within which an application for remitting the award has to be made. This period runs from the date the award was made and published with the court having a discretion to extend the time limit.

112   CK Tay Sdn Bhd v Eng Huat Heng Construction & Trading Sdn Bhd (1989) 1 Malayan Law Journal 389.

113   See Ong Guan Teck & Ors v Hijjas (1982) 1 Malayan Law Journal 105 where the judge decided to grant relief as there was nothing to show that the award or any part of it, contained any error on the face of it and that the applicant having "doubts" however genuine could not constitute sufficient grounds to remit the award.

114   Lian Hup Manufacturing Co Sdn Bhd v Unitata Bhd (1994) 2 Malayan Law Journal 51.

115 Sharikat Pemborong Pertanian & Perumahan v Federal Land Authority (1971) 2 Malayan Law Journal 210.

116 Chung & Wong v C.M. Lee (1934) Malayan Law Journal 153; KS Abdul Kadeer v M.K. Mohamed Ismail (1954) Malayan Law Journal 231.

117 Turner East Asia Pte Ltd v Builders Federal Hong Kong Ltd & Anor (1988) 2 Malayan Law Journal 502.

118 Halfdan Greig & Co c Sterling Coal & Navigation Cor (1973) 2 ALL ER 1073 applied in Yee Hoong Loong Corp. Sdn Bhd v Kwong Fook Seng Co (1993) 1 Malayan Law Journal 163.

119 The Council of five persons consists of two Board Members, one representative from the Federation of Malaysian Consumers Association (FOMCA), one representative from the University of Malaya and one other person.

120 IMB Annual Report 1996, p. 4.

121 As at the end of 1996, the backlog of civil cases was reduced from 8,0000 to 25,000.

NETHERLANDS

A G Dorée and M H B Nelissen
University of Twente, Twente,
The Netherlands

1.   BACKGROUND

1.1   Economic

In 1995 the gross turnover in the Dutch construction sector (residential and non-residential (R&NR), civil engineering (CE) and installation enterprises) amounted to about 90 billion guilders (approx. $48 billion), no VAT included. Turnover is expected to increase until 2002 at the average growth rate of 2 percent a year, to more than 100 billion guilders. About half of the industry's turn-over in construction industry is made in housing (residential), a third in non residential buildings, and a small percentage (20%) in civil engineering related investments (roads, bridges, dikes, canals etc). The industry consists of many small and medium enterprises, with a few large firms. The industry accounts for about 10 percent of the Dutch gross national product and approximately 12 percent of employment.

	Nr. of enterprises [-]	Turnover [guilders]	Employees [-]
Small (1-10 employees)	33.500 (82%)	31 billion (35%)	110.000 (28%)
Medium (10 –100 employees)	7000 (17%)	40 billion (46%)	193.000 (49%)
Large (100 or more employees)	400 (1%)	17 billion (19%)	91.000 (23%)

Table 1: % to size of enterprises[CBS 1997, Bouwcijfers 1995-1996]

1.2   Legal

The Netherlands is a constitutional monarchy with a democratic parliamentary system (two chambers). Through three levels of decentralised government, the local authorities also have their own designated tasks and competence. This structure is provided in the constitutional law; a section of the Dutch Public law. The two other parts of the Public law are the Criminal law and the Administrative law. The line between Constitutional Law and Administrative Law is not always clear. The Administrative law can be conceived as the regulations governing a public entity's interference with peoples' daily lives. The Administrative law contains regulations related to matters such as town and country planning, education and welfare.

The Private law is another part of Dutch law. It contains not only Family- and Marital law but also Contract law dealing with the way agreements have to be made.

The legal system, being part of the public domain, is arranged under the Constitutional law. It provides for the Dutch System of Courts. Jurisdictions within the system of courts are defined in both geographical and functional senses:

- Canton Courts [62] are small courts of law that deal with cases under Fl 5.000,- (= $2,500.-). Canton Courts all have chambers of one judge.

- District Courts [19] deal with cases that are not dealt with by the Canton Court. The District Courts have chambers with one (80% of the cases) or three judges (20% of the cases). Every District is divided into departments for administrative reasons.

- Courts of Appeal [5] deal with appeals to a higher court after a judgement by the Canton Court or the District

Court. This court, in the same way as the District Court, has a one and a three judge chamber.

- The Supreme Court [1] is the highest court in the hierarchy. It has chambers with three or five judges. Most cases are decided by a three judge chamber. This Supreme Court is an appeal court only. It reverses decisions because of violations of the law only. The Court exclusively engages in judicial questions.

The courts not only decide upon the written law but also take the unwritten law into account. This "unwritten" common law is illustrated by the Common Principles of Proper Government (Abbb). These principles can be purely written, purely unwritten or a combination of both. These principles relate to common rules of conduct for public organisations/institutions, governing their actions towards the citizens and society. Interests of all parties for example have to be considered according to the Abbb. This code develops by a way of jurisprudence.

1.2.1   Construction law

There is no specific part in national law acknowledged and referred to as "construction law". So all law relevant for acts and conduct within the construction industry is dispersed within various areas of the national law.

Over the years professional bodies, associations of contractors and the large public clients (especially the Ministry responsible for matters of Traffic, Transportation and Water and Flood control) have drafted standard contract documents (model or reference contracts). These documents are obligatory for public sector clients and are furthermore widely used in the private sector. These documents, often referred to as "common conditions", are drafted to improve the clarity of the contractual relationship (also in case of a conflict). Professional bodies and associations play a prominent role in developing these common conditions. See Section 1.2.2.; Construction Contracts.

With a simple reference in a contract, parties can make the common conditions apply. If a conflict arises then, as a consequence of these conditions, court will not interfere. The two parties will seek arbitration first. This can be seen as an extension of the law. Although it is not acknowledged as Construction law, and although the national Contract law prevails, these documents and subsequent conditions can be seen as an institutionalised and acknowledged specific extension of the law. It has even resulted in a formally installed system of arbitration, recognised within the legal system. The arbitration system is explained in the Sections 2.1 and 2.2.

Furthermore, public clients are obliged to arrange procurement as provided for European regulations, The EU procurement works directive is especially relevant for the construction industry. This directive is translated and implemented as the Uniform Procurement Directives (in Dutch the UAR-92).

One particular rule (code/"law") is the so called obligation to warn the opposite party to the contract. This code comprises that contractual parties are obliged to notify each other if one suspects something has gone wrong (Although any action and its' consequences are the responsibility of the other party). For example, when party A expects a fault within the domain of party B's responsibility, A is obliged to inform B about the expected mishappening. Where B is not warned despite A being reasonably aware of the fault, will result in A being held liable because of the negligence. This "warning" code has profound impact on arbitration decisions and subsequently on conduct in the construction process.

1.2.2 Construction contracts

Most contracts drafted between clients, architects, engineers and contractors use a set of common conditions as base. Although clients and companies have their own modifications and amendments. Since Dutch law is expressed in Dutch, foreign sets of Common conditions (JCT, etc) are not used. Contract terminology has to be the same as the language used in the law in order to avoid interpretation problems. This is true almost without exception. The most frequently used common conditions are:

- for construction contracts: Uniform Administrated Conditions 1989 (Dutch: UAV-1989).
- for engineering professionals: the Arrangement for the Relation between Client and Advising Engineering-bureau 1987 (RVOI 1987)
- for architects: the Standard conditions in the Relation between Client and Architect 1997 (SR 1997)

Common conditions deal with subjects as reward or fee systems; compensation for work not primarily included in the contract; tasks, responsibility and authority of contract partners; several procedural issues; and, of course, how to act in case of a conflict (see sections 2.1 and 2.2).

1.3   Literature

In the Netherlands several Dutch magazines aimed at, and dealing with, the construction industry are published. The magazine "Bouwrecht", freely translated as Construction law, edited and published by the Institute of Construction Law [Instituut Bouwrecht]. Also, the Foundation Construction Research [Stichting Bouwresearch (SBR)] publishes on issues relevant for practitioners in construction industry. The following books about construction conflicts have been published in the last couple of years. Unfortunately these publications are written in Dutch (without a written English summary).

- Aansprakelijkheid voor gebreken aan bouwwerken (SBR 179): deals with post-construction failures and liabilities;
- Algemene voorwaarden; betekenis voor de bouwpraktijk (SBR 214): explaining the relevance, contents and practical use of common conditions;
- Geschillen in de bouw; Beheersing en Beslechting (SBR 302): deals with disputes, conflicts and differences in the context of contracts.

2.   DISPUTE MANAGEMENT AND CONFLICT RESOLUTION

Introduction

In the Netherlands the difference between a conflict and a dispute is not clear. Of course it is inevitable that parties will have different views and opinions. If these differences cause problems, the first action is to try to solve them in a normal, rational and orderly fashion. These normal situations are not considered to be conflicts, although they can be treated or approached as such. Neither theory, practice, nor law gives a general accepted definition of the concept. What is agreed upon is that "conflict" consists of several stages. To avoid confusion this Section will be restricted to the stage where the conflict is usually described and recognised as manifest.

A polarised situation exists where parties, who have a mutual need for each other, acknowledge the differences in points of view between them, and signal that it is not their intention to give in. This is already a deadlock situation or can be expected to become so rapidly.

To explore the Dutch approach to conflict and its resolution we start at the worst case scenario: parties turn to the legal system. In practice, owing to the use of common conditions in the

Netherlands, this means resorting to arbitration. Section 2.1 describes the Dutch arbitration system. Section 2.2 considers what happens when arbitration was not agreed upon in the contract, and the case is brought directly to court. Section 2.3 describes the alternative ways to solve manifest conflicts; often with third party intervention such as Conciliation and Mediation. Both sections are, as far as known and where relevant, illustrated with the facts and statistics of practice.

2.1   Arbitration under Dutch law

The most prominent arbitration institute, in size and operation, is the Council of Arbitration for construction contractors. This Council was established by the professional associations of the three mayor player in the construction industry (engineers, architects and contractors). The members of the institutes' board are nominated by the Royal Institute of Engineers (KIvI), the Dutch Bond of Architecture (BNA) and the Common Bond Construction Companies (AVBB). Each of the first two organisations nominates two members of the board and the AVBB nominates four. The chairman of the board is chosen by the board. The appointment of this person requires approval of the Ministry of Traffic Transportation and Flood-control. The Council of Arbitration consists further of members, extraordinary members and a supporting staff. In 1995 the Council of Arbitration had a total of 75 members and extraordinary members.

Under Dutch law it is possible for parties to seek arbitration. The law is very flexible on how to arrange this arbitration. It is possible that both parties appoint one arbitrator and that these two arbitrators together appoint the third, but it is also possible that both parties appoint one arbitrator together and that this person has to make a decision.

If two parties decide to seek arbitration they have to describe what the dispute is about. After that they are free to choose the method of arbitration. In practice it is usual that parties refer to the rules of an Arbitration Institute. Such reference makes only this institute competent for resolving the dispute.

If a party goes to a different arbitration institute from that designated by the contractual arrangements, or if a party goes to the District Court while the other party claims arbitration as was agreed upon, then an appeal on the grounds of the incompetence of the arbitrator or judge should be made at the beginning of the case. If this is omitted then implicitly the result is accepted.

In case of an appeal on the ground of incompetence the arbitrator is the first one to decide whether or not he is competent. The possibility exists that an arbitrator may decide that he is competent while the defendant maintains his incompetence. In this case the whole arbitration procedure has to be concluded and after that the defendant can go to the District Court to ask for annulment of the decision of the arbitrator. Should the arbitrator decide that he is incompetent this is the final decision in the arbitration procedure and the plaintiff can go directly to the District Court to ask for annulment of this decision. If the judge decides that the arbitrator is competent then the arbitrator must conclude the procedure.

The opposite is also possible; the plaintiff could go to the District Court and the defendant could appeal on the ground of the incompetence of the District Court. In this case the judge has to decide his competence in the case. This decision, however, is a decision which a party can appeal to a higher court of law. This results in a long delay before a final decision is reached.

There are four main differences between the procedure before an arbitration institute and before court.

- Before the court of law both parties need a lawyer. In case of arbitration parties can appear themselves or they can have a representative, which may be anyone. There are no professional restrictions on this representative.

- The arbitration procedure is much more informal than a judicial procedure. This can be seen, for example, in the way it is possible to send the summons. In a judicial procedure a summons has to be brought by an official of the court, while in an arbitration procedure a summons can be sent by ordinary mail.
- In the ordinary judicial procedure one of the two parties can go to a higher court, in the arbitration procedure this is not possible. This has the disadvantage that neither party can object to the decision but it has the advantage that a final decision is obtained very quickly and that both parties can move on. An exception to this is the Council of Arbitration where it is possible to go to a higher authority.
- Hearings by a Court of law are public and hearings by an arbitration procedure are not. Sometimes, if there is a lot of media interest or the accused has to remain anonymous the judge will declare a closed hearing but these are exceptions.

A judge must decide a case according to the law but the rules for an arbitrator are much more flexible but they are also justified by law. An arbitrator operates in a context which he understands, so he knows what the practices are in a particular industry and he can take them into account where the law leaves an opportunity to interpret the rules. Also because of these arbitration rules a lot of procedural formalities can be ignored.

The decision of an arbitrator can be enforced in the same way as a decision of a judge can. A financial penalty or custody are ways to enforce a decision. Before these kinds of measures can be used the arbitration decision needs the approval of the President of the District Court in which it was pronounced. The only reason for the President to refuse to sign the decision is that elementary legal principles have been violated. If the president refuses to sign then the successful party can appeal to the Court of Appeal to enforce the judgment.

If the unsuccessful party does not accept the decision of the arbitrator then there is no way to appeal to a higher court. There is, however, a possibility to appeal to a higher court if the Council of Arbitration has pronounced a decision. If the unsuccessful party does not execute the judgment of the arbitrator and there is no way to enforce the judgment then the successful party has to go to a court of law because the other party has committed an unlawful action (UA) and an UA falls within the jurisdiction of the court of law. The court of law will not re-examine the case but will decide whether the action is lawful or unlawful.

2.2   Arbitration in the construction industry

In the previous section the most used set of common conditions was already mentioned. These conditions also prescribe how to act in case of a conflict:

- The UAV89 and the UAR designate to the Council of Arbitration of construction companies as the official arbitration institute in case of a conflict. The UAV89 starts with the explicit rule that all parties give up their right to go to a Court of law.
- The RVOI states that conflicts have to be solved by a friendly arrangement as far as possible. When this is not possible the case has to be referred to the Committee of Disputes from the Royal Institute of Engineers. This Committee exists of a chairman and four members.
- The SR 1997 states that conflicts have to be solved by the Foundation Arbitration Institute Architecture.

If a private party acts as a client, the common conditions may be used also (in fact this is often the case). But in case of a conflict where they were not, it is possible for both parties even then to decide, with hindsight, to ask for arbitration.

Most organisations such as the Royal Institute of Engineers and the BNA have their own dispute committee. When a conflict between two members of such an organisation arises this committee can pronounce a decision. As this judgement has no judicial basis a party can always ignore the decision but the organisation can then inflict disciplinary measures on one of the two parties. If this party still fails to obey the dispute committee then the other party has to go to the Court of law to enforce their right.

It is clear that almost any profession and specialisation in construction industry has its own method and institute of arbitration. One of the reasons underlying this situation is that every sector has its own special, branch and product-tied knowledge. The disadvantage of this diversity is that the arbitrators are not tied to the decisions of other arbitrators so it is possible that a client will lose a case just because the differences in arbitration competence. To reduce this problem, for several years it has been possible to combine these procedures. However, before combining they have to be started separately. When several procedures are started then the President of the District Court can order the combining of these procedures. If combinations of procedures are formed then both parties have to agree on which common conditions are valid for a certain combination and which arbitrator has to reach a decision. If both parties disagree, then the President of the District Court in Amsterdam will appoint an arbitrator and identify the common conditions that are valid. Both parties can, however, agree upon not combining by contractual arrangement. Recently initiatives have been taken to investigate the possibility of combining all the arbitration institutes into one big overarching arbitration institute for the entire construction industry.

The Council of Arbitration is one of the major sources of jurisprudence. In 1995 866 cases were referred. Almost 90 percent (777) were taken into consideration. In 324 cases it actually came to a decision. About 40 cases resulted in an settlement during the process. One third of the 777 were retracted before the judge produced a decision. Sixty percent of the cases were decided by one judge only and 40% of the cases were decided by a council. In 61 cases a member-lawyer was one of the three arbitrators in the council. The choice between one or three arbitrators is based on the amount of money involved. When this amount exceeds $ 35,000.- [nl: Fl 65.000,-] then three arbitrators are assigned. Notwithstanding this rule, parties may also agree amongst themselves to assign just one arbitrator. Only a few examples of cases have actually gone to court. This happens because no arbitration has been agreed upon, no competent arbitration institute was available, or the decision of an arbitration-institute was not executed.

The members of the council are people with a construction industry background. In some cases referred, the contractual arrangement between parties already anticipated that, in case of a conflict, a member lawyer had to be one of the arbitrators. Also the chairman of the Council may want a member-lawyer to advice him in the process. In both cases the member-lawyer is asked to join the council. These member-lawyers, having a judicial background, are extraordinary members of the Council of Arbitration.

The essence of arbitration lies in the fact that on the one hand the arbitrators, being professionals in the business themselves, know the technical complexities of construction problems, while on the other hand regulations in law about conflicts in construction industry are sparse.

As long as the UAV 1989 is valid, parties have to ask for arbitration in case of a dispute. When the UAV 1989 is not valid, then parties have to go to court. The judge's decision is binding and can be enforced. If arbitration is asked for, the decision of the institute can only be enforced with the permission of the President of the District Court. If one party does not fulfil its part of the decision then the other party can go to court and ask the judge to enforce the decision of the arbitration institute. Then the action is a criminal procedure because one party commits an unlawful action and these actions fall within the jurisdiction of the court of law. What happens before a court of law is described in the next section.

Also in this arbitration system a method of appeal is created. In case of an appeal on a higher court the Council of Arbitration exists of 3 or 5 members and these are different from those who have decided the case at first instance.

2.3   Court involvement

If two parties have not agreed on the validity of the UAV and a dispute exists, the case, most of the time, has to be referred to the District Court because usually the amount of money involved exceeds 5000 guilders. This section will therefor focus upon the procedure of the District Court and will not go into the procedure of the Canton Court.

The normal procedure of going to court involves lawyers. Party A writes a summons which is sent to party B and which contains information on: the court of law before which the case is referred; the date at which party B is expected to appear before the court: and the reasons why the case is brought to court.

After this both parties send two written conclusions to the court of law about the case. Party A writes a conclusion (refers to the summon that explains the reason why this case has been brought). Party B reacts to the conclusion. Party A then counter-pleas and finally party B reacts to the counter-plea. Before the judge can pronounce a decision lawyers of both parties have the opportunity for a final plea. In practice this final plea happens rarely to save money and time.

After the exchange of conclusions the court of law can pronounce a final judgment, but it is also possible that the judge pronounces an interim decision because he wants to hear specialists or witnesses. If one of the two parties pleads incompetence or if the court of law or one party wants to involve a third party then the judge has to decide about these problems before he can resolve the actual conflict. This decision is also an interim decision.

Another way for the judge to reach a decision is by a report from an expert. This report advises the judge to which decision should be made given certain circumstances. If both parties agree on the adviser then the judge will appoint this person. If parties cannot agree on who should advise the judge, then the judge will decide for himself who will be the adviser. Often this is someone who is well known by the court of law because they have done this advice work before. The report from the adviser is sent to both parties and to the judge and both parties can write a reaction to the report.

The final judgement pronounced after this procedure means the end of the case, unless one party appeals to a higher court of law. Until the higher court of law has pronounced a judgement, the judgement of the court of law cannot be enforced. In most cases the judge of a court of law allows parties to execute the judgement but at their own risk. If the higher court of law decides that the other party is right then the first party has to demolish the building or he has to pay compensation.

If one party does not agree with the judgment of the District Court than that party can ask a higher court, the Court of Appeal, to review the case. The process by the Court of Appeal is the same as by the District Court but now both parties only have one opportunity for a conclusion instead of two. After hearing the parties, the Court of Appeal can reach a decision and the only way to challenge this decision is by a notice of appeal [nl: *Cassatieverzoek*] to the Supreme Council.

An interim judgment can also be appealed as above.

2.4   Mediation

Mediation is seen as a way to help parties to solve a problem by intervention of a person. This person guides the parties towards a mutually reasonable solution. Mediation should prevent the parties from seeking arbitration. There are professional and trade associations that have their own mediation committees. Members can call upon their association to mediate in case of a problem.

The Dutch Mediation Institute (DMI) reports about 150 mediation attempts a year. Only 10 of these were related to conflicts in construction industry. The DMI has 15 mediators specialised in construction industry. Some mediators have a background in construction industry but most of them have a juridical one with experience in construction industry. This may seem a contradiction because a juridical procedure is the opposite of a mediation procedure. In a juridical procedure the goal is to get everything possible and the goal in a mediation procedure is to make both parties content. However, in practice it turns out that these juridical mediators also do a good job.

Problems between landlords and people who rent a house problems have been solved very effectively by mediation. This is the reason that the DMI has introduced a working-group for social mediation.

3.   THE ZEITGEIST

Given the language and entry barriers, the Dutch construction market is a rather closed market. In major projects, capacities and qualities are often drawn together. This might be seen in the concerted production in the struggle against the sea, the reconstruction after World War II, land reclamation projects, and the big effort put into social housing. Both the demand and supply side are well organised. The national spirit is co-operative, tolerant, inquisitive, down-to-earth and pragmatic. The national maxim is: "Be normal, that is different enough already". The social and income structure is relatively flat. The work ethic (Calvinistic) is high and subsequently - so is productivity. The political situation is stable. The economy is characterised as a "deliberation-economy". It is not exceptional that a court judge, when confronted with a case, orders the conflicting parties "back to the table" to re-negotiate and try to solve the problem amongst themselves.

The Netherlands is a small country with a population around 15 million people (highest population density of the European countries). The country measures some 400 kilometres North-South and some 150 kilometres across. One could travel from one end of the country to the other in four to five hours. The Dutch construction market must be assessed on the same scale. Operating on this scale in this market means you can't afford to make enemies. There is no place for hit and run strategies, because you're bound to bump in to each other eventually.

This context (market and culture) create checks and balances in project situations preventing conflicts from escalating. "There must be a reasonable way to solve the problem without the high fee for the lawyers"; "since we're all grown-ups we must be able to solve the problem in a adult fashion". Statements of such tenor are often heard in times of crises, and are acted upon.

Given all this it won't come as a surprise that in the construction industry parties seldom choose to come to the level of third party involvement. In the contractual arrangements of major projects often special intra-project conflict solving procedures and mechanisms are installed (project internal dispute committees etc.). Construction law is only a small specialism in education at the Dutch Universities.

However, since the ratification of the European directives on procurement the number of conflicts rose, and the hostility in projects increased. The drift towards more "market-oriented" policies make project relationships more adversarial. The number of procurement disputes brought before the council of arbitration has risen. It has to be seen whether this is a temporary rise or a sign of a trend for the long term.

4.   GLOSSARY

- Cassatieverzoek:
  Notice of appeal at the Supreme Council

- Algemeen Verbond Bouwbedijf (AVBB):
  Common Bond Construction Companies

- Koninlijk instituut van Ingenieurs (KIvI):
  Royal Institute of Engineers

- Algemene Beginselen Behoorlijk bestuur (Abbb)
  Common Principles of Proper Government:

- Commissie van Geschillen (CvG):
  Committee of Disputes from the KIvI

- Bond van Nederlandse Architecten (BNA):
  Bond of Dutch Architecture

- Arbitrage Instituut Bouwkunst (AIBk):
  Arbitration Institute Architecture from the BNA

- Onrechtmatige daad (OD):
  Unlawful action

- UAV 1989 Uniforme Administratieve Voorwaarden:
  Uniform Administrated Conditions 1989

- RVOI 1987 Regeling voor de Verhouding tussen Opdrachtgever en Adviserend Ingenieursbureau:
  Arrangement for the Relation between Client and Advising Engineering-bureau 1987

- Standaardregeling tussen Opdrachtgever en Architect 1997 (SR 1997):
  Standard conditions in the Relation between Client and Architect 1997

- Uniforme Aanbestedings Regelement 1992 (UAR-EG 1992):
  Uniform Procurement Directives 1992

- Ministerie van Verkeer en Waterstaat:
  Ministry responsible for matter of Traffic, Transportation and Water- and Flood-control

- Kantongerecht:
  Canton Courts

- Arrondissementsrechtbank:
  District Court

- Gerechtshof:
  Court of Appeal

- De hoge raad:
  The Supreme Court

Bibliography

Haan P. de, Voorst van Beest M.A. van, Bregman A.G. en Langendoen H.: "Bouwrecht in Kort bestek", 3^e druk, Stichting Instituut voor Bouwrecht, Kluwer 1996 Deventer, ISBN: 90-268-2721-0.

Huismn A.M., Groot A. de: "SBR 302: Geschillen in de Bouw; Beheersing en Beslechting", Stichting Bouwresearch, 1993 Rotterdam, ISBN: 90-5367-094-7.

"Oxford Dictionary of Law" 3rd edition, Oxford University Press 1994 Oxford, ISBN: 0-19-280000-0.

Kaplan M.J.G.P., Rinnooy Kan A.H.G.: "Onderhandelen; Structuren en Toepassingen", Academic Service Economie en Bedrijfskunde, 1991 Schoonhoven, ISBN: 90-5261-031-2.

Buur A.P.: "De Verwachtingen voor de Bouwproduktie en de Werkgelegenheid in 1997:, Economisch Instituut voor de Bouwnijverheid [EIB], 1997 Amsterdam.

Cliteur P.B.: "Inleiding in het Recht", Wolters-Noordhoff, 1992 Groningen, ISBN: 90-01-19271-8.

Gerritzen-Rode P.W.A., Vlies I.C. van der: "Beginselen van Bestuursrecht" 2^{e} druk, Samson H.D. Tjeenk Willink, 1994 Alphen aan den Rijn, ISBN: 90-6092-756-7.

"Bouwcijfers 1995-1996: Bedrijfseconomische resultaten van de Bouw en Bouwinstallatiebedrijven", Centraal Bureau voor de Statistiek [CBS], 1997 Voorburg/Heerlen, ISBN: 90-3572-443-7.

"Regeling van de Verhouding tussen Opdrachtgever en adviserend Ingenieursbureau [RVOI 1987]", Koninklijk Instituut van Ingenieurs [KIvI], 1993 's-Gravenhage, ISBN: 90-73331-02-1

"Uniforme Administratieve Voorwaarden voor de uitvoering van Werken [UAV 1989]",

"Uniforme Aanbestedings Regelement EG [UAR-EC 1991]",

"Standaardvoorwaarden 1997 Rechtsverhouding opdrachtgever-architect [SR 1997]", Koninklijke Maatschappij tot Bevordering van de Bouwkunst: Bond van Nederlandse Architecten, 1997 Amsterdam, ISBN:

"Jaarverslag 1995", Raad van Arbitration voor de Bouwbedrijven in Nederland.

Dorée A.G.: "Voortbrenginsprocessen in de Civiele Techniek", Universiteit Twente, 1995 Enschede.

OMAN

Richie Alder
Trowers and Hamlins, Sultanate of Oman

1.   BACKGROUND

1.1   Economic

The Sultanate of Oman has undergone a period of dramatic change and development since His Majesty Sultan Qaboos bin Said came to power in 1970. The economy remained dependent on oil reserves, and continued oil and gas exploration remains a national priority. Natural gas reserves have recently been located, in relation to which ambitious export plants are presently being formulated.

1.1.1   Trends

Oman's fifth Five Year Development Plan began in January 1996, and focuses on new measures to boost the private sector and foreign investment, by considerable expenditure on matters such as human resource development and civil development, while aiming to reduce the annual deficit by reducing spending in areas such as defence and national security. Oman has already embraced a concept of privatisation and use of private finance and can now be recognised as a leading light for private sector funded infrastructure projects in the region. With great reliance having been placed, in the past, on Oman's oil and gas reserves, future expansion of the economy is now planned to take place in the private sector outside these industries. The government has indicated moves to provide for incentives for industry outside this traditional area of wealth.

1.2   Legal

The Sultanate of Oman's legal system is founded on the precepts of the Shariah. The Shariah Courts still hold jurisdiction over personal and family matters. However, a large body of legislation has been enacted relating to all aspects of commercial law which takes the form of Royal Decrees and subordinate legislation, of which the more important are Ministerial Decisions. All such decrees and decisions are published in the Oman government's Official Gazette and generally only come into effect from the date of publication. This legislation has been drawn from the precedents established by other countries in the region and is based, to a significant extent, on French and Egyptian sources of jurisdiction.

To the extent that Omani law is not covered by the codified law, reliance is still placed on custom and practice and the Shariah, although the Shariah Law and its court do not generally play any part in the resolution of commercial disputes.

Selected decisions of the Commercial Court, illustrating legal principles which it has followed, are published in an annual yearbook, but although the court does tend to follow its previous decisions, there is no binding rule of precedent in Oman.

Until 1997, the judicial body responsible for disputes concerning commercial matters was the Authority for the Settlement of Commercial Disputes. Whilst independent, the Ministry of Commerce & Industry had the power to exercise an administrative role over its operations. With effect from July 1997, however, this body has been formally reconstituted as a court of law under the Ministry of Justice, and is known now as the Commercial Court. The court sits in Muscat, although summary courts of jurisdiction are established throughout Oman to settle claims up to R.O.15,000.

It is presumed that the Commercial Court's proceedings will retain the existing nature of the Authority for the Settlement of Commercial Disputes' proceedings - that they will be inquisitorial not adversarial with a requirement, therefore, on the part of the litigants, that detailed written pleadings, supported fully by documentation, be submitted. It is frequent practice for the court to appoint experts in the relevant field of the dispute, to prepare a report to the court, although the court is not bound to accept the findings or recommendations of the expert.

There is a right of appeal to the Court of Appeal, whose judgment is final. We are aware of one or two instances only over the last seventeen years in which a direct petition to His Majesty the Sultan has resulted in a request to the Court to review its decision.

Advocacy may only be undertaken by a licenced practitioner. A licence will only be granted to an Arabic speaker. All proceedings are undertaken in Arabic; any documents in a language other than Arabic upon which the parties wish to rely must be translated into Arabic.

Access to the court is limited to those whose dispute is "commercial", "as defined in the laws of the Sultanate". A non-exhaustive list of examples of such disputes are contained within the Royal Decree governing the Commercial Court, and in reality so long as the court is satisfied that the underlying dispute between the parties can be regarded as "commercial" in the broad sense, rather than "private", it will not decline jurisdiction. In any event, "Activities related to building and construction" fall within the definition of "commercial activities" in the Oman Commercial Law.

The newly introduced Lawyers Law governs the relationship between the lawyer and client, including a duty of confidentiality, prohibition upon acting in circumstances where there would be a conflict of interest, and an entitlement to fees. The lawyer must be in possession of a formal power of attorney, and the Lawyers Law provides that a lawyer must be instructed where the value of the claim exceeds R.O.5,000.

1.2.1   Construction Law

There is no specific legislation relating to construction law, and indeed no specific reference to construction contracts within the Oman Commercial Law, which is the principal legislation dealing with trade and business within the Sultanate.   The Omani commercial laws are broadly based on the system used in Kuwait, and it is believed that the laws' jurisprudential roots stem from the principles of the French Civil Code and Egyptian Law.

The current major players in the construction sector are Tarmac/Wimpey, Mott MacDonald, Costain, Foster Wheeler and Strabag.

Under a new Royal Decree relating to engineering consultancy offices, the owner of a consultancy office designing or supervising the execution of a structure is jointly responsible with the contractor for any defects in the buildings designed by him or executed under his supervision, for a period of ten years, as from the date upon which the client receives the structure. This liability applies even if the defect was due to the condition of the land on which the project was constructed, or if the owner has approved the construction of defective installations.   Any condition or clause in a contract which seeks to release this warranty, or to limit it, is considered null and void. If a party wished to make a claim against a consultancy engineering office under the decennial liability provisions, (see 1.2.2. below) it must do so within three years from the date of discovery of the alleged mistake or error.

1.2.2   Construction Contracts

A standard form of construction contract is used by the government. Although drafted in English, this has been translated into Arabic. This documentation is based upon FIDIC standard forms. By virtue of the application of the Oman Commercial Law, it is possible to argue that the standard Omani terms and conditions for construction contracts comprise the "local custom and practice" for such contracts for the purposes of defining the obligations of parties in the absence of any written contracts. The Oman Commercial Law provides that the limitation period for the obligations of merchants towards each other is ten years from the date when performance of such obligation lapses, unless the law provides for a shorter period. This time period is consistent with clause 62 (4) of the Standard Omani Building and Civil Engineering Works Conditions which also provides for decennial liability in relation to defects in construction (as distinct from design) which lead to "failure" or "collapse" of the relevant structure. This liability did not, by its terms, depend on the existence of fault on the part of the contractor, but it is excluded if there is any other apparent cause and unsound construction. The liability lasts for ten years from the date of issue of the maintenance certificate.

The standard terms provide for the reference of disputes to a sole arbitrator to be agreed upon by the parties; failing such agreement on the application of either party to the Commercial Court, which shall appoint an arbitrator. Thereafter, the arbitration will proceed as a private arbitration, subject to the provisions of the Arbitration Law (see 3.2.2. below).

2. CONFLICT MANAGEMENT

As a general rule, other than arbitration or proceedings before the Commercial Court, there are no non-binding conflict management systems in place in Oman.

As for quality matters, ISO 9000 applies, and has been adopted as the national standard in Oman. There is a tender board system for government organisations letting any major contracts. Only those companies which are registered at the Oman Tender Board can submit a tender for public works, other than works directly concerned with palace contracts. Certain large contracts are open to foreign companies who are invited to tender.

There are regulations regarding products of national origin, which effectively create a restriction on imports. Such products have to be given preferential treatment over imported goods, to the extent of a 10% price margin. Products originating in GCC member states attract a 5% margin of preferential treatment. Therefore, the ruling encourages procurement of products (where available) firstly from Oman, secondly from GCC states and finally from overseas, unless those imported products more than 10% cheaper than similar Omani or GCC products.

As for the value of procurement, this depends upon the type of project, but as a rule of thumb for petrochemical contracts approximately 40% of the value of the contract is allocated to procurement, 40% to sub-contractors and 15 to 20% to the engineering services/management fee.

3.   DISPUTE RESOLUTION

A commercial conciliation and arbitration facility was created under Article 7 of the 1979 Statutes of the Oman Chamber of Commerce & Industry, but has done little in the way of a sustained volume of business.

3.1   Non-binding

There is no developed concept of non-binding dispute resolution within Oman.

3.2   Binding

3.2.1   Adjudication

Whilst a party to a construction contract would be free to agree to the appointment of an adjudicator, we have no experience of such a process in Oman.

3.2.2   Arbitration

Arbitrations in Oman can take place either through the Commercial Court or by way of private arbitrations.

There is no fetter upon the parties agreeing the method and course of arbitration proceedings. Any agreement to arbitrate is likely to be upheld by the Commercial Court, even in circumstances where such arbitration is to take place abroad, and may be subject to foreign law, in which circumstances the ultimate award may not, in any event, be enforceable in Oman.

An Arbitrator will usually require that his fees be paid by the parties in advance.

A new arbitration law, the Law of Arbitration in Civil and Commercial Disputes (the "Arbitration Law"), came into effect on 1st of July 1997, and sets out formal rules and regulations governing any arbitration undertaken within Oman. The Arbitration Law appears to have been based, in the main, on the UNCITRAL model law. The law extends to any international commercial arbitration which is taking place abroad if the parties have agreed that the Arbitration Law shall be the governing law. The law enshrines the freedom of the parties to agree upon the format and procedure of arbitration undertaken within Oman, but permits the interference of the Commercial Court in certain instances, such as the failure by the parties to agree upon an arbitrator, or if the arbitrator fails to discharge his responsibilities.

The Arbitration Law sets out the rules relating to the exchange of pleadings, which must be in Arabic unless otherwise agreed; the arbitrator is entitled (but not obliged) to hold hearings; witnesses and experts are entitled to deliver testimony without taking an oath. The arbitrator may seek the opinion of an expert, upon which the parties to the arbitration are entitled to reply. Unless the parties agree on the number of arbitrators, they shall be three, and there must always be an odd number in any event.

The Commercial Court has the power to fine any witness who refuses to co-operate.

The arbitrator's statutory obligation is to apply rules agreed upon by the parties; if no such rules have been agreed, to apply the "most suitable rules in the law", or to apply the "common rules of justice, without strictly adhering to the articles of the law". A time limit of twelve months is imposed for the award to be reached in the absence of any other period agreed between the parties. This period can be extended by the Commercial Court.

The arbitration award should be issued in writing and signed, and should be based on reasons and justifications. A copy of the award is to be made available to the parties within thirty days from its date of issue. All arbitration awards are to be referred to the Commercial Court, and a copy of the text of the settlement must be deposited by the successful party with the court. It shall not be published until the approval of the parties involved. It is presumed that this reference to publication is to publication in the yearbook of the Commercial Court (see 3.2.4. below).

Arbitration judgments reached in accordance with the Arbitration Law are not capable of appeal to the Commercial Court. The court will however entertain an application to set aside the arbitration judgment in specific cases, such as the failure of the arbitrator to comply with terms agreed upon by the parties or if he exceeds the terms of his appointment. As already mentioned, the judgment will automatically be considered invalid by the court if it contains anything contrary to public order. Any application to the Commercial Court must be made within ninety days of the declaration of the judgment.

Arbitration awards granted in Oman may be enforced through the Commercial Court, but enforcement may not take place by the Commercial Court until it is satisfied that it does not contradict a rule of the Omani court on an issue of dispute; that it does not violate the *ordre public* in Oman and that it has been properly notified to the losing party.

As for the enforcement of foreign awards, the recent amendments to the Commercial Court's rules permit the enforcement of foreign judgments so long as there is reciprocity, and the judgment is not contrary to public order.

Oman is a party to the GCC Agreement for the Enforcement of Judgments, Judicial Assistance and Notifications of 1995, and the Convention on the Settlement of Investment Disputes, but not to the New York Convention 1958.

3.2.3   Expert Determination

We have no experience of this procedure having been adopted in Oman.

3.2.4   Litigation

The Commercial Court has jurisdiction to hear disputes relating to construction contracts, even if one party has no address or residence within the Sultanate, so long as the case relates to property in the Sultanate or to an obligation arising, implemented or which is enforceable in the Sultanate.

A legally binding agreement to arbitrate is likely to be upheld by the Commercial Court although whether such an arbitration award will be enforceable in Oman will depend upon those factors referred to in 3.2.2 above. The Commercial Court has jurisdiction to intervene in private arbitrations as set out in the Arbitration Law referred to 3.2.2 above, but the court's jurisdiction to review a decision of an arbitrator is limited to a failure by the arbitrator to have followed the rules of the arbitration, whether agreed between the parties or as set out in the Arbitration Law.

Insofar as construction disputes which come before the Court are concerned, there are no specialist judges who would be appointed.   However, the court will frequently appoint an independent expert to analyse any technical issues and to provide an opinion, upon which each party is given an opportunity to comment. The court is not bound by the opinion of the expert. No fetter can be placed upon the court's power to appoint an expert by the parties.

The court procedures are inquisitorial. Primary importance is placed on documentary evidence and oral evidence is rarely given although legal counsel do present oral submissions at hearings before the Commercial Court in addition to their written submissions. The parties' lawyers do have the right to cross examine themselves, but this right is exercised through the court.

There is no concept of without prejudice in Oman. A party to the proceedings may submit in evidence to the court any documents that it wishes, and there are no formal rules with regard to discovery of documentation or interrogation of parties. The court does have the power, upon the request of one of the parties, or upon its own initiative, to direct any of the parties or any other person to produce any documents, records or books relevant to the case.

Whilst there is no prohibition upon the Commercial Court applying foreign law to a contract said to be subject to a foreign legal system, we believe it unlikely that such foreign law would be upheld to the extent that it conflicts with Omani law.

Prior to the amendments to the regulations governing what was previously known as the Authority for the Settlement of Commercial Disputes, issued on the 26th of March 1997, constituting the Commercial Court, foreign judgments or arbitration awards were not normally recognised, and the parties were required to relitigate the matter before enforcement could take place. There are now, however, express reciprocal enforcement provisions, subject to a number of conditions being fulfilled, including that both parties (or their representatives) must appear before the Commercial Court, that the judgment must not be contrary to Omani law, and that it must have been obtained in a jurisdiction in which Omani Judgments or arbitration awards are themselves enforceable.

Other than filing fees and any expert's fees, a successful litigant is unlikely to recover his legal costs in proceedings before the Commercial Court. Local lawyers are permitted to act on the basis of a contingency fee, although this may now be called into question by Article 48 of the Lawyers' Law, which provides that "a lawyer's fees may not be an interest or share in the disputed rights".

3.2.5   Negotiation

See above - there is no formal process of negotiation in Oman.

3.2.6   Ombudsman

There is no equivalent in Oman of the Ombudsman system.

4.   THE ZEIGEIST

4.1   Economic Overview

Oman has succeeded, over the past 27 years, in achieving significant socio-economic progress in all economic sectors.

Political stability and the judicious use of oil revenues have been instrumental in achieving the current growth rate of 6% p.a. and in providing the base for the development of a modern infrastructure and general economic development. The economy has moved from a low-income economy into a modern, free market economy, and is currently aiming to diversify its resource base away from reliance on oil towards renewable resources including commodity production and services. Oman implements 5 year plans, of which 1996-2000 is the fifth, representing a transitional stage in which the government is striving to achieve a balanced budget by the year 2000. Oman's Vision 2020 is based on diversification of the economy away from reliance on oil; one of the basic supports of this diversification strategy is the Oman LNG Project, the largest and the most ambitious project of its type in Oman. Other projects which are being encouraged include petrochemical, aluminium smelter and fertiliser plants. His Majesty Sultan Qaboos bin Said has recently declared 1998 to be the "Year of the Private Sector", in order to encourage the privatisation investment programme and the continued growth of the Muscat Stock Market. Inward investment has been encouraged by the promulgation of the Foreign Capital Investment Law and the establishment of the Omani Centre for Investment Promotion and Export Development. Human resource development is also a major strategy, with the government launching an ambitious vocational training programme in order to be able to fit Omanis into private sector jobs currently occupied by non-Omanis. It is envisaged that this process of Omanisation should reach 95% and 75% in the government and private sectors respectively, by the year 2020. Oman is also taking advantage of its attractions for tourists; tourism is high on the agenda for development and diversification of the economy. The hotel infrastructure is being expanded to meet the demand, although having carefully studied other countries' tourism policies, mass tourism is being discouraged.

4.2   Legal Overview

1996 saw the promulgation of the Basic Law of the State, establishing important principles relating to succession, governance, rights of citizens and the formation of principles for a new judicial system. New laws will be introduced over the next two years in order to bring the legislative and legal system of Oman in to line with the principles established in the Basic Law. Also in 1996 the policy of liberalisation of the economy and the encouragement of foreign investment resulted in key changes to the Commercial Companies Law and the laws relating to Commercial Agencies, Taxation and Foreign Investment all of which are primarily focussed towards the encouragement of greater inward investment in Oman's industrial and non-oil based industries, as referred to above.

The Basic Law represents the first attempt in Oman to put in place legislation which deals specifically with the basic constitutional framework under which government is carried on. Part I confirms that Islam is the religion of Oman and that the Islamic Shariah is the basis of legislation. Part II sets out, inter alia, the economic principles which enshrine the free market economy, interpreted as a constructive cooperation between public and private sectors. Part III deals with public rights and duties, Part IV with the status and powers of the Head of State. Part VI deals with the Legal System and the Independence of the Judiciary.

Other relevant developments include the specification of the responsibilities of the Ministry of National Economy and the Ministry of Finance, which had previously been one Ministry. Royal Decree 42/96 sets out a series of privatisation policies and controls, giving priority to privatisation of service industries such as sanitation, drainage, electricity, water, communications, highways and postal services. Recent changes to the Commercial Companies Law have simplified the procedure for establishing and expanding joint stock companies. The Law of Income Tax on Companies has also recently been amended, reducing the rates of tax applicable to Omani registered companies with foreign ownership of 90% or less, and concessionary rates for public joint stock companies with at least 51% Omani shareholding, in which at least 40% of the company's capital must have been offered for public subscription.

PORTUGAL

José Filipe Abecasis
A M Pereira Sárragga Leal,
Oliveira Martins, Júdice e Associados,
Lisbon, Portugal

1.   BACKGROUND

1.1   Economic[1]

Although historically small Portugal's construction industry has been growing rapidly ion the last few years, supported by foreign investments, EU funds for the building of infrastructures and some major public projects like Expo '98. The estimated total gross value of production of the Portuguese construction industry for 1997 is of Esc. 2.638.400.000.000$00 (approximately US.$ 14.822.471.910), representing an estimated growth of 6.9% over 1996.

There is a high number of EU contractors established in Portugal, most of whom operate by forming close association with Portuguese contractors in consortium or by buying up Portuguese contractors.

Until recently, the local industry was mainly composed of small size companies owned by entrepreneurs of the "self-made-man" type[2]. Such companies employed, for the most part, poorly skilled workmen, mainly in production and administration areas, little or no attention was paid to management or legal requirements.

The situation arose where only public works were subject to formal contracts. Private undertakings were governed only by proposals and estimates of costs presented by the contractors. Occasionally correspondence might be exchanged, but contracts were seldom entered into (except in major projects) and, where they did exist, they were quite rudimentary, scarcely exceeding the indispensable legal requirements and the remission for the public works regime in any other aspects.

In addition, employers did not have technically qualified experts to inspect the works of contractors effectively and often lacked the necessary knowledge and time. Work was therefore poorly supervised. Defects were often not detected in time to enable the parties to resolve their conflicts before resorting to court.

Under these circumstances, legal certainty was obviously affected by the vagueness of any contractual terms regarding the parties' rights and obligations and by the breaches of contract that were remedied by the owner's acceptance (i.e. inspections that were carried out before delivery or that were improperly done, claims not laid down for non-observance of deadlines, absence of contractual provisions in respect of penalties, lack of documents relating to meetings held between the employer and contractor, etc.). In addition, there were judicial and doctrinal disagreements about whether private contracts should be subject to the same legislation as public contracts. The prevailing opinion was that they should with regard to the regulation of procedures, but not with regard to provisions granting special powers to the public entity party.

At present, the situation described above has already undergone, and continues to undergo, significant changes. International competition resulting from the integration of Portugal into the European Union; the relaxing of economic, social and cultural barriers; a higher level of education; the industrialisation of the former almost artisan construction companies; and a profound restructuring of the Portuguese economy, are among several factors which have resulted in the formation of completely different types of companies. Employers are now more experienced and more capable of understanding the need for contractual devices to protect their rights and interests, and are more apt to resort to such devices.

Therefore, the Portuguese construction industry is now dominated by an ever growing number of national companies, which are gradually becoming more international[3] by intervening in foreign markets or by associating with foreign companies (whether temporarily or permanently), as well as aiming at the domestic market. These companies increased their knowledge of business management, legislation, contracting techniques and the supervision and execution of works.

Even the man in the street is now more aware that, if he intends to engage a building contractor, it is not enough simply to talk, exchange correspondence and obtain an estimate to regulate the relationship which will be established, or to guarantee the protection of his interests.

1.1.1   Trends

The gross value of production of the Portuguese construction industry was PTE 1.706.400.000.000$00 (US $9.596.516.854) in 1992 and PTE 2.373.900.000.000$00 (US $13.336.516.853) in 1996, representing a 1.2% growth of this industry in 1993; 0,5% in 1994; 5,9% in 1995; and 4,5% in 1996; a 6,9% growth is forecasted for 1997.

Likewise the construction industry employed 346,200 people in 1992; 340,200 in 1993; 330,800 in 1994; 340,300 in 1995; and 343,100 in 1996.

Between 1990 and 1994, which is the last year in respect of which this type of data is available, companies with 5 to 9 employees represented around 10.5% of all companies in this industry and companies with over 100 employees increased from 39.9% to 36.2%, being affected by the crises of the construction industry during those years, by which large companies were more deeply affected.

Mortgage credit interest rates fell from 24.6% in 1985 to 13.9% in 1994, and is, at present approximately 9%.

The inflation rate of the Portuguese economy was 13.3% in 1990; 13.3% in 1991; 12.1% in 1992; 9.2% in 1993; 7.5% in 1994; 6.0% in 1995; and 4.3% in 1996; and it is expected to be 3.4% in 1997.

1.2   Legal

1.2.1   Portuguese Legal System

The Portuguese legal system stems from the Roman or Latin systems, whose distinctive trait is the strict definition of the sources of law and the separation of powers amongst the different sovereign powers, in particular the legislative and the judicial power. Therefore, in addition to being based on the codification of the various areas of Law containing the relevant supporting rules, Roman law systems are characterised by the fact that the judicial precedent is not a source of law[4].

According to article 6 of the Constitution, Portugal is a unitary state, although it recognises the autonomy of local powers, both in respect of self-governing regions (Açores and Madeira) and city councils. These local powers enjoy different levels of autonomy which is broader in the case of the self-governing regions as a result of their political and administrative Statutes, whose terms empower such self-governing regions to establish their own regional government[5] and to legislate in respect of matters of specific interest for their population, without prejudice to the Constitution and to the laws applying to the whole of the Republic (General Laws of the Republic)[6]. To a smaller degree, local powers too, have a certain degree of autonomy as far as the definition and protection of the interest of their populations are concerned and have decision-making and executive bodies (*Assembleias Municipais* and *Câmaras Municipais*, respectively), which are empowered to issue regulations in respect of matters of their own concern, within the limits set forth by the Constitution, by the laws and by the regulations issued by the central administration[7].

At present, local powers of an intermediate degree, known as *Regiões Administrativas* (administrative regions) are being established; these shall not, in principle, enjoy such autonomy as the self-governing regions, in particular with regard to legislative powers, but shall have more autonomy than the local powers as they shall be entrusted with the co-ordination of the action taken by the various Municipalities belonging to such administrative regions.

Therefore, the Portuguese legal system, based on the 1976 Constitution as amended in 1982, 1989 and 1997, may be characterised by its unity.

Immediately below the Constitution, the Laws of the Parliament and Decree-Laws of the Government, rank *pari passu*, although certain matters are, in whole[8] or in part[9], reserved for the Parliament. Should the above mentioned statutes (General Laws of the Republic) affect the interest of the self-governing regions, they may be regulated or even, in some cases, adapted by the legislative bodies of the self-governing regions, by means of regional laws (*"decretos legislativos regionais"*) which, however, shall uphold the setting provided by the law they regulate[10].

At a lower rank there are regulations developing the general principles set forth in the above mentioned statutes; such regulations may be issued by the Government as Rulings (*"Decretos Regulamentares", "Portarias"*)[11] or by the legislative bodies of the self-governed regions (*"Decretos Regulamentares Regionais"*)[12].

Lastly, with the nature of administrative regulations (which are therefore subject to the laws they regulate), there are the orders issued by the local powers (*"posturas", "regulamentos"*) in respect of matters of specific interest to the pertaining population, amongst which the creation of taxes for the provision of services, the granting of licences in respect of utilisation of municipal properties and the plans governing municipal planning and urban development (*Plano Director Municipal*)[13] are of particular interest for the purpose of this monograph. Because such Plans must obviously co-exist with those of the abutting Municipalities, they must follow national and regional land use policies and are subject to ratification by the Government.

Furthermore, as a result of the entry of Portugal to the European Union on January 1st 1986, the national legal system incorporates Directives and Regulations issued by the European Community (in the form of laws or decree-laws); incidentally, when expressly set forth, some of the above mentioned regulations are automatically applicable in Portugal and in the remaining member States.

Likewise, international treaties and conventions of which Portugal is a signatory State, are incorporated in the Portuguese legal system, after approval by the Parliament and enactment by the President[14].

1.2.2   Portuguese Court System

The judicial system in Portugal comprises the Constitutional Court, the courts of first and second instance, the Supreme Court of Justice, the Court of Accounts, the military courts, tax courts, administrative courts of first instance, the Central Administrative Court, the Supreme Administrative Court, maritime courts and penal courts.

As mentioned above, unlike common law jurisdictions, courts in Portugal are not bound by judicial precedent.

The judicial process, mainly in civil and administrative courts, is organised under the adversarial principle which is complemented by a certain degree of inquisitorial power granted to the judge. That is to say, courts may not examine any question nor any facts therein unless a party renders it to the case and the other party is offered the opportunity to oppose[15]. However, once the questions are submitted to the court, the judge must endeavour to ascertain the material truth and seek to settle the dispute[16].

Portugal is divided into judicial districts, which are in turn divided into counties ("*comarcas*"). These counties are grouped into judicial circles.

The Constitutional Court is empowered to determine whether a law or decree-law is constitutional. The court exercises its power at three different levels:

a)   it may determine whether a statutory bill or an international treaty to be enacted or ratified by the President of the Republic or signed by one of the ministries, is constitutional;

b)   it is empowered to decide appeals in which a law or a decree-law has been held to be unconstitutional;

c)   it may determine whether a law or a decree-law already enacted is constitutional, whether in the abstract or in the context of any material case.

The Constitutional Court comprises 13 judges, ten of whom are appointed by the Parliament. These judges then choose the remaining three from judges of the other courts. Judges are appointed for a tenure of six years, which is not renewable.

The courts of first instance ("*tribunais de comarca*" or circle courts) have jurisdiction over a particular county and are empowered to hear all cases not specifically reserved for other courts. They may be organised as specialised courts, depending on the matter they are to examine, be it civil, labour, family, minors, penal law, bankruptcy or maritime law. There is no court specialised in construction law.

The courts of second instance ("*tribunais de relação*") have jurisdiction over a particular judicial district and are empowered to decide appeals in cases where the amount involved exceeds the limit of the competence of the courts of first instance, namely cases involving more than PTE 500,000.00. They may also try judges and district attorneys, decide jurisdictional disputes arising amongst judges of the relevant judicial districts and review foreign judicial decisions. There are special sections for civil, criminal and labour jurisdictions. Courts of special competence may be organised in judicial districts and circles.

The Supreme Court of Justice has jurisdiction over the entire Portuguese territory. It is empowered to decide appeals concerning questions of law - as opposed to cases involving questions of fact - where the amounts involved exceeds the limit of the competence of the courts of second instance (i.e. cases involving more than PTE 2 million). It may also uniformise civil courts jurisprudence, try the President of the Republic, judges and public attorneys of the Supreme Court or of the courts of second instance, settle jurisdictional disputes arising between other courts and decide habeas corpus matters.

There are four types of special courts:

a)   Court of Accounts, empowered to supervise the public expenditure of the Portuguese Administration and to render its opinion on the public budget (*"Conta Geral do Estado"*);

b)   military courts, empowered to hear cases involving military matters;

c)   tax courts, empowered to hear cases involving a violation of tax laws;

d)   administrative courts, empowered to hear cases involving matters of administrative law, such as public contracts, challenges decisions of the Administration and responsibility of the Administration for damages caused in the exercise of the irrelevant powers.

Finally there are arbitration courts, whose existence is provided for in the Constitution[17] and whose creation and operation is governed, in its fundamental principles, by Law 31/86 dated August 29th; this statute further sets out the rules for the setting up of the court and process, where the parties did not establish such rules in their arbitration clause. There are a few institutional centres of arbitration, whose establishment is subject to the authorisation by Order of the Ministry of Justice[18], some of them specialised in questions of consumers protection and one specialised in questions of construction law (A.I.C.C.O.P.N., in the city of Porto). This matter will be further developed under 3.2.4.

1.2.3   Legal Professions

The system of legal profession in Portugal is organised as follows:

a)   The judges are magistrates having tenure of the sovereign power of the courts[19] and appointed to try the cases submitted by the parties as well as to administer justice; they are independent from the remaining powers of the State.

b)   The magistrates of the Public Prosecutor's Office are an independent body of magistrates, entrusted with representing the State in court, conducting penal prosecution and, in general, defending legality[20];

c)   The lawyers are independent professionals entrusted with providing legal counsel to their clients, in and out of court, advising and protecting the interest of the mandator[21];

d)   The "*Solicitadores*" are specially trained technicians who essentially discharge bureaucratic duties relating to the preparation and application for documents and who may assist their clients in certain proceedings of a technical rather than legal nature;

e)   The Notaries Public, who are public servants who oversee certain kinds of legal acts for which the law requires special formalities (notarial deeds, wills, certain powers of attorney, etc.); they confer legal form and attest to the genuinity of any documents thus executed[22];

f)   The Heads of Real Estate or Companies Registries are public servants entrusted with the registration of all acts for which registration is required by law, in particular in respect of real property and any rights attaching thereto or in respect of companies and other legal entities and any relevant information thereon (registered capital, partners, registered office, binding of the company, etc.)[23].

The relation between client and lawyer need not be formalised save for the performance of any official acts in which the lawyer is to represent the client. Such formalisation is usually made by means of a power of attorney whereby the client appoints the lawyer as his attorney.

However, it is becoming increasingly customary for both parties to specify the terms of their relations in particular the lawyer's fees and payment conditions, as well as how any contacts are to be made and, as far as possible, the anticipation of the time such services are expected to take. These matters are sometimes put down in writing, in the form of letters exchanged between the Parties.

1.3   Consumers Protection

The basis of the legal system governing consumers' protection in Portugal is set out in Law 24/96, dated 31st July. Furthermore, the Constitution sets out the consumers' rights to the quality of goods and services, to the protection of health and safety and the right to be indemnified for damages, in addition to empowering the relevant associations to protect their members' rights in court[24].

Pursuant to article 2 of the above mentioned law, a consumer is understood to be anyone who purchases goods, services or rights for non professional purposes and as such, although no distinction is made between individuals and legal entities for the purposes of providing such protection, this description applies, in the majority of cases, to individuals.

Consumers are entitled to the quality of goods, services and attaching rights being supervised by the State and ensured by economy operators, so as to protect consumers' health and in order to make such products safe and suitable for the purposes they serve; they are further entitled to be clearly and efficaciously informed of the characteristics and functions of such goods, services or rights and to that effect misleading advertisement is not permitted; consumers are entitled to be effectively indemnified for the damages sustained through accessible and expeditious legal processes; they are entitled to have their rights and interests duly represented by consumers' associations upon the legal or administrative definition of any matters which concern them.

In this respect, the establishment of several rules extending the period during which economic operators may be held responsible for the quality of goods, services and rights[25] and are under the obligation to indemnify any damages sustained, is particularly relevant.

On the other hand, as far as the settlement of conflicts in this area is concerned, several institutional arbitration centres have been established, whose competence is in general restricted to a city or region; such arbitration centres adopt simple and swift proceedings in order to render prompt decisions.

1.4   Construction

1.4.1   Construction Law

In respect of construction law, the Portuguese legal system contains several statutes governing many aspects of this issue, not only in respect of contracts and the performance thereof but also in respect of the technical rules to be complied with and of the planning of the territory whereupon constructions is to take place.

With regard to the compliance with technical rules for construction, the most important statute is Decree-Law 38382 dated 07/08/51 (*"Regulamento Geral das Edificações Urbanas"* - General Regulation for Urban Construction). This Regulation sets out minimum construction sizes and features to be complied with, as well as the urban parameters which, in the absence of any other provisions[26], must be taken into account by the constructor in respect of standing and volume. On the other hand a number of other statutes, listed in Ministerial Order 338/89, dated 12/05, regulate and detail various aspects of construction - safety, quality, environmental protection, soundproofing and heat insulation, specific purposes of constructions, etc.[27].

Mention must be made of Decree-Law 167/97, dated 04/07 and of Regulations 36/97 dated 25/09 and 34/97, dated 17/09 on account of their importance to the Portuguese economy and of the comprehensiveness of their provisions. The first of the said statutes sets out the legal provisions governing licensing of tourist developments (by Municipalities, having heard the Directorate General of Tourism); the second one sets out quality requirements for construction, finishings and installations of hotels and the third similarly, but in respect of real estate developments.

With regard to construction contracts and their performance, the Portuguese legal system makes a distinction between the relations of private entities between themselves and the cases in which a public entity is also a party.

Relations between private entities are governed by the provisions of the Civil Code, namely the legal provisions governing contracts, from the negotiation and formation thereof, to the performance, exercise of rights and fulfilment of obligations, changes to the contracts, unfulfilment of the obligations of the Parties thereunder as well as the applicable penalties. These general provisions are complemented by a group of rules, also contained in the Civil Code, concerning construction contracts in particular, set out in articles 1207 to 1230. These rules set out the obligation of the constructor to ensure the quality of the construction and to comply with the relevant technical rules (article 1208); the supervision of the construction by the Owner (1209); the transfer of ownership of the works and of the materials installed, during the performance thereof (1212); the performance of any additional works or the reduction of works in respect of the original project (1214 to 1217); and the provisions governing the guarantee provided by the constructor in respect of the quality of the works performed, possible claims in respect of construction defects and the repair thereof (articles 1218 to 1226).

As far as construction contracts to which public entities are parties, such contracts are essentially governed by Decree-Law 405/93 dated 10/12; this statute is often used to supplement construction contracts entered into between private entities as it provides a detailed regulation of the relations between the parties.

Decree-Law 405/93 is divided into three main parts. In the first part the different kinds of contracts are characterised, according to their extent and to the amount of work involved and sets out the rules governing public procurement; the second part deals with the relation between the parties during the performance of the works, payment conditions and provisional and final acceptance, as well as the guarantee period; the third part deals with litigation arising out of the failure of the parties to fulfil their obligations under the contracts.

As far as land use planning is concerned, it should be noted that, to date, there is no basis law (*"lei de bases"*) of national scope, establishing land use policies to be adopted by plans at the regional or local level. The corresponding bill is still being prepared and shall be soon discussed in Parliament.

However, Decree-Law 176-A/88, dated 18/05 sets out the provisions governing the preparation of regional land use plans. As a result of the said Decree-Law, a number of regional land use plans were drawn up and cover, at present, the following areas: Rio Douro, Outer Lisbon, Aguieira, Coiço and Fronhos, central coasts, Algarve, Alto Minho; coast of Alentejo[28] and a few others are being prepared.

At the local level, Decree-Law 69/90, dated 02/03 sets out the obligation of all Municipalities to draw up the so called land use municipal plans for the pertaining areas. These plans are of three different categories according to their being more or less detailed:

a)   The General Municipal Plans (*"Planos Directores Municipais")* which cover the whole of the Municipality and set forth the guidelines for the organisation of the land, specifying which areas are intended for various purposes (habitation, industrial, trade and services, agriculture, etc.); the overall urban parameters and rules for the protection of the environment, landscape and cultural heritage.

b)   The urbanisation plans (*"Planos de Urbanização"*) which cover a specific area of the Municipality and provide more detailed rules governing the development of such area;

c)   The Detail Plans (*"Planos de Pormenor"*) which cover a limited area of the Municipality and provide the details of the pertaining urban regulation;

Lastly, with regard to the licensing of construction works to which all private works are subject, the most important statutes are - in addition to Decree-Law 38382 dated 07/08 referred to above - Decree-Law 448/91 dated 29/11 and Decree-Law 445/91 dated 20/11.

The first of the above mentioned statutes (Decree-Law 448/91) sets out the legal provisions governing the licensing of urban plot division for the purposes of construction, including the construction of the pertaining infrastructures. The licence is granted by the Municipality where the plot is located. Should the plot concerned not be covered by a Detail Plan or by a Urbanisation Plan, the Municipality shall request the opinion of the regional or central authorities before rendering any decision. Obviously the plot division plan shall comply with the urbanistic rules set out in the relevant Plans and shall describe the buildings' features and the use for which they are intended as well as provide the construction projects of the corresponding infrastructures, of which the constructor shall be in charge.

Decree-Law 445/91 sets out the legal provisions governing the licensing of private works and the submittal of projects, the appreciation thereof, the consultation of entities outside the Municipality and the approval or rejection of the application. Again, these licences are granted by the Municipalities after consultation of the regional or central authorities before any decision is rendered if the construction project is not covered by any plot division plan, detail plan or urbanisation plan. Only private works to be performed inside already existing buildings, which do not affect the structure or the use thereof, do not require any licence; in such case it shall be enough to merely inform the Municipality of the performance, characteristics and location of such works.

The applications referred to above both in respect of plot divisions and of construction of buildings, must be submitted by whoever is entitled to dispose of the property. The pertaining designs and projects may only be signed by properly trained technicians - architects or engineers, depending on the works to be performed - who undertake any responsibilities for the quality of the construction and for the compliance with all applicable legal provisions and regulations. Likewise, the performance of the work itself must be monitored by a properly trained technician - an engineer - who shall be responsible for the quality of the works performed in compliance with the project and with the relevant licence. Such a technician is usually *Director Técnico* appointed by the Constructor.

1.4.2   Construction Contracts

In Portugal there is only one standard form for construction contracts in which the Owner is a public entity. This form was published by Government Ruling 428/95 dated 10/05 which regulates the above mentioned Decree-Law 405/93 dated 10/12. The said Decree-Law contains the legal provisions governing public construction works. This form is used for all public construction works unless the extent or the particular complexity of the works require special contractual conditions to be expressly set out, which are also published in the form of administrative regulation.

With regard to construction contracts for private works there is no published standard form and the provisions of the above mentioned Decree-Law 405/93 are usually applied as far as the performance of the works, payment conditions and acceptance of the works are concerned, or alternatively the use of clauses of the form established in Government Ruling 428/95.

However, it is increasingly frequent, in particular as far as contracts of higher value are concerned, to adopt the form of construction works contracts issued by FIDIC[29], as is the case, for example, for the construction of the new River Tejo crossing.

Construction contracts concerning public works are governed, as referred to above, by Decree-Law 405/93 dated 10/12 and by Governmental Ruling 428/95 dated 10/05. The main characteristic of this kind of contracts is essentially that the choice of the contractor is subject to a number of very strict rules[30] - amongst which the establishment, from the outset, of a material part of the clauses of the prospective contract -; the contractor is not entitled to justify the suspension of the works bringing forward the failure of the other party to comply (*"Excepção de não cumprimento"*) unless such has been previously approved by the public entity which is a party to the contract or has been ordered by a judicial court; the public entity which is a party to the Contract may, in cases of repeated unfulfilment by the contractor, take possession of the site and of the materials and equipment of the contractor, by an act of authority. However, in all other matters, contracts governing public works ensure the balance of the positions held by the Parties and set out adequate mechanisms of compensation for cases of breach of contract or alteration of the conditions originally foreseen.

Another distinctive trait of public works contracts is the fact that they are regarded as administrative contracts, not only because one of the parties is a public entity but also because the relation between the Parties is subject to specific regulations of the Administrative law - as the one summed up in the preceding paragraph - and to the public interest of the contract concerned. As such the interpretation and performance of these contracts are the jurisdiction of the Administrative Courts[31].

However, both the general law governing the activity of the Administration[32] and articles 229 and 230 of Decree-Law 405/93 entitle the parties to agree on the submittal of the relevant conflicts to arbitration courts which shall operate in accordance with the general terms of the civil procedural law[33], although a somewhat simplified process in that parties may only submit a plea and opposition and may only present two witnesses for each of the facts under discussion; final pleadings must be put down in writing.

Pursuant to articles 229, nr. 2 of Decree-Law 405/93 the awards of the arbitration are rendered according to equity[34] and therefore cannot be appealed to courts.

The above mentioned award shall be kept in the files of the *"Conselho Superior de Obras Públicas e Transportes"*[35] (Higher Board for Public Works and Transport) whose Chairman shall specify the terms under which the award is to be enforced by the public entity; on the other hand, enforcement by the contractor is entrusted to judicial courts.

Private works contracts, on the other hand, as referred to above, are essentially governed by the Parties themselves and by the general provisions of the Civil Code governing contracts. Articles 1207 to 1230 of the Civil Code regulate certain particular aspects of the contractual relations in construction contracts such as: the duty to perform the works according to the applicable quality requirements; the right conferred to the Owner to promote the supervision of the works, directly or through a representative; the transfer of ownership of the works and of the materials therein installed; changes to the agreed works schedule; claims in respect of construction defects and the obligation of the contractor to repair such defects during the 5 year guarantee period[36]; the transfer of the risk during the performance of the contract; the extinction thereof.

Decree-Law 445/91, dated 20/11 (legal provisions governing the licensing of private works) and Decree-Law 155/95 dated 01/07 (legal provisions governing safety at work on building sites) are also sources of law in this respect and set out obligations of the contracting parties with regard to licences, preparation and performance of the construction works.

Any litigation relating to the interpretation or performance of private construction contracts, not provided for by the parties, shall be submitted to the court of first instance having jurisdiction over the place where the obligation concerned should have been fulfilled or over the place where the party at fault is domiciled, at the discretion of the plaintiff[37].

The parties may agree upon the territorial jurisdiction of the court to which litigation arising out of the contract[38] shall be submitted, which, as a rule, is the one having jurisdiction over the domicile of any of the parties or over the place where the works are to be performed, or may agree upon the submittal of such litigation to an arbitration court provided such litigation respects such a disposable right.

Law 31/86 dated 29/08 regulates the setting up and operation of the arbitration court and enables the parties to appoint the arbitrators and specify the rules according to which the court is to operate and the proceedings are to be followed; it also sets out, several supplemental provisions which shall be applicable in case none of the above have been established by the parties.   The arbitration court shall settle any litigation which may arise under a contract or any litigation already existing which the parties agree to submit to arbitration.

The awards of the arbitration court on any litigation submitted by the parties is, for the purposes of the Portuguese legal system, equivalent to a decision rendered by a court of first instance. Therefore, unless the parties have waived their right to appeal or have set out that arbitrators shall decide according to equity[39], they may appeal in respect of arbitration awards to courts of second instance having jurisdiction over the same jurisdictional division where the arbitration court was set up.

2.   CONFLICT MANAGEMENT

2.1   Non-binding

2.1.1   Dispute Review Boards

This particular process of settlement of disputes is not traditional in Portugal. It was not until very recently that a mechanism of pre-litigation settlement with similar characteristics was seen in respect of some of the construction contract and contracts for the purchase of plots entered into by PARQUE EXPO '98, S.A.[40] and other contracts based thereon.

Such process is merely intended for the settlement of litigation over technical matters. According to the above mentioned process a committee, generally consisting of three expert technicians, is set up, two of such technicians being appointed by each party (the constructor and the owner) and the third one being co-opted by the first two or agreed upon by the parties.

Such committee is only set up for the express purpose of settling any litigation. As a rule, only the third expert, appointed by the parties to preside the said committee, is independent.

The committee is to settle the question raised within a short period, normally one month and the party which is not satisfied with the decision rendered may submit the question to a court which is normally an arbitration court.

2.1.2   Dispute Review Advisers

This process of settlement of conflicts is not traditional in Portugal either and there is no knowledge of any contracts setting it forth.

2.1.3   Negotiation

Although there is long-standing tradition in Portugal in respect of this process of settlement of disputes, the same has never been subject matter of any in-depth study enabling it to be included in university tuition. Therefore, the practice of negotiation is essentially learnt on the basis of personal experience.

As a matter of fact, as the Portuguese market for construction is small (if compared with the majority of the markets of other European countries) and is dominated by a relatively small number of companies (medium size companies, by European standards), and also the Administration is the most important client, given the amount and scope of the public works awarded, the major private companies on the market work in close contact and are well known to each other.

This being so, in addition to the contracts being performed in a fairly informal fashion (even in the case of public works, although to a lesser degree) the resolution of disputes too is, at first, subject to mechanisms of negotiation at various different levels (technicians, heads of department and directors) and it is not until it becomes impossible to resolve the question by negotiation that the parties resort to arbitration or to courts.

These mechanisms, in respect of which no special formalities are set forth[41], are, at first, triggered by the individuals directly involved in the works who then refer whatever questions they have been unable to settle to their superiors who shall, in turn, try to reach an amicable solution.

2.1.4   Quality Matters

The public administration department in charge of controlling and promoting quality development in Portugal, is the *"Conselho Nacional de Qualidade"* (Quality National Board), wherein the *"Comissão Sectorial para a Construção"* (Construction Board) operates as a representive of various bodies and associations of entrepreneurs which operate in the construction market.

The Construction Board is currently examining suitable forms of certifying a larger number of construction companies in Portugal, according to NP EN ISO 9001 requirements, as the number of firms so certified is currently extremely small. Portugal has adopted the norms of ISO 9000, under the name NP EN ISO 9000.

At present, quality control relating to construction consists basically in setting out quality patterns required for the works in the Schedule of Conditions attached to the invitation of constructors to submit their proposals, and in the monitoring of the works, as they are being carried out, entrusted to entities appointed by the owners.

It should be noted that the new statute regulating the licensing of private works does not require the same to be examined from a technical point of view if such works have a certificate of quality and compliance with the applicable legal provisions and regulations. The said certificate is granted by duly qualified entities whose technical capability[42] is recognised by the Ministry of Planning and Management of the Territory.

Public works contracts may be awarded by two different processes:

(i)   public invitation;
(ii)   closed invitation.

The first category includes:

(i)   public tenders;
(ii)   closed tender with applications;
(iii) tenders by negotiation.

As a matter of fact, any of the above mentioned processes is opened by the publication of the relevant notice and all entities meeting the relevant requirements may be admitted to tender.

The form of public tenders, in particular, is applied whenever the value of the contract to be awarded exceeds the amount fixed by the European Union for the application of the directives in respect of the co-ordination of awarding procedures and whenever the specific characteristics of subject matter of the contract to be awarded do not set out any other process.

Whenever the value of the contract is lower than the amount referred to above or whenever its characteristics so require, it is possible to resort to the form of closed tender by submitting an application or of tenders by negotiation with notice. In this case the bidders who are in the best position, from those who responded to the notice of the tender, shall be chosen and any extremely complex proposals and/or projects are then developed further as it would be impracticable to do so at an earlier stage or in documents of a more general nature.

The following are closed tender process:

(i)   closed tender without applications;
(ii)   tender by negotiation without publication of notice;
(iii)   direct arrangement.

The processes have the same scope of application and are only applicable for contracts of a lower value or regarding the award of contracts whose object is the continuation of a contract previously entered into; if a public tender previously launched for the award has been deserted; if the object of the contract may only be performed by a specific entity (for example, patent rights or copyrights); or if the execution of the contract, being a matter of urgency, may not be subject to the formalities of a public tender.

The processes applicable to the establishment of public works contracts, concession of public works and procurement of public works are regulated by the legal system governing public works, as approved by Decree-Law 405/93 dated 10/12 through which Directive 89/440/CEE dated 18/07/89[43] was incorporated into the Portuguese legal system.

The procedure for the establishment of public contracts for the rendering of services and the lease and purchase of movable assets are regulated by Decree-Law 55/95 dated 29/03, whereby Directives 92/50/CEE   dated 18/06/92 and 93/36/CEE dated 14/06/93 were incorporated into the Portuguese legal system.

Lastly the covenanting of construction works, purchase of goods and services and the concession of exclusive contracts, of public works and services by the bodies of the local administration (Municipalities) and the regional administration (Self-governing Regions) is regulated by Decree-Law 390/82 dated 17/09, of which articles 2,3,4,5,7,8 and 9 (concerning public tenders for public works and the procurement of goods and services) were revoked by the above mentioned Decree-Law 55/95.

The criteria adopted for the choice of the best proposal as far as the processes open to various candidates are concerned, must be mentioned in the notice and in the tender programme or in the invitation to tender, depending on it being a public tender or a closed tender, and listed in order of importance.

Of course, after being assessed, the proposal chosen shall be the one which best meets the criteria mentioned above, for which purpose a report showing the ranking of each proposal shall be drawn up.

Between the assessment of the proposals and the award, the draft contract shall be drawn up, in accordance with the conditions set out in the tender and those of the above mentioned proposals.

Should the preferred bidder consider that the draft contract does not strictly reflect the said conditions, the same may lodge its complaint with the owner and may desert the contract should its complaint not be attended to. After completion of the process described above, the award is formally granted and the contract entered into, for which purpose the preferred bidder shall provide a performance bond in the amount of 5% of the total value of the contract.

The proposals are examined in two stages. In the first one, during the public tender itself, the documents presented are formally examined in order to ascertain that bidders are duly qualified and fulfil the conditions required to be admitted to tender as well as any other documents and relevant attachments which shall comply with the required formalities. During the second stage, the period may be more or less depending on the complexity of the contract to be awarded, the proposals are assessed from the technical point of view in accordance with the criteria set out in the notice and tender program[44].

The two stages referred to above shall be carried out before two separate committees[45], both designated by the owner and consisting of no less than three members. The committee presiding over the public tender shall draw up the relevant minutes, which shall be delivered to the bidders at the latter's request. The committee set up to examine the proposals, which may or may not request the assistance of experts without voting rights, shall further draw up a report of its works grounding the appraisal of the proposals, strictly in compliance with the criteria set out in the tender documents.

The participation in closed tendering is, by the nature of the process itself, limited to the bidders invited by the owner to submit a proposal, for which reason the examination of the technical and financial capabilities of the bidders, being prior to the choice of the entities invited to tender, is dispensed with.

In public tenders, on the other hand, bidders must produce evidence that they meet the conditions required to be admitted to tender, attaching to their proposals, for such purpose, whatever documents are expressly required under the terms of the Tender Notice and in the Tender Programme, which shall be referred to as capability documents.

Applications shall be submitted in various opaque envelopes, all of which shall be closed and sealed. One of the envelopes shall contain the required capability documents; another envelope shall contain the price proposal as well as any other relevant documents, such as any documents which account for the price offered, unit price list, works and payment schedules; the third envelope may, at the discretion of the bidder, contain any other documents not required by the owner but providing further clarifications.

To the extent such is foreseen in the tender documents[46], the bidder may further submit alternate proposals, prepared strictly in accordance to the above mentioned documents, which may alter any clauses of the schedule of conditions (conditional proposal) or a variation of the project put out to tender (proposal with variation). Such proposals must then be clearly identified as such and assessed separately, the corresponding price offer being also submitted.

In 1996, 3049 public works projects were put out to tender in Portugal, in the total value of PTE 489,019,000,000.00, 2128 of which were actually awarded in the total value of PTE 336,668,000,000.00. 31.1% of the said awards were in respect of the construction of buildings; 4.7% to hydraulic works; 32.2% to communication routes; 24.9% to urban construction and 6.9% to other kind of works[47]. The comparison between the value of the contracts awarded and the net value of the product of Portuguese construction companies - i.e. PTE 2,638,400,000,000.00 - shows that public works accounted in

1996 for approximately 12% of the Portuguese construction market.

2.1.5   Partnering

As already mentioned, the construction market in Portugal is rather small; the market is controlled by a small group of medium size firms; the most important client is the Administration with public works contracts.

Therefore, Portuguese construction companies are always in contact with each other, be it by associating with each other for the performance of contracts[48], be it by competing for the award of other contracts, or even by participating in associations for the protection of their interests.

As a result of the above, the environment is one of co-operation, where companies seek to settle whatever conflicts may arise between themselves or to define policies of market growth and of protection of common interests.

It is increasingly frequent for smaller companies to "gravitate" around larger companies; some of such smaller companies specialise in the performance of certain kind of works while other, less specialised, perform general tasks. There is no formal relation between these two kinds of companies which merely operate together in the market by means of construction subcontracting; thus the larger companies, upon being awarded the contracts, appoint such smaller companies as subcontractors for the performance of tasks or the procurement of equipment and man power.

The larger companies profit from this kind of association as it enables them to cut down expenses with personnel on payroll and construction equipment, which would otherwise be much more significant representing an excessive burden in terms of budget and management and would thus impair their commercial agility.

The smaller firms benefit from having a certain number of works guaranteed, which it would be difficult to secure alone.

2.2   Binding

2.2.1   Partnering

Another consequence of the specific traits of the construction market in Portugal, referred to in 2.1.5 hereabove, is that construction companies often become associated either in consortia[49] (as it is usually the case) or under the form of Complementary Groups of Companies[50] or European Economic Interest Groups[51] (not so usual) be it for the purposes of the performance of any specific construction contract or in order to establish a long lasting form of carrying on their activity.

Any of the above mentioned forms of association implies the existence of a board wherein all members are represented and have equal voting rights or pro rata to their relevant holdings, as set out in the contract. The above mentioned board is the natural, albeit informal, place were disputes amongst associates may first be settled.

The majority of joint venture agreements (under any form) set out that, should the settlement of disputes not be possible under the terms of the preceding paragraph, the various boards of directors should seek to reach an agreement, for which purposes a period of time of 1 to 2 months is foreseen. The provisions governing this kind of the settlement of disputes, although foreseen in the agreement, are not yet formally established.

The majority of joint venture agreements provide for the submission to arbitration court of any disputes which remain unsolved.[52]

3. DISPUTE RESOLUTION

3.1 Non Binding

3.1.1 Conciliation

In Portugal there is no distinction between the terms conciliation and mediation[53], as employed in this monograph, nor are they formally established or regulated processes of settlement of disputes.

As a matter of fact, in this respect, in addition to the process referred to in 2.1.1 hereabove, which is still extremely new and very little used, or to the negotiation processes mentioned in 2.1.3. 2.1.5 and 2.2.1, which are, as referred to hereabove, basically technical processes and therefore have no uniform characteristics and are not regulated separately, no other material content may be found for these terms.

Mention is made in 3.2.4 hereafter - outlining the procedural terms of litigation[54] - that the judge must, in all action taken before a decision is rendered, endeavour to cause the parties to reach an amicable settlement over the litigation, and seek the conciliation of such parties.

3.1.2 Executive Tribunal

This form of settlement of conflicts is, again, not implemented in Portugal and least of all established in formal terms. There is a slight similarity to pre-litigation settlements amongst the boards of directors, usually foreseen in consortium agreements, as mentioned in 2.2.1 hereabove. However, such processes are essentially based on negotiation and are informal and do not imply the participation of third parties unless specific circumstances of the case in question lead the parties to request the assistance of someone in a better position to promote the resolution of the dispute.

3.1.3   Mediation

About this term see 3.1.1 hereinabove

According to the Portuguese legal system no binding decision can be rendered by the mediator in this form of processes.

As a matter of fact, in accordance with the proposed terms, the mediator does not correspond to arbitration courts.[55] In Portugal, only decisions rendered by arbitration courts[56] or the agreement between the parties,[57] are binding.

Therefore, because the mediator is not an arbiter and because the parties agree to require the intervention of such mediator merely in order to help them find a resolution and not to render any decision,[58] no binding force could be ascribed to such decision.

3.1.4   Negotiation

Given the informality which, in Portugal, characterises the negotiation process referred to in 2.1.3 hereabove, no process of resolution of conflicts by negotiation with different traits can be broached in this chapter.

3.2   Binding

3.2.1   Adjudication

The only forms of binding decisions legally foreseen in Portugal for the settlement of disputes are the agreement between the parties and the arbitration, under the terms referred to in 3.1.3 hereabove, or lawsuits.

On the other hand, contrary to the Anglo-Saxon practice, Portuguese construction contracts do not provide for the intervention of any entity, independent from both the owner and the contractor,[59] entrusted with the settlement of any disputes arising out of the contract. Therefore this form of resolution of disputes is not used in Portugal.

3.2.2   Arbitration

Arbitration is permitted by Portuguese law to resolve disputes in respect of contracts. It is either domestic or international, depending on the parties and the law governing the agreement. Portugal is a party to the 1927 Geneva Convention for the Execution of Foreign Arbitration Awards and also of the 1958 United Nations Treaty on the Recognition and Execution of Foreign Arbitration Awards.

Domestic arbitration rules apply to cases where the parties to an agreement are residents in Portugal and their relation has no relevant connection with a foreign country. Pursuant to these rules the parties may include a clause whereby they agree to submit to arbitration any disputes arising out of the agreement, or they may resort to arbitration after a particular dispute has already arisen.

In case of agreement between a resident and a non resident governed by foreign law, the parties may further agree to submit any disputes arising out of the agreement to arbitration. In such cases, the international arbitration rules apply. In general, any clause in international agreements providing for foreign arbitration is valid and enforceable in Portugal, provided that the connection between the agreement and the applicable foreign law is a reasonable one. The parties may freely choose the place of the arbitration, regardless of the agreement being domestic or international.

Arbitration courts usually comprise three arbitrators, one appointed by each party and the third arbitrator by agreement, or by the Court of second instance if the parties fail to reach an agreement within one month. Furthermore, arbitration institutions have recently been established.[60]

Law nr. 31/86 dated August 29th is the fundamental statute regarding voluntary arbitration. The said statute states that with regard to arbitration proceedings which take place in Portugal:

- the arbitrators must be Portuguese citizens with full legal capacity;
- the parties to the arbitration may confer upon the arbitrators the authority to judge *ex aequo et bono*, in which case the arbitrators have a wide discretion as to the procedures to follow;
- despite the said discretion of the arbitrators with respect to procedure, the parties to an arbitration must receive exactly the same treatment;
- a summons must be served upon the defendant;
- both parties to the arbitration must be heard before a final decision is rendered; and
- the arbitration award must be in writing.

The parties may freely choose the procedure rules for the arbitration, as regards the hearing of the parties, the production of evidence and the powers of the arbitration to seek any further evidence. To that purpose, the parties may resort to the regulations of a permanent arbitration entity duly set up. Law 31/86 states that, should the parties fail to reach an agreement upon this issue, such rules shall be set out by the arbitrators.

Arbitration awards are generally appealable in Portugal to the Courts of Second Instance (*"Tribunais de Relação"*), unless the parties agree to waive their right of appeal. However, arbitration awards rendered by arbitrators who were granted authority to decide *ex aequo et bono* are not appealable under any circumstance.

An arbitration award rendered in Portugal has the same authority and may be enforced under the same terms as a judgement of an ordinary court.

As regards arbitration awards rendered abroad, Resolution 37/96 passed by the Portuguese Parliament and Decree 52/94 of the President of the Republic finally ratified and approved for ratification the New York Convention of 1958, on the Recognition and Enforcement of Foreign Arbitration Awards. The text of the Resolution defines arbitration awards: these are not only the decisions rendered by arbitrators appointed for certain cases, but also the ones rendered by permanent arbitration bodies chosen by the parties.

Upon signing, ratifying or joining the Convention, any state may declare, on the basis of reciprocation, that it will only apply the Convention to the execution of awards rendered in any of the contracting states.   It may also declare that it will apply the Convention only to cases considered commercial by the state's law, whether derived or not from a contract.   According to article III of the Resolution, any state that is a party to the Convention shall enforce the award in accordance with the laws of its procedural code.   This means that, before being enforced, a foreign arbitration award needs to be confirmed by a Court of Second Instance in Portugal, the one having territorial jurisdiction over the place of residence of the person against whom it will be enforced.

The confirmation and execution may only be denied in cases where the party against whom it will be enforced produces to the competent court evidence:

- of the incapacity of the parties which signed the arbitration clause;
- that the arbitration clause is not valid according to the law chosen by the parties, or, should no law have been

chosen, according to the law of the country where the award was rendered;

- that the party against whom the award was rendered was not duly notified to choose an arbitrator, nor notified of the arbitration process, or that, because of any other reason, he was prevented from opposing the proceedings;
- that the award concerns a dispute that was not covered by the arbitration clause;
- that the award goes beyond the scope of the arbitration clause;
- that the setting up of the arbitration board, or the arbitration process were not in agreement with the arbitration clause, or in case no such clause was signed, in agreement with the law of the country where the arbitration took place;
- that the decision was appealed against and was annulled or suspended by a competent authority.

Confirmation and performance may also be denied by the court in the country where they were sought in cases where the object of the process may not be settled through arbitration in that country, or the confirmation and performance of the award are against the country's mandatory rules.

If the arbitration is not submitted to the regulations of a permanent arbitration entity, which has a table of costs and rules for the apportion to the parties, nor is this subject agreed upon by the parties, then each party shall bear its own costs and the equitative partition of common costs.

The number of permanent arbitration entities in Portugal has recently been increased and are at present the following:

- Portuguese Delegation to the ICC;
- Portuguese Industry Association;
- Faculty of Sciences of the Catholic University;
- Association for Conciliation and Arbitration;

- Arbitral - Sociedade de Arbitragem;
- Portuguese Bar Association;
- Institute for Conciliation and Arbitration;
- Associação Comercial de Lisboa - Câmara de Comércio e Indústria Portuguesa and Associação Comercial do Porto - Câmara de Comércio e Indústria do Porto;
- Associação Centro de Arbitragem de Conflitos de Consumo da Cidade de Lisboa;
- Associação Centro de Informação de Consumo e Arbitragem do Porto;
- Associação de Arbitragem de Conflitos de Consumo do Distrito de Coimbra;
- Instituto de Autodisciplinada Publicidade;
- Associação dos Industriais da Construção Civil e Obras Públicas do Norte;
- other minor associations

Portugal is a signatory of the following multilateral conventions:

- 1923 Geneva Protocol on Arbitration Clauses;
- 1927 Geneva Convention on the Enforcement of Foreign Arbitration Awards;
- 1958 New York Convention on the Recognition and Enforcement of Foreign Arbitration Awards;
- 1965 Washington Convention on the Settlement of Investment Disputes between States and Nationals of Other States.

3.2.3   Expert Determination

As already referred in 3.2.1 hereabove this system of resolution of conflicts is, in practice, not used in Portugal.   The only comparable system is the one described in 2.1.1 hereabove.

3.2.4   Litigation

Portuguese courts are not subject to the judicial precedent and are completely free to render their own decisions on the basis of the facts presented and under the terms of the applicable law. Also, as already mentioned, courts may only render their decisions with regard to matters and facts submitted to them by the parties[61].

With regard to lawsuits, the following rules specify the competence of Portuguese courts to hear and decide the questions raised by the parties.

As far as international matters are concerned - involving more than one country's legal system, be it on account of the parties' nationality, be it because the facts which are relevant for the case took or are to take place in different States - the jurisdiction of Portuguese courts is subject to the following conditions[62]:

- the defendant or any of the defendants must be resident in Portugal, unless the lawsuit refers to real property located abroad;
- the lawsuit must be lodged at a Portuguese court, according to the Portuguese legal provisions on territorial jurisdiction;[63]
- the fact which is the cause of action must have been performed in Portugal;
- the right brought forward can only take effect through Portuguese courts, provided there is a relevant term of connection between the subject matter of the lawsuit and the Portuguese law.

Portuguese court have exclusive jurisdiction in the following cases[64]:

- lawsuits concerning real property located in Portugal;
- bankruptcy or assets-stripping proceedings in respect of companies having registered offices in Portugal;

- process for the appreciation of the validity of the incorporation or dissolution of companies having registered offices in Portugal;
- processes for the appreciation of the validity of the entry of any rights into Portuguese public records.

At the domestic level, judicial courts are generally competent and are only prevented from hearing cases entrusted to special courts.[65] There are no special courts dealing with construction law or other essentially technical matters to which the former relate.

As mentioned hereabove,[66] the parties may freely submit to arbitration any questions which may come to arise in respect of the legal relation established between themselves or any other party which may already have arisen, unless they relate to indisposable rights. The establishment of such clauses prevent courts from rendering a decision in respect thereof and causes the lawsuit to be extinguished if the defendant brings forward and produces evidence of the establishment of such a clause.[67]

The judicial process, mainly in civil and administrative courts, is organised under the adversarial principle, which, however, is complemented by a certain degree of inquisitorial power granted to the judge. That is to say that the court may not examine any question nor any facts therein unless a party renders it to the case and the other party is offered the opportunity to oppose[68]. However, once the questions are rendered to the court, the judge must endeavour, as he deems fit, to search the material truth and seek the just composition of the conflict.[69]

On the other hand, it is also the judge's duty to promote - whenever he deems feasible and adequate and at any stage of the process - the conciliation of the parties, serving, in a way, as mediator.[70] Such duty however, despite it not being subject to the parties' consent but only dependent on the judge's own initiative, is discharged within the scope of the judicial process and can therefore not impose a solution which has not been freely agreed upon by the parties. In particular, the judge can not summon the parties in order to seek a conciliation more than once during the process.

As regards the inquisitorial powers of the judge - it should be reminded here that such powers are always limited to the facts brought forward by the parties, which are, indeed, the only ones the court can hear - the following highlight a number of outstanding aspects:

- the power to request the assistance of an expert for the examination and trial of facts, in lawsuits concerning questions essentially technical[71];
- the power to order that expert evidence be produced[72];
- the power to order the judicial inspection of things or persons[73].

Evidence to be produced in judicial processes is strictly subject to the adversarial principle[74] and is usually documentary, save for testimonial evidence or personal deposition. Documentary evidence must be attached to the written pleadings of the parties in which mention is made of such evidence; legal or technical opinions may be attached at any time until closing of the discussion; expert evidence must give rise to a written report to be attached to the process. Whenever new evidence is attached to the process the other party is notified to reply.

When producing evidence, parties may bring forward facts relating to earlier negotiations between themselves,[75] although the will to covenant any kind of undertakings does not bind the parties in case of unsuccessful negotiations. Therefore, any representations made by the parties in the course of negotiation must clearly state that such representations are only binding for the purposes of the negotiation itself and do not constitute a waiver of the parties' rights in case of litigation.

Once the adversary proceedings of the discussion concerning the facts and the applicable law, is over, the court renders its decision which shall settle all questions raised in respect of the legality of the parties, jurisdiction of the court and feasibility of the process; the court must further specify which facts it deems proven and by what evidence as well as settle all matters of law broached in the lawsuit, convicting or acquitting the parties as deemed justified.[76]

The expenses incurred shall be borne by the parties pro rata to the admission of their relevant positions,[77] according to a table of judicial tax, depending on the value of the lawsuit.   Such expenses do not, however, include those incurred due to expert evidence or other similar evidence, whose amount depends on the complexity of the evidence to be produced, the fees of the experts being in accordance with the current practice.

Decisions rendered by first instance courts on the subject matter of a lawsuit may be appealed to second instance courts of the relevant judicial district,[78] which is competent to examine the opposed questions of fact and of law. The second instance court may request that evidence be again produced before it, as deemed necessary for the purposes of reaching a decision, or order the total or partial repetition of the first instance trial, should it consider that the decision opposed is insufficiently grounded or explained or is contradictory as regards the questions of fact.[79]

Decisions rendered by second instance courts (*"acórdãos"*) on the subject matter of the appeal, may be in turn appealed to the Supreme Court of Justice, which may only decide questions of law. The Supreme Court of Justice establishes which law is applicable to the subject matter of the lawsuit or orders the same to be addressed to the second instance court, should it deem that any decision rendered on questions of fact is incomplete, unclear or contradictory, thus avoiding the decision of the case.[80]

Portuguese courts competent to render a decision on any lawsuits shall apply thereto the Portuguese or foreign law according to the rules on conflict of law contained in articles 25 to 65 of the Civil Code. In this respect, the following rules are especially important to the foregoing:

- individuals are governed by the law of their nationality, which defines the legal personality and capacity of such individuals;
- legal entities are governed by the law of the state where their principal place of business is located, which governs the incorporation, legal personality and capacity and duties of the boards thereof;
- as far as their perfection and form is concerned, business proposals are governed by the law applying to the matter thereof or to the place where the same is executed, unless the former states that the business proposal is null and void if certain formal requirements are not complied with;
- contracts are governed by the law chosen by the parties provided such choice reflects a serious interest,[81] or by the law of the place where both parties are resident or in the absence of any such place, the law of the place where the contract is entered into.

The nationality or place of residence of the parties is not relevant for the purposes of the jurisdiction of Portuguese courts for the trial, with the exception of the place of residence of the defendant in Portugal which is relevant for the international jurisdiction of Portuguese courts. Thus, a decision rendered by a Portuguese court against a non resident may be enforced under the same terms as any other decision, provided the defendant against whom such decision was rendered has assets located in Portugal which may account for the debts thereof.

Decisions rendered by foreign courts may be enforced in Portugal after being reviewed by second instance courts ("*Relação*") having jurisdiction over the location of the individual against whom the decision is to be enforced. The requirements of such review are the following:

- no doubts arise on the validity of the document containing the decision;
- such decision must be *res judicata* in the country where it was rendered;
- the jurisdiction of the foreign court must not have been established by fraud nor can the matter of the lawsuit be the exclusive jurisdiction of Portuguese courts;
- no similar lawsuit between the same persons may be pending in any Portuguese courts unless, according to the provisions on pendency, the foreign court is competent.
- the defendant must have been duly summoned and the process must comply with the adversarial principle and the principle of equality of parties;
- the decision rendered must be consistent with international public order principles of the Portuguese state.[82]

3.2.5   Negotiation

Not used in this country. See the remark made on 3.1.4.

3.2.6   Ombudsman

The Ombudsman corresponds in Portugal to the officer known as *Provedor de Justiça*. The *Provedor de Justiça* is an independent entity having no executive power, entrusted with attending to the citizens' complaints in respect of actions or omissions of any public powers, with analysing such complaints for which purpose he shall be assisted by the bodies against which the complaint is lodged and with addressing whatever recommendations he deems necessary in order to prevent or redress injustices.

The definition of the preceding paragraph may be found in Article 23 of the Constitution which refers to this officer; he is elected by the Parliament. On the basis of this constitutional charge, other *Provedores* have been appointed having similar duties, only specialised in certain areas, such as the *Provedor* for the users of telecommunications services.

The Conselho do Mercado de Obras Públicas e Particulares (CMOPP) (Committee for the Market of Public and Private Works), although having different characteristics, has also been set up for the protection of individuals and focuses on real estate and construction market.

The CMOPP is an entity of the public Administration and depends on the Ministry of Public Works.

It is relevant that the above mentioned entity issues the certificate ("*alvará*") of the licence of public or private works contractor for the purpose of carrying on the construction business, specifying the kind and value of the works any contractor is qualified to carry out. Such licence is based on the analysis of certain aspects such as the technical and economic capacity and financial standing of the contractor as well as its experience in the performance of works. The CMOPP also attends to any complaints lodged by any private or public entities in respect of defects in the works performed by contractors and may investigate such complaints as a result of which the type (scope and value) of the licence held by the contractor may be changed.

4.   THE ZEITGEIST

Ever since its entry into the European Union and as a result of the consolidation of the Common Market, with its increased competition resulting from the free circulation of persons and goods, services and establishments amongst member countries, Portugal has undergone and continues to undergo a deep and increasingly fast economic, industrial and administrative restructuring coupled to a considerably high economic growth rate, of 2% to 3% during the 90s.

In respect of industry in general and of the construction industry in particular, such restructuring is the result of the general awareness that companies need to be re-organised, at the operational, financial and administrative level, enabling such companies to adjust to a whole new market, a larger, but much more competitive, market, where protective mechanisms no longer have a place.

As a result, an increasingly larger number of companies, which did not know or were not able to adjust to the new circumstances, are going or have gone bankrupt or are undergoing asset-stripping procedures, under agreements entered into with the relevant creditors.   At the same time, on the other hand, many new companies, more flexible and aggressive, endowed with a sounder financial capacity[83], have been established.

The Portuguese construction industry, in particular, involved, as are many others, in this restructuring process, has benefited not only from the general economic activity and the resulting need of built areas, but also from the execution of major public and private projects such as Expo '98 and the two River Tejo crossings as well as the construction of a variety of urban and communication infrastructures, which have created a durable situation of full employment (over-employment sometimes), reflected in the growth rate referred to in the first part of this monogragh.

With regard to the public Administration, however, this process has been troubled by the typical passiveness of extremely complex bureaucratic mechanisms and the inbred centralisation of the power.

However, also in the case of public Administration certain steps have been taken in the direction of modernisation.   For such purpose the new (1991) Administrative Procedural Code was published, regulating in simpler terms the attendance by the Administration of complaints lodged by individuals and setting out the fundamental principles governing the administrative process. In the context of construction law, the whole legal system governing planning and zoning consents and the construction of buildings was also redrafted in 1991; procedural norms and deadlines have clearly been set out for each phase; processes have been made simpler and further guarantees have been provided in respect of the rights of individuals. In that same respect, the publication of the General Municipal Plans, which regulate the planning of plots within each Municipality, further protects individuals against possible arbitrary action taken by the Administration.

On the other hand, powers and duties have progressively, albeit insufficiently, been transferred to Municipalities and the creation of local or regional departments has given rise to the decentralisation of the central Administration. For this same purpose, the establishment of the *Regiões Administrativas* (administrative regions) is in progress; such administrative regions shall be intermediate centres of power - above the local powers of Municipalities and under central state power - much more autonomous and resourceful.

With regard to legislation, in addition to the aspects referred to above mention must be made to the deep redrafting of the Civil Procedural Code in 1995 and 1996, which came into effect in 1997. Changes focused on speeding up judicial processes by simplifying some of the relevant phases; deadlines have been shortened; the principle of the co-operation between the parties and the court has been enhanced in order to facilitate the examination and understanding of the matters under disputes as well as the pursuit and definition of material truth. Likewise, many unnecessary special processes have been extinguished and the process as a whole has gained clearness and simplicity.

With regard to Tax Law, VAT began to be levied in 1986 and Individual and Corporation Income tax and Council (collected by Municipalities on real property) tax were created in 1989 thus unifying - and simplifying - direct taxes, and in 1991 the Tax Procedural Code, containing procedural rules to protect individuals against tax authority was published; however the truth is that taxes continue to be high in particular because of the existence of several indirect taxes levied on consumers and on the economic activity, which are acknowledged as one of the main sources of income for the state.

In particular Stamp Duty, levied on many acts and contracts; Oil Products Tax, various taxes on electricity and water consumption, the accrual of VAT at each stage of production and the *sisa* levied on the conveyance of property, account for approximately 50% of the market cost of property. This situation is common to several other economic activities and poses many problems to Portuguese companies in terms of competitiveness in comparison to foreign companies, in particular, European industries which are subject to much lower taxes and production costs.

Lastly, with regard to international relations of the Portuguese construction industry, its remarkable growth, in particular the major public works now in progress, have attracted many foreign companies to the Portuguese market and especially companies from the European Union which have become associated with Portuguese companies for the performance of such projects, in particular, under the form of consortia, but also through the acquisition of holdings in the registered capital of such Portuguese companies.

Portuguese companies have also, although to a smaller degree, extended their activity abroad, in particular to the other countries of the European Union, to Northern Africa and to Portuguese speaking African countries.

1   We would like to thank the *"Associação de Empresas de Construção e Obras Públicas"* (AECOPS) (Building and Public Works Companies Association) for its co-operation in preparing this monograph, in particular, for the statistic data provided, which were obtained from "Construction Report - 1996-1997".

2   Between 1990 and 1994 companies with a turnover of less than PTE 10,000,000.00 (US $56.178) went from 42.6% to 32.6% of all construction companies; companies with a turnover of over PTE 5,000,000,000.00 (US $ 28,089,887) went from 0.2% (24) to 0.4% (81).

3   In 1989, Portuguese construction companies effected sales of goods and services abroad for an overall amount of PTE 32,454,000,000$00 (US $ 182,352,842), which went to PTE 73,200,000,000$00 (US $ 411,235,955) in 1995.

4   The only exception to this principle is the power of the Constitutional Court (developed hereunder) as regards the control of laws' compliance with the Constitution (judicial review), in the abstract, in material cases and prior to enactment, as set forth in articles 278 to 282 of the Constitution and in articles 51 and following of Law 28/82 of 15/11.

5   Articles 6 and 233 of the Constitution.

6   Articles 229 and 234 of the Constitution.

7   Articles 237 and following of the Constitution and Decree-Law 100/84 dated March 29th (setting forth the powers and duties of city councils).

8   Matters in respect of which the Parliament has exclusive powers to legislate, such as those concerning elections and referendum, organisation of the political power, relations between sovereign powers, national defence and security and participation into international bodies (article 164 of the Constitution).

9   Matters in respect of which the Parliament can delegate to the Government powers to legislate, setting forth the relevant principles and terms, such as those concerning fundamental rights and duties of citizens, basis of the economic organisation, health,

education, environmental protection, tax system and penal law (article 165 of the Constitution).

10   Articles 112, nr. 4, 169 and 227 of the Constitution.

11   Article 112, nr. 7 of the Constitution.

12   Article 227 of the Constitution.

13   Article 242 of the Constitution and article 39 of Decree-Law 100/84 dated March 29th.

14   Articles 8, 135 and 161 of the Constitution.

15   Article 3 of the Civil Procedural Code stipulates that the court may not resolve the conflict of interests that the lawsuit supposes, unless such resolution is required by one of the parties and the other is duly called to lodge its opposition.

16   Article 265 of the Civil Procedural Code sets forth that the judge must take or cause to be taken, even on his own initiative, all action necessary to ascertain the truth and to resolve the dispute with justice, in respect of the facts over which he has jurisdiction.

17   Article 209 of the Constitution.

18   Decree-Law 425/86, dated December 27th and Government Ruling 639/95 dated June 22nd.

19   Articles 110 and 202 of the Constitution.

20   Articles 219 of the Constitution.

21   Article 1st of the Deontoligical Code of Lawyers.

22   Article 1st of Decree-Law 207/95, dated August 14th (Notarial Code).

23   Also respectively Decree-Law 224/84 dated July 6th and Decree-Law 403/86 dated December 3rd.

24   Article 60 of the Constitution.

25   In particular, for the purposes of the foregoing, the recent change to article 1225 of the Civil Code, whereby the term of the guarantee period provided by the constructor for the quality of the construction was extended to at least five years, as well as article 916 of the above mentioned Code, whereby the same responsibility is also undertaken, for the same period, by the vendor of real estate.

26   See above, the municipal plans for the planning of the territory.

27   Article 66 of Decree-Law 445/91, dated November 20th (see
      hereinabove) requires the Minister for Public Works, Transports
      and Communications to publish periodically a list of legal
      provisions to be observed by the authors of construction projects.
      The said Government Ruling 338/89 is the more recent statute
      published under this provision.

28   Respectively: Decree-Regulatory 60/91, November 11th; Council
      of Ministers Resolution 21/89, May 15th; Decree-Regulatory
      22/92, September 9th; Council of Ministers Resolution 38/90,
      September 14th; Decree-Regulatory 11/91, March 21st; Council
      of Ministers Resolution 45/93, June 7th; and Decree-Regulatory
      26/93, August 27th.

29   Conditions for contracts of civil engineering produced by the
      Fédération Internationale Des Ingénieurs-Conseils, FIDIC 4th
      Edition, published by Sweet & Maxwell, London, 1991 (FIDIC's
      "Red Book").

30   See 2.1.4 below.

31   Articles 3rd, 9th and 51st of Decree-Law 129/84, April 27th
      (*Estatuto dos Tribunais Administrativos e Fiscais*) and article
      10th of Decree-Law 267/85, July 16th (*Lei de Processo nos
      Tribunais Administrativos*).

32   Article 188th of Decree-Law 442/91, November 15th (*Código do
      Procedimento Administrativo*).

33   Law 31/86, August 29th, see 3.2.2 below.

34   We will come back to this point later (3.2.2 below), as it is not
      quite clear if it is presumed that the court will decide *ex eaquo et
      bono* or if the court must decide on those terms.

35   This Board is a consulting body of the Ministry of Public Works
      for matters concerning public works and was established by
      Decree-Law 37,015 dated 16/08/48.

36   According to the majority of the doctrine, such guarantee term
      (specifically set forth for the construction of real estate) must be
      understood as being in of public interest, no shorter term being
      acceptable. We quote Prof. Vaz Serra "*Revista de Legislação e
      Jurisprudência*" issue 106, page 300: "it may be inferred from
      article 1225 of the Civil Code which reads «within five years as

of acceptance or within the covenanted guarantee term» that the parties may freely covenant a guarantee term of more or less than five years ... They may without a doubt covenant a term longer than the legal term established, however the possibility to covenant a shorter term is not so acceptable as, if the responsibility of the contractor set forth in article 1225 is established to protect the public interest and the stability of the buildings or other real estate, which for their own nature are supposed to be long-lasting, as well as the interest of the owner which may be the victim of its own inexperience failing to detect construction defects in time, no clause reducing such five year term can be valid."

37   Article 74 of the Civil Procedural Code.

38   Articles 99 and 100 of the Civil Procedural Code.

39   Setting out that arbiters are to decide according to equity, pursuant to articles 29, nr, 2 of Law 31/86, corresponds to the automatic waiver of the right to appeal.

40   The company PARQUE EXPO '98 was incorporated by the Portuguese state in 1993 to organise and set up the Lisbon International Fair of 1998 (Expo '98).

41   Mention should be made, however, that the majority of construction contracts (public and private) sets out that the Technical Director, representing the contractor, and Supervisor, representing the Owner, should liaise with each other on a on-going basis such liaison being the starting point for the negotiations.

42   Article 5 of Decree-Law 445/91, dated November 20th and Decree-Law 83/94, dated March 14th.

43   Directives nr. 71/304/CEE and 71/305/CEE, both dated 26/07 had already been incorporated in the Legal provisions governing Public Works, approved by Decree-Law 235/86 dated 18/08, before this one.

44   Article 65 of Decree-Law 55/95 and article 97 of Decree-law 405/93, as clarified by the Order of the Ministry of Public Works, Transport and Communications, nr, 63/94-XII, dated 30/12/94, published in the official journal, II series, dated 17/01/95.

45   Although none of the above mentioned statutes makes any
      express distinction between the committee presiding over the
      tender and the one entrusted with the technical examination of the
      proposals, such distinction is current practice.

46   See Jorge Andrade da Silva, "Regime Jurídico das Empreitadas
      de Obras Públicas, 3rd Issue, Almedina, page 216.

47   Data supplied by the "*Relatórion Anual da Construção - 96/97*",
      (Building and Public Works Companies Association), Lisbon,
      1997.

48   See 2.2.1 below.

49   Companies associate on the basis of a contract whereby no new
      legal entity is created, such contract being merely subject to being
      put down in writing and not to the record thereof at the companies
      registry. The said contract does not set out the solidarity of the
      members of the consortium for the debts thereof, unless
      otherwise provided for in the contract.

50   This kind of association is more formal, the pertaining contract
      being subject to notarial deed and record at the companies
      registry. Members are solidarily responsible for the debts of the
      Group but only after the assets thereof have been used up. The
      Group thus formed is a new legal entity.

51   Established by Regulation (EEC) 2,137/85 dated 25/07 and
      regulated, in Portugal, by Decree-Law 148/90 dated 09/05.
      Although the pertaining articles of incorporation is not subject to
      notarial deed, the incorporation must be recorded at the
      companies registry and a new legal entity is created. The
      members of the Group are solidarily responsible with no
      limitation but only to the extent the Group itself fails to honour its
      obligations after being expressly summoned.

52   See 3.2.2 below.

53   See 3.1.3 below.

54   See 3.2.4 below.

55   While Portuguese law sets out arbitration courts, in practice no
      such recognition exists for the mediator to confer it judicial
      existence. His assistance is *ad hoc* and expressly required by the
      parties themselves; the mediator discharges his duties for as long

as the parties deem fit, neither party being under the obligation to comply with the decisions thereof.

56 Article 26, nr. 2 of Law 31/86, dated 29/08 sets forth that «for enforcement purposes, arbitration awards are as effective as first instance courts decisions».

57 Article 406, nr. 1 of the Civil Code sets forth that «the contract must be timely performed and may only be altered or terminated by agreement of the parties or as foreseen in the law».

58 Indeed, as a result of the foregoing, on account of the mediation being a mere step in the negotiation process whereby the parties seek an agreement, strictly speaking, the mediator merely assists the parties in reaching an agreement rather than render a decision.

59 Mention is made here to the Independent Engineer of the FIDIC contract form, 4th edition.

60 See 1.2.2 hereabove.

61 See 1.2.2 above.

62 Article 65 of the Portuguese Procedural Code.

63 E.g., the competent court on the basis of the location of the property or on the basis of the place where the obligation is to be fulfilled.

64 Article 65-A of the Portuguese Civil Procedural Code.

65 See 1.2.2 above.

66 See 1.2.2 and 3.2.2 above.

67 Article 493 and 494 of the Portuguese Civil Procedural Code.

68 Articles 3 and 264 of the Civil Procedural Code. Article 3 states that the court may not resolve the conflict of interests that the lawsuit supposes, unless such resolution is requested by one of the parties and the other is duly notified to lodge its opposition.

69 Article 265 of the Civil Procedural Code states that the judge must take or cause to be taken, even on his own initiative, all action necessary to ascertain the truth and to resolve the dispute with justice, in respect of the facts over which he has jurisdiction.

70 Article 509 of the Civil Procedural Code.

71 Article 649 of the Civil Procedural Code. The assistance of an expert in court is increasingly frequent in respect of the technical aspects of construction contracts.

72   Article 579 of the Civil Procedural Code.

73   Article 612 of the Civil Procedural Code.

74   Save for injunctions in which non hearing of the defendant was authorised, in case it is reasonably feared that the purpose of the injunction may be impaired.

75   Save for negotiations established for out of court resolution of pending litigation in which a lawyer has taken part, which are subject to confidentiality.

76   Articles 658 to 665 of the Civil Procedural Code.

77   If A files a lawsuit against B and requests that the latter be ordered to pay 100 and B is ultimately only ordered to pay 60, the parties shall bear the expenses in the proportion of 60 and 40 by, respectively B and A.

78   See 1.2.2 above.

79   Articles 691 to 720 of the Civil Procedural Code, in particular, article 712.

80   Articles 721 to 732 of the Civil Procedural Code.

81   Serious interest is understood as the law of the citizenship of either party or the law governing the execution of the deal or if the latter is, in any relevant way, in relation with the law chosen.

82   Article 1096 of the Civil Procedural Code.

83   In 1983, 53.3% of the Portuguese construction industries were individual entities and 40.8% were organised as companies (joint stock companies or quota companies). In 1994 this ratio was of 45.5% for the former and 53.5% for the latter. In December 1993 construction companies held 12.95% of bank credit granted to non financial companies, and 14.34% in December 1996; this shows an increase of companies credit and a larger use of the available financial facilities and the same occurs in the hotel industry and with service companies.

QUÉBEC

Daniel Alain Dagenais
Lavery de Billy, Montreal, Quebec

1.   BACKGROUND

1.1   Economic

In Québec, approximately 15 billion C$ is spent annually on total immobilisation expenditures, in both the residential and non-residential sectors. Following the trend of the North American economy, the Québec construction industry experienced a surge in activity in the 1980s and a high of 74,179 housing starts was reached in 1987. However, construction activity slowed in the late 1980s because of weakening economic growth and a general oversupply of commercial buildings and office space. The construction industry has stagnated in the 1990s[1] and this has been particularly true in Montreal, which is Québec's economic centre.[2] With the return of favourable economic conditions, building activity has grown somewhat in 1996 and 1997, but remains well off its pre-recession highs. However, the price of new housing has remained flat since the beginning of the 1990s and building activity has been stagnant in Montreal. Industrial construction outside of the urban centres has fared well in 1996-97, as a number of large projects are now underway *(e.g.* construction of smelters and other industrial facilities).

1.1.1 Trends

Robust growth in both industrial and residential construction in the 1980s has given way to a stagnant market in the 1990s, characterised by increased competition among builders and continuing low prices, as supply continues to outpace demand both in the industrial sector and the housing market. Thanks to low interest rates and resurgent economic growth, building activity should improve in the coming years, albeit at a slow pace.

1.2   Legal

Canada is a confederation of ten provinces, including Québec, whose governments exercise relatively broad powers over matters which fall within their jurisdiction, subject to the overarching jurisdiction of the Federal Government. The *Constitution Act, 1867*[3] provides for a separation of powers between the Federal and provincial governments. Generally speaking, the Federal Government has authority over topics which affect the citizens of every province, as well as issues of national importance, whereas the provinces have general powers over "property and civil rights in the province".

With respect to the organisation of the Courts and the administration of justice, there is considerable jurisdictional overlap between the Federal and provincial governments. The "superior courts" are the courts of original general jurisdiction in the province, without limit over the subject matter. Within these "superior courts", there is a trial division (called the "Superior Court", in Québec) and an appellate division (the "Court of Appeal"). The "provincial court" (the "Court of Québec") has a jurisdiction which is similar to that of the Superior Court but it can only hear cases where the value of the plaintiff's interest is less than $ 30,000.

The Small Claims Court, a division of the Court of Québec, hears claims not exceeding $3,000. Plaintiffs represent themselves in the Small Claims Court and legal persons counting more than five employees cannot present a claim. This forum is used especially for complaints related to consumer protection issues.

Note that the Government of Québec names the judges of the Court of Québec and determines their compensation, whereas the Federal Government has the authority for naming and remunerating the judges of the "superior courts".[4] Because the province is responsible for the administration of justice, it pays the costs related to the operation of the courts

The Government of Canada has created a parallel system of Federal Courts which have (non-exclusive) jurisdiction over existing and applicable Federal law. This Court has a trial division and an appellate division. The Government of Canada has the responsibility for naming the judges *and* for assuming the costs of the administration of justice for the Federal Courts.

Finally, whether the case originated from the superior courts, the provincial courts or the Federal Courts, a party may seek permission for leave to appeal from the Supreme Court of Canada, which sits in Ottawa. The Supreme Court is a national tribunal, which has jurisdiction over all questions, whether criminal, civil or constitutional and is the highest appellate Court in the country.

The 1980s and 1990s have seen a growth in administrative tribunals, both in the provinces and at the Federal level. However, there are no specialised Courts to decide conflicts in the construction industry. If litigation is required, these cases must proceed through the general Court system.

In Québec, there are two types of legal professionals: lawyers and notaries. Only lawyers can plead before the Courts. Notaries have a limited field of exclusive expertise, which relates to questions of private law, including mortgages, marriage contracts and the transfer of personal property, and can authentify documents. Because the field of construction law often involves litigation and representation before the Courts, building owners, contractors, engineers and other industry actors are generally represented by lawyers.

The lawyer's relationship with his client has a dual nature. On the one hand, the lawyer acts as a *mandatary* for his client when he *represents* him, notably during negotiation, mediation or arbitration sessions with another party and in pleading cases before the Courts. The *Civil Code of Québec* defines *mandate* as a contract by which the mandator empowers another person, the mandatary, to represent him in the performance of a ** juridical act with a third person; and the mandatary, by his acceptance, binds himself to exercise the power.[5] The mandatary must act in the best interests of the mandator and must avoid placing himself in a situation that could give rise to a conflict of interest.[6]

On the other hand, when the lawyer provides advice to his client or drafts contracts or other documents for his benefit, he does not act as the client's representative. Instead, the lawyer is engaged in a contract for *services*, which the *Civil Code of Québec* defines as a contract by which the provider of services undertakes to carry out physical or intellectual work for another person, the client, or to provide a service, for a price which the client accepts to pay.[7] The lawyer is bound to act in the best interests of the client, with prudence and diligence, and to act in accordance with usual practice and the rules of art, and to ensure that the service provided is in conformity with the contract.[8] The *Code* also includes specific articles on the parties' obligations with respect to the price of the services provided, and its possible modification in certain circumstances.[9]

1.2.1   Construction Law

Construction law touches upon a number of different areas, such as contracts, civil liability, insurance and bankruptcy. In a civilian jurisdiction such as Québec, the *Civil Code* establishes the general legal regime which governs the parties' relationship, but matters dealing with bankruptcy and insolvency are governed by a specific Federal statute.[10] The private law of Québec was governed by the *Civil Code of Lower Canada* [hereafter the *CCLC]* from 1866 to 1993, and on January 1, 1994, the *Civil Code of Québec* [hereafter the *CCQ]* came into force thus replacing the old *Code*. A transitional regime applies to matters which straddle these two periods.[11]

The *Civil Code* contains both imperative and suppletive provisions. The chapter on the "contract of enterprise" (articles 2098 to 2129 *CCQ)* includes provisions which apply more specifically to construction contracts and outlines the parties' respective obligations. Article 2118 *CCQ* provides that the contractor, the architect and the engineer who directed or supervised the work are "solidarily" *(i.e.* jointly and severally) liable for the loss of the work occurring within five (5) years after the work has been completed. Article 2119 *CCQ* indicates possible grounds of exoneration for the parties involved in the construction project: essentially, they must prove that another participant in the construction project committed a fault,[12] or that the defects in the work resulted from decisions imposed by the client in selecting the land or materials, or the subcontractors, experts or construction methods.

Because the CCDC standard-form contracts are widely used throughout Canada, practitioners in Québec sometimes use common law cases as authorities (such as cases reported in the *Construction Law Reports)*. The *Recueil de droit immobilier* is a specialised reporter which includes the most recent Québec case law on issues relevant to real estate.

1.2.2 Construction Contracts

The Canadian Construction Documents Committee [hereafter the CCDC] includes contractors, professionals and other participants in the construction industry, which develop model contractual documents for use in construction projects across Canada. One of the most widely used CCDC documents is the CCDC 2 form of the Stipulated Price Construction Contract.[13] The current version of the contract was developed in 1994, as the result of a five-year review process. The use of the CCDC 2 standard contract is not mandatory and the parties may use it as a basis for further negotiations or may devise their own agreement. They may also adapt the standard-form contract to their particular needs and objectives.

The CCDC 2 contract basically has two parts: the Articles of Agreement (which provide a basic outline of the contractual relationship between the parties, including elements such as the price of the contract and the amount of the holdback) and the General Conditions, which provide the basic terms of the contract.

The CCDC 2 contract has a new section on "Dispute Resolution", which provides a comprehensive framework for the settlement of conflicts.[14] The parties first submit their problem to the "Consultant", an independent professional (usually, the Project Engineer or Architect), whose functions are essentially those of a Disputes Review Advisor, as we will see in Section 2.1.1 below. However, a party may, within fifteen (15) working days after receipt of the Consultant's decision, send a notice of dispute to the Consultant and the other party, who then has ten (10) working days after the receipt of notice to set out the particulars of his response (article 8.2.2). The parties must make all reasonable efforts to resolve their dispute by amicable negotiations and to disclose the information necessary to facilitate the negotiations (article 8.2.3). After a period of ten (10) working days following the receipt of the responding party's reply, the parties shall request the assistance of a Project Mediator.[15]

The mediation is conducted according to the Rules for Mediation of CCDC 2 Construction Disputes, which emphasise the importance of confidentiality.[16] Rule 9 gives the parties much latitude in proceeding with their mediation session: they must make a frank disclosure of all the information which may be necessary to resolve the dispute (Rule 9.1), the Project Mediator may caucus privately with the parties to assist the process (Rule 9.3) and he will assist the parties to isolate points of agreement and disagreement and to consider acceptable compromises or accommodations (Rule 9.4). Furthermore, the time period for the mediation may be extended upon agreement of the parties Rule 9.5).

If the parties proceed to arbitration, the CCDC 2 Rules for the Arbitration of Construction Disputes will be applied. The process is more formal than mediation, and the final arbitration decision is binding on the parties (Rule 18.6). The Arbitrator must be an experienced and skilled commercial arbitrator and preferably has some knowledge of relevant construction industry issues (Rule 8.5). He enjoys broad discretionary powers, not unlike those of a judge, to make interim orders, to inspect documents, exhibits or other property and to extend or abridge any time periods (Rule 10.3). However, he is not required to follow the legal rules of evidence (Rule 14.1). Within five (5) days after having been appointed, the Arbitrator conducts a Procedural Meeting with the parties to reach a consensus on the procedure to be followed (Rule 9). The parties participate in an exchange of statements which outline their respective positions (Rule 11.1) and attach a list of documents thereto (Rule 11.3). Each party has access to the other's documents (Rule 12.2), and the Arbitrator may order the disclosure of any additional documents he considers relevant (Rule 12.1). Not less than 21 days before any hearing, each party must provide the other with a list of his witnesses as well as a written summary of their evidence; in the case of expert witnesses a copy of the expert's report must be provided (Rule 12.4).

Within 30 days after the hearing has closed, the Arbitrator renders his final decision, accompanied by his written reasons (Rule 18) and makes certain orders with respect to costs (Rule 19). The arbitration process provided for by the CCDC 2 1994 contract therefore resembles the judicial process, but the Arbitrator has more flexibility in his application of the rules of procedure and the contents of all hearings and meetings are to be kept strictly confidential by the participants (article 13.2).

Note that if the parties do not request an arbitration within ten (10) working days after the end of the mediation, the arbitration agreement will no longer be binding on them and they can refer the matter to the Courts, to a non-CCDC arbitration or to any other dispute resolution forum they might agree upon (article 8.2.7).

2.   CONFLICT MANAGEMENT

2.1   Non-Binding

2.1.1   Disputes Review Board

The parties to a construction contract may require the intervention of a single person (the Disputes Review Adviser) or of a group of people (the Disputes Review Board) to help them resolve problems or disagreements which arise on the project. The Board is constituted of a representative from each party as well as a neutral adviser. Its members monitor the evolution of the project closely and tour the site when necessary. The Board intervenes quickly when differences occur and attempts to mediate the dispute without slowing down the project. Its effectiveness, like that of the Disputes Review Adviser, is limited because it does not make binding recommendations.[17]

2.1.2   Disputes Review Adviser

The "Consultant" in the CCDC 2 1994 contract documents plays a role somewhat similar to that of the Disputes Review Adviser. First of all, the Consultant has important responsibilities for the administration of the contract (article 2.2.1 of the CCDC 2 1994 contract). Although he does not exercise control over the Contractor, Subcontractor or Suppliers who carry out the work (article 2.2.5), the Consultant interprets the provisions of the Contract Documents (article 2.2.6), he has the authority to reject work which in his opinion does not conform to the contractual requirements (article 2.2.8), and he conducts reviews of the work to determine the date of substantial performance (article 2.2.12). If, however, the Consultant's decision does not resolve the parties' conflict or if he does not have authority over a question, the parties must avail themselves of the Dispute Resolution process set forth in Part 8 of the CCDC 2 1994 contract and in the Rules of Mediation and the Rules of Arbitration in annex.

In practice, even if the parties do *not* use the CCDC 2 1994 contract, the engineer or the architect who oversees the project will often act as an intermediary between the parties and will attempt to resolve disputes which may arise during the project.

2.1.3   Negotiation

In 1992, the Government of Québec held the *Sommet de la Justice,* which gathered lawyers, academics, government officials and representatives of citizens' groups to examine and evaluate the adequacy of the civil justice system in Québec. Since that time, there has been a growing impetus for the development of conflict management and alternative dispute resolution methods, such as negotiation, conciliation, mediation and arbitration. For example, the

Professional Education course sponsored by the Québec Bar, which all future lawyers must follow before their articles, includes several mandatory classes on negotiation techniques and students participate in simulated negotiating sessions. The Québec Bar also organises seminars on the "reasoned" negotiation method, to promote more effective negotiation techniques. There is an emphasis on the fact that conflict management yields a "win-win" outcome, as opposed to the "win-lose" outcome of litigation. However, the development and use of conflict management systems in Québec is still in its early stages, as compared with the United States and the other provinces of Canada, where it has become widely known, accepted and put into practise by the judicial community.

Honesty, good faith, flexibility and creativity are the qualities of a good mediator. The lawyer who acts as a mediator must abide by the *Code of Ethics of Advocates*,[18] which governs his conduct in all his professional activities. He cannot make a statement *de facto* or *de jure* knowing it to be false (s. 4.02.01 d), nor can he conceal or knowingly omit to reveal what the law obliges him to reveal (s. 4.02.01 f), nor can he abuse a colleague's good faith or be guilty of breach of trust or disloyal practices towards him (s.4.03.03).

Of course, the art of negotiation depends on the parties' objectives and the lawyers' individual style. There is no hard and fast rule on what makes for a successful negotiation in any given case. The Québec Bar suggests four (4) basic strategies: the "contributory method", the "reasoned method" the "competitive method" and the "mixed method"[19]. If the negotiator uses the "contributory method", he will seek to co-operate with his counterpart in order to arrive at a solution that is mutually acceptable. The "reasoned method", inspired by the celebrated American authors Fisher and Ury, aims to separate the people from the problem, focuses on the parties' interests not their positions, sets forth a broad range of possible solutions to the conflict and requires that any proposals be based on objective criteria. The negotiator keeps in mind the "big picture" instead of haggling over details. When the parties' interests are too far apart for a more co-operative technique, the negotiator may use the "competitive method", by which a party seeks to impose its own solution and obtain concessions from the other side. Finally, the negotiator can apply a combination of the three above methods (the "mixed method"), and adapt his strategy according to the other party's stance and depending on the progress of the negotiations.

Negotiation is an essential part of any practise in construction law. From Day One, where the parties agree on the contract price, the holdback and the conditions under which the work will be performed, to the day where post-construction conflicts must be resolved, months (and sometimes years) after the work has been completed, the parties attempt to arrive at a mutually satisfactory agreement. Because of the high costs of litigation in the construction industry, negotiation must at least be the first step in the process of conflict management.

2.1.4   Quality Matters

Quality Standards Councils have appeared across Canada, both in the private and public sectors, in order to provide information on quality standards and to encourage businesses to apply those standards. In Québec, a growing number of contractors have received ISO 9002 certification (installation and production), mainly because certain important public bodies which engage in procurement on large projects (namely Hydro-Québec, the province's hydro-electrical monopoly) give preference to firms with ISO certification. Other contractors obtain ISO 9002 certification for the purpose of doing business abroad, particularly in Europe where ISO is a prerequisite for competing on projects.

In Québec, many construction contracts are awarded through the Québec Bid Depository System (QBDS), whose members are the Québec Construction Association (CQA), the Corporation of Master Electricians of Québec (CMEQ) and the Corporation of Master Pipe Mechanics of Québec (CMMTQ). These professional corporations and trade associations have established a Tendering Code which governs most aspects of the tendering process and which are revised periodically (the last revision took place in 1996). The Code seeks to establish objective criteria for the awarding of contracts. It applies if it is voluntarily required by the owners or if certain conditions are met, independently of the will of the parties, *i.e.* if the work must be performed in Québec, if there is more than one tender, if the price of the tender of the governed speciality exceeds $ 10,000 and if the tendering documents make the presentation of comparable tenders possible. Owner-clients (who are generally not members of any speciality) are not bound by the rules of the Tendering Code, unless they request its application.

Lawsuits arising from the application of the Tendering Code are generally taken by third parties who submit an unsuccessful bid for a project, while a competitor bypasses the BSDQ rules and obtains the contract. In such cases, the Courts exonerate the owner-client from liability if he did not request the application of the QBDS rules and did not *induce* the successful contractor to derogate from the requirements of the Tendering Code. However, the specialist who knowingly bypasses the rules of the Tendering Code, which he is bound to observe as a member of his corporation or professional association, will be held liable in damages towards his aggrieved competitor. The *Master Electricians Act*[20] and the *Master Pipe-Mechanics Act*[21] both provide that any person who obtains a contract for the execution of work covered by one of these *Acts* without abiding by the rules of the Tendering Code is subject to a penalty equal to 5 % of the contract price, payable to the QBDS. (This penalty does not preclude a civil liability suit by the aggrieved subcontractor).

The Tendering Code includes provisions on the commitments which must be taken by the bidder and the recipient contractor (chapter C). A bid bond and a letter of intent or tender security must be submitted by the tendering contractor to the QBDS (chapter D). The Code also describes the procedure for the deposit of tenders at the QBDS or at the client's offices (chapter E). Note that the owner may authorise the QBDS to accept and distribute official envelopes to a "short list" of invited persons (section E-5). In general, contracts which are awarded through the QBDS must be given to the bidder who has submitted the lowest price (section J-2); this principle can only be set aside under limited circumstances (section J-3).

As we have mentioned above, the purpose of the Tendering Code was to introduce objective rules in the procurement process, particularly when the owner-client is a governmental entity, to ensure that an impartial decision will be made in awarding the contract. Also, since the owner-client can "invite" a limited number of contractors to submit a bid on the project, he can avoid dealing with certain firms which are financially shaky or which have undesirable business practices. In spite of these features, many owner-clients in the private sector prefer not to use the QBDS, because it tends to make the tender process more cumbersome. This may explain why the number of projects registered with the QBDS by owner-clients has declined from 5,223 in 1995 to 4,805 in 1996, even if the construction industry experienced growth in 1996.

2.1.5   Partnering

In Québec, "partnering" is not a technique for conflict management but rather a method for avoiding disputes altogether. A partnering arrangement involves the creation of a relationship characterised by good faith, confidence, co-operation and team spirit between a contractor and his supplier or between other participants on a construction project. Although partnering does not give rise to any new contractual obligations between the parties, it aims to achieve certain objectives, such as an optimal utilisation of the parties' resources, improved efficiency and cost control and a higher quality of work.[22] It seeks to provide the parties with the flexibility required to complete the job on schedule and on budget. At the beginning of the project, the parties get together and define their respective objectives and also try to anticipate the potential problem areas. Certain officials chosen by the parties (e.g. an engineer or architect) deal with dispute resolution on a punctual basis. As soon as problems arise, the parties address the difficulty and attempt to find a solution which will minimise the disruption to the project.

Although "partnering" is a promising means for the prevention of disputes, it does not always work well in practice. For example, in a large construction project in which the author's firm was involved, the owner, a multinational corporation, had contracted with a consortium of firms to build its new plant, which was to have world-class technology for the production of paper products. A number of changes were required by the owner as the project went on, but no formal, *written* requests were ever made. The contractors nonetheless accepted to make the changes so as to accommodate the client's needs. As the project neared completion, the consortium had incurred substantial losses (in the range of several million dollars) and the client refused to re negotiate the terms of the contract. The members of the consortium had to raise the threat of walking off the site to force the client to revise its contract price, but ended up losing millions of dollars anyway.

This example highlights the fact that even if the parties must show goodwill in the application of alternative dispute resolution methods, they must be careful to preserve their contractual rights.

2.2 Binding

2.2.1 Partnering

See 2.1.5 above.

3. DISPUTE RESOLUTION

3.1 Non Binding

The popularity of non-judicial dispute resolution methods has grown considerably in recent years, not just in legal circles, but also in the business community. For example, the Montreal Chamber of Commerce has devised an Agreement relative to the resolution of conflicts. The companies which sign the Agreement accept not to take any judicial action against another signatory before having jointly undertaken with the other party to examine whether the problem could be resolved through alternative dispute resolution methods. The Chamber has pledged to take the necessary steps to inform its members of the available options and to create a framework (such as a directory of mediators and arbitrators) to assist them in resolving their disputes extra-judicially.

3.1.1   Conciliation

There is little difference between conciliation and mediation, as applied to resolve disputes in the construction industry in Québec. In labour law, however, conciliation is a somewhat more flexible mechanism, because the conciliator works to bring the parties together and to give them the opportunity to talk, but he does *not propose* solutions if the parties reach an impasse.[23]

3.1.2   Executive Tribunal

Although mediation and arbitration are by far the most widely known and applied dispute resolution techniques, the "mini-trial" method may be used in certain cases. The parties voluntarily constitute a three-person panel, with a senior representative from each party and an impartial moderator. The panel hears the testimony of a key witness from each side, followed by a brief legal argument by the party's lawyer. At the end of the "mini-trial", an executive from each side meets with the mediator and they attempt to reach a solution in light of what has been said.[24]

3.1.3   Mediation

Mediation is one of the most widely used dispute resolution techniques in Québec, not only for construction disputes but for all types of civil conflicts. The Québec Superior Court, the Court of first instance in Québec, has sponsored a voluntary mediation programme which has proven to be quite successful. In half the cases scheduled for a hearing of three (3) days or more which are before the Superior Court and are ready for trial, the Co-ordinating Justice for Mediation in Civil and Commercial matters sends a letter to the parties' lawyers and invites them to an information session on alternative dispute resolution techniques. Once the information session is over, the parties can start their mediation immediately. The judge who was given the case then withdraws from the file and leaves his place to the mediator who was chosen by the parties. All discussions which are held during the course of the mediation remain strictly confidential and cannot be used for the ends of litigation.

The parties may contact a mediator together, or one party may contact a mediator, who then approaches the other party. If the contract provides for *voluntary* mediation only, the parties may be less inclined to follow the route of mediation if their differences run too deep.

The mediator has a number of responsibilities in the execution of his mandate and must always be aware of potential ethical concerns. The mediator (especially if he is a lawyer[25]) must avoid any situation in which he is (or seems to be) in a conflict of interest. Since the mediator does not *represent* a party, but acts as an intermediary between the parties to a dispute, he cannot be said to be representing opposing interests. However, the mediator should disclose to the parties at the outset any past or present situations *(e.g.* business, professional or social relationships) which may affect his impartiality and obtain the parties' approval before conducting the mediation.[26] This does not mean that the mediator should remain passive; he should take the initiative, when necessary, to bring the parties to make compromises, but must always act fairly and equitably. He should clearly establish the "rules of the game" for the mediation and may provide the parties with a mediation mandate form, which invites them to consult an independent legal adviser and to act in good faith and disclose all the information which might be useful for successful negotiations to ensue.[27]

The mediator should have some practical expertise in his field and should have an established reputation among his peers. He should also have the capacity to evaluate the strengths and weaknesses of the parties' respective arguments. Furthermore, he must be able to quickly identify the problem areas in a given file and to invite the parties to address these points.[28] The mediator may be required to make tough decisions: for example, if one party is not acting in good faith and "stalls" the mediation merely to buy time, the mediator should take the responsibility of ending the talks so as to save the opposing party aggravation and unnecessary expense. Finally, if the parties arrive at a settlement, should the mediator draft the text of the agreement? The parties will often ask him to do so. However, this presents an ethical dilemma for the mediator because he must put in writing *his* understanding of the agreement, which may not correspond to the parties' understanding. Also, a party who has second thoughts on concessions made during the mediation may invoke the "ambiguity" of the agreement to obtain a reduction of his obligations. For these reasons, the mediator would be well advised to include an exoneration clause in the final text, which provides that he cannot be held personally liable for the content of the negotiated agreement.[29]

There are some essential steps to be followed for a successful mediation:

1) Creation of an appropriate climate for the discussions and outline of the mediation process
2) Factual research and identification of the matters in dispute
3) Creation of options and alternatives
4) Dialogue, negotiation and decision-making - hammering out the agreement
5) Negotiating the details and drafting the agreement
6) Revision and approval of the agreement by the parties; ratification of the agreement, which has the effect of a transaction (ends the dispute)

7)   The parties act upon the agreement; follow-up and evaluation of the process.[30]

At the first stage of the process, the mediator will canvass his role and help the parties initiate the dialogue. At the second stage, the parties, together with the mediator, will discuss all the factual elements which have some importance to their eyes, so as to better identify each side's objectives and constraints. During the negotiation phase, the mediator will hold "plenary" negotiation sessions and may confer with each party separately; he may also propose alternatives and solutions to the parties' problems. At some times, the mediator will merely act as a moderator for the talks and at other times, he will direct the debate and propose solutions or alternatives. Mediation is a flexible process, which functions according to the parties' needs. The specific rules of a given mediation may be provided by the mediator (mediation groups typically set out certain rules on the basis of which the mediation will be conducted). As we have seen above at 1.2.2, the CCDC 2 contract provides a more structured procedure for a mediation which takes place in a construction context.

Co-mediation is an interesting option for the resolution of construction disputes because it allows two (or more) professionals from different fields to act as mediators. Co-mediation has a number of advantages: it improves the appearance of impartiality, it takes some weight off a mediator's shoulders and allows the mediation to proceed more easily in spite of differences in the parties' availability. The mediator with specialised expertise (e.g. the engineer or the architect) may preside the sessions where more technical questions are involved. The co-mediation must confer on a regular basis and share their impressions on the progress of the mediation and may hold caucuses with each party separately. Co-mediation offers interesting possibilities for the resolution of construction disputes, because it provides the parties with a multi-disciplinary approach to their problem, which corresponds to the complex nature of certain problems which arise on projects.[31]

In Québec, the mediator does not enjoy "judicial immunity" as judges do (although suits against mediators are a very rare occurrence). There is a very fine line between the acts committed solely in the context of a person's mandate as a mediator and the acts which may be considered as professional actions (and which may therefore entail his professional liability as a lawyer). This is especially true in more complex cases which require considerable legal expertise. For those reasons, the mediator would be well advised to acquire professional insurance, even if his practice is devoted to mediation.[32]

The parties may agree to apply the "med/arb" method in case the mediation is unsuccessful. In such a case, the mediator will become an arbitrator and actually *decide* the dispute. This method has a number of drawbacks: the parties may be reluctant to disclose all the relevant information they have in relation with the dispute, out of fear that the mediator will use this information against them if he must decide the matter as an arbitrator. The med/arb technique also raises questions on the confidentiality of information. These problems may be alleviated if a different person is appointed to act as the arbitrator of the dispute.

3.1.4   Negotiation

See 2.1.3 above.

3.2   Binding

In Québec, arbitration and litigation are the main dispute resolution techniques available to the parties. Both processes take place in a well-established legal framework, which is provided by the *Civil*

Code of Québec and the *Code of Civil Procedure* [hereafter the *CCP*]. Although arbitration is more flexible from a procedural point of view, both methods focus on the parties' *rights* and consider the dispute resolution process as a zero-sum game *(i.e.* one party gains what the other party loses).

3.2.1 Adjudication

Although there is no formal process known as "adjudication" in Québec, the "Consultant" in the CCDC 2 contract plays a role that is similar to that of the adjudicator. The Consultant is an independent specialist who has the responsibility for interpreting the contractual documents and for overseeing the execution of the work and its conformity with the contractual documents (see section 1.2.2 above). Section 8.1.3 of the CCDC 2 contract provides that if a dispute is not resolved promptly, the Consultant will give such instructions as are necessary to ensure the proper performance of the work and to avoid any delays pending settlement of the dispute. The Consultant's orders in such cases do not jeopardise the parties' rights.

3.2.2 Arbitration

The *Civil Code of Québec* has a separate chapter on "Arbitration Agreements", as did the *Civil Code of Lower Canada*. The articles on this topic were added to the *CCLC* and the *Code of Civil Procedure* in 1986,[33] as the provinces of Canada implemented the provisions of the model law adopted by the UNCITRAL on June 21, 1985.[34]

The arbitration agreement is defined in article 2638 *CCQ:*

An arbitration agreement is a contract by which the parties undertake to submit a present or future dispute to the decision of one or more arbitrators, to the exclusion of the courts.

The Québec Courts have historically hesitated to recognise the validity of the *clause compromissoire* (by which parties accept to submit future disputes to arbitration), but the question was definitively settled (in the affirmative) by the Supreme Court of Canada in 1983.[35] In addition, article 2638 *CCQ* clearly states that the parties may choose to submit their disputes to arbitration to the exclusion of the Courts.[36]

Note that the Courts have considered that a dispute can be submitted to arbitration if the parties have agreed on a "compromise" *(compromis)* or if they have included a *clause compromissoire* in their contract. A compromise is an acceptance by the parties to submit a current and existing dispute to arbitration, while a *clause compromissoire* is a contractual clause by which the parties agree to submit future disputes to arbitration.[37]

Arbitration agreements must be in writing (article 2640 *CCQ),* but must not confer an advantage to one of the parties with respect to the selection and designation of the arbitrators (article 2641 *CCQ).* Article 2642 *CCQ* provides for the severability of the arbitration clause. Not only is it distinct from the other provisions of the parties' contract, the ascertainment by the arbitrators that the contract is null does not entail the nullity of the arbitration agreement.

The *Code* thus lends considerable importance to the arbitration agreement by giving it autonomy from the rest of the contract.

Finally, article 2643 *CCQ* recognises that one must refer to the parties' contract for the rules applicable to the conduct of the arbitration. The parties may set forth certain specific rules (as is the case in the CCDC 2 construction contract, with the Rules for Arbitration of Construction Disputes), they may refer to the rules of a domestic or international arbitration body (the Québec Centre for National and International Commercial Arbitration, the International Chamber of Commerce, *etc.)* or they may devise their own rules for the purposes of the arbitration. If the parties do not include or refer to specific provisions, they will be governed by the *Code of Civil Procedure*, which has a separate book on arbitration, which was adopted in 1986.

The *Code of Civil Procedure* provides a comprehensive set of rules on the conduct of an arbitration, on the award and its homologation and on the recognition and execution of awards made outside Québec. First of all, article 940 *CCP* recognises the parties' freedom to establish the terms of the arbitration, but also sets forth certain provisions which are of public order. For example, the parties cannot set aside provisions which indicate that judicial intervention is warranted *(e.g.* when the Court intervenes to name an arbitrator (article 941.3), to revoke the appointment of an arbitrator (article 942.7) or to homologate or annul an arbitration award rendered in Québec (articles 946 to 947.4)).

Article 940.6 *CCP* truly modernises the law of Québec with respect to arbitration by making a direct reference to international legal sources. It provides that where matters of extra-provincial or international trade are involved, the interpretation of the provisions which are relevant to arbitration proceedings shall take into consideration (1) the model law on international commercial arbitration adopted by the UNCITRAL on June 21, 1985; (2) the UNCITRAL's Report on the work of its 18th Session held in Vienna

(June 3-21, *1985);* and (3) the Analytical Commentary on the draft text of the model law on international commercial arbitration included in the Secretary General's Report to the 18th Session of the UNCITRAL. Article 940.6 *CCP* applies if three conditions are met: there must be an *arbitration,* on a *commercial* matter, and there must be an international facet to the parties' relationship.[38] (An international construction contract would be one such situation, since it is commercial in nature and involves contractors, engineers, architects, *etc.* from different jurisdictions (or who must perform work in a foreign country)). In the author's opinion, this is a positive development because article 940.6 *CCP* provides more certainty and predictability to foreign parties who are engaged in an arbitration to be decided in Québec, since the UNCITRAL rules will serve as a source of reference for the arbitrators.

Three arbitrators are named: one is named by each party, and the arbitrators selected choose the third person (article 941 *CCP).* (However, the parties often take exception to this provision, because of the expense of a three-person arbitration panel.) A party may ask that the mandate of an arbitrator be revoked if that person does not have the necessary qualifications or for one of the grounds of revocation of a judge's mandate (which apply, as adapted, to arbitrators) (article 942).[39] The *Code* also requires that arbitrators reveal to the parties any grounds which may justify the revocation of their mandate (article 942.1). If a party decides to ask for the revocation of an arbitrator's assignment, he must do so in writing within fifteen (15) days after the nomination of the arbitrator or of his realisation that a cause of recusation exists. If the arbitrator does not resign, the other arbitrators will come to a decision on the issue (article 942.3). If the recusation still cannot be obtained, a party may apply to the Courts to have the arbitrator's appointment revoked (article 942.5). The judge's decision on revocation or recusation of an appointment shall be considered final (article 942.7).

In practice, article 942 *CCP* may cause difficulties for the practising lawyer who acts as an arbitrator. Section 234(3) *CCP* provides that a judge's mandate can be revoked if he has given advice upon the matter in dispute, or has previously taken cognisance of it as an arbitrator, if he has acted as attorney for any of the parties, or if he has made known his opinion extra-judicially.

Since the same provisions also apply to arbitrators, this means that the practitioner who has been involved in the case himself or whose partner has been involved should be reluctant to arbitrate the dispute. The arbitrator, just as the lawyer, should avoid any appearance of conflict of interest.[40]

Article 943 *CCP* provides that the arbitrators may decide on their own jurisdiction (the *kompetenz-kompetenz* principle).[41] This leads to an uneasy balance between the arbitrators' powers and those of the Courts. On the one hand, the arbitrators have the power to trace the limits of their jurisdiction but on the other, a Court may refuse to homologate an arbitration award if it deals with a matter which was not contemplated by the parties or if the award contains decisions on matters which lay outside the scope of the agreement (article 946.4 (4) *CCP*).[42]

After some initial reluctance, the Québec Courts have generally refrained from intervening in a matter which has been submitted to arbitration or which may be submitted to arbitration in the future by virtue of a contractual clause to that effect.[43] For example, in *CIM4, société d'ingérnene V. Immeubles Marton lteé.*,[44] a case which dealt with a contract between an engineering firm and a building contractor, who was to prepare certain plans and drawings and provide services during the construction of a building above a metro station in Montreal, the parties had agreed on an arbitration clause applicable to all conflicts which could arise with respect to their rights and obligations during the term of the agreement. In the course of the works, the contractor made a claim of $ 1,244,235 against the engineering firm, on the grounds it had failed to provide for adequate supporting works to maintain the arch of the Metro tunnel in place. The engineering firm presented a motion to the Superior Court to obtain a declaration that the arbitrators had no jurisdiction to hear the contractor's claim because it was related to delictual (i.e. extra-contractual) matters and because the engineering firm would be precluded from calling third parties in warranty during the arbitration since they were not parties to the *clause compromissoire*. The Superior Court dismissed the engineering firm's motion, on the grounds that the arbitration clause had been perfected and that it should be given a liberal and generous interpretation; the merits of the delictual claim would have to be determined by the arbitrators. Furthermore, the Court found that the arbitrators did not lose jurisdiction because the conduct of third parties was at issue.[45]

The arbitrators conduct the arbitration according to the procedure they determine. They have all the powers necessary to exercise their jurisdiction, including the power to name an expert (article 944.1 *CCP*). The arbitrators may require each party to produce a written statement of claim, with the necessary supporting documents. These must also be communicated to the other party (article 944.2 *CCP*). If one of the parties does not present himself, the arbitrators shall record the default and may continue the proceedings nonetheless; however, if the party who submitted the dispute to arbitration fails to state his claim, the arbitrator shall terminate the proceedings unless another party objects (article 944.5). Witnesses are summoned in the usual manner provided for in the *Code of Civil Procedure*, i.e. by *subpoena* and *subpoena duces tecum* (article 944.6) and the arbitrators have the power to administer oaths and receive solemn affirmations (article 944.7).

The arbitrators decide matters on a majority vote, but one arbitrator may decide procedural questions, with the approval of the parties or the other arbitrators (article 944.11).

Article 944.10 addresses the legal regime applicable to the arbitrators' decision:

The arbitrators shall settle the dispute according to the rules of law which they consider appropriate and, where applicable, determine the amount of the damages.

They cannot act as amiable compositeurs except with the prior concurrence *[sic]* of the parties.

They shall in all cases decide according to the stipulations of the contract and take account of applicable usage.

Therefore, the arbitrators will only decide *ex aequo et bono* if the parties specifically provide so. This possibility could be especially interesting in commercial arbitration, where the parties might prefer that the matters be decided according to the prevailing rules of commercial practice.[46] To date, parties seem reluctant to have arbitrators decide as *amiables compositeurs*.[47]

The arbitrators are bound to keep silence on a case while the outcome is pending, but may state their conclusions in the final decision (article 945). If the parties settle the dispute, the arbitrators will record the agreement in an arbitration award (article 945.1). The arbitrators decide the matter by a majority vote and must provide written reasons; if an arbitrator refuses to sign or cannot sign, the other arbitrators must note that fact, but the decision has the same effect as if all the arbitrators had signed (article 945.3). The arbitration award binds the parties (article 945.4). The arbitrators may, within thirty days after deciding the matter, correct any clerical mistake in the decision (article 945.5); they may also, upon the application of a party within thirty days of the arbitration award, correct mistakes, interpret a specific part of the award or render a supplementary award on a part of the application which was omitted in their decision and these changes form an integral part of the original award (article *945.6)*. This decision must be rendered within sixty (60) days after the application (article 945.7).

A party will then apply by motion to the Superior Court for the homologation of the arbitral award (article 946.1 *CCP)*. Furthermore, the Court will not have the power to enquire on the merits of the dispute (article 946.2 *CCP)*. The Court cannot refuse homologation except in limited cases, *i.e.* because of some major procedural irregularity, because one of the parties was not qualified to enter into the arbitration agreement or if the award is invalid according to the law elected by the parties (article 946.4).[48] The Court cannot refuse homologation of its own motion unless it finds that the matter in dispute cannot be settled by arbitration in Québec or that the award is contrary to public order (article *946.5)*.

Finally, once the award has been homologated, it is executory as a judgment of the Superior Court (article 946.6). The only possible recourse against an arbitral award is an application for its annulment (articles 946.2 to *946.5)*.

The *Code of Civil Procedure* also includes a Title on the Recognition and Execution of Arbitration Awards made outside Québec. Article 948 provides that the Title will be interpreted by taking into account, where applicable, the 1958 New York Convention on the Recognition and Enforcement of Foreign Arbitral Awards. The Québec legislator has thus sought to implement the principles of comity and reciprocity which are at the heart of the New York Convention. Article 949 *CCP* provides that the foreign arbitration award will be recognised and implemented in Québec if the matter at issue may be settled by arbitration in Québec and if its recognition and execution are not against public order.[49] To make a foreign arbitration award executory, a party must present a motion for homologation before the Court which would have jurisdiction to decide the matter in dispute that was submitted to the arbitrators, and the motion must be accompanied by an original and a copy of the arbitration award and the arbitration agreement, which must be authenticated by an official representative of the Government of Canada or a delegate of the Government of Québec working outside Québec or a public officer of the jurisdiction where the award was made (article 949.1). Article 950 *CCP* identifies a series of circumstances in which a party may validly object to the recognition and execution of a foreign arbitration award, notably if the party against whom the award is to apply was not given proper notice of the arbitration proceedings (article *950* (3)) or if the decision dealt with a dispute which lay outside the scope of the arbitration agreement (article *950(4))*.[50]

Article 951.1 *CCP* provides that a court examining an application for recognition and execution of a foreign arbitration award cannot inquire into the merits of the dispute. However, as we have seen above, even if the arbitrator has jurisdiction *ratione materiae,* the Courts may be tempted to refuse the recognition of a foreign arbitration award if the subject matter lay outside the arbitrator's jurisdiction.[51] Finally, once the foreign arbitration award has been homologated, it is executory as a judgement of a Québec Court (article 951.2 *CCP)*.

Finally, the Courts have decided that arbitrators enjoy immunity from suit, just as judges do. In the *Zitfrer* v. *Sport Maska inc.* case, the Québec Court of Appeal found that:

[Immunities] issue from the rules of public law and not from those of private law, because of the relationship between arbitration and judicial functions even in cases where the conclusion of a compromise is related to contracts of a private nature... In the absence of fraud or bad faith, the arbitrator enjoys immunity from civil liability that lawyers might want to impose on them. [52]

The Supreme Court of Canada overturned the Québec Court of Appeal's decision,[53] but not on this point. Parties who believe that they have been wronged by an arbitration award should instead apply to the Superior Court to have the decision annulled (articles 947 to 947.4 *CCP)* or to have the Court refuse homologation (article 946.4).

In short, the *Code of Civil Procedure* sets forth a cohesive regime which incorporates a number of references to international law, such as the 1958 New York Convention and the model law on international commercial arbitration adopted by the UNCITRAL. Parties who have not provided for specific rules regarding their arbitration will be well served by the general regime of the *Civil Code of Québec* and the *Code of Civil Procedure.*

3.2.3   Expert Determination

The parties may designate an expert to decide a specific issue. As the Supreme Court of Canada explained in the *Sport Maska* case,[54] the powers which are conferred to the third party will determine if this person decides as an arbitrator or as an expert. If the person hears arguments from both sides and is required to remain strictly neutral, then he is acting as an arbitrator. However, if the person only decides on a specific part of the contract and *can* apply his personal knowledge to arrive at a decision, he is providing an expert opinion.

In the context of construction, the Consultant (in the CCDC 2 contract) has much of the same role as the expert, since he renders opinions periodically, on problem matters within his jurisdiction *(e.g.* interpretation of the contract). Of course, the Consultant's decisions are not final and the parties may choose to go to mediation and then arbitration if they are not satisfied with his holding.

3.2.4   Litigation

3.2.4.1 Preparation of the Case until the Hearing

We will make hereafter a cursory presentation of the judicial process in Québec. As we have seen in section 1.2, a case can be heard in first instance in the Court of Québec or the Superior Court, depending on the amount of the claim. There are no specialised tribunals which hear cases related to construction; all civil cases proceed through the Courts of general jurisdiction.

Furthermore, at the present time, the Superior Court[55] only proposes *voluntary* mediation in cases set for a hearing of three (3) days or more where the certificate of readiness has been issued. There is no *obligatory* recourse to mediation or any other alternative dispute resolution method, except in family law matters, where the parties *must* attend a pre-hearing information session on the mediation process.[56] Note however that the use of mediation, conciliation and arbitration are widespread in the field of labour relations.

As we have seen in section 3.2.2 above, article 2638 *CCQ* recognises the parties' right to include a legally binding arbitration clause in their agreement. Furthermore, so long as the case has not been inscribed on the roll of the Superior Court or the Court of Québec (and provided the arbitration agreement is valid), the Court shall refer the matter to arbitration on the application of either party (article 940.1 CCP). Before or during the arbitration proceedings, the Court may nonetheless grant provisional measures (article 940.4) or intervene on specific points at the request of one of the parties.

Although the Québec Courts do not exercise "inquisitorial" powers (and generally do not actively intervene in the conduct of proceedings), a tribunal may, after the issue is joined, order of its own initiative that any fact relating to the case be investigated, verified and determined by an expert whom it designates, or may refer to an accountant the establishing or auditing of accounts in any matter where accounts have to be rendered or settled (article 414 CCP). Furthermore, if the case presents exceptional difficulty or is particularly important, the Court may refer to three (3) experts (or accountants, depending on the nature of the problem) (article 415). The judgement appointing the expert will clearly state his duties and the delay he has to file his report (article 416). The expert's mandate may be revoked on the same grounds as for judges (articles 234 and 235 CCP), upon the presentation of a motion by one of the parties (article 417). The expert has broad powers to fulfil his mandate: he may examine any object or visit any place which is necessary for the preparation of his report, and may summon witnesses by means of subpoenas (article 420). A party may request that the expert's report be rejected on the grounds of irregularity or nullity; if there is no such contestation, the report forms part of the evidence in the case, though it does not bind the Court (article 423).

Every judicial proceeding is introduced by a declaration (article 110 CCP), which explains the object of the claim as well as the facts the claimant seeks to invoke (article 76) and must be accompanied with a notice to the defendant to appear, within ten (10) days (article 119).

As soon as the time provided for an appearance has expired, the Clerk may inscribe the case for proof and hearing before the Court (article 192). If the defendant appears, he has ten (10) days following the expiry of the time provided for filing an appearance to make his defence (article 173). The defendant may plead any ground of law or fact to reflite the plaintiffs claim in whole or in part and may within the same proceeding make a cross-claim if it arises from the same source as the principal demand or from a related source (article 172). If the plaintiff does not reside in Québec, the defendant may require that the plaintiff give security for the legal costs which may be incurred in consequence of his suit (article 65).

The *Code of Civil Procedure* includes a chapter which discusses Preliminary Exceptions. If the defendant believes that he has been summoned before the wrong Court, he can ask that the case be referred to the competent Court within Québec, or that the action be dismissed if there is no such court (article 163 *CCP*).[57] The motion for declinatory exception must be decided first by the competent Court (article 161(2)).

What are the other exceptions which a defendant may raise? He may make a motion for the dismissal of the action (1) if there is *lis pendens* or *resjudicata*, (2) if one of the parties does not have the necessary capacity, (3) if the plaintiff clearly has no interest in the suit or (4) if the suit is unfounded in law, even if the alleged facts are true (article 165 *CCP*). The defendant may also ask for the suspension of the suit, notably if he wishes to implead a third party (article 168 (5)), if he makes a motion for particulars (article 168 (7)) or if he requests the communication of a document by the plaintiff (article 168 (8)); the defendant may also ask for the striking out of allegations which are immaterial, redundant or libellous (article 168 *infine*).

The plaintiff may, if he so chooses, respond to the defendant, within ten (10) days after the filing of the defence. The defendant may only reply if he obtains leave from the judge in chambers (article 182 *CCP*). The issues are then joined (article 186) and a party can inscribe the case for proof and hearing (article 274). This is the act of procedure which indicates to the Clerk of the Court that the case is ready for discovery.[58]

The parties must take a number of steps before the case is ready for the proof and hearing. They must comply with Rules 15 and 17 of the *Rules of Practice of the Superior Court of Québec in Civil Matters*.[59] Rule 15 provides that a case will only be placed on the roll for hearing if the Clerk issues a certificate of readiness and files it in the court record. For this to occur, the parties must serve and file a declaration for inscription on the roll for hearing which indicates the nature of the case, the number of witnesses (expert and ordinary) and the expected length of that party's representations. This declaration shall be accompanied by a list of the communicated exhibits. The party must also attest that all documents such as depositions, introduced in evidence in whole or in part, expert reports and/or medical reports which he intends to use at trial and which are in his possession have been communicated to the other party and that no further preliminary motions will be made with respect to those documents (article 331.8 *CCP*). Once all the parties to the case have issued a declaration to this effect, the certificate of readiness will be issued by the Clerk. Rule 17 provides that after the production of the certificate of readiness, no other document, testimony, report or other exhibit may be produced without permission of the Court, which will be granted only when considered necessary in the interest of justice.

At least ten (10) days before the date set for proof and hearing, the parties must serve on the other attorneys and must file in the record of the Court a summary statement of the questions of fact and law at issue in the case Rule 18).

In cases of some length or complexity (usually scheduled for three (3) days or more), the trial judge or another judge may call the parties to a pre-trial conference before the certificate of readiness is issued. At this conference, the judge and the parties will discuss whether the questions at issue can be simplified and how to expedite the hearing. Admissions are usually made at pre-trial conferences and some discussion is made of the questions of fact and law rule 18.2). The parties will then be summoned by the Clerk to set a trial date, according to the length of the case and the availability of judges. The Clerk will send to the parties a notice of the date fixed for proof and hearing at least thirty (30) days and not more than (60) days before it occurs (article 278
CCP).

3.2.4.2 Testimony and Evidence

There are generally three parts to the proof: the plaintiffs evidence, the defendant's evidence and the plaintiffs rebuttal evidence. The plaintiff will adduce his evidence, using testimony, writings, admissions, material evidence and any other elements which are allowed by the *Civil Code of Québec.* During the testimony of the plaintiffs witnesses, the defendant's lawyer will make the necessary objections to prevent suggestive questions or questions which do not deal with the facts at issue (article 306 *CCP).* Witnesses only benefit from limited immunities.[60]

What of communications that were made in the context of settlement negotiations? The Québec Courts, inspired by the common law of England, have considered that these communications are privileged.[61] Both the party making and the party receiving the communication may raise the privilege. The fact that the statements were made "without prejudice" is not determinative; the Courts will recognise a privilege if the parties *intended* that the contents of their discussions not be revealed. However, it might be more difficult to argue that the communications were privileged if they were made to a non-lawyer. Royer, a noted expert on evidence, is of the opinion that a person could only claim that his discussions with a mediator or a conciliator are privileged by extension of the privilege which is recognised for communications made for the purpose of settling a dispute.[62]

Once the plaintiffs lawyer has finished questioning each witness, the defendant may cross-examine on all the facts at issue and may establish in any manner the grounds he may have for challenging a witness's testimony (article 314 *CCP*). Although there is no formal examination of the witness by the Court, the judge may ask the witness any question he deems useful according to the rules of evidence (article 318).

The defendant will adduce his evidence in a similar manner, and the plaintiffs lawyer may proceed to a cross-examination of each one of the defendant's witnesses. Finally, the plaintiff may avail himself of the right to present rebuttal evidence. Strictly speaking, rebuttal evidence should only be used to contradict the defendant's allegations and not to support the plaintiffs case. However, in practice, the Courts are quite flexible in allowing the plaintiff to make his case, namely by allowing the examination of other witnesses (article 289 *CCP*). The defendant could, for his part, ask the Court's authorisation to present additional evidence once the plaintiff has finished his rebuttal.

The Court may, on application by a party, appoint a commissioner to receive the testimony of any person who resides outside Québec or in a place which is too far distant from the place where the case is pending (article 734 *CCP*). The party who requests a rogatory commission must prove that the person resides in a place which is located too far from the place where the case is to be heard, and must also show the usefulness of such a commission and establish that it will have some influence on the questions at issue in the case.[63] The judgement will name a commissioner (article 429) and the parties who wish to be represented at the rogatory commission must give notice to the commissioner of the name and address of the person who will represent them (article 432). The examination will take place following the ordinary procedure, *i.e.* an examination followed by a cross-examination; in addition, a party may ask to have other examinations and cross-examinations admitted by the Court and attached to the commission. The commissioner himself may ask any questions which he deems to be relevant to the case (article 433). Within the delay specified in the judgement, the commissioner shall return to the clerk a certificate attesting that he has carried out his duties as specified in his mandate, to which he will attach the depositions and exhibits which the witnesses have produced (article 436).

The parties may file their exhibits in the record of the Court only at the time of the hearing (article 331.7 *CCP*). In proceedings introduced by declaration, exhibits must have been communicated to the other party not later than sixty (60) days after a declaration of inscription on the roll for hearing is served by any party, failing which exhibits may only be filed with the authorisation of the Court (article 331.8). In proceedings introduced by motion, communication of exhibits is performed by providing a copy of the exhibits to the parties upon service of the motion (article 331.6).

Once the parties have presented their evidence, they will make their pleadings before the Court, in which the attorneys present their arguments of fact and law. After the defendant has made his argument, the plaintiff may reply and the defendant may answer if the plaintiff has raised a new point of law. No other address will be made, except with permission of the Court (articles 291 *CCP)*. Note, however, that a party may ask at any time for the presentation of additional evidence; the Court may also of its own motion order the re-opening of the hearing (articles 292 and 463).

3.2.4.3 Legal Costs

A word on costs. The losing party must pay the costs of the case, including the costs of the stenographer, unless the Court by decision giving reasons decides otherwise (article *477 CCP)*. In Québec, costs are established according to the *Tariff of Judicial Fees of Advocates*.[64] On the one hand, they include outlays such as the costs of certification and service of proceedings and the expert's fees for the preparation of his report (but not for his testimony at trial). On the other hand, they also include taxable fees, such as (relatively minimal) amounts for incidental motions and for the decision on the merits.[65] If the amount of the action exceeds $ 100,000, a fee of 1% on the excess of $ 100,000 may be taxed (article 42 of the *Tariff)*. Furthermore, the Court may, of its own initiative or at the request of a party, grant a special fee to that party, in addition to any other fees, if it is an "important case" (article 15 of the *Tariff)*. The "importance" of the case will be assessed according to the seriousness and complexity of the questions at issue, the number of witnesses, the length of the trial, *etc.*

Unlike the Courts in the common law provinces of Canada, the Québec Courts cannot exercise their discretion to establish the amount of the costs; they are bound to follow the *Tarrif* (the award of the "special fee" is an exception to this rule). Note however that the Courts may bypass this problem to a certain extent by considering lawyers' fees as a heading of damages, especially in cases where a party had to defend against a vexatious complaint.[66] These cases remain an exception to the rule.

3.2.4.4 Appealing a Decision

As we saw in section 1.2, decisions of the Court of Québec and the Superior Court can be appealed to the Québec Court of Appeal. Generally, final judgements of a court of first instance are heard as of right by the Court of Appeal, except in cases where the value of the object in dispute is less than $ 20,000. The appellant must obtain the permission of the Court of Appeal in other cases where a final judgement is involved. Note that an appeal lies from an interlocutory judgement[67] only on leave granted by a judge of the Court of Appeal if the pursuit of justice requires that leave be granted (article 511(1) CCP). Article 29 states that an appeal lies from an interlocutory judgement if one of the three following conditions are realised: (1) when the judgement decides the issue in part; (2) when the judgement orders the doing of anything which cannot be remedied by the final judgement; or (3) when the judgement delays the trial unnecessarily. The right to appeal from interlocutory judgements is particularly important in the context of a Court's decision on objections which are made during the hearing.

3.2.4.5 Annulment of a Transaction or an Arbitration Award

On the other hand, if the parties settle their dispute as a result of their negotiations, or through mediation, conciliation or arbitration, a party who believes that he has been aggrieved by this decision must seek relief from the Superior Court or the Court of Québec depending on the value of the object in dispute. If any of the dispute resolution methods (other than arbitration) is used, the parties will sign a "transaction", as it is called in the law of Québec, to evidence their settlement. To obtain the annulment of a transaction, a party cannot invoke an error of law; however, annulment is possible if a party's consent was vitiated for one of the causes of nullity of contracts, *i.e* fear, lesion or an error on an essential element of the contract (articles 1399 and 2634 *CCQ*).[68] A transaction may also be annulled if it is based on writings which are later proven to be false (article *2635 CCQ);* or if documents were withheld by one of the parties (or were held by a third party with the knowledge of one of the parties) and were subsequently discovered (article 2637 (1) *CCQ);* or if the transaction related to only one object and the documents later discovered proved that the person had no rights in it (article 2637 (2) *CCQ)*. Of course, the party requesting the annulment has the burden of proving that one of the causes of nullity applies in the case at hand, given that a transaction has the authority of a final judgement between the parties (article 2633 *CCQ)*.

If an arbitration award is made, articles 947 to 947.4 *CCP* set out the procedure for annulment. Article 947 *CCP* reads:

The only possible recourse against an arbitration award is an application for its annulment.

Annulment is made by a separate motion to the Court or by opposition to a motion for homologation (article 947.1). Article 946.4[69] enumerates the grounds which may justify a Court in ordering the annulment (or in refusing to order the homologation).[70]

3.2.4.6 Proof and Application of Foreign Law

The Québec Courts are receptive to applying foreign law to contracts if such was the will of the parties. Article 3111 *CCQ* is a decidedly modern provision to the extent that it allows for the application of foreign law. The article reads as follows:

A juridical act, whether or not it contains any foreign element, is governed by the law expressly designated in the act or the designation of which may be inferred with certainty from the terms of the act.

A juridical act containing no foreign element remains, nevertheless, subject to the mandatory provisions of the law of the country which would apply if none were designated.

The law of a country may be expressly designated as applicable to the whole or a part only of a juridical act.

Furthermore, if no law is designated in the contract or if the law designated invalidates the juridical act, the é Courts will apply the law of the country with which the act is most closely connected, in view of its nature and the surrounding circumstances (article 3112 *CCQ*). Article 3113 *CCQ* establishes that a juridical act is presumed to be most closely connected with the law of the country where the party who must perform the prestation has his residence or establishment, if the act is performed in the course of its business.

Note also that article 3121 *CCQ* provides the following:

Failing any designation by the parties, an arbitration agreement is governed by the law applicable to the principal contract or, where that law invalidates the agreement, by the law of the country where arbitration takes place.

This article thus refers back to the law applicable to the main contract and emphasises that in any case, the arbitration agreement survives and is governed by the law of the jurisdiction where the arbitration takes place.

How does a Québec Court acquire knowledge of foreign law? Foreign law is assimilated to any other fact, which must be proven at the Court's request. Article 2809 *CCQ* is an innovative provision on the proof of foreign law:

Judicial notice may be taken of the law of other provinces or territories of Canada and of that of a foreign state, provided it has been pleaded. The court may also require that proof be made of such law; this may be done, among other means, by expert testimony or by the production of a certificate drawn up by a jurisconsult.

Where such law has not been pleaded or its content cannot be established, the court applies the law in force in Québec.

This article makes it clear that a simple reference to the law of the foreign jurisdiction will not be considered sufficient. The foreign law must be established by a writing, by testimony or by an admission. Literal proof of the law of the foreign jurisdiction can be made by filing a copy of a foreign law or judgement which has been issued by the official foreign authority or a copy which has been authenticated by a foreign official.[71] Testimonial evidence will usually be given by a law professor or a practitioner who has sufficient knowledge of the law of the foreign jurisdiction.[72] Admission must be clear, precise and voluntary; the Court may nonetheless require that this evidence be completed with an expert's testimony.[73] A written opinion prepared by a "jurisconsult"[74] may also serve as proof of the foreign law. However, in such a case, since the jurisconsult will not testify before the Court, the judge may be more demanding with respect to the expert's qualifications or to the probative value of the writing.

3.2.4.7 Jurisdiction and Private International Law Issues

The Court may, on the application of a party, decline jurisdiction if it considers that the authorities of another country are in a better position to decide the matter; this is the *forum non conveniens* rule (article 3135 *CCQ*). Also, on the application of a party, a Québec Court may stay its ruling on an action which is brought before it if another action, between the same parties and based on the same facts is pending before a foreign authority, if that suit may lead to a decision that could be recognised in Québec or if the foreign Court has already rendered its decision (article 3137 *CCQ*).

The *Civil Code of Québec* also provides intricate rules for the recognition and enforcement of foreign decisions.[75] Article 3155 *CCQ* establishes as a presumption that a Québec authority will recognise and declare enforceable any decision rendered outside Québec, subject to certain exceptions. These exceptions include cases in which the authority of the state where the decision was rendered did not have jurisdiction (1); the decision was not final or enforceable in the jurisdiction where it was rendered (2); the decision was made in contravention of the fundamental principles of procedure (3); a dispute between the same parties, based on the same facts and having the same object has given rise to a decision rendered in Québec, or which is pending before a Québec authority, or has been decided in another foreign country and meets the conditions for recognition in Québec (4); or the outcome of the foreign decision is manifestly inconsistent with public order as understood in international relations (5). However, the Québec Courts are generally respectful of the foreign Courts' exercise of their jurisdiction: a Québec authority will limit itself to verifying whether the decision for which recognition is sought meets the conditions set out in the *Civil Code of Québec*, without entering into an evaluation of the merits (article 3158 *CCQ*). The *Code* also provides that a transaction which is enforceable in the place of origin may be declared to be enforceable in Québec on the same conditions as a judicial decision, to the extent that these conditions apply to the transaction (article 3163 *CCQ*).

Pursuant to article 3155, the *Civil Code of Québec* applies the "mirror principle", *i.e.* the jurisdiction of foreign Courts is established in accordance with the jurisdictional rules which apply to the Courts of Québec, at articles 3141 to 3154. Furthermore, the foreign Court may validly exceed this jurisdiction in situations where a Québec Court would have done the same, for example if it was a forum of necessity (article 3136), if provisional or conservatory measures were ordered (article 3138), or to afford the protection of a person or property located in the jurisdiction (article 3140). However, the Québec authority may also refuse to accept the jurisdiction of the foreign Court if a Québec Court would have declined to exercise its jurisdiction in similar circumstances *(e.g.* in application of *the forum non conveniens* principle (article 3135) or in case *oflispendens* (article 3 137)).[76]

Article 3164 *CCQ* establishes the additional requirement that "the dispute is substantially connected with the country whose authority is seised of the case". Furthermore, the foreign Court's jurisdiction may be excluded for any of the grounds set forth in articles 3165 to 3168 *CCQ*.

Article 3165 *CCQ* provides that the jurisdiction of a foreign authority will not be recognised by Québec authorities if the law of Québec gives exclusive jurisdiction to its authorities to decide on that subject matter (1); if Quebec law recognises the authority of another jurisdiction, because of the subject matter or because of an agreement concluded between the parties (2); or if Québec law recognises an agreement by which exclusive jurisdiction has been conferred upon an arbitrator (3).

Also, where a personal action of a patrimonial nature *(i.e.* involving some monetary consideration) is involved, article 3168 *CCQ* states that the jurisdiction of a foreign authority will be recognised in six (6) situations *only:*

(1)   the defendant was domiciled in the country where the decision was rendered;

(2)   the defendant possessed an establishment in the country where the decision was rendered and the dispute relates to its activities in that country;

(3)   a prejudice was suffered in the country where the decision was rendered and it resulted from a fault that was committed in that country or from an injurious act which took place in that country;

(4)   the obligations arising from a contract were to be performed in that country;

(5)   the parties have submitted to the foreign authority disputes which have arisen or which may arise between them in respect of a specific legal relationship;

(6)   the defendant has recognised the jurisdiction of the foreign authority.[77]

Paragraph 5 is particularly important for parties who have willingly submitted their disputes to a foreign Court. Unless the decision violates another provision of the *Civil Code (e.g.* it violates public order), the foreign decision will be recognised in Québec.

Although article 3164 *CCQ* states that the jurisdiction of foreign authorities is submitted to the *same* rules as the Courts of Québec, article 3168 *CCQ* has a more restricted scope than article 3148 *CCQ,* which sets out the cases where a Québec authority has jurisdiction in personal actions of a patrimonial nature. For example, 3168 (3) requires the prejudice *and* the fault to have been committed in the foreign jurisdiction for that Court to have authority over the matter, whereas it is sufficient that *either* the fault *or* the prejudice occurred in Québec for the Québec Court to have jurisdiction (article 3148 (3) *CCQ)*.   Also, article 3168 (4) provides that *"the"* contractual obligations must be performed in the foreign country (it is understood that *all* the obligations flowing from the contract must be performed in the jurisdiction), while 3148 (3) only requires that *one* of the obligations imposed by the contract be performed in Québec.[78]

In summary, the *Civil Code of Québec* has established elaborate rules for the recognition of foreign decisions and each situation must be evaluated separately to determine which rules of private international law should apply. In general, though, one can say that the "mirror principle" applies and that a Québec Court will generally be respectful of the jurisdiction of a foreign Court if it would also have assumed jurisdiction in a similar case.

See 3.2.2 for a discussion of the *Code of Civil Procedure's* principles on the recognition of arbitration awards made outside Québec.

3.2.4.8 The "Fast Track" Procedure

On January 1, 1997, the Québec National Assembly approved major amendments to the *Code of Civil Procedure.*[79] These amendments were designed to introduce a "fast track" procedure into the law of Québec, to simplify the system of civil justice and make it more cost-efficient. As a result, most cases where the amount claimed or the value of the object in dispute is equal to or less than $ 50,000 must proceed by the new rules. In addition, a number of matters are submitted to the "fast track" procedure independently of the amount at stake. Article 481.1, paragraph 2 b) *CCP* reads:

> These special rules also apply to the recovery of a claim, irrespective of the amount in issue, in any matter concerning
>
> b)   the price in a contract for services or of enterprise, other than a contract pertaining to an immovable work, if the value of the object of the dispute is more than *$50,000,* or in a contract of leasing or a contract of carriage;

In practical terms, this means that the "fast track" procedure will apply to contracts related to immovable works (engineering, design, building *per se, etc.),* but only if the amount of the dispute is less than or equal to $ *50,000.* This provides some relief for professionals who work in the construction industry, because an accelerated procedure is not well suited to multi-million dollar construction law cases which require extensive expert evidence.

Now, for a few characteristics of the "fast track" procedure. There are two absolute rules: in all cases, the defendant must file his defence within ninety (90) days after service of the declaration and notice by the plaintiff (article 481.9); and the inscription for proof and hearing must be made not later than 180 days after service of the declaration and notice (article 481.1 1).[80] The defendant must file an appearance within ten (10) days after service of the declaration (article 481.6); within ten (10) days after expiry of the time for appearance, the defendant must present *together* any demand for security, dilatory or declinatory exception or exception to dismiss the action (article 481.7). A cross-demand forms part of the defence and is subject to the same rules as the principal demand *(i.e.* to the "fast track" procedure) (article 481.10). A separate roll is kept for "fast track" cases and a declaration for inscription on the roll for hearing must be filed not later than thirty (30) days after the inscription for proof and hearing (article 481.14).

When the rules of practice provide for the issuance of a certificate of readiness (i.e. for most cases which proceed through the Superior Court), a declaration of inscription on the roll for hearing must be filed not later than 30 days after the inscription for proof and hearing. The party to which the declaration and inscription are served has 30 days in which to serve and file a declaration of inscription to the same effect, on pain of foreclosure (article 481.14 *CCP*). Exhibits must be communicated to the other parties within 30 days after a declaration for inscription on the roll is served by one of the parties. The clerk then sets a date for proof and hearing and gives notice to the parties at least 30 days in advance (article 481.15) Finally,. the judge who hears the case has four (4) months (rather than six (6)) in which to render his decision. All in all, from the date the declaration is served to the date the judgement is rendered, the process should take from 14 to 16 months if the "fast track" rules are applied.[81]

The judiciary has been receptive to the new procedure and has generally refused requests to opt out of the "fast track", except in very limited circumstances. The accelerated procedure is advantageous for a plaintiff who has prepared his case and has gathered the elements necessary to complete his file, such as expert reports, because it imposes a strict 90-day delay to the defendant to make his plea. Practitioners have certain reservations with respect to the new procedure because it imposes rigid deadlines, which cannot be adapted to the parties' circumstances, even with their mutual consent. Because the new provisions have been in force for less than one year, there are few statistics to indicate if the "fast track" procedure has allowed (a) the parties to proceed more quickly or (b) the Courts to hear more cases (and thus improve the access to justice). We must give the Québec Bar and the judiciary some time to adapt their practices and to learn the workings of the new system before drawing any conclusions on the success (or failure) of the new accelerated procedure.

3.2.5   Negotiation

See 2.1.3 above.

3.2.6   Ombudsman

The National Assembly adopted legislation creating the office of "Public Protector" in 1968.[82] Although the title has more *panache*, the Public Protector performs functions which are quite similar to those of the ombudsman in other jurisdictions. In a nutshell, he represents the interests of citizens against the Administration. The Protector will intervene whenever he has reason to believe that a person or group of persons has suffered or may likely suffer prejudice as the result of an act or omission of a public body, its chief executive officer, its members or a person holding an office, employment or position accountable to the chief executive officer of that public body (section 13 of the *Public Protector Act* [hereafter the *PPA]*). The Protector may act on his own initiative or on the behalf of another person.[83]

However, the Protector is an institution of last resort. He cannot intervene in respect of an act or omission of the Administration if the person concerned has a legal remedy that can adequately correct the situation within a reasonable time, or if that person has omitted or failed to pursue such a remedy within the available time, or if the public body in question "is bound to act judicially" *(e.g.* if it is a Court or an administrative tribunal) (section 18). Furthermore, the Protector will not intervene if a request is presented more than one year after the person acquired knowledge of the situation complained of (article 20).

When the Protector deems it appropriate to intervene, he gives the author of the action (or the CEO of the public body, if such is the case) an opportunity to be heard and, where appropriate, shall give him the opportunity to remedy the situation (section 23). The Protector's interventions are private and may include an investigation (section 24). If the Protector is of the opinion that a person or a public body has not complied with the law, has acted in an unreasonable manner or has generally failed in its duties, he shall notify the CEO in writing of that fact (article 26.1) and shall make any recommendations he deems necessary (article 26.2). If no action is taken, the Public Protector may notify the Government in writing in a special report or in his annual report to the National Assembly (article 27). If he deems it to be in the public interest, the Protector may comment in public on any report he has submitted to the National Assembly or on any intervention he has made (article 27.4). The Protector and his assistants enjoy immunity from suit for official acts done in good faith in the performance of their duties (article 30) and no extraordinary recourse (e.g. judicial review) may be taken against the Public Protector (section 31).

The Public Protector has tended to be a high-profile, controversial figure and, because of the nature of his functions, he has often been at odds with the Government, whatever the party in power. His functions, however, are limited to the relations between citizens and the Administration and do not extend to private disputes.

Although there is a Consumer Protection Office in Québec, which deals with the application of the *Consumer Protection Act*, there is no official who acts as an ombudsman in construction matters.

4.   THE ZEITGEIST

The initiative which lead to the adoption of the "fast track" procedure which came into force on January 1, 1997, is a prime illustration of the legal *zeitgeist* in Québec. The Minister of Justice at the time, the Honourable Paul Bégin, made the following comments before the Commission des institutions, a permanent body of the National Assembly, upon presenting the legislation:

[Translation] This bill, Mr. President, includes a number of measures which will improve the quality of civil justice in Québec. I am persuaded that the adoption of these measures will provide new avenues to deal with the problems of costs and delays associated to the judicial process in civil matters. Finally, Mr. President, I reiterate that the changes brought to the procedure by means of declaration... aim to simplify the procedure in first instance and to make civil justice more accessible and less costly for all citizens.[84]

There is a consensus among lawyers, judges and government officials that changes need to be made to the civil justice system, to reduce costs and delays and to make the Courts more accessible to the people. The Government of Québec's enactment of legislation to create the "fast track" procedure is an important step towards achieving these goals.

The growing popularity of alternative dispute resolution methods is another promising development for improving access to justice. Mediation should become especially popular, in all fields of law, from commercial disputes to family law cases, because it allows the parties to resolve their differences in an informal and non-confrontational setting.

The Québec economy is growing, albeit at a slower pace than the provinces of Ontario or Alberta. However, outside Montreal and the Saguenay region, which has experienced a construction boom in the wake of last year's flooding of the area, the peripheral regions have not experienced much job growth in 1997. Housing starts have grown modestly but the vacancy rate remains high everywhere, which has contributed to depress prices.

Québec is currently in a state of transition. The Government has made substantial budget cuts, which has resulted in a sharp decline of employment in the public sector. The Greater Montreal region has bright prospects for the year 2000 and beyond, because of the growth of sectors with "high-tech" expertise, such as pharmaceuticals, biotechnology, aeronautics and software imaging. There has also been considerable foreign investment in these sectors. However, the city (and the regions of Québec) have to deal with the post-industrial *malaise* which grips most Western economies and which has translated into a steady decline of the manufacturing sector. The peripheral regions depend on cyclical sectors of the economy *(e.g.* forestry and mining) which have suffered a downturn in the second part of 1997. Therefore, slow growth seems to be the outlook for the near future.

The construction industry has fared poorly in the 1990s, with more competition, low prices and high vacancy rates. Given Québec's probable economic outlook, construction will continue to remain relatively stagnant for the next two or three years.

5.   TAXONOMY

I would like to make a few remarks on the use of the terms "conflict" and "dispute". In Québec, practitioners use both terms interchangeably (the French words *"conflit"* and *'differend"* are used). Professor John E.C. Brierley, a noted expert on the topic of alternative dispute resolution methods, uses the terms "conflict" to designate a difference of opinion or interests and the term "dispute" when the differences are so great that the parties are on the verge of taking judicial action. Brierley makes the following remarks on the usage of the term "dispute" *("differend")*.

[Translation] The concept involves a conflictual relationship which opposes two adversaries, who both have different arguments to make.

One must ask whether arbitration may be used to settle an opposition of interests which has not yet degenerated into a dispute of a judicial nature and which, for those reasons, could not be the object of a claim.[85]

I believe that the distinction between the *two* terms is necessary because they refer to two different stages in the relationship between the parties. In fact, they will only seek the assistance of a mediator or an arbitrator or will resort to the Courts if they are engaged in a *dispute*.

1   A low of 21,885 housing starts was recorded in 1995 (Commission de la construction du *Québec*, *Analyse de la construction du Québec 1995* (Montréal: Commission de la construction du Québec, 1996) at 24).

2   "New Housing Price Index" in Government of Canada, *Canada Year Book 1997 at* 372; "Région de Montréal: moins de locaux vacants dans un contexte de stagnation" Les Affaires, 27 September 1997 at 88.

3   (U.K.) 30 & 31 Victoria, c. 3, included in Revised Statutes of Canada *1985*, Appendix II, No.5. The Canadian Constitution, formerly a statute of the United Kingdom, was patriated to Canada through the *Constitution Act, 1982*, being Schedule B to the *Canada Act 1982* (U.K.), 1982, c. 11.

4   See articles 92(14), 96 and 100 of the *Constitution Act, 1867*.

5   Article 2130 of the *Civil Code of Québec* [hereafter the *CCQ*].

6   Article 2138 *CCQ*.

7   Article 2098 *CCQ*.

8   Article 2100 *CCQ*.

9   Articles 2106 to 2109 *CCQ*.

10   *Bankruptcy and Insolvency Act*, Revised Statutes of Canada *1985*, c. B-3.

11   *An Act respecting the Implementation of the Reform of the Civil Code*, Statutes of Québec 1992, c. 57.

12   The architect or the engineer must prove that the defects in the work did not result from an erroneous or faulty expert opinion or plan he may have submitted or from any failure to direct or supervise the work. The contractor, for his part, may he relieved from liability only by proving that the defects resulted from an erroneous or faulty expert opinion or plan of the architect or engineer selected by the client. The subcontractor may he relieved from liability only by proving that the defects resulted from decisions made by the

contractor or from the expert opinions or plans furnished by the architect or engineer (article 2119 *CCQ*).

13   See H.J. Kirsh and L.A. Roth, *The Annotated Construction Contract (CCDC 2-1994)* (Aurora (Ont.):Canada Law Book, 1997) for a comprehensive presentation of the CCDC 2 standard form contract.

14   See P. Shaposnick, "Mandater le mecanisme de resolution de litige" in Insight Conferences, "La nouvelle formule normalisée de marché à forfait du CCDC", Montréal, June 14, 1994.

15   The parties shall appoint a Project Mediator within thirty (30) days after the contract was awarded, or if they neglected to do so, within fifteen (15) days after either party requests that the Project Mediator be appointed (article 8.2.1).

16   For example, Rule 3.2 provides that the Project Mediator cannot disclose documents or information provided by one party to the other unless authorised to do so by the party providing the information; Rule 4.1 requires that all information communicated during the mediation shall not be disclosed by the Mediator or the parties. Finally, Rule 13.1 provides that parties who participate in a mediation session will not use any information or documents which were used or disclosed during the mediation process for a subsequent arbitration or trial.

17   S. Roy, "Definitions" in La *médiation en matières civiles et commerciales, 1995* (Montreal: Barreau du Québec, 1995) at 5-3.

18   Revised Regulations of Québec, C. B-i, r. 1 (as amended).

19   See generally Barreau du Québec, *Négociation* (Montreal: École du Barreau du Québec, 1996) at *56-58*.

20   Revised Statutes of Québec, c. M-3, s. 28.

21   Revised Statutes of Québec, c. M-4, s.27.

22   S. Roy, "Médiation commerciale, prévention des conflits et "partnering" in *Développements récents en médiation (1995)* (Cowansville (Qué.): Yvon Blais, 1996) 191 at 213.

23   S. Roy, "La médiation commerciale" in *La médiation en matières civiles et commerciales, supra* note 17 at 29-10 and 29-11.

24   S. Roy, "Définitions" in *ibid.* at 5-5.

25   Note that only lawyers may act as mediators in the context of mediation which is sponsored by the Superior Court.

26   Article 3.06.06 of the *Code of Ethics of Advocates* provides that the lawyer must avoid any situation in which he may he in a conflict of interest. In addition, mediation and arbitration groups usually have their own "Code of Ethics".

27   Clairmont, "L'avocat et la mediation" in *Développements récents en médiation (1996)* (Cowansville (Qué.): Yvon Blais, 1997) 155 at 16O-61.

28   S. Roy, "La médiation commerciale" in *La médiation en matières civiles et commerciales, supra* note 17 at 29-13.

29   See generally D.F. Gauthier, "Déontologie en médiation" in *Developpements récents en médiation (1995), supra* note 22, 83 for an interesting and in-depth discussion of the mediator's qualities and his obligations towards the parties.

30   S. Roy, "Les étapes et le processus de médiation" in *La médiation en matières civiles et commerciales, supra* note 17 at 8-1.

31   See generally S. Roy & H. de Kovachich, "La co-médiation: ses multiples facettes" in *Les développements récents en médiation (1996), supra* note 27, 1.

32   P Daignault, "Quelques réflexions sur la responsabilité professionnelle de l'avocat-médiateur et la couverture d'assurance" in *ibid.,* 73 at 86-gg.

33   *An Act modifying the Civil Code and the Code of Civil Procedure in matters related to Arbitration,* Statutes of Quebec 1986, C. 73.

34   The provinces' legislative action followed the adoption of a *Commercial Arbitration Code* by the Government of Canada (which was based on the model law adopted by the UNCITRAL) in *An Act relating to Commercial Arbitration,* Statutes of Canada 1986, c. 22, s.5.

35   *Zodiak International Productions Inc. V. The People's Republic of Poland,* [1983] 1 Supreme Court Reports 529.

36   This principle also appears in article 940.3 of the *Code of Civil Procedure.* Article 2639 *CCQ* provides an exception to the principle of article 2638 *CCQ,* for matters related to the status and capacity of persons, family matters or other matters of public order, which cannot he the object of arbitration.

37   J.E.C. Brierley, "La convention d'arbitrage en droit québécois interne" (1987] Cours de perfectionnement du Notariat 507 at 548-550; *Sport Maska Inc. V. Zitfrer,* [1988] 1 Supreme Court Reports 564 at 613-14 [hereafter *Sport Maska].*

38   L. Marquis, "La notion d'arbitrage commercial international en droit québecois" (1992) 37 McGill Law Journal 448.

39   Articles 234 and 235 *CCP* set forth a number of grounds which may justify the revocation of a judge's assignment, notably if he is directly interested in an action in which one of the parties will be called upon to act as a judge (article 234(4)); if he has any interest in favouring any of the parties (article 234(8)); or if he is related to an attorney or counsel or to the partner of any of them (article 234(9)). There is some controversy in the case law as to whether the enumeration of cases in article 234 *CCP* is exhaustive. The guiding principle is that a judge should recuse himself *(i.e.* withdraw from the case and refer the matter to another judge) if he cannot hear the matter with impartiality.

40   M. Lalonde, "Nomination des arbitres et procedure d'arbitrage" [1987J Cours de perfectionnement du notariat 573 at 585.

41   In fact, the arbitrators decide on the limits of their powers *(e.g.* whether they can decide on the validity of the contract which confers powers on them) rather than their jurisdiction (S. Thuilleaux & D. Proctor, "L'application des conventions d'arbitrage au Canada: une difficile coexistence entre les competences judiciaire et arbitrale" (1992) 37 McGill Law Journal 470 at 481).

42   In *Télébec ltée*. V. *Société Hydro-Québec*, Jurisprudence Express
      97-1061, the Québec Superior Court found that the arbitrators'
      declaration of partial lack of jurisdiction was an overly restrictive
      interpretation of their own jurisdiction, and that the parties had not
      had the opportunity to make adequate representations on the issue.
      The arbitration decision was therefore annulled.

43   See *Condominiums Mont St-Sauveur inc.* c. *Les Constructions
      Serge Sauvé ltée.*, [1990] Recueil de jurisprudence du Québec 2783
      (Que'. Ct. App.); *International Civil Aviation Organization* V.
      Tripal Systems Pty. Ltd., [1994] Recueil de jurisprudence du Québec
      2560 (Qué Sup. Ct.); see also in the context of homologation of an
      arbitration award, *Moscow Institute of Biotechnology* c. *Associés de
      recherche medicale canadienne (A.R.M C.)*, Jurisprudence Express
      94-1591 (Qué. Sup. Ct.) (settled out-of-court, March 30, 1995)
      [hereafter *Moscow Institute*].

44   Jurisprudence Express 96-385 (Qué. Sup. Court) [hereafter *CIAM*].

45   *Ibid.* at pages 10-12.

46   Brierley, *supra* note 37 at *557-58*.

47   Lalonde, *supra* note 40 at 579.

48   In the *Moscow Institute case*, *supra* note 43, the Superior Court
      dismissed a party's claim that it had been deprived of its right to be
      heard, because the parties had presented all their evidence.
      Furthermore, the parties' will to use arbitration had to be upheld.

49   Article *949 CCP* thus resembles para. 2 of Section V of the
      *Convention on the Recognition and Enforcement of Foreign Arbitral
      Awards*.

50   This provision mirrors para. 1 of Section V of the *Convention on the
      Recognition and Enforcement of Foreign Arbitral Awards*.

51   Thuilleaux & Proctor, *supra* note 41 at 482-87.

52   *[1985]* Court of Appeal 386 at 393.

53   *Sport Maska, supra note 37.*

54   *Ibid.* at 603-605.

55   The Québec Court of Appeal has just recently intitiated a voluntary mediation programme for parties who are set to proceed before that ꙏCourt.

56   Article 814.3 *CCP.* A party who has a valid reason may request not to attend the information session (such reasons may include the inequality of the power relationship between the parties, the disability of the party or the considerable geographical distance between the residences of the parties) (article 814.10). Note also that at any time *during* the hearing of a contested application, the Court may order the adjournment of the proceedings and the referral of the parties to the Family Mediation Service of the Superior Court or, at their request, to the mediator of their choice, for the settlement of one or more of the elements in dispute (article *815.2.1).*

57   Note that a Court's lack of jurisdiction *ratione materiae* can be raised at any time, and can even be raised by the Court of its motion (article 164 *CCP).*

58   Barreau du Québec, *Preuve etproce'dure* (Montréal: École du Barrcau, 1994) at 249.

59   Revised Regulations of Québec, C. *C-25,* r.8 (as amended). There are no equivalent rules for the Court of Québec, which allows a less formalised approach to the proof and hearing.

60   Communications made to one's partner during the marriage (article 307 *CCP)* and those made by government officials on information revealed to them in the exercise of their functions (article 308) are privileged. A witness cannot refuse to testify in a civil case on the grounds that his reply might incriminate him or expose him to legal proceedings; however, if he objects on those grounds, his reply cannot be used against him in peuni proceedings in Québec (article 309). Note also that a professional may not disclose confidential information revealed to him by reason of his profession, unless he is authorised to do so by the person who confided the information to him or by an express provision of law (section 9 of the *Charter of Human Rights and Freedoms,* Revised Statutes of Québec, c. C-12).

61   See generally on this issue J.-C. Royer, *La preuve civile*, 2d ed. (Cowansville (Qué.): Yvon Blais, 1995) at 686-90.

62   *Ibid* at 689.

63   D. Ferland & B. Emery, *Précis de procédure civile du Québec*, 3d ed. (Cowansville (Qué.): Yvon Blais, 1997) at *523*.

64   Revised Regulations of Québec, c. B-1, r. 13 (as amended) [hereafter the *Tariff*].

65   The amounts for a decision on the merits range from $75 to $1,000 in the Superior Court and from $450 to $1,250 in the Court of Appeal (articles 25 and 59 of the *Tariff*).

66   J.-L. Baudouin, *La responsabilité civile délictuelle*, 4th ed. (Cowansville (Qué.): Yvon Blais, 1994) at 142.

67   Article 29 *CCP in fine* gives the following definition for an "interlocutory judgment":
Any judgment is interlocutory which is rendered during the suit before the final judgment.

68   Following mediation or negotiations between the parties, it is unlikely that lesion could be invoked, because this requires the exploitation of one party by another, which may be evidenced by a serious disproportion between the consideration given by each party (article 1406 *CCQ*).

69   Article 947.2 *CCP* provides that articles 946.2 to *946.5* apply to an application for the annulment of an arbitration award.

70   See section 3.2.2 above for our discussion of the annulment of an arbitration award.

71   Royer, *supra* note 61 at 62-64.

72   Ibid at 65-66.

73   Ibid. at 67.

74   This term has not been defined in the new *Civil Code*, but presumably designates an expert in the law of the foreign jurisdiction.

75   Note that the term "Decisional" is used to designate a decision of a judicial, administrative or religious authority (in the context of family law). Foreign arbitration awards are subject to their rules of

recognition in the *Code of Civil Procedure* (articles 948 to *951.2)* (J.A. Talpis & J.-G. Castel, "Le Code civil du Québec: Interprétation des règles du droit international privé" in *La réforme du Code civil*, vol.111 *(Québec:* Presses de l'Universite' Laval, 1993) 807 at 912).

76   H.P. Glenn, "Droit international privé" in *La réforme du Code civil*, *ibid.,,* 671 at 769-78.

77   If a party defends an action on the merits before a foreign Court without any reservation and without making a declinatory exception, this constitutes a recognition of the foreign jurisdiction (Glenn, *ibid.* at 778).

78   Talpais & Castel, supra note 75 at 918 - 19.

79   *An Act to Amend the Code of Civil Procedure, the Act respecting the Régie du logement, the Jurors Act and other Legislative Provisions*, Statutes of Québec 1996, c. 5.

80   Furthermore, the 180-day delay for the inscription may not be fatal, if the lawyer was in the impossibility of acting any earlier (and if the other party did not suffer any prejudice).

81   R Mongeon, "Le fonctionnement de la procedure allégée" in *Amendments to the Code of Civil Procedure: How Will They Affect your Professional Life?, Montréal, McGill University, April 4 & 5, 1997 at 23.*

82   See the *Public Protector Act*, Revised Statutes of Québec, C. P-32 (hereafter the *PPAJ.*

83   Persons can request the Protector's intervention by giving their name and address and explaining their grievances in writing to the Protector (section 20 *PPA).*

84   *Québec,* National Assembly, Commission permanente des institutions, *Journal des débats,* no.20 (23 May 1996) at 3.

85   Brierley, *supra* note 37 at 534-35.

ROMANIA

Magda Brosteanau
Universitatea Technica Iasi, Iasi, Romania

1.   BACKGROUND

1.1   Economic

The Romanian construction industry outputs are: buildings, civil engineering work, and plant. It lies with other branches of industry, such as building materials, metallic constructions and products etc.

Some characteristics of the Romanian constructions industry expressed statistically are[1]:

Employment by branches	%	'90s	'80s
Construction		4.3	8.3
Municipal services and housing		6.5	3.8
Transport		6.4	6.1
Administration		0.8	0.6
Construction output			
Total economy		55	100
Gross domestic product			
Construction		70.2	100
Index of investments			
Construction		25	100
Fixed assets			
Construction		2.1	4.3
Structure of investments			
Building and assembly works		47.2	41.0
Equipment		39.0	49.3
Geological works		5.7	2.0
Other investments expenses		8.1	7.7

Depending on the governmental decisions and according to the strategic importance of the domain we have in Romania: national companies (rail, oil, post and telecommunications etc.); headquarters, as black holes for funds (for energetic industry, natural gases etc.); trade companies, private, State and mixed capital, national and multinational.

Setting up and operation of the following Romanian companies: Unlimited Partnership, Limited Partnership, Limited Liability Company, Public Limited Companies, Equity Joint - Venture, Contractual Joint - Venture, and Holding are enforceable by law.

The perception of capital market is favourable. Foreign investors are already interested in buying stock in Romanian companies and are looking for business opportunities in Romania. The potential financing forms are: from the budget, from personal funds (investors', partners', and shareholders' by public bidding), and crediting. The art of doing business is to make money with somebody else's. Although the absence of a favourable loan line is being felt, there are some crediting opportunities. The banks can credit investors on demand. During the negotiating of investment credits banks audit the investors' requests (current and non-current assets, and the technico - economic documentation of the investment) in view of ensuring its opportunity and profitability, assess the level of own financial standing, and ensure the reimbursement of the credits in due time. The large interest policy leads to loss of capital in small and average firms. Only a relaxed fiscal policy can be stimulating.

The inflation rate is operating by higher costs of constructions, as well. There are many difficult problems: power conservation, earthquake safety, constructions maintenance, too many dwellings in blocks of flats, giant plants, rehabilitation, lower performance levels, a few highways and international airports, absence of high speed trains etc.

1.1.1   Trends

Until 1990, the Romanian State was the only investor that would impose a credit level. After 1990 the investment works have collapsed. We are witnessing a period of transition towards a decentralised economy in which the law of demand and offer operates by costs (the higher the prices, the higher the production). This stage has not been, however, reached. Now the State has a very low requirement. The private sector does not have the necessary economic power to finance important contracts - small and average sized company buildings, banks and insurance company buildings, churches, individual flats, roads, highways are now being built.

We are witnessing a lent recovery in investment policies, especially in the private sector.

As compared to the countries with a highly competitive market economy, Romania does not have, with timid exceptions, institutions specialised in developing small businesses, as well as associated adequate legislation. The business incubator centres, flexibly organised in research institutes, technical universities, the Chamber of Commerce and Industry, in collaboration, sometimes, with similar Western institutions, are an alternative to indirect financing of a small business. They offer investors specially developed space, birotics facilities, managerial consulting, legal advising, training opportunities, and exchange of experience.

The irreversible reform that the Romanian economy is undergoing at the moment, imposes substantial reassessments and reshaping in the constructions industry as well. The quality of constructions is a key component of the free market contractual relations. Wording and applying quality conditions as performance requirements in agreement with world practice will facilitate the free circulation of products, services, persons, and capital in this field.

1.2   Legal System

The legal sector includes construction law, decrees of regulatory agencies, Court decisions, and tariffs. Co - determination in the construction industry acts at the level of the establishment (e.g. of construction legislation or of business contracts), and at the board level of controlling.

The Romanian constructions industry operates within the civil legal system by contractual relations in respect of law, governmental bills and other specific regulations' concerning investments and construction, published in the Monitorul Oficial of Romania.

The construction industry is subject to the regulation of the Ministry of Public Works and Land Development, Ministry of Transports, and Ministry of Finance. They ensure the guidance, loans, provision, and management as well as the check - up, approval and arbitration in construction and installation works. The main participants are: the specialised departments; the State Inspection in Construction, Public Works, Urban Planning and Land Development; Romanian Technical Experts Corporation; the Institute of Research in Construction and Construction Economics; Environment Protection Agency; and the local bodies (Technical and Investment Departments within District Councils; the Urban Planning and Land Development within the Town Councils; the Chamber of Commerce and Industry; the Cadastral and Land Development Office etc.). These bodies are the instruments of State control of the unitary application of legal stipulations on quality matters, comprising all the stages and components of constructions quality system, as well as on exposing contraventions, applying punitive sanctions and, if necessary, discontinuing the inadequate works.

The legal system co-ordinated by the Ministry of Justice is applied locally by the Court of Law, Appeal Courts, Notary Offices, and Attorneys' Offices. There is no legal system with special reference to constructions. A jurist's profession is a distinct one. The large firms have their own solicitors. The individuals and legal entities (companies) can address the Romanian Court of Law by appointing a lawyer. There are ombudsman institutions and district Consumers' Protection

Offices and Assurance Agencies. There are no differences in applying the legal system between individuals and legal entities, they have the same legal rights and obligations.

The external control of quality in constructions, inspired from the German system, is ensured through licensed experts.

The Ministry of Public Works and Land Development organises examinations for civil engineers and architects as design supervisors, supervisory managers, and technical experts in constructions for various fields in legal related works.

The Ministry of Justice organises examinations for civil engineers as technical experts in legal expert determinations.

The examination boards are made of highly reputed professionals from Academia, research, and other professional associations (Association of Romanian Civil Engineers, Association of Structural Engineers, Romanian Concrete Society, Romanian Association of Construction Entrepreneurs, "Anton Sesan" Academic Society, Professional Association of Road and Bridge Workers etc.).

These specialists work within the branch of the Romanian Technical Experts Corporation, the Local Legal Offices for Expert Determination, Technical Experts Company LTD, with banking, and in freelance works. The legal frame of supervisory management on the quality in constructions is governed by law.

Contractual relationships between clients and professional advisors are established within extra - legal works. When a legal expert determination is carried out the People's representative, and/or the Court of Law which requires it nominate the technical expert from a list provided by the Local Office for Expert Determination. If the list does not contain licensed experts in the field, other professionals, PhD holders with well known experience may be appointed. The expert may be challenged by the parties involved in the trial in case of incompatibility or in the absence of a letter of abstention from the expert. At the same time, each party has the right to request a specific expert to take part in the expert examination.

1.2.1   Construction Law

Construction law, officially recognised as an academic subject in education curricula, contains the technical prescriptions regarding quality criteria in constructions as a whole, components life span stages, elements, building materials and assessment ways. The quality conditions are worded as performance requirements which reflect the interests of the community to which the State is held responsible for ensuring individual and global protection; they are compulsory and are legislated and sanctioned by the public authority.

Construction law operation is affected by a great deal of legislation (Safety Measurements, Environment Legislation, Fire Protection etc.). Codes and standards have a direct relevance to construction method assessment. The legislation undergoes large, fast modifications, and it is, often, confusing and contradictory. There are a small number of laws translated in foreign languages.

The technical prescriptions in constructions offer answers to repetitive, technical, and business questions that occur between users' needs, the general social interest (by State administration bodies), technical and economic opportunities for all partners. The wording as performance requirements gives partners on contract the opportunities to establish any quality level in free will agreement with the users' needs and available funds; a free competition is, thus, promoted.

The compulsory performance requirements are: structural safety, users' security, fire resistance, hygiene and environment protection, acoustic protection, comfort and energy conservation. The performance criteria are given at the minimum levels.

Depending on complexity, level, and endorsement ability, the structure of technical prescriptions in constructions has three levels:

(i)   The present - day standards are documents dictate the characteristics of materials, products, components, elements, and installations, and basic data for design, erection, acceptance, operation, maintenance, rehabilitation, and post - use. The endorsement ability is one of the Romanian Standardisation Institute.

         After 1990 some standards have been replaced with technical agreements of the new products and procedures. The competence lies with the licensed testing laboratories, and Romanian Board for Technical Agreements.

(ii)   The technical regulations contain technical codes, instructions and specifications that stipulate the methods and procedures applied in all stages of building investment, for ordinary objects, case studies, and special works.

(iii)   The workshop manuals with an advisory aim are used for implementing technical standards and codes; they suggest technical and economic solutions.

After 1990, the reshaping of construction law has aimed at unifying it to the international codes. The technical regulations requiring quantification under specific Romanian conditions have been set up on the bases of own research results in correlation with the international evaluation methods.

There is a published collection on following subjects:

         Quality Assessment
         Building Investment

> Road and Railways
> Technical Municipal Constructions
> Land Development
> Standards
> Labour and Payment
> Trade Company
> Taxes, Tariffs, Rates

The descriptive construction law will co-exist for a time with the qualitative one.

1.2.2 Construction contracts

The contract is, legally, an agreement between parties in accordance with private law norms[2].

The norms[3] give the frame content for a project in terms of stages of design, auction documents, tenders, and construction contracts; they correspond to the present day constructions market in Romania, the experience of using the auctioning-bidding procedure in concluding contracts, and the recommendations from international financial institutions.

There are contractual procedures, national and international: by public auction, by price bidding, and by direct entrusting. The contract nature depends on the phase and form: commitment, standard contract, supply contract, and entrusting contract.

The object of commitment is to provide consulting services and feasibility studies; it is concluded between investors and contractors (consultant, designer, auction committee manager, and investors' representative).

The object of a designing contract is to provide the feasibility study, the technical project on a stage basis, and the auction documents; it is concluded between investor and the successful contractor (designer).

The object of a construction phase contract is either a turnkey contract which includes the operation or a separate contract; it is concluded between investor and the successful contractor (general undertaker and specific entrepreneurs).

The object of the sub-contract is either the erection of some parts or materials, goods and services supply; it is concluded between contractor and sub-contractors (professionals, suppliers, traders, other legal entities).

Depending on the method of payment, there are three types of construction contracts: measurement contracts, lump sum contracts, and cost plus a fixed fee contracts.

The employment of labour force is made by individual and group contracts, over a limited or unlimited duration.

The minimum level of training on the job is required for all contracts. The contractual procedures are enforceable by the Romanian construction law.

The contract form is not part of the bid; it is filled in and signed by the winning contractor. The annex documents, partially contained in the bid, ranging from particular to general, are parts of the contract.

The list of contract parts contains:

Contract form

> Nature of Contract, Number, Date
> Parties to the Contract
> Object of Contract
> Cost of Contract
> Duration of Contract in calendar months
> Parties Responsibilities

Litigation
Final Clauses
Authorised signatures

Annexed documents

Letter of Acceptance
Bid
Technical Project
Special Conditions
Additional Conditions
General Conditions

The general conditions contain the following chapters:

Construction Phase Documentation
Contractor and Specific Contractors
Partners' Obligations. Guarantees for Construction. Risk
Sharing Insurance
Labour
Contract Terms
Commencement of Works. Delays. Acceptance
Mended Works during the Guarantee Period
Modifications. Additions. Omissions
Materials. Tools and Equipment. Provisional Works
Measurements
Terms of Payment. Provisional and Final Payments.
Bonus and Penalties
Price Alternatives
Sub - contractors. Hourly Paid Works
Major Force Cases
Litigation
Contract Cancellation
Hard Currency and Exchange Rate
Other Provisions. Legal Effects

The additional conditions are specified in each construction work. Some articles read as following:

> Art. The contractor will design the details of the technical project annexed to the contract; the details will be endorsed by designer and awarded by a project supervisor.

> Art. The contractor will present the letter of guarantees for construction issued by a bank agreed on by investor (bank name and address).

> Art. The investor acknowledges his obligation to deposit in the contractor's account the negotiated advance of money within "n" days from contract conclusion.

> Art. The contractor will ensure transportation, accommodation, lunch, and payment of labour force.

> Art. The contractor will observe legal and religious holidays and traditions when using labour force etc.

The special conditions modify, complete and cancel the provisions of general conditions. Some articles read as following:

> Art. The contract documents will be written in two languages. The correspondence between parties will be in Romanian and English simultaneously.

> Art. The payment of importing materials, equipment, and services will be made in hard currency, convertible at the exchange rate on the day of purchase.

> Art. The contractor will present the final invoice within 10 days from works finishing.

Art. The value of payment for construction works will be up - dated depending on indices of price and tariff evolution for materials, tools, for transport by road, railway etc, and on indices of salary increase for budgetary staff.

The evolution indices increase with the price increase indices of industrial production in branches and classes communicated by the Romanian Statistics Commission, with the transportation tariffs practised by the Ministry of Transports, with the price of fuel, and salaries issued by governmental bills. The coefficients representing the weight of pay-offs, salaries, fuel, tyres, spare parts etc. are introduced by contractor in the bid, depending on their ownership, i.e. contractor's, associated partners' or sub - contractor's.

The parties to a contract convene to settle litigation either by mutual agreement, by arbitration or in the Court of Law operating in the area of construction placement. In case of disputes on quality of materials, tools, supervisory procedures, accuracy of sampling, each party may ask an expert determination after a previous notification of the other party. The costs of investigation and damages will be supported by the party whose guilt was proved, and the litigation will be settled according to contractual provisions.

2.   CONFLICT MANAGEMENT

Conflict might be considered as a gradual escalation of a disorder state[4].

The Romanian perspective on conflict versus dispute is pertinent[5].

Conflict means: misunderstanding; an incompatibility of interest; disagreement; antagonism; clash; difference; argument (violent) etc.

Dispute means: contradictory discussion between two or more people or groups of people; controversy; quarrel, by extension, fight for supremacy, for solving rivalry in one's favour; sports contest etc.

Conflict means an incompatibility of interest, and dispute means the fight for supremacy. If we use the terms of modern game theory, the techniques of solving a conflict are win - win or lose - lose, and solving a dispute ends up win - lose (either victory or defeat). The question that not even Harold Nicolson could answer in "Diplomacy" (1939) is how can a warrior be turned into a merchant[6].

The social causes of conflict might be worker versus supervisor, line versus staff, and Union versus Company.

The main economic cause of conflict is responsibility versus accountability.

2.1 Dispute Review Boards

A Board is established to evaluate disputes, and to make settlement recommendations to the parties. The Board members become knowledgeable about the object by periodically visiting the site. There are formal and independent Boards that advise, analyse and arbitrate disputes, usually of 3-7 members, depending on the importance of construction.

(i)   A formal dispute review Board process comprises mandatory inspections to these sites to ensure the existence and observance of construction quality system, e.g.:

- Appeal Review Auction Board - with the endorsement of Ministry of Finance and Ministry of Public Works and Land Development or according to situations stipulated in the Law of Disloyal Competition concerning the participants in a public auction;
- Sampling Board;
- Running Inspection Board due to the State Inspection in Constructions;
- Final Acceptance Board;
- Monitoring Board in Operating etc.

(ii)   An independent or formal dispute review Board process comprises binding or non-binding inspection on applying the law in the domain of quality criteria in construction by sharing of responsibilities, e.g.:

- Supervisory Board of construction works quality due to investor by supervisory manager or consulting firms;

- Follow - up Board of site application of designing solutions due to designer on investor's request;
- Distributing Board of responsibilities among participants in construction stages - own professional staff, associated partners, sub - contractors, bidders - due to contractor, in agreement with own strategy of quality management and the law;

- Expert Examination Board of documents, projects, tests, direct construction works, and interventions in time due to technical experts upon request from one party, the State Inspection in Constructions, the civil or penal Court of Law etc.

The State Inspection in Construction elaborates the inspection methodology, the instructions and procedures which will be used with investors, designing - construction - operation and post - use companies, and with specialised offices of local public administration that grant construction permit as well as technical regulations for the protection of dwellers. The District Inspection in Construction collaborates with other State control bodies; with prefectures, and local public administration (district and mayor's councils); with Universities, and Institute of Research in Constructions and Economics of Constructions; with the Romanian Technical Experts Corporation, and Technical Experts Company LTD; with the licensed laboratories, and technical departments staff of quality control and monitoring of participants etc.

The acceptance of construction works has two stages: the acceptance at the end of determined stages, and the final acceptance when the guarantee period expires.

The acceptance Boards will be able to assess if the obligations assumed in the contract are met. The acceptance Boards members are nominated by the investor, usually of 5 - 7 members, for public works, and of 2 - 3 members, for private works; they are made up of the investor's representative, another one on the part of the local administrative bodies, possibly one from the Chamber of Commerce and Industry, and some specialists in the field (possibly from Fire Brigade Headquarters, Zonal Commission for the Protection of Monuments and Historic Sites etc).

The designers will present, in written form, their point of view on construction. If the constructor's representative does not take part in the acceptance Board, the investor may invite a neutral technical expert who will note down in situ conditions.

The acceptance may be admitted with or without objections, postponed or rejected. If there are doubts as to the written stipulations in the technical book of the construction or the quality of works, the Final Acceptance Board may require an expert examination, other documents, additional testes, samples etc.

Dispute Review Boards operate as counsellors and make recommendations to the parties on observing authority limits, local autonomy, and parties' autonomy.

2.2   Dispute Review Advisers

Dispute Review Advisers are either a third party, neutral or work on behalf of one party, who simply advise on a problem or potential dispute: either the consulting civil engineer, design supervisor, supervisory manager, technical expert, consulting firm, and business incubator centre.

The consulting civil engineer works usually, for consulting firms or in freelance works licensed within one or more fields: Civil Engineering, Architecture, Project Management, Supervisory Management etc.; he/she can be authorised by investor to represent him/her in building the project. Sometimes he/she is described as an early settlement adviser who offers consulting, pre-feasibility studies, field studies, auction documents, technical assistance.

Consulting services may be: legal advice from an attorney, engineering advice from a civil engineer, accounting data and banking from the financial controller.

The design supervisor is employed by the investor.

The technical expert, usually a member of the Romanian Technical Experts Corporation and of Technical Experts Company LTD is the third, neutral party, irrespective of which party pays him/her.

The supervisory manager is employed by the general undertaker.

The clients go to the financial consulting firms to draw the business plans and credit, insurance documents, brokerage, mutual funds, and quality control.

2.3   Negotiation

Negotiation is a process in which two or more partners with common interests, convergent or divergent but solvable, agree to discuss an acceptable compromise, an agreement and/or a favourable transaction.

Negotiation is a process in which individuals or groups, as partners not adversaries, seek to reach goals by making concessions; the aim is to harmonise the common interests of the parties between which, in the beginning, there seemed to be an incompatibility of interests.

Negotiation can be considered one of the great discoveries of humanity.

Negotiation starts at the supplier - manufacturer level, and ends at the producer - user level, at mutual favourable prices within the company. The supply - sale price will be highly conditioned by the demand - offer ratio, the fluctuation of influence factors, and by mutations. The demand - offer ratio expresses, on the one hand, the quality - price link, and the competitive capacity, on the other; it is shown in the stock exchange quotations, and on the extra - exchange market - RASDAQ. The influence factors are: geographic location of business; volume of transaction; nature and quality of objects of discussions; company image; desired return; clauses on contract; originality degree; special relations between partners etc.

Within the company internal relations, negotiations develop either at manager - employees or patronate - Union levels; it is a way of communication for an optimisation of wages/productivity or wages/profit levels.

The Romanian negotiating strategies are piecemeal approach, and total approach or a combination of them.

The appropriate theory has three steps: pre-negotiation; negotiation, and post-negotiation. The most used negotiating techniques are: limited authority technique, feed-back technique, statistic game technique, detente technique and surprise technique.

2.4   Quality Matters

The Romanian building quality system is composed of: technical regulations; quality of used products; technical agreements for new products and procedures; experts' opinion; test laboratories accreditation; metrology; constructions acceptance; life-span behaviour and inspection; constructions post-use; State control of quality and quality management. There is a strategy of Total Quality Management and many procedures of Quality Assurance.

The Romanian construction law contains the performance requirements identified by ISO; ISO codes are not compulsory; they are of an advisory nature; competition makes companies comply with quality standards.

The qualitative level of constructions is assessed in a file consisting of: users' requirements; essential performance requirements; performance criteria; performance levels; estimating methods; performance specifications, norms, standards, and technical codes, for all stages of engineering structure (materials, products, equipment, components, sheath, structure, construction as a whole, and built environment).

The stages of a construction project are: promotion, engineering, and operation.

The engineering stage comprises design, construction, and commissioning. In terms of performance levels these mean:

> promotion - to - desired level;
> design - to - required level by technical codes and
> designed level;
> construction - to - achieved level.

Commissioning is the period devoted to testing and treating all the elements as a unit.

The responsibility in each of the stages is as follows:

> to designers in the design phase;
> to contractors and suppliers in the construction phase;
> to suppliers and caterers for the quality of deliveries;
> to the Final Acceptance Board in the commissioning phase; operation and maintenance are the client's responsibility.

The major specialist involvement is found even in the smallest construction project such as: in estimating and procurement; in engineering; in supervision; in general cost accounting; and in personnel relations. The major procurement systems used for design are: typified projects, re-usable projects, and custom-tailored projects. The legal designing share is 5 % of the contract value, and 2 - 3 % after auctioning. The use of typified projects makes it possible to reduce the designing share by 30 %.

Romania acknowledges the copyright only for the custom-tailored architecture projects and software in structural engineering. Clients pay taxes in obtaining construction licences, and authorisations, and construction permits.

The procurement systems used in construction are provided by major materials and installations supplies (as speciality contractors in ready-mixed concrete delivery or in reinforcing steel bars or pipes etc.), and major equipment and plant suppliers (manufacturers of generators or cranes, excavators or concrete pumps suppliers etc.).

The target of a construction process can be the construction as a whole, parts of construction or construction works (earthworks, formworks, finishing works, infrastructure, superstructure etc.). In Romania, the construction processes are listed in the Labour Performance Rates in Constructions defined as the time allowance necessary for an optimum size gang to carry out a unit of measure of a given construction work under specific conditions. Romania has only one method of measurement of construction works, based on the collection of Construction Work Items (34 books for new works and 7 for repairing and maintenance works), according to work classification. Each item has enclosed the usage rates of resources for materials, labour, and equipment, defined as the amount of resource used to build a unit of measure of the specified work item. Note that the material resource usage rate allows for wastage due to the handling and transport, and the material usage rate for formworks represents the percentage of wastage per a single use which is meant to last many cycles of usage.

The quality of materials is controlled by standards. Technical agreements of limited duration will be obtained for new and imported products and procedures. Special norms and instructions stipulate the technical and technological conditions of manufacturing, transport, placing, and protection of mixed materials construction works. Suppliers' deliveries are accompanied by formal quality certificates containing specifications about transport, handling, storage, and guarantee. Contractors have to review by sampling certificate, a formal number of evidence specimens per accepted delivery, and check the state of stored products. When the works are accepted, the contractor will present the quality certificates of deliveries, and the evidence on specimens analysis certificates.

Locally, the research facilities provided by Universities can be used in this respect. For instance, in Iasi, quality tests can be carried out, depending on the investigated item, in the laboratories providing: boundary layer wind tunnel, roads testing pilot unit, geotechnics and foundations testing stand, facilities for mechanical and chemical testing of building materials, steel structures testing stand etc.

2.5   Partnering

Partnering is a long commitment of co - determination at the level of partnership.

One of the most important features of a building investment is the distribution of responsibilities among partners who operate by partnership relations: investor, financier, bidder (consultant, designer, contractor or undertaker, supplier or caterer), client (owner or tenant). In other departments, e.g. in industry, the manufacturer is or may be the only responsible for market prospection, comparing various alternatives, doing research for a competitive prototype, designing objects and technologies, home trading and exporting.

Napoleon's Code, included in the Romanian legislation, stipulates the liability. Designer and contractors (leader, associated partners, and sub - contractors) are held responsible for unknown structural defects occurring within ten years of final acceptance, and during the life span specified in the project, for losses in loadbearing capacity or stability under earthquake loads. The civil - penal -

moral responsibility of civil engineers is higher as compared to others, and it may be partly covered by insurance contracts. This is not enough. There is a difference between partnership and partnering.

To make the argument more plausible, let us use some scenarios:

(i)   Co - determination at the level of construction law

The tree - like partnership chart that aims to obtain a quality level and a return on construction stages does not provide the transparence of the initial aim: a high performance level of construction as a whole. If the total quality indices, as, for example, the total cost of investment - operation on square metre of built surface or the total consumption of primary energy on square metre of sheath area of a heated building found themselves as a percentage in the required performance level per stage of construction, then we could speak of partnering.

(ii)   Partnering in a small construction project

The attic-roofing development of a block of flats with real architectural and comfort disfunctions could become reality only by a partnering between the flat owners' association, the local public administration, one or more construction companies, one or several investors.

The advantages are: improvement of the image and functionality of the block of flats and of town planning; increased comfort and functionality for dwellers together with a dignified identity; means and prestige for constructors; functionally, 40 % cheaper space as against a new building. The disadvantages arise from the owners' subjectivity.

(iii)   Co - operation

Romania has been designated to co-ordinate the project "Financial policies on promoting small and average business by micro-guaranteed crediting programmes", within the South East European Co-operation Initiative, as a consequence of the U.S., European Union, Switzerland, and South East European States partnering.

Partnering is a way of building profitable understanding between different partners in order to build a common objective: by maximising the effectiveness of each participant's resources.

3.   DISPUTE RESOLUTION

Disputes may arise between parties as competitors over quality of documents, of works, end products, responsibilities for delays, appropriate payments due to changed conditions, parties' rights and responsibilities by contract and in law etc.:

(i)   Challenges facing the public administrative sector including finance (e.g. for the conformity between the technical project and the agreement from the Ministry of Finance emitted on the grounds of an endorsed feasibility study);

(ii)   Challenges confronting the construction industry and trade led by procurement system;

(iii)   Challenges confronting the involved parties;

(iv)   Challenges confronting investor's requirements and design development for additional works;

(v)   Challenges confronting contractors, suppliers, and ADR Procedure etc.

The systems of dispute resolution can be specified in the contract between parties or decided upon when the dispute arises, and they will comprise: informal arbitration (i.e. mediation and conciliation); administrative arbitration; and arbitration in the Court of Law.

3.1   Informal Arbitration

This is the process whereby a third neutral party gives a decision which can be binding by contract unless the parties wish to proceed to arbitration in the Court of Law. The aim of informal arbitration by mediation that it is the parties' conciliation.

3.2   Conciliation

The terms mediation and conciliation are parts of the same process that may recur until a dispute is resolved if at all. Conciliation is a mutually acceptable resolution made by mediation between parties.

3.3   Mediation

Mediation is a free-will private dispute resolution process in which a neutral third party, being knowledgeable of the dispute topic, helps the parties to make endeavours to reach a negotiated settlement; the neutral third party is, usually, chosen by mutual agreement of the involved parties.

In Romania mediation is perceived as an informal arbitration which is binding if it is specified in the contract under the chapter "Litigation".

3.4   Negotiation

Here, this term is used as a mutual agreement for distributing responsibilities among parties for caused damage, unless contractual clauses stipulate otherwise.

3.5   Administrative Arbitration

The administrative arbitration is a formalised method of ADR consisting of one executive from each party in dispute and a neutral, formal one; the executive must have power to achieve settlement.

This term may be translated into Romania as State Arbitration ('80s), Economic Tribunal ('90s), and State Inspection in Constructions, Public Works, Urban Planning and Land Development which exert control, observe contraventions, apply punishments and, if necessary, discontinues inadequate works. In this case, the dispute resolution is binding unless the parties wish to proceed to arbitration in law.

3.6   Adjudication

Adjudication is a procedure whereby the Auction Board gives an adjudicative award which is binding on the parties to the contract unless they wish to appeal.

3.7   Arbitration in Law

Arbitration in law is a trial whereby formal disputes are determined by the Court of Law. There are no separate Courts for constructions litigation with specialised procedures. The construction disputes may be dealt with under the jurisdiction of the Economic Section of the Court.

Various functions are fulfilled by specific law offices such as the Auditing Office, Expert Determination Office and, more recently by the Sale and Financial Administrations Corporation in view of accelerating the technical, banking or accounting expert examinations in assessing damages.

If the informal arbitration is unsuccessful the parties may submit the case to the Court. This is the main default dispute resolution system for individuals. The legal action is the last resort of dispute resolution for firms since some of the consequences can be high fees, fines, and bad public relations. The multinational firms prefer legal action because of arbitrators' immunity.

The disputes become civil litigation, and/or penal cases (criminal offences). The law breaking elements are punished if expressly stated. The civil law code arbitrates litigations within the legal procedure. The criminal law code is the procedure used in solving criminal cases. Recourse may be taken to the Court against a judicial ruling, in civil or criminal cases. If a final sentence cannot be reconciled the revision procedure is applied. In the most instances the Court enforcing awards are compulsory unless one of the parties or both of them wish to proceed to superior instances such as the Appeal Court, the Supreme Court of Law or International Arbitration Court (e.g. the International Court of Law in the Hague).

The executory procedure is the competence of the Court (e.g. the Court, by its executory officer, advertises the public sale of assets for the payment of debts to the bank). The Court of Law is not controlled by the State.

The Romanian arbitration in law is based on two principles:

the principle of objective truth;
the principle of arbitrator's active role.

The arbitrators' responsibilities are:

> to discover the truth;
> to protect the rights and interests of involved parties;
> to appoint technical experts;
> to rule out legal decisions and/or sentences;
> to abstain from and/or be challenged in case of incompatibility.

The challenge is the possibility offered to parties to remove arbitrators or experts for partiality in equity causes. If the whole panel of judges is challenged, the file will be sent to an instance of the same rank. If the challenge is denied, the file is returned to the initial instance.

Evidence is given in writing and orally and it encompasses: parties' declarations; witnesses' testimonies; written documents; expert determination; and material evidence. Cross examination may be used. Judges or parties can ask witnesses and/or experts questions only by means of the president of the panel who can approve or not direct questioning.

The law ensures protection against contraventions, defends the judicial order, the State, ownership, the citizens and their rights. The limits of law enforcement for deeds perpetrated on the Romanian territory and abroad by a Romanian, against a Romanian, and/or against the State are: territoriality; legal personality; reality; universality; international conventions; legal immunity; and extradition (e.g. business transactions are governed by the law of the country in which they are operating). In time limits of law enforcement for actions and / or retro - actions are: partial enforcement; compulsory one; optional one; and temporary enforcement.

Contraventions are penalised by administrative or penal fines, and penalties. The claimant pays the legal costs. The losing party is charged with all the legal costs.

3.8   Expert Determination

Expert determination is evidence used by a physical person, contract parties, judges' panel to determine an issue which necessitates professional advice.

There are legal and extra-legal expert determinations, compulsory and conditioned (by damage, functional transformation or on demand). The legal expertise is optional.

Fees depend on the expert's qualification, the complexity of works, and the required time. The expert can be nominated in the legal instance by one, both parties or by the Court. The expert's conclusions can by freely assessed by the judge. The interested party can require another expert.

Technical Experts Company LTD instructs extra-legal expert determinations at the clients' request by a commitment contract. The experts provide quality, promptness, and confidentiality.

The conventional procedure of expert determination refers to Assessment Decisions and Quality Assurance Decisions.

3.8.1   Assessment Decisions

The clients request Assessment Decisions for:

    Privatisation
    Establishment of Company Patrimony
    Increasing of Social Capital
    Establishment of Real Estate Value
    Setting the Collateral Value for Bank Credits
    Auctions. Mergers. Liquidations. Splitting etc.
    Establishing the Damage Value
    Stamp Duty for Notarial Documents
    Evaluation of Intravillan Land. Surveying Works

International Transport Expenses
Post Labour Performance Rates of Design Works etc.

The assessment decisions methodologies are in connection with the construction legislation, inflation index, data provided by the National Commission of Statistics, mean exchange rate of hard currency, direct information provided by designing and construction firms, materials - products - energy suppliers, and owners.

(i)   There are five Assessment Decisions Methods for establishing the technical cost of the existing constructions:

- Standard Cost Method - uses price lists from typified projects, correction frames, up - dating and indexations;
- Assimilation Method - makes a comparison between the costs of two similar constructions with equivalent facilities of which one with a known construction cost;
- Coefficients Method - applies only for the accounted (administrative) value which reflects in situ (commercial) value;
- Direct Cost Method - evaluates the current assets from group 1, Buildings, and group 2, Civil Engineering Works, on the grounds of bill of quantities; it is an accurate method;
- Substitution Value Method - uses price/sqm indices (standard prices), correction frames, up-dating, and indexation.

All construction features are input data.

(ii)   For the reimbursement of the cost of the works under construction, the Price Forming as in technical project auctioning is used.

(iii)   The Standard Currency Method for establishing the up-
          dating indices generated a conflict between specialists.

(iv)   Individualising Index Method applied to price determines
          the free market price of the existing constructions. Some
          input data are: technical features, location, proximity of
          commercial areas, output, mean statistic tenders in the
          area for similar constructions (some owners overvalue
          their properties).

          The legal instance recommends the evaluation of the
          commercial value.

(v)   There are four Assessment Decision Methods for land:

          •   Patrimony Value Method
          •   Gross Capacity of Hire - Value Method
          •   Balanced Value Method
          •   Goodwill Method

          The differences between the administrative value of land
          and the commercial one found by expert determination
          are considerable. The land value is 15 - 45 % of the entire
          value of a firm's fixed assets. These two aspects may alter
          the economic - financial diagnosis. The short term
          consequences are: difficulties encountered in
          privatisation, association with the Romanian and foreign
          partners, stock sales etc. On a longer term, the land value
          pay - off influences the cost of products or services,
          favouring or harming firms depending on their location.

(vi)   The commercial value of constructions erected on
          guaranteed land (without ownership rights) is found by
          means of two methods influenced by management:

          •   Benefit Capacity Method
          •   The Method of Gross Capacity of Hire - Value

(vii)   Renewing Payments Method in splitting with mortgaged buildings has generated a conflict between specialists. The procedure of incurring payments between parties depending on their contribution to the advance payment and mortgage is logical under monetary stability conditions. Wherever the Roman law is the source of the legal system the solutions are the same: the quota remain the same, in nature or money and cannot be mistaken by the payments in collateral credits, without forced payments to harm the adverse party. This is an equity clause.

3.8.2   Quality Assurance Decisions

The goal of Quality Assurance Decisions procedure is to measure or compute the achieved performance level of construction outputs after the setting standards. It may be a sampling supervisory or a total one, in each or all stages of a building investment.

Supervisory management is an official responsibility. It entails own controlling, commissioning, and State inspection, and expert determination. There is an expert's custom-tailored participation.

The methodologies for Quality Assurance Decisions are in connection with construction legislation, protection level, type of expert examination, occupancy importance class, location, features of construction/works, technical project/technical book of construction/documents, technology, time, and direct information.

(i)   The goal of Commissioning procedure is to observe the standard due to investor, or designer or own controlling department; the standard will be the first made element (reinforcing bars, foundation element, pre-cast concrete panel, the first apartment etc.):

- Establishing, by measurement, of correspondence between in situ elements and technical project;
- Checking the existence of formal quality certificates of suppliers' deliveries;
- Checking the existence of specimens analysis certificates of contractors' controlling and partial acts acceptance especially for hidden works;
- Sampling. Running Tests. Monitoring.

(ii)   Expert determination of Seismic Protection Level of the existing constructions may be made by following methods[7]:

- Qualitative Expert Determination
- Non - Destructive Tests
- Load - Carrying Capacity
- Biographical Method
- Non - linear Dynamic Method

(iii)   The intervening decision is made depending on the index of seismic protection, priority, decay, damage, duration of intervention, owner's requirements (i.e. modifying the function of buildings, attic development, decrease of floor number, retrofitting etc.), designer's opinion, and investor's option:

- Repairing
- Retrofitting
- Rehabilitation
- Partial Demolition
- Demolition

All decisions may be made under certainty, risk, and uncertainty.

3.9 Litigation

The disputes submitted to the Court of Law become civil litigations (causes) and / or penal ones.

3.10 Negotiation

This term is used as a mutual agreement among parties for all accepted or rejected Court procedures and decisions (with reference to expert, judge, damages etc.).

3.11 Ombudsman

The Ombudsman institution, lately introduced in Romania, investigates complaints against public bodies. Through the ombudsman institution the legal entities and physical persons can appeal to the international Court of Law. On the other hand, there are public and private agencies and offices that operate within consumer protection, insurance, banking, and legal services.

4. THE ZEITGEIST

The spirit of the time for the Romanian nation is an optimistic one. The political and legal zeitgeist is operating in the aims of reviewing and introducing proper laws (acts) carried out by both Chambers of the Parliament, Senate and Deputies House; the aims of the review are:

- to reduce the complexity of the privatisation rules;
- to harmonise and to unify the construction legislation to the international codes;
- to offer the same opportunities to the foreign and Romanian investors;
- to improve the access to justice;

- to decentralise the formal responsibilities;
  to give a chance to everybody.

The economic zeitgeist is depressing: despite the increased construction prices these lag behind constructions costs. The mean income is almost US $ 110, and the cost of living is similar to Europe. We need a lot of money for building investment. It would be a big mistake if we felt impressed by the world general economic recession and we went out of the construction competition, where there is an exceptional potential.

The liberalisation of the Romanian market is irreversible and will not be subjected to any restrictions. The new law of legal land circulation will eliminate the State's pre-emption right in buying land. The foreign investors will begin to notice the Romanian potential.

The educational and research aim of Civil Engineering Speciality is to provide the graduates with a sound university training in order to operate all over the world. The Romanian higher education system in Civil Engineering is adjusting to competition and activities abroad. Moreover, there is a need to provide a sound training to adapt to new demands that go beyond traditional visions of Civil Engineering, and to offer young Romanians new job opportunities in order to support the Romanian enterprises, and to increase Romania's chance to balance her foreign trade exchange in constructions. As a professor of Building Engineering at the "Gheorghe Asachi" Technical University of Iasi (Romania) I begin my lectures like any civil engineer who, after having got free from inertia, faces his/herself and time: "Where are we going to go in Civil Engineering?" Who knows? Things always change. The essence of dispute avoidance is quality.

1   See Romanian Statistics Commission, (1992) *Romanian Statistical Yearbook*, Printing House Coresi, Bucuresti, p 125, 275, 279, 282, 293, 461.

2   See Civil Roman Law Code, Book 3, Title 3, *On Contracts and Conventions*, Article 942 - 1222.

3   See Monitorul Oficial no 232, (1996) *Methodological Norms of projects in terms of stages of design, auction documents, tenderds, and construction contracts*, annex to Common ordinance of Ministry of Finance, Ministry of Public Works and Land Development.

4   For further information see Kinard, J., (1988) *Management*, Western Kentucky University, D.C. Heath and Company Lexington, Massachuttes, Toronto, p. 308 - 311.

5   See DEX, (1996) *Explanatory Dictionary*, Academy (Ed), Bucuresti.

6   See Zartman, W., (1992) *Regional Conflicts Resolution*, in Sinteza No 93, U.S. Information Agency, Washington, D.C., p 13.

7   See P 100, (1991) *The Romanian Anti - seismic Design Norms*, Printing House of Public Works Department, Bucuresti, p 88 - 104.

SCOTLAND

Duncan Cartlidge
The Robert Gordon University, Aberdeen, Scotland

1.   BACKGROUND

1.1   Economic

While Scotland is often referred to as a region of the United Kingdom, it is a quite distinct country in its own right and exhibits some very different characteristics from the rest of the UK. Apart from having its own separate administrative, legal and educational systems, it differs in terms of industrial and economic performance.

The total population in Scotland was estimated in 1996 to be 5,148,900 of which just over 59% are domiciled in the Glasgow/Edinburgh conurbation and accounts for only 9% of the total population for the UK, making it one of the most sparsely populated countries in the EU.

The GDP of Scotland is on average approximately £42,000.000 per annum of which, in 1997, the total output of the construction industry was £4,900 million, that is 11.7 % of GDP. The most important sector in terms of output being Electrical Engineering which accounts for 37% of total GDP.

Scotland's output accounts for 8.5% of the total UK GDP and looking at per capita figures the Scottish economy performs well in comparison with the UK. GDP per head in Scotland was 97.4% of the UK average in 1995 and Scottish per capita GDP was the third highest in the UK behind the South East and the East Midlands but with the second highest growth ranking during 1993 - 96 .[1]

Scotland's recent impressive growth is forecast to continue, based on its bias of output towards faster growing industries, in particular Financial and Business Services.

The growth rate for the construction industry in Scotland during 1991 - 94 was 2.3% compared with a - 0.5% growth rate for the UK as a whole [2], while during 1996 growth fell slightly to 2.1% and currently construction employs approximately 127,000 people, which represents between 4.9% - 10.6%, depending on the region, of the total working population of 2,900,000.[3]

The Scottish construction industry output for 1997 can be broken down as follows: [4]

New Build -   £2,900 million

Repair and Maintenance -   £2,000 million

1.1.1   Trends

Of the total of new build work, £1,240 million was public/infrastructure works, leaving £1,660 million of private sector works. Of the Repair & Maintenance work, £318 million was public works with private work output standing at £1662 million[5]

During the 1990s Scotland has become one of the United Kingdom's strongest growing and prosperous regions, however, there are many uncertainties surrounding developments in the immediate future. Not least is the potential for more autonomous government and the establishment of a Scottish Parliament in 2000, which could have significant impacts on the economy. In particular, autonomy may see the end of the so called 'Barnett Formula ' which is a method established in 1978 by Lord Barnett for giving Scotland billions of pounds extra for public services - 23% extra in fact compared with England. Scotland will have to rely instead on its newly founded tax raising powers to make good the expected short fall. Independent forecasts of the Scottish economy expect GDP growth in 1997 to be around 2.8 percent, which is somewhat less than the UK as a whole. [6]

Protagonists of total independence from the United Kingdom claim that Scotland is a net contributor to the UK economy.[6] However, even the income from North Sea oil is insufficient to prevent Scotland being in deficit to the UK Exchequer in the sum of approximately £8 billion per annum.

Despite these clouds on the economic horizon the July 1997 Confederation of British Industry Quarterly Industrial Trends Survey indicated that optimism in Scotland's [7] business situation was 16% greater compared with the previous years.[8] Certainly the long term growth prospects are further supported by having received 15% of all UK direct inward investment in 1990 - 95. Scotland is also a major beneficiary EU subsidies, according to the Department of Trade and Industry, Scotland received ECU 214 million or 17% of the UK's allocation of EU Structural Funds in 1996.[9]

1.2   Legal

Within the territorial limits designated as Scotland the state has developed an organisation of courts and legal practitioners, a body of principles, rules and precepts and a body of knowledge and thinking about law which amounts to the Scottish legal system. It is largely indigenous having grown and developed naturally within the country but having strong historical connections and affinities with legal systems with continental Europe.[10]

An industrial tribunal, sitting in Edinburgh upheld Lord Simon of Glaisdale's decision in 1972 that ' *Scots are a nation because of Banockburn, Flodden and Culloden..............* ' , accordingly the tribunal found that for the purposes of the Race Relations Act 1976 the English and Scots are separate racial groups and there was preserved to Scotland, under the Act of Union its church, its education system and its legal system.[11]

After the Wars of Independence, culminating in the Battle of Bannockburn in 1314, Scotland's external political and economic ties were mainly with France and the Low Countries. Many Scottish lawyers obtained their legal education in France, Germany and Flanders where they studied and practised Roman Law.

In the late 17th century, Viscount Stair (1619 - 1695), Lord President of the Court of Session and the first of the so-called ' "institutional writers" published his *Institutes of the Law of Scotland.* In this Stair set out the whole of Scots law as a rational, comprehensive, coherent and practical set of rules deduced from common-sense principles.[12] Stair was guided in his task by Roman cannon law or Roman-Germanic systems as well as reported Scottish decisions and statutes. Other notable institutional writers include : Sir Thomas Craig (1538 - 1608), Sir George Mackenzie (1636 - 1691) and Professor George Joseph Bell (1770 - 1843).[13]

In the 18th and 19th centuries, the practice of studying in Europe declined, due in part to the Napoleonic Wars and gradually English Law replaced Roman Law as the main external source of Scots Law. There was a move towards English legal methods and the doctrine of *judicial precedent* came to be more strictly applied in Scotland. The House of Lords became the final court of appeal for Scots civil cases despite the fact that judges in the House of Lords were likely to be English Lawyers. Currently, Scottish and English courts are not bound by each other's decisions but they do consider them persuasive.

While Scottish legal methods are now influenced by the English legal system, nevertheless in Scotland, civil law rests more on generalised rights than in England. It remains to be seen however what influence Scottish autonomy has on future attitudes.

The two most important branches of Scottish law are civil and criminal law. Civil is separated into two parts - public and private. Public law regulates and controls the exercise of political and administrative power within the state whereas, private law deals with the rights and obligations of citizens.

The major areas of private law are :

> Family law.
> The Law of contract.
> The law of delict.
> The law of property.
> Mercantile law

The formal sources of Scots Law are said to be: [14]

> Primary :   Legislation
>   Judicial Precedent
>   Institutional writings
>   Custom
>   Equity
>
> Secondary :   Extraneous sources

It is worth noting that unlike England, Scotland has never had a distinction between law and equity. The term equity therefore does not have the same associations as in England

Litigation in Scotland is generally an adversarial or contentious, rather than an inquisitorial or investigative procedure; the judge presides as it were over a contest, ensuring that the rules of law are correctly applied and followed. A decision is based on the legal arguments and facts submitted by the pursuer and the defender or by their legal representatives.

The main characteristic of the Scottish judicial system is separate courts and tribunals for civil, criminal and administrative matters. A further characteristic is the division of courts into inferior and superior courts. Inferior courts have jurisdiction locally, while superior courts have jurisdiction over the whole of Scotland. In addition, courts and tribunals are divided into those which have original jurisdiction, which hear cases for the first time and those of appellate jurisdiction, which hear appeals from other courts and tribunals.

The legal profession in Scotland has two branches : advocates and solicitors. A practitioner may move from one branch to the other but may not practice in both fields simultaneously.

Members of the Faculty of Advocates may practice before the Court of Session, the House of Lords and the Court of Justice of the European Union in addition to all the lower Scottish Courts and are briefed by solicitors. Advocates may be senior or junior and after 10 to 15 years practice may apply to be a senior or Queen's Counsel. The Faculty has its own constitution and entry criteria.

To practice as a Solicitor in Scotland it is necessary to be a member of the Law Society of Scotland which regulates the conduct and standards of the profession. Solicitors undertake most of the work of the sheriff courts and unlike advocates are allowed to form partnerships. Since 1993 solicitors with at least five years continuous court work experience may apply to practice in their own right in certain Scottish courts. Some solicitors are also *Notaries Public* and have certain functions connected to bills of exchange and foreign business which can not be performed by an ordinary solicitor.

1.2.1   Construction Law

The Law Society of Scotland has a Specialist Panel where accreditation is given to lawyers with five years continuous experience in construction law. There are no specialist chambers for lawyers who practice construction law, but several members of the Faculty of Advocates have expertise in construction law.

Two Scottish Universities: Robert Gordon University, Aberdeen and Strathclyde University, Glasgow offer MSc/LLM in Construction Law and Arbitration.

Construction law in Scotland is drawn from a number of sources there is no specialist construction law report, however, reports on cases involving construction law may be found in :

Scots Law Times and

Green's Weekly Digest

1.2.2   Construction Contracts

The first standard form of building contract was issued in Scotland in 1915. Currently, the Scottish Building Contract Committee, established in 1964, is one of the main sources of construction contracts in Scotland, where contracts are normally in a standard form. The contracts from this source are based upon the standard forms of contract drafted in England by the Joint Contracts Tribunal and then adapted for use in Scotland by the addition of a Scottish Supplement. The major difference between the Scottish SBCC forms and English JCT forms being the clause relating to Arbitration which is the preferred method of dispute resolution in SBCC contracts. It should be noted that although the use of standard forms is wide spread, they are frequently amended.

The SBCC is composed of representatives of professional bodies, building contractors organisations, local and central government and industry and commerce generally. Only one contract issued by the SBCC, the Minor Works Contract, is printed in a unique Scottish form i.e., without the need for a supplement.

There follows a list of the SBCC standard forms of contract:

Scottish Building Contract - With Quantities or Without Quantities Edition - July 1997 ; with Scottish Supplement

Scottish Building Contract - With Approximate Quantities Edition - July 1997 ; with Scottish Supplement.

Scottish Building Contact - Contractor's Designed Portion - July 1997 with Scottish Supplement

Scottish Building Contract - With Contractor's Design - July 1997 ; with Scottish Supplement.

Scottish Minor Works Contract (1986 Edition - Sept. 1996 Revision)

Scottish Prime Cost Building Contract (July 1997 Revision)

Scottish Measured Term Contract (Sept. 1996)

Other Standard Forms of Contract in use in Scotland are :

Institution of Civil Engineers Conditions of Contract : 6th Edition.

General Conditions of Government Contracts for Building and Civil Engineering Contracts (GC/Works/1) - 1997.

General Conditions of Government Contracts for Building and Civil Engineering Minor Works (GC/Works/2) - Edition 2, 1980.

Institute of Chemical Engineers Standard Forms of Contract.

The British Property Federation - ACA Standard Form of Contract 1983

Institution of Civil Engineers - The Engineering and Construction Contract - 1993

It can be seen from the above that the building client in Scotland can be faced with a bewildering range of forms of contract.

2.   CONFLICT MANAGEMENT

2.1   Non-binding

2.1.1   Disputes Review Boards

There is very little evidence to suggest that DRB are being adopted as a means of managing conflict in Scotland. There are no published figures, as this type of ADR is a private procedure, but the author's own research has been unable to find reference to a dispute where a DRB was used.

2.1.2   Disputes Review Advisors

Once again there is little evidence to suggest that the use of such a person in practice. However, to the author's knowledge there are persons within the industry, who are respected and trusted who are acting in this capacity, but rather as an 'add on' service for existing clients than a stand alone service.

Conflict is an issue in the Scottish construction industry, however the past five years has seen a move away from the traditional procurement paths that have so often lead to conflict in the past.

Major contractors operating in Scotland claim to have changed from a *'claim preparation'* approach to *'managing a commercial process'* approach in an attempt to soften their image and several of the major contractors operate courses for their managers in negotiation skills. Traditionally in Scotland the building industry, as distinct from the civil engineering industry, has been the subject of very little conflict. The major reason suggested for this is that, the industry is so comparatively small, that contractors are reticent to *'bite the hand that feeds them'* and gain a reputation for having an aggressive and confrontational approach. However, the civil engineering sector has not been so fortunate in generating this ethos.

There has been a move away from traditional procurement, that is competitive selective tendering, within the past ten years. It is thought by many contractors that this procurement path engenders a mind set of *'lowest price - biggest claim'* and should be avoided. Instead there has been a move towards establishing long term win/win relationships with clients which use certain of the principles of partnering. See 2.1.5.

2.1.3   Negotiation

As previously referred to in 2.1.2 there is a movement to training construction and commercial managers in negotiation skills in a attempt to manage any potential trigger points where conflict has traditionally occurred and several major contractors have recently gone public on their change of approach to conflict management.

2.1.4   Quality Matters

Until the latter half the 1980's the primary procurement path in the Scottish construction industry was single stage competitive tendering based on the Scottish Building Contract, with quantities, with Scottish Supplement. Recently, there has been a move away from this procurement path, by both clients and contractors towards design and build type contracts and partnering. Figures are difficult to obtain but it is thought that the split between traditional and 'new' procurement paths is now around 50:50. The clients' move was in the belief that design and build/non-traditional procurement provided better value for money, for contractors the shift was prompted by excessive losses from the preparation of abortive tenders.

2.1.5 Partnering

Partnering is a post-Latham[15] buzz word which seems to be used by many but truly understood by few. In this section of the monograph it can be taken to mean a non-binding agreement in order to maximise returns and reduce the potential for conflict and in that respect partnering has been in operation for many pre-Latham years.

2.2 Binding

2.2.1 Partnering

Commercial clients building in Scotland, for example the supermarket chains; Asda and Sainsbury have post-Latham, proclaimed the virtues of partnering. Also, in the oil and gas industries, in which Scotland is a world player, partnering has been used for many years to the advantage of clients and suppliers. There is however, still a tendency in the construction industry to use partnering like a stick to beat a dog, i.e. if you don't perform there will be no more work. It obviously makes commercial sense for large organisations building a standard product, over a period of time to use partnering, the rest of the industry remains sceptical and ill-informed for the time being.

3.   DISPUTE RESOLUTION

3.1   Non-binding

3.1.1   Conciliation

Conciliation is not recognised as distinct from mediation. The UNCITRAL model has not been adopted in Scotland.

3.1.2   Executive Tribunal

There is no evidence of usage in Scotland although any mini-trial would be conducted in private with confidentiality surrounding the outcome.

3.1.3   Mediation

Mediation to produce a non-binding decision is not apparently used a great deal in Scotland

3.1.4   Negotiation

Negotiation has historically taken place at a number of trigger points during the course of a contract, the outcome can be binding if accepted by both parties.

3.2   Binding

3.2.1   Adjudication.

There have been major ground breaking developments in the referral of disputes to adjudication during 1996/97 in the form of :

The Housing Grants, Construction and Regeneration Act 1996 (Part II)[16], and

The Draft Scheme for Construction Contracts (Scotland)
Part I 1997[17].

The Act was a direct result of The Latham Report (1994)[18] -
See 4 The Zeitgeist , and gives a party to a building contract the
right to refer a dispute for adjudication. In addition the Act
provides the right, in the event of a contract not complying with
the Act, for a dispute to be referred to the provisions of Scheme
for Construction Contracts (Scotland) for adjudication. At the
time of writing the Scottish Scheme is in the consultation phase
and it is expected that it will not be fully implemented until
1998/99. It must be reported that the initial reaction by lawyers to
the legislation has been critical, nevertheless, this development is
of primary importance, as for the first time it acknowledges the
value of adjudication in dispute settlement in the construction
industry. The decision of the adjudicator will be binding and may
be accepted by the parties as final, or alternatively will only bind
the parties until referral to arbitration or legal proceedings,
usually after the contract works have been completed.

3.2.2   Arbitration

Arbitration is the preferred method of dispute resolution in the
Scottish Construction Industry. Figures are difficult to obtain, as
arbitration proceedings are private and the awards are not
published, but it is thought that approximately 5 - 8 cases are
heard each year, with many more disputes being referred but
settled prior to the arbiter's award being made. The traditional
reasons for the use of arbitration in Scotland have been: speed,
informality, cheapness, privacy and expertise, although in recent
times most, if not all of these reasons have been called into
question. See 3.2.4. The principle references in Scotland
Arbitration are :

Articles of Regulation 1695

The Arbitration (Scotland) Act 1894

The Administration of Justice (Scotland) Act
1972

The Arbiter has a wide discretion to determine how proceedings are conducted in Scotland[19]. At the more formal end of the spectrum there is the adversarial type of procedure patterned on the ordinary Sheriff Court, with counsel instructed to plead the case of the parties. At the informal end there is the personal examination of the evidence by the Arbiter. Which approach is chosen will depend upon the complexity of the issue, the agreed view of the parties and the arbiter's experience.

Scottish Law of Arbitration differs from most other European countries in that it is effected mainly by means of judicial precedent with some assistance from juristic writings. It may be considered therefore that it does not travel well, and this basis discourages references to Arbitration in Scotland in international commercial cases.

A major risk when parties decide to opt for arbitration is that the arbiter may incorrectly interpret questions of fact and questions of law - on these grounds there is no right of appeal, the arbiter's decision is final. There are however limited grounds on which the Court of Session may set aside an award. These are :

> Corruption, bribery or falsehood.
> Improper procedure.
> Ultra fines compromissi and
> A defective award.

It has been common practice in the Scottish Construction Industry to refer disputes to a third party for resolution. Of the available procedures reference to an Arbiter is the one most usually referred to in the dispute resolution clause(s) of a contract. Arbiters may be a named person, a member of the Institute of Arbitrators or perhaps a person recommended by the Royal

Institution of Chartered Surveyors in Scotland Arbitration Service.

In Scotland the Arbiter's jurisdiction is derived solely from the agreement of the parties and the Arbitration clause in the contract may :

> Lay down the conduct of the proceedings, which have been established beforehand and which are binding on the parties. The Arbiter, must in these circumstances respect this form of agreement as failure to do so would render the outcome invalid . e.g. A Statutory Arbitration, or

> More usually it is the duty of the Arbiter in Scotland to establish the conduct of the proceeding, e.g. Non-Statutory Arbitration.

In the case of Statutory Arbitration it can be necessary for the arbiter to express an opinion on a point of law pertinent to the proceedings, also during a non-statutory arbitration a party may apply to the arbiter, unless expressly excluded from the arbitration agreement, to apply to the Court of Session to give an opinion on law arising out of the proceedings.

In addition, in the author's experience it is relatively common practice for disputes between parties to a construction contract to be referred to an eminent person, with extensive knowledge of construction law and who is respected by both parties, for a quicker and less formal, but non-binding dispute resolution procedure. Such a procedure may take half a day and be held in the informal offices of a neutral party.

3.2.4 Litigation

The traditional reasons for the adoption of arbitration as the preferred method for settling disputes in Scotland , see 3.2.2 , have been repeatedly called into question during the past few years[20]. Arbitration is now seen by some as a lengthy and expensive procedure in need of updating. The impetus for reform was a working party, established in 1993 to consider what changes should take place in the handling of commercial litigation to establish a fast, efficient, economic and reputable system, in the light of increasing criticism of the existing arbitration procedures. The findings / recommendations of the working party can be summarised as follows :

• The existing system was slow and lacking in technical expertise. It should be recognised that the working party looked at all arbitration procedures and therefore it was not necessarily the expertise of the construction arbiters that was called into question.

• Much more control was required by the court, together with a hands-on approach by the judge. Traditionally, in the Court of Session there has been a non-interventionist tradition in the Scottish judicial procedures, the neutral position of the judge being seen as involving a passive and reactive role.

• There was a need to introduce simple pleadings.

• A recommendation that commercial cases should have their own allocation of judges on a full-time basis.

Following the working party, in March 1994, Lord Penrose was appointed as a full-time commercial judge to the Court of Session, together with another senior judge with commercial experience, who would be available when required. New rules for commercial actions in Scotland were introduced in the Court of Session on September 1994.[21] The intention of the commercial court was to accommodate the needs of the business world and to reduce the delays that had been experienced with the existing system. Commercial action has a very wide definition, it means any action arising out of, or concerned with, any transaction or dispute of a commercial business nature[22] and includes disputes relating to Building and Engineering Contracts. In order for an action to be heard in the Commercial Court the words 'Commercial Action' are inserted in the summons by the request of a party. A commercial action may be withdrawn from the commercial roll only on the application of a party The procedure will then involve preliminary and procedural hearings which are held in chambers and are relatively informal and could take the form of a round table discussion on how the cases are to proceed to achieve a speedy solution. During these hearings the focus will be on identifying issues and time tabling for optimum speed.

It is difficult to say whether or not the Commercial Court has been accepted by the Scottish Construction Industry, but research carried out during the first two years of its operation show that out of a total of 286 registered cases only a small proportion were related to the construction industry[23]. As to one of the primary objectives of Commercial Courts, that is to speed matters up, there are signs that the system is becoming a victim of its own success. However, one of the outcomes of the Commercial Court will be a body of case law, long absent in Scotland which may assist parties in dealing with disputes.

Scottish Courts do not have the power to direct the use of mediation without the consent of the parties, however, a procedure known as judicial references allows the parties to withdraw the action from a court and refer the question to a judicial referee, this process leaves the action in the court. If the parties agree to go to arbitration, any court action would stop, the subsequent award however may be enforced by the courts.

3.2.5   Negotiation

See 2.1.3

4.   THE ZEITGEIST

The spirit abroad in the Scottish Construction Industry seems to
be a move away from conflict towards managing the commercial
process for the benefit of the client. This is to some extent
historical but impetus was added with the publication in 1994 of
the Latham Report [24]and the subsequent legislation[25]. Initially,
there was the fear that the Latham Report would go the way of
many other construction industry reports and have little impact,
however there are signs that there are changes in the way that the
construction industry conducts business and the hope is that this
momentum will be maintained.

1   Earley,F.et al., 1997. *Scotland,* London : The Council of
        Mortgage Lenders.
2   Johnson,J.et al., 1997 (Jan). *Scotland : High Profile Growth.*
        London : Credit Lyonnais Laing.
3   Scottish Office, et al., 1997 (March). *Scottish Economic*
        Bulletin Number 54. Edinburgh : HMSO.
4   Scottish Office et al ., 1997 (March). *Housing and Construction*
        Statistics - Great Britain Parts 1 & 2   Edinburgh : HMSO.
5   Scottish Office et al ., 1997 (March). *Housing and Construction*
        Statistics - Great Britain Parts 1 & 2 Edinburgh : HMSO.
6   Frazer of Allander Institute, 1996 (Dec.). *GDP Growth*
        Forecast from 1995-1997. Scotland and the UK.
7   Scottish National Party et al., 1997 (March). *The Economic*
        Case for Independence. Edinburgh : SNP.
8   Confederation of British Industry , 1997 (July). *Quarterly*
        Scottish Industrial Trends Survey. Glasgow : CBI Scotland.
9   Johnson,J.et al., 1997 (Jan). *Scotland : High Profile Growth.*
        London : Credit Lyonnais Laing.
10   Walker.D.M., 1992. *The Scottish Legal System, 6th Edition.*
        Edinburgh : Sweet & Maxwell.
11   Bowditch,G. 1997. The Scots and English are different races,
        tribunal decides. *Times.* 28 March :p4.
12   Manson-Smith D., 1995.*The Legal System in*
        Scotland. Edinburgh : HMSO.
13   Marshall,E.A., 1995.*General Principles of Scots Law, 6th*
        Edition. Edinburgh : W.Green, Sweet & Maxwell.
14   Marshall,E.A., 1995.*General Principles of Scots Law, 6th*
        Edition. Edinburgh : W.Green, Sweet & Maxwell.
15   See 3.2.
16   The Department of the Environment et al., 1996. *The Housing*
        Grants, Construction and Regeneration Act 1996. London :
        HMSO.
17   The Scottish Office et al., 1997 . *The Scheme for Construction*
        Contracts (Scotland) Regulations 1997 , Edinburgh : HMSO.
18   Latham Sir .M., 1994. *Constructing the Team.* London : HMSO.

19   Hunter R.L.C., 1987. The Law of Arbitration in Scotland.
        Edinburgh: T & T Clark Ltd.

20   Clancy R. et al., (1997). The New Commercial Cause Rules.
        Scots Law Times. 6 (14 Feb) : p 45 - 49.

21   Lords of Council and Session, 1994 (Sept.). *Act of Sederunt
        (Rules of the Court of Session 1994 Amendment No. 1)
        (Commercial Actions) 1994.* HMSO. Edinburgh.

22   Morrison,N. (1994).Scotland's New Commercial Court. *Journal
        of the Law Society of Scotland* (Oct.) : p354 -356.

23   Clancy R. et al., (1997). The New Commercial Cause Rules.
        Scots Law Times. 6 (14 Feb.) : p 45 - 49.

24   Latham . Sir M., 1994. *Constructing the Team.* London : HMSO.

25   The Department of the Environment et al., 1996. *The Housing
        Grants, Construction and Regeneration Act 1996.* London :
        HMSO.

SWEDEN

Ulf Franke
Stockholm Chamber of Commerce
Gunnar Lindgren
Nyström & Partners
Lars Ranhem
Ferax Contract Management

Editorial assistance from
Jens Pedersen and Peter Dyer
Advokatfirman Foyen & Co

1. BACKGROUND[1]

1.1 Economic

During the last two decades, construction investment in Sweden has contributed annually between 6 and 13 per cent of the Gross Domestic Product. Repairs and maintenance of constructed facilities correspond to about 5 per cent of GDP, subject to less variation than the volume of investment. The construction industry provides jobs directly or indirectly to about one ninth of the Swedish labour force.

There are four large general contractors operating on the national market together with several thousand smaller firms, which are local and often specialised. A similar structure is found among architectural practices and technical consultants, where a few large companies compete with a great number of small firms. Although all major contractors and consultants have operations abroad, the number of projects in Sweden with foreign firms as participants is limited.

1.1.1 Trends

After a rapid expansion during the second half of the 1980s, culminating in 1990, the total volume of construction in Sweden has declined. However, taking the various subsectors into consideration, a more complicated pattern emerges.

Thus, investment in new dwellings peaked in 1991. A withdrawal of government subsidies and changes in tax legislation have depressed housing demand to an unprecedented low level. Moreover, there was a sharp reduction of investment in new offices.

After the sudden depreciation of the Swedish krona in 1992, construction for the manufacturing industry soon revived. Nevertheless, building for industry clients accounts for no more than 10 per cent of the total volume.

To some extent, investment in heavy civil engineering work has compensated for the reduction in building. During the 1991-95 period, investment in roads and railways increased considerably and reached a record level in 1995. Major projects were initiated. New financial arrangements were put into use for the rail link between Stockholm and its airport, Arlanda. The Swedish government initiated proposals for a Build-Operate-Transfer concession in 1993, and work is in progress. The 16.2 km tunnel and bridge complex for the fixed link across Öresund is another example of a complex, large-scale investment in infrastructure, jointly backed by the governments of Denmark and Sweden.

1.2   Legal

Like the civil law in many other countries the Swedish civil law is founded on the principles of private ownership for natural and legal persons and a high degree of freedom of contract. The laws as they exist today are the result of a long historical development marked by continuity rather than by abrupt changes.

Ownership is protected by the constitution and nobody can be forced to abandon property by means of expropriation and need not put up with the community restricting the use of land or buildings if it is not called for in order to satisfy a very urgent public interest. Whoever is forced to abandon his property or suffers considerable restrictions on the use of his land is entitled to compensation in accordance with grounds fixed by law.

The freedom of contract is of very great importance in Swedish contract law, but the judicial system is not content with referring to what the parties themselves have agreed. Since, in practice, the contracts are rather often incomplete, legislation often gets involved, stating what shall apply in a certain situation when the parties themselves have failed to reach an agreement. The Swedish contract law is largely an optional type of legislation, which does not only have to submit to what has been expressly agreed but also to commercial customs or established practice in a certain line of business.

Mandatory provisions, which are unconditionally binding irrespective of what has been agreed by the parties, are found in Swedish contract law, in principle, only in the field of consumer legislation. In the case of contracts between an producer and a private individual regarding a purchase or the performance of a service, on the other hand, there are a number of mandatory statutory provisions which, accordingly, restrict the freedom of contract.

Like the corresponding Danish, Norwegian and Icelandic laws, the Swedish civil laws can be said to be founded on the North Germanic sources of law which are also, to some extent, influenced by Roman law. In Sweden, however, the principles of civil law have not been compiled in large systematic codes such as the French Code Civil or the German Bürgerliches Gesetzbuch. Instead the law-maker has intervened in such areas where practical requirements have been considered to exist. Thus, the Swedish legislation in the field of civil law is not systematically built up on the basis of general principles as is the case in Germany and in some other countries. Nowadays, Nordic law can be said to constitute a special judicial system which is influenced more by continental than by Anglo-American law, since Nordic law is, to a greater extent, based on written law. Accordingly, the courts have by no means as large space for law-creating activities as in countries applying Common Law. For the court's adjudication process, however, earlier decisions, precedents from the Supreme Court, are of great importance. The European Community Law has lately gained a great and growing importance to Swedish law and has had an influence on many civil laws.

The Swedish civil procedure is based on the fundamental principle that the court is separated from the parties; thus, the parties themselves are responsible for the inquiry of the case and are also themselves responsible during the session before the court. The court's most important function is to try the case and the Swedish procedure is not at all inquisitorial. However, the court has the authority to pose questions to parties and witnesses and shall, according to the law, work for a conciliation between the parties.

Proceedings can only be initiated by a party and the parties define, by means of pleadings and grounds, the compass of the court's trial. Each party is entitled to plead his own case before the court. The court may not pass judgment unless the defendant has been notified about the plaint or unless either party has received summons to attend the main hearing. It should be noticed that the principle that "nobody shall be passed judgment on untried" only means that the parties shall be afforded the "opportunity" to plead their case. If a party does not make use of his right, the court can still, in case of disputes amenable to out-of-court settlement, decide the case by means of a so-called judgment by default. The court sessions are open to the public and this applies both to preparatory meetings and the main hearing in civil cases.

The general courts are district courts (tingsrätter), courts of appeal (hovrätter) and the Supreme Court (Högsta domstolen). There are also special courts for certain cases requiring special expert knowledge. No special construction court exists, however.

The district court is the court of first instance with competence to pass judgments within a certain geographical area, called a judicial district (domsaga). There are about one hundred district courts and in the Swedish judicial system the district courts play the dominant role.

The function of the court of appeal is to be superior instance to the district courts. There are six courts of appeal in the country, of which the oldest and largest is the Svea Court of Appeal (Svea Hovrätt) in Stockholm, established in 1614. The chief judge of a court of appeal carries the title Court of Appeal President. The preparatory work is to a large extent carried out by junior judges undergoing training.

The Supreme Court (Högsta domstolen) is the final court of appeal in cases taken there from the courts of appeal. The main purpose of the proceedings in the Supreme Court is to serve as a guide for the court's decisions in future, similar cases. In order to have a case tried by the Supreme Court it is necessary for a party to apply for and obtain a review permit (certiorari). Review permits may be granted by the Supreme Court only if it is of importance to the adjudication process and, therefore, the Supreme Court is a typical "precedents" instance.

To become a law-earned judge with the right to pass judgment in civil cases a bachelor of law degree and Swedish citizenship are required. A law-earned judge is appointed by the Government and has an irremovability protected by the constitution. Apart from the obligation to retire at pensionable age, the judge can only be dismissed from his service if he has committed a crime or by repeated malpractice has proved himself to be clearly unfit to hold an office as judge.

A consequence of the legal capacity is that each and everybody can be a party in a judicial process, i.e., have a party capacity. The party capacity according to civil law corresponds to the procedural capacity in procedural law. In the case of a legal person it is the representative, e.g. the board of directors in a limited liability company, who has the procedural capacity.

By using a power of attorney a party can commission a proxy to handle a judicial process. It is common but not necessary that the proxy is a "lawyer"[2]. What is required by law for a person to act as proxy at court is that the proxy is suitable for a certain case with regard to "integrity, knowledge and earlier activities". A distinction must be made between the proxy and the counsel. The counsel is not authorised to take any action on behalf of the party. The function of the counsel is to be of help when the party pleads his own case.

"Lawyers" are regularly found in the courts. They often function as proxy or counsel for a party in cases of dispute. To be able to be a "lawyer" it is necessary to have been elected to the Swedish Bar Association (Sveriges Advokatsamfund), which is an independent semi-official professional organisation of lawyers with university education. The Swedish Bar Association is officially sanctioned and only those who have been elected to the association have the right to use the title of "lawyer". The number of "lawyers" is small in Sweden compared with many other countries.

To be elected to the association is required, among other things, a bachelor of law degree and five years of practical work, of which three years in a law firm. The person concerned is also required to have shown integrity and suitability for the profession of "lawyer".[3]

1.2.1   Construction Law

Swedish legislation does not contain any rules which are directly applicable to construction contracts.

The provisions of the Contract Act of 1915 are generally applicable to agreements in industry and commerce; naturally, therefore, they are also applicable to construction contracts. However, the Contract Act mainly covers the question of how agreements are brought about, when tenders and orders are binding, in which situations agreements become invalid, etc. and, accordingly, does not give any information regarding the material contents of an agreement.

In Swedish judicial doctrine the construction contract has often been characterised as an assignment or a contract for services. Since also for these two types of agreement there are no legal provisions, the legal position is still uncertain in the case of construction contracts with a firm as employer, unless the parties have themselves regulated the contents of the agreement. However, ever since the late 1800s, conditions of contract have been prepared, on different occasions, for construction work for the government as well as for private employers. Moreover, ever since the beginning of the 1900s, within different organisations common to employers and contractors, General Conditions have been prepared which have been widely used and from time to time been revised in the light of the development. The latest edition of General Conditions was published in 1992 and is called "General conditions of contract for Building, Civil Engineering and Installation Works, AB 92". Thus, AB 92 is a so-called agreed document and has been prepared by the Construction Contracts Committee (Byggandets Kontraktskommitté, BKK), which is an association on which all important organisations on the employer and contractor side are represented. On the employer side the government as well as local authorities and private clients are represented, whereas the contractor side is represented by organisations for building and civil engineering contractors and specialist contractors, etc.

As far as construction contracts between firms regarding the performance of building and civil engineering works are concerned, the use of AB 92 is very frequent. In these agreements, the application of AB 92 is a rule with practically no exceptions. Even if the parties should occasionally have failed expressly to refer to AB 92, a reference may therefore often be considered implied between the parties. For natural reasons, this is the case especially if the parties, on several earlier occasions, have concluded construction contracts with one another and then always been using AB 92 as contents of the agreement. In other situations, too, AB 92 for traditional contracts and ABT 94, which is the corresponding standard document for design-and-build contracts, must be considered largely to be an expression of what is considered to be the existing Swedish optional construction law. In particular, this may be said about certain typical contract conditions, for instance regarding the scope and execution of the contract, whereas other conditions - especially, perhaps, conditions regarding time limits and the manner of solving disputes, etc. - might probably never rightly be said to reflect the existing Swedish construction law.

Thus, as appears from the above, AB 92 in particular but ABT 94 and their predecessors as well, have so long been generally applied between professional employers and contractors that, in important parts, they must be considered to be an expression of what, according to common opinion, is the existing construction law between such parties. Therefore, the agreed documents AB 92 and ABT 94 must be considered an important source of law as regards the existing Swedish construction law.

Another important source of construction law is a limited number of precedents in the Supreme Court and a number of arbitration awards which have been published, in anonymous form, in law reports. The Swedish construction law has also been described in a number of books, written by lawyers specialising in construction law. The expositions in these books must also be considered to be an important source of law to those who have to acquaint themselves with the existing Swedish construction law.

In journals of a general legal nature as well as in trade journals articles are occasionally found on problems relating to construction law. On the other hand, there are no particular specialist reports published.

In Sweden there are a number of law firms with altogether some 20 lawyers specialising in matters and cases of construction law. In addition, all major law firms always have one or more lawyers who, besides other types of cases, also deal with construction conflicts and who also act as arbitrators in arbitration tribunals.

Several organisations representing employers, contractors or consultants have lawyers on their staff specialised in construction law. Major clients and contractors also employ lawyers who, in addition to company matters and matters relating to real property law also, to a large extent, deal with matters of construction law. Leading project management companies have university-trained engineers who, based on experience, have acquired both technical and practical knowledge of construction law and the application thereof in the individual case.

1.2.2   Construction Contracts

As has already appeared under section 1.2.1 Construction Law there is in Sweden a very well developed system of standard forms for construction contracts. These standard forms consist of national General Conditions which, to some extent, have been translated into other languages to enable them to be used in connection with international procurement of large-scale contract works in Sweden.

All contract works of reasonable size are, in principle, procured with the General Conditions as contract terms. Only minor, craft-type contract works are procured without the use of any kind of such conditions. Actually, bespoke contracts are not at all used in connection with competitive procurement in the construction sector, unless it is a question of special kinds of agreements, such as agreements regarding Build-Operate-Transfer projects, agreements regarding land distribution competitions and the like.

The system with General Conditions within the construction sector in Sweden is based on the attitude that it is rational and economic to bring together, in packages of provisions, such provisions as can regularly be the same, irrespective of type of contract and irrespective of the particular design of the individual object.

In Sweden there is the attitude that, by using General Conditions, we get more clear-cut conditions that give a safer application and interpretation than bespoke contracts. The fact that the conditions recur in everyday procurement practice also means that the most important conditions get an established meaning for the parties involved. By that the General Conditions also contribute to the fundamental principle that all tenderers shall be given the opportunity to submit tenders on equal terms will be easier to uphold.

In connection with the preparation of our General Conditions, which are all agreed documents, the parties have been guided by certain principles. As far as possible, economic consequences of contract provisions shall be easily estimated by tenderers so that the projects are not unnecessarily burdened with costs due to obscurities in the provisions. A fundamental principle is that each party to an agreement shall be responsible for the measures taken by him. It has been generally stated that, in different respects, the provisions shall contain a reasonable distribution of risks between the parties.

The General Conditions most widely used are the following.

General Conditions of Contract for Building, Civil Engineering and Installation Works, AB 92

These conditions, which have existed in several editions since the beginning of this century, are revised from time to time, the latest edition dated 1992. AB 92 is intended to be used for all kinds of traditional (execution) contracts, by which is understood contracts where the employer is responsible for the design work and the contractor is only responsible for the actual execution. The conditions are used in the agreement between a client and a general contractor, but also in the agreement between a client and, for instance, a building contractor or a specialist contractor in connection with a divided contract. AB 92 has been prepared by the Construction Contracts Committee (Byggandets Kontraktskommitté BKK), consisting of representatives of central government, local authorities and some fifteen other organisations representing clients, building and civil engineering contractors and specialist contractors. AB 92 is undoubtedly the most important conditions of contract within the system of construction law.[4]

General Conditions of Turnkey Contracts for Building, Civil Engineering and Installation Works, ABT 94

These conditions are intended for design-and-building contracts where the employer enters into an agreement with a contractor regarding the entire or a substantial part of the design work in addition to the actual execution. These conditions, too, have been prepared by the Construction Contracts Committee. The conditions are used in the agreement between a client and a turnkey contractor regarding an entire project, but is also used in agreements between a client and a contractor, who has only to execute a special contract and who has undertaken also to be responsible for the design work and to bear the so-called functional responsibility. The conditions originate from AB 92 and several of them are identical in AB 92 and ABT 94. Still, in the practical situation, the parties' responsibility may be different, even if the provisions are identical in wording, since, when ABT 94 is being used, the contractor is responsible for the design work.

The following clauses are of special interest when it comes to the extent of the Contract Works.

Clause 1
The extent of the Contract Works is determined by the Contract Documents. The Contract Documents are mutually complementary, provided that circumstances do not give occasion for another procedure.

Clause 2
If the Contract Documents do not show requirements or commitments concerning usability or property, the work shall be executed in a manner which satisfies what can be required with regard to the planned use of the object, specified by the Employer or otherwise known to the Contractor.

General Conditions for Sub-Contracts, AFU 96

These conditions are mainly intended to be used in the agreement between a general contractor and a sub-contractor - irrespective of type of sub-contractor - but can also be used when a contractor other than the general contractor enters into an agreement with a sub-contractor. AFU 96 has been prepared by the Swedish Construction Federation (Byggentreprenörerna) representing the building and civil engineering companies, and some thirty associations representing the very large number of specialised companies that are found in the construction industry. The most important provision in AFU 96 is the one stating that AB 92 shall apply for the contract. In addition, AFU 96 contains a number of stipulations supplementing or modifying AB 92 in order to attain co-ordination of defects liability periods, inspections, etc. between the different contracts forming part of a project.

General Conditions for Sub-Contracts on a Turnkey Basis, AFTU 96

AFTU 96 is intended to be used for sub-contracts on a design-and-build basis, i.e. when the entire or a substantial part of the design work as well as the actual execution of the sub-contract rests with the sub-contractor. Thus, the conditions refer to the relation between a turnkey contractor as employer and his sub-contractor in those cases where the sub-contractor is also a turnkey contractor. The most important condition in AFTU 96 states that ABT 94 is applicable to the contract with the variations and additions specified in AFTU 96. The variations and additions to ABT 94 mainly relate to provisions intended to attain co-ordination between the responsibilities of the turnkey contractor and the sub-contractor, defects liability periods and inspections, etc.

General Rules of Agreement for Architectural and Engineering Consulting Services, ABK 96

The rules with appurtenant commentaries are the result of negotiations between the Construction Contracts Committee (Byggandets Kontraktskommitté, BKK) representing clients, and the Swedish Federation of Architects and Consulting Engineers (Arkitekt & Ingenjörsföretagen) representing architects and consulting engineers. The commentaries form an integral part of ABK 96 and shall be used as a guide in connection with the interpretation and application of the rules. ABK 96 is intended to be used for assignments within the professional field of architects and consulting engineers. In the introduction to the rules it is stated that important prerequisites for a fully satisfactory result of the assignment are that the client and the consultant

- have a unanimous conception of the purpose, scope and level of quality of assignment;
- maintain a dialogue, before and during the performance of the assignment; and
- show each other confidence and openness in general.

According to ABK 96 the consultant shall perform the assignment without undue influence from suppliers or others who might have an effect on his objectivity and also otherwise observe sound professional practice.

General Conditions for Purchases of Goods intended for professional use in construction, ABM 92

These conditions have been prepared by the trade organisation of the building and construction companies and the Building Materials Section of the Federation of Swedish Industries (Industrins Byggmaterialgrupp), which consists of representatives of suppliers of a large number of commodity groups within the building materials industry. ABM 92 and its predecessors have, for the most part, replaced the terms of delivery drawn up by individual companies and various trade organisations on the supplier side. ABM 92 has attained a wide field of application and, by its standardisation, has facilitated for sellers and buyers to prepare individual delivery agreements for the construction industry.

A fundamental principle in the work with ABM 92 has been that the provisions shall be designed in such manner that they represent a suitable balance between the interests of the buyer and the supplier. In order to attain uniformity of provisions in the employer-contractor/buyer-seller chain the provisions have, as far as possible, been adapted to the conditions in AB 92.

General Conditions for Single-family home Contracts where the buyer is an individual consumer, ABS 95

These conditions are the result of negotiations between the National Board for Consumer Policies and representatives of the construction industry. It has been explicitly stipulated in the conditions that agreements beside these general conditions must not disturb the balance in the agreement between the parties.

According to ABS 95 the contractor shall carry out his undertaking in a professional manner and, with due care, look after the consumer's interests and confer with him to the extent required. If, during the contract period, the contractor has realised or should have realised that the data or results of investigations provided by the consumer are incorrect, the contractor is obliged to inform the consumer thereof and confer with him regarding suitable measures to be taken.

Moreover, the contractor is obliged, during the contract period, to carry out the variations and additions required by the consumer.

Defects are considered to exist if the result of the contractor's work differs from what was agreed regarding design or execution. Defects are also considered to exist if the result of the contractor's work does not satisfy the demand for professionalism in connection with the handing-over of the works, even though the difference is due to an accident.

According to the provisions the defects liability period is two years. However, the contractor is liable for substantial defects which become apparent after the expiry of the defects liability period, if the defects are shown to be due to negligence on the part of the contractor. The latter liability expires after ten years, calculated from the handing-over of the works.

The handing-over of the works takes place in connection with the completion inspection. The contractor is obliged, without cost to the employer, to remedy defects noted in the inspection report.

Besides the protection given to the individual consumer according to ABS 95 he is also protected in accordance with the Act on consumer services, which contains mandatory provisions which, thus, cannot be set aside by means of an agreement.

According to the Act on consumer services the building company is obliged to see to it that the service is not performed in conflict with statutory provisions intended to ensure that the object of the service is reliable from the point of view of security.

If, with regard to the price or the value of the object of the service, a service cannot be considered to be of reasonable use to the home buyer, the building company shall advise him against having the service performed. If only after work has begun to be carried out, it turns out that the work cannot be considered to be of reasonable use to the home buyer or that the price for the work may become higher than the home buyer was able to expect, the building company shall inform the home buyer thereof and ask for his directions. If a defect is not remedied correctly or if it is not remedied within a reasonable period of time, the home buyer is entitled to a reduction of the price.

The building company is obliged to compensate the home buyer for damage suffered by him due to delay and he is also liable to pay compensation to the home buyer for damage caused to him because of defects. The building company is also liable to pay compensation for damage caused to the home buyer, if the damage has been caused through negligence on the part of the building company.

If the price does not follow from the contract, the home buyer shall pay what is considered to be reasonable with regard to the nature, extent and workmanship, or the prevailing price for corresponding services at the date of the agreement. If the company has indicated an approximate price, the price stated must not, however, be exceeded by more than 15%.

Since a couple of years back the law prescribes a ten-year warranty for defects and damages in buildings. According to the law, a party who erects a building which is intended to be used as a dwelling for permanent use is obliged to take out a ten-year insurance covering building defects.

The building defects insurance follows the building also in case there is a change in ownership. The building defects insurance also applies irrespective of what happens to the parties concerned, e.g., if the building company goes bankrupt. A great advantage with the building defects insurance is that no investigations regarding responsibility or negligence have to be made before the repair work can be commenced.

The insurance covers reasonable costs for remedying defects in the design of the building, in materials used or in workmanship, as well as reasonable costs for remedying damages to the building cased by the defect. By defects are to be understood deviations from a professionally acceptable standard at the date of the execution of the work.

In the Act on consumer services it has been specially stated that the act is not applicable in those situations where the ABS 95 contract applies together with the building defects insurance, since, by that, the consumer protection is stronger than is afforded by the provisions of the Act on consumer services.

According to ABS 95, disputes arising from agreements between building companies and home buyers shall be settled by court. A number of cases, decided by the Supreme Court, have shown that an arbitration clause in agreements between building companies and home buyers shall be considered undue.

Dispute Resolution Clauses

In AB 92 and ABT 94 the following stipulations regarding the settlement of disputes are found:

"Disputes arising from the contract agreement shall be settled by arbitration in accordance with the Swedish Arbitration Act unless otherwise stipulated in other contract documents."

Notwithstanding this arbitration clause either party has the right to refer to a court any matter which manifestly does not relate to a sum larger than ten so-called base amounts (equivalent to approx. USD 5,000). Moreover, either party has the right to refer to a court any indisputable, mature claim in respect of the contract.

General Observations

It is the general view of the parties on the Swedish construction market that the system of construction law with well developed standard forms for construction contracts is working very well. The General Conditions are considered to be well balanced and enjoy a high reputation with the companies.

In certain individual cases, by deviating from the General Conditions employers have transferred risks and responsibilities to the contractors to such an extent that the reasonable and balanced allocation of responsibilities and risks aimed at by the conditions has been disturbed. In such cases, the contractors still submit most tenders without making reservations, which is probably due to the competitive situation which is a consequence of the pronounced buyers' market which is at present prevailing in Sweden.

To sum up, the Swedish system of construction law with standard forms for construction contracts can be said to have the following advantages:

(a)   The General Conditions have been agreed by representatives of the parties which, as in the case of other agreed documents, is a guarantee that the conditions contain a reasonable distribution of responsibilities and risks between the parties.

(b)   The General Conditions are regularly used, which means that the contents become well known and supported by the parties on the market.

(c)   All links in an agreement are covered by General Conditions with a corresponding distribution of responsibilities and risks between the parties, which is bound to give justice in individual agreements.

(d)   The various General Conditions are co-ordinated from the point of view of responsibility.

(e)   The General Conditions with risks and responsibilities are co-ordinated with specially developed insurance solutions, which is bound to guarantee the parties security. Thus, for instance, according to the General Conditions the contractor shall always have a third party insurance and an all-risks insurance. Then there is an agreement with the insurance companies stipulating the minimum terms and conditions that always have to be met when the concepts of third party insurance and all-risks insurance are being used. According to ABK 96, a consultant is always obliged to take out and to maintain during the period of responsibility a third party liability insurance for consultants with an insurance amount corresponding to the agreed liability to pay damages.

(f)   Moreover, the General Conditions are co-ordinated with standardised procurement instructions, instructions for the contract, technical specifications and measurement methods.

2.   CONFLICT MANAGEMENT

This section deals with tools and techniques to avoid conflicts as well as systems of conflict management. It discusses sources of possible conflicts, strategies to avoid conflicts and - if they occur - procedures for conflict management and dispute resolution.

Frequent sources of conflict are:

*   Obscurities and discrepancies in the contract documents.
*   Lack of co-ordination between various technical documents.
*   Late or defective drawings.
*   Schedule delays due to unrealistic scheduling or mismanagement from the contractor.
*   Cost overruns due to underestimation of the contract price or the contractor's mismanagement.
*   The site or other circumstances differ from what they should be assumed to be.
*   Defects in the permanent works.

The list above is not considered to be complete.

The Swedish approach to construction conflicts may be summarised as follows:

*   Firstly, avoiding conflicts.

*   Secondly, if conflicts nevertheless arise, managing them
         promptly and in a professional manner so that they have
         the smallest possible effect on the works, with regard to
         time schedule, budget and the required end product.
*   Thirdly, if a conflict has become irreconcilable, a well-
         known set of procedures for dispute resolution enables
         the parties to place the dispute in a proper forum.

Avoiding conflicts

Avoiding conflicts is substantially a matter of employer's
procurement and contract strategy and the quality of the contract
documents.

A professional employer generally adopts a strategy with the
following characteristics:

*   Managing the process of procurement and contract
         administration firmly, fairly and professionally, thus
         taking care of his interests in a co-operative and
         constructive way.
*   The distribution of risks between the employer and the
         contractor is adapted to the best interest of the project
         objectives.
*   Information regarding the contract works is given in the
         enquiry documents to such an extent that the tenderer can
         judge the conditions prior to his tender and estimate the
         contract works.
*   The responsibility for the correctness of data, documents
         and designs rests with the party providing them.
*   The criteria for award of contract are clearly stated in the
         enquiry documents and implemented during the
         evaluation process.

> *   Use of standard conditions for the contract and other
>   standard documents when preparing the enquiry
>   documents.

The employer's strategy aims at preparing for a successful completion of the contract works. However, the contractor's ability and his determination to do a good job are crucial for a contract performance without conflicts. This fact indicates the importance of procedures for the selection of tenderers and award of contracts.

Managing conflicts

Managing conflicts and handling claims involves implementation of procedures in order to avoid possible escalation and facilitate resolution of conflicts and settlement of claims. The aim is to handle all kinds of conflicts in such professional and managerial ways that conflicts have the smallest possible influence on the works. The following strategy is normally accepted as appropriate:

> *   The procedures are indicated in the enquiry documents
>   and agreed upon in the contract documents.
> *   Handling of conflicts and claims at the lowest suitable
>   organisational level.
> *   Regular meetings to review the contractor's performance
>   reporting including non-conformities, schedule and cost
>   performance.
> *   Negotiations with the contractor to reach agreements on
>   any change and the subsequent influence on schedule and
>   cost. These agreements are added to the contract
>   documents.

According to Swedish practice the parties handle and settle, by using conflict management techniques, the great majority of differences and conflicts which arise between them in connection with the execution of the works or otherwise the performance of the contract.

Dispute resolution

If the parties fail in reaching an agreement they have a set of non-binding or binding procedures for dispute resolution at their disposal. Dispute resolution is here identified by a third party intervention. Thus, the various procedures involving a third party are discussed in section 3. The following procedures are applied in Sweden:

Non-binding
* Expert
* Mediation
* Conciliation
* Dispute Review Advisers
* Dispute Review Boards (unusual)

Binding
* Expert
* Arbitration
* Litigation

2.1   Non-binding

2.1.1   Dispute Review Boards

Discussed in section 3.1.5.

2.1.2   Dispute Review Advisers

Discussed in section 3.1.6.

2.1.3   Negotiation

The character of the construction contract and the practical situation in which it is applied means that, during the performance of every contract, a number of negotiations must take place regarding the scope of the contract as well as discussions regarding cost adjustments for variations and additions, extensions of times, etc. For natural reasons, the parties have different interests to look after and defend in these matters and, as a rule, it is by means of negotiations between the parties at different levels that agreements are eventually made.

In view of the enormously large number of conflicts of this kind it can easily be established that negotiation is the most dominant form for dispute resolution in the construction industry. Even if the concept of conflict is "soft" term, it may possibly still be considered too "hard" as a description of the clashes of opinion now referred to which, to a greater or lesser extent, occur in connection with every contract and are settled by negotiations.

2.1.4   Quality Matters

Quality Assurance (QA) and Quality Control (QC) address the management of the construction process as well as the product of the contract. Construction contracts primarily focus on the completion of the works and the end product. This fact reflects a common idea, especially among contractors; the employer just has to specify the product that he wants to procure, the contractor will deliver the specified product and take full responsibility for compliance with specifications and requirements. It is the task of the contractor to direct the works and manage the construction process. Consequently, the Swedish standard forms of contract conditions (AB 92, ABT 94) include only the following statements with reference to the construction process and the performance of the works: the contractor shall carry out his undertaking in a professional manner, the contractor alone is authorised to direct the execution of the works.

The introduction of the 1987 edition of the ISO 9000-series in the construction field promoted new thinking; employers are interested not only in the product - and the price - but also in the performance of the contractor. The employer wants to be assured of the contractor's capability to deliver the specified product at the agreed price. Consequently, implementation of quality assurance and quality control requirements in construction contracts could be looked upon as an instrument of conflict management - i.e., to avoid conflicts.

The tendency today in engineering and construction is:

*   Employers require that consultants and contractors
        operate quality assurance systems.
*   Consultants and contractors show an increasing interest in
        demonstrating, in procurement situations, their capability
        to fulfil contractual undertakings.
*   Employers implement quality assurance and quality
        control in their own project management operations.

Characteristic for construction is the involvement of many organisations in every construction project; furthermore, the co-operating organisations differ from project to project. A common language facilitates communication between organisations. The ISO-series provides such a language and is the most wide-spread quality system. The European Standard EN ISO 9001:1994 (as well as 9002 and 9003) has the status of Swedish Standard. A limited number of contractors and consultants have their quality assurance systems certified according to ISO 9001 or ISO 9002.

The management of quality matters in contractual situations is developing. Many employers, public as well as private, include requirements on QA/QC in the enquiry documents. The tenderer is requested to show his capability to maintain a system for QA/QC in conformance with the requirements. Sometimes the tenderer has to submit his quality plan in his tender; in other cases the contractor has to prepare the quality plan within a certain time after award of contract. During the execution of the works it is the contractor's duty to perform inspections and testings and to document acceptable results. The employer monitors the contractor's adherence to approved quality plans and procedures, e.g. by quality audits. The monitoring normally includes the contractor's documentation of achieved quality and random sampling.

Procurement Systems

Three procurement procedures are commonly used in the private sector and for public procurement of large projects (according to the Procurement Directives of the EC and the corresponding Swedish law):

*   Open procedures - all interested contractors may submit
        tenders.
*   Restricted procedures - only those contractors invited by
        the employer may submit tenders.
*   Negotiated procedures - contractors submit tenders and
        the terms of the contract are negotiated with one or more
        of them.

These procedures are applicable to both traditional and design-and-build contracts.

Open or restricted procedures are most frequently used, while negotiated procedures are common in the private sector but are applicable in the public sector only by exception.

2.1.5   Partnering

In this context partnering is defined as a long-term arrangement under which employers and contractors decide to collaborate closely in order to deliver requirements as cost reduction, improved quality and innovative solutions. *Long-term* indicates a project-to-project approach.

Partnering, in the strict sense of the term, is not applied to a large extent today in Sweden. In the fifties and sixties long-term co-operation between clients in the manufacturing industry and contractors was not uncommon. The employer and the contractor worked close together in the planning and design stage. The contract works were performed under a design-and-build contract.

In addition to successful experiences from a single-project partnership agreement for the construction of a newspaper office building, completed in 1992, there are a few cases where recurrent clients with a US background have initiated partnership agreements recently.

2.2 Binding

2.2.1 Partnering

Discussed in section 2.1.5

3. DISPUTE RESOLUTION

As has appeared above under 2 Conflict Management, in our presentation the concept of dispute has been reserved for situations where some form of third party intervention has become necessary. The fact that, at a certain stage, the parties are of different opinion regarding, for instance, the required size of a variation or an addition in a contract is thus not enough for the parties to be considered to be in conflict with one another. However when the parties cannot themselves solve the problem but have engaged lawyers, who pursue strongly their principals' points, it is correct, according to Swedish linguistic usage, to use the concept of *tvist* (dispute(. The large number of these conflicts are solved after negotiations between the parties, where the lawyers handle the negotiations in the presence of their principals.

If the lawyers together with their principals do not succeed in solving a construction dispute by means of negotiations, the dispute is regularly referred for settlement by arbitration or by court.

Disputes to which the state is a party shall always, according to the construction contract, be subject to court trial, since it is thought that the state cannot or should not, from a general pint of view, disregard the public judicature. It does occur, however, that in construction disputes the state makes an agreement in the individual case that the dispute shall instead be referred to arbitration if this is considered to be suitable with regard to the character of the dispute.

The General Conditions applied, such as AB 92, ABT 94, etc., contain a clause that disputes shall be settled by arbitration. But the clause has been given the addition "unless otherwise stipulated in other contract documents". A number of major local authorities, among them the city of Stockholm, have taken advantage of this and stipulated that disputes, if any, shall be settled by court instead of arbitration.

It has been stated that this is not based on the attitude that the courts would be more skilful than arbitration tribunals in settling construction disputes. Instead, the attitude is based on the idea that the contractors prefer to refrain from trial when a court is involved, because the proceedings are then public and because it takes too long to get a final decision, since it may be a matter of trial in several instances.

Thus, like most foreign legal systems, the Swedish legal system offers two forms of dispute resolution leading to an enforceable decision: court action and arbitration.

As appears from the above, arbitration clauses are common in the construction industry, except in agreements with consumers, since arbitration clauses in such agreements are considered undue and are therefore set aside by the courts.

It can be noticed that many of the companies are of the opinion that the arbitration procedure is an expensive form of dispute resolution and that, in practice, this opinion also counteracts the use of arbitration.

In the light of the wide-spread use of arbitration clauses in the construction industry, one would also think that arbitrations are very common. The use of arbitration clauses are not, however, a direct reflection of the number of arbitrations that actually take place.

In 1995, the Central Bureau of Statistics made an investigation on "Arbitration in Sweden". All lawyers and assistant lawyers in law firms, permanent judges, company lawyers and professors of a jurisprudential subject were asked, altogether 5,000 persons.

The result of the investigation showed that there are maximum 300-400 arbitrations per year in this country. In 13 per cent of the cases a sole arbitrator handled the procedure. In the remaining cases the arbitration tribunal consisted of three arbitrators. After treatment of the cases on their merits arbitral awards were made in 55-60 per cent of the disputes. The construction disputes accounted for some 17 per cent of the arbitrations, i.e., some 60 arbitrations per year.

Our assessment is that, in principle, there is no abuse of dispute resolution, neither by court nor by arbitration. The above mentioned investigation seems to confirm, in actual figures, the correctness of our assessment.

3.1   Non-binding

3.1.1   Conciliation

The term conciliation is used in Sweden as a concept for the agreement often concluded by the parties regarding the point at issue after a dispute has been referred to court or arbitration.

Conciliation can be effected at any time between the date of the institution of the dispute and the decision by the court. Conciliation means that the dispute is settled in respect of the part to which the agreement refers. To use the concept of conciliation for the type of agreements that are often made between the parties before a construction dispute has been referred to court, or an arbitral procedure has been started, is alien to Swedish linguistic usage. Conciliation can be said to be the result at which the parties aim by entering into negotiation, see 3.1.4, and mediation, see 3.1.3.

According to an investigation made by the Central Bureau of Statistics in 1995, 40-50% of all arbitrations entered into in Swedish end up in conciliation.

3.1.2   Executive Tribunal

It can be said that this procedure is not being used as a model for dispute resolution in Swedish construction.

3.1.3   Mediation

Mediation is a very frequent form for the parties to try to reach a negotiated settlement in Sweden. Usually, a single mediation is appointed by the parties.

Practically always this form is being used as a way of trying to solve a situation of conflict without having to refer to it to court or arbitration. A form used in the construction industry is also that the parties each appoint an expert and that the two persons so appointed are instruction to suggest, in a common report, a solution to the dispute between the parties.

According to the Code on Judicial Procedure it is possible for a court to order mediation and to appoint a mediator. In practice, a condition for this is that the parties, when asked by the court, agree that mediation shall take place. This procedure is infrequent.

Experience shows that mediation is an effective way of solving disputes.

3.1.4   Negotiation

As has appeared above, sec 2.1.3 and 3.1.1, all agreements after negotiations can be made, in practice, at all stages from the very first clashes of opinion until the court decision has been given.

3.1.5   Expert Report

It quite often occurs that the parties, in a situation of conflict due to clashes of opinion jointly instruct a third party, a lawyer or an engineer, to submit a report regarding the dispute and the way to solve it. The report is non-binding but entails strong pressure on the parties, and the parties often feel morally obliged to make an agreement in accordance with the report. The procedure is regularly used in order to try to avoid judicial trial. The experience of this model is very good.

3.1.6   Dispute Review Boards

Formation of a non-binding DRB at the beginning of construction aims to assure timely and equitable dispute resolution. DRB is used in international civil engineering contracts, which involve complex technical features that might lead to substantial claims or disputes. Up to the present DRB has not been used in Sweden, leaving the Öresund Link out of consideration. The Danish-Swedish consortium has introduced DRB in three design-construct contracts for the coast-to-coast link between Denmark and Sweden. The experiences so far are too limited to evaluate the effect of DRB.

3.2   Binding

The decision of the court, after treating the case on its merits, takes the form of a judgment. The disputes in the construction industry always relate to matters which are for the parties themselves to decide without having them referred to court (cases amenable to out-of-court settlement). This means that the parties have a great influence on the proceedings and decide, for instance, the compass of the process by their pleadings and grounds. In principle, the court may not deviate from the parties' agreement.

According to the Code on Judicial Procedure the court session shall be oral and it may be said that, in principle, it is forbidden to read out written pleas. By this reading prohibition the court has a better chance to exercise an active direction of the procedure so that the process is concentrated on the matters of dispute.

In the Code on Judicial Procedure it is stipulated that the court shall base the judgment on what has occurred in the main hearing and that only those judges who have been present during the entire hearing are allowed to pass judgment.

In order to satisfy the demand for speed, fairness and legal security in the trial the process has been divided into two parts: preparation and main hearing. The purpose of the preparation is to prepare the case in such a manner that it can be decided at one concentrated main hearing.

The plaintiff presents his claim in a plaint ant the defendant presents his view on the claim in a defendant's plea. Both the plaintiff's and the defendant's plea shall contain pleadings, grounds, and information regarding evidence (information regarding the forms of evidence referred to and what shall be proved by each evidence).

The court always summons a meeting, an oral preparation, at which the court works for a clarification of the matters of dispute. Usually, the court then tries to find out if the parties are inclined to make a conciliation. It is thought that the court should be cautious in the discussions regarding conciliation since the confidence in the court's objectivity can otherwise be questioned. The normal thing is that a case is decided at the main hearing.

In a statement of facts the parties present their grounds. The statement of facts includes a number of data and is usually done in such manner that the parties, in a "narrative" form, develop their claims and pleas and account for the origin and contents of the dispute.

After the statement of facts follows the taking of evidence. The parties as well as witnesses and experts are heard. Written evidence is presented. It is common, though, that the written evidence is presented already during the statement of facts at the place in the narration where, as regards contents, it belongs.

Normally, the taking of oral evidence begins with party examinations. Then follows the examination of witnesses. This order is based on the wish to avoid the parties' narrations to be influenced by the statements of the witnesses.

The main hearing is concluded by the pleading. During the pleading the parties try to convince the court that the ground for the claim and the opposition, respectively, is proved by the evidence referred to and that the objections raised by the opposite party is of no significance. The parties usually also, although it is not necessary (*jura novit curia*) emphasise the legal provisions that are relevant, by referring to sections of law, cases. legislative history, etc. An ambition in connection with the formulation of a pleading ought to be that is shall be so clear and convincing that it can be used as grounds for the court's decision.

Before the main hearing is concluded the parties are afforded the opportunity to present their claims for compensation for litigation costs. It is too late to claim compensation for litigation costs when the hearing has been concluded.

In connection with the conclusion of the hearing the court decides when and how the judgment shall be passed. After that the court proceeds to deliberation.

According to the main rule the losing party is liable also to pay the opposite party's litigation costs. In the case of conciliation the parties' litigation costs are usually offset against one another.

According to the Code on Judicial Procedure the court shall , ex officio, try the reasonableness of the amounts claimed, unless such a trial is unnecessary, for instance, if the party losing the vase has consented to the opposite party's claim regarding litigation costs.

In principle, the court procedure described above is applied, in similar form, in connection with arbitration.

3.2.1   Adjudication

This form of dispute resolution is not applied in Sweden.

3.2.2   Arbitration

Introduction

Domestic Arbitration

Arbitration is widely used in Sweden in the commercial field in general and in the construction field in particular. Although no figures are available because of the confidential character of arbitration, it is considered that a great majority of business and construction disputes are solved by arbitration. Most written contracts include an arbitration clause, and such clauses are common in standard form contracts.

Institutional arbitration plays an important role in Sweden. It is mainly practised by the Arbitration Institute of the Stockholm Chamber of Commerce (the "SCC Institute") which is the centre for institutional arbitration in Sweden. The SCC Institute was established in 1917. It administers both domestic and international arbitration and has in recent years emerged as one of the leading arbitral institutions in the world.

International Arbitration

Over the past decades there has been an ever-growing trend to designate Sweden as the location of choice in arbitration clauses in international contracts, and many international arbitrations take place in Sweden. In most such cases both parties are from other countries than Sweden and it is not rare for one party to be a State or state agency and the other a commercial corporation.

The parties are from virtually all parts of the world. There are particularly many cases with parties from China and the former USSR on one side and US and Western European corporations on the other.

Most international arbitrations in Sweden are administered by the SCC Institute.

The Law on Arbitration

Arbitration as means of solving disputes has a long tradition in Sweden. Statutory provisions on arbitration appeared in Swedish legislation as early as in the 14th century.

The present law on arbitration is mainly contained in two statutes: the *Arbitration Act* and the *Act on Foreign Arbitration Agreements and Awards*, both adopted in 1929. The Arbitration Act is the main statute on arbitration, dealing with arbitration proceedings in Sweden generally, while the latter Act deals with arbitration agreements and proceedings having a foreign element. It also regulates the enforcement of foreign arbitral awards.

Only minor amendments to the 1929 legislation have been made. The most important ones were adopted in 1971 in order to implement the *Convention on the Recognition and Enforcement of Foreign Arbitral Awards* of 1958 (the "New York Convention"). In response to the increasing number of international arbitrations in Sweden, some amendments were also made in 1976 to exclude certain provisions, which were inconvenient in conducting cases involving parties from outside Sweden, such as the statutory time limit for making the award.
Although rather old the Arbitration Act has proved to be well adapted to modern arbitration, domestic as well as international.

One reason for this is that the autonomy of the arbitral process is fully recognised in Sweden. The result of that process, i.e., the arbitral award, is overturned only if the procedure followed by the arbitrators is in contradiction with the parties' agreement or otherwise fails to meet minimum standard of fairness.

Another reason is that the Arbitration Act deals with the procedure very summarily and in general terms, thus giving the parties and arbitrators a considerable freedom to adopt the procedure best suited to the circumstances of the particular case.

Also, most provisions in the Arbitration Act are non-mandatory and to that extent parties may allow the procedure to be governed by other rules, such as the Rules of the SCC Institute.

The Government has, however, initiated a review of the Arbitration Act and the Act on Foreign Arbitration Agreements and Awards and a committee has presented a proposal which will comprise both Acts. The new Act is expected to be adopted by the Parliament in 1998 and come into force on 1 January 1999.

The new Act is not expected to change any of the overriding principles of flexibility and autonomy inherent in the present Act. Instead it will mean a modernisation in language and structure and a further internationalisation of the arbitral process in Sweden by taking into account useful provisions found in other modern arbitration laws, notably the UNCITRAL *Model Law on International Commercial Arbitration.*

Conventions

As appears above Sweden has ratified the New York Convention 1958. It is noteworthy that Sweden has ratified the Convention without making any of the reservations which are open to the contracting states.

A part from the New York Convention Sweden has ratified the *Protocol on arbitration clauses,* 1923, *Convention on the execution of foreign arbitral awards,* 1927, and the *Convention on the settlement of investment disputes between States and Nationals of other States,* 1965.

The Arbitration Institute of the Stockholm Chamber of Commerce

(1)   Codification of the doctrines of separability and competénce de la competénce.

(2)   A clarifying rule as to the standing of the arbitration agreement in case of assignment of the main agreement by either party.

(3)   Detailed provisions as to disqualification of arbitrators and the consideration of such issues.

(4)   Clarification as to the parties' right to submit new or altered claims and counterclaims.

(5)   Limitation of the grounds on which an award can be held void by a court.

The Arbitration Review's findings have not yet been transformed into a government Bill and it remains to be seen to what extent the proposed changes will be brought forward and accepted by the Swedish Parliament (*Riksdag*). However, since there is little or not political controversy surrounding this field, it is safe to assume that the new act will by and large correspond to the Arbitration Review's suggestions.

The SCC Institute was established in 1917. Although mainly domestic arbitration was envisaged at that time the Institute has always been prepared to assist in international arbitration. Over the past few decades this work has become an especially important part of the Institute's field of activity. In recent years the number of international cases filed with the SSC Institute has been close to a hundred per year, and the number is on the increase.

The present Rules of the SCC Institute (SCC Rules) came into force in 1988 and are particularly adapted for the administration of international arbitration proceedings.

The recommended arbitration clause reads as follows:

> Any dispute, controversy or claim arising out of or in connection with this contract, or the breach, termination or invalidity thereof, shall be finally settled by arbitration in accordance with the Rules of the Arbitration Institute of the Stockholm Chamber of Commerce.

If such reference is made it means that the arbitral tribunal will consist of three arbitrators, unless otherwise agreed by the parties. Each party appoints one arbitrator and the SCC Institute the third, who will act as chairman. There is no list of arbitrators but the parties may appoint anyone of any nationality as arbitrator as long as he is impartial and independent.

The parties are usually represented by counsel, who may be of any nationality.

The parties may agree on any law to govern the substance of the case as well as on any language or languages for the proceedings.

The arbitrators act on the basis of presentations - both oral and written - submitted by the parties. The award shall be given within one year and no appeal is permitted on the merits.

Apart from these Rules the SSC Institute also in 1988 adopted Conciliation Rules and Procedures and Services under the UNCITRAL Arbitration Rules. In 1995 the SCC Institute adopted Rules for Expedited Arbitrations and Insurance Arbitration Rules.

The SCC Institute acts as appointing authority apart from under the UNCITRAL Arbitration Rules also, for instance, in cases where the arbitration agreement does not contain reference to any set of rules.

The Arbitration Agreement

No particular form is prescribed for the arbitration agreement. Accordingly the arbitration agreement may be oral although in practice it is almost always in writing.

Arbitration agreements may be of two kinds, viz. those which relate only to existing disputes and those which provide for the resolution of future disputes. The latter ones, which by far are the most common, are often included as a clause in the parties' underlying agreement   although such a clause may also be contained in a separate writing.

Standard form contracts are widely used in Sweden, not least in the construction field. As indicated above most such contracts contain an arbitration clause.

Although no particular form is prescribed for the arbitration agreement, it must specify the subject matter and state that such matter should be referred to arbitration. When future disputes are referred to arbitration they should, according to the Arbitration Act, arise "from a particular legal relationship specified in the agreement". A reference to "any dispute arising under this agreement" is acceptable.

Arbitrability

The arbitrable field is considered to be very wide in Sweden and includes all matters on which the parties may themselves determine. By way of case law the scope of arbitrability has even been extended beyond that limit. For instance, it has been considered that the task of deciding on a pure question of fact may be qualified as arbitration if it is relevant to a question which can be subject of a civil action. Valuations often qualify as arbitrations even if the arbitrators are not asked to make any order based thereon.

Arbitrators may also be given the task of filling gaps in an agreement, e.g., to settle the price to be paid for goods or services at some future date under a long-term contract. Swedish law seems to be quite generous in this respect, so that the procedure is arbitration if the decision is to be taken according to some principles laid down (expressly or by implication) by the agreement. Arbitrators are, of course, always entitled to fill gaps in the course of interpreting a contract, by implying terms which are necessary but have not been expressed therein. To this extent no express authorisation is necessary.

Doctrines of Separability and Kompetenz-Kompetenz

The doctrines of separability and kompetenz-kompetenz are fully recognised in Sweden. This means that an arbitration clause is considered as being separated from the contract of which it forms part, thereby ensuring its existence when for all other purposes the contract itself is effectively at an end.

A separate arbitration clause gives an arbitral tribunal a basis to decide on its own jurisdiction, even if it is alleged that the main contract has been terminated by performance or by some intervening event. It is this power of the arbitral tribunal which is frequently referred to as kompetenz-kompetenz.

Effects of the Arbitration Agreement

A valid arbitration agreement constitutes a bar to court proceedings. The defendant must, however, invoke the agreement before pleading to the substance, as a court will not take judicial notice of the agreement on its own motion. If, on the other hand, the agreement has been invoked, it constitutes an absolute bar, and then the court has no discretion but to refer the parties to arbitration. The respondent will lose his right to invoke the arbitration agreement in certain cases, especially where the claimant has previously attempted to initiate arbitration, but the defendant has failed to co-operate by neglecting to appoint an arbitrator.

Apart from constituting a bar to court proceedings, the arbitration agreement has certain other effects. The most important one is that it entitles the parties to the assistance of the courts and authorities to effectuate the agreement if such assistance proves to be necessary.

Multi-party

Although neither the Arbitration Act nor the SCC Rules particularly deal with multi-party arbitration many such arbitrations have been satisfactorily carried out. In several such cases the parties have grouped themselves on two sides whereby a multi-party problem does not arise. In other cases they have negotiated, sometimes in consultation with the Secretariat of the SCC Institute, a multi-party arbitration clause adapted to the circumstances of the particular case. Frequently in such cases the parties have provided for the appointment of all three arbitrators by the SCC Institute.

Composition of the Arbitral Tribunal

The parties are free to compose the arbitral tribunal according to their own wishes, both as regards the number of arbitrators and each arbitrator's standing and qualifications. The statutory provisions will be applicable only if the parties have made no provision in this respect.

The Arbitration Act provides that disputes may be referred to "one or more arbitrators". Sole arbitrators are therefore fully recognised regardless of subsequent statutory reference to "arbitrators". However, three arbitrators is the most common number, and this is also the number provided for in the Act. It is provided that the tribunal shall consist of three arbitrators, "one appointed by each party, and the third by the two arbitrators so appointed". The third arbitrator acts as chairman. If a party fails to appoint an arbitrator or if the two arbitrators fail to agree on the third arbitrator, such appointment shall be made by the District Court unless otherwise agreed between the parties, for instance by reference to the SCC Rules whereby the SCC Institute will fulfil such functions.

Qualifications

There are no restrictions, with respect to nationality or otherwise, regarding the eligibility of arbitrators. In construction disputes parties often appoint arbitrators knowledgeable in construction matters, he may be lawyer or engineer. However, all arbitrators, even those chosen by parties, are required to be impartial and independent. The Arbitration Act contains a list of grounds disqualifying a person from acting as an arbitrator, with the general purpose of ensuring impartiality and independence.

Swedish law does not confer immunity from liability upon arbitrators. Obstruction of the proceedings, in the form of refusal to act or otherwise, may give rise to an action for damages, although such cases are very rare. There are no reported cases in recent years.

The Arbitral Procedure

Initiation of the Proceedings

Arbitration proceedings are commenced by a party requesting "the application of the arbitration agreement". The request shall be directed to the other party and shall contain "notice...of the question or questions as to which an arbitration award is requested".

Although not a necessary part of the request for arbitration, it is advisable that the claimant notifies his choice of arbitrator in the request, because then the respondent, as noted above, must notify his choice within fourteen days.

General Framework of the Procedure

As noted above the Arbitration Act deals with the procedure very summarily and in general terms. Arbitrators are directed to act impartially and to give each party sufficient opportunity to present his case.

Apart from this the main provision of the Act only stipulates that the arbitrators, as far as possible, are to act in accordance with the instructions of the parties and otherwise deal with the case in an impartial, practical and speedy manner.

Parties and arbitrators thus have considerable freedom to adopt the procedure best suited to the circumstances of the case. A purely written procedure is possible, but normally oral hearings are arranged. The oral part of the proceedings often takes the form of one or several preparatory meetings and a main hearing. Efforts will be made to concentrate the presentation of all oral evidence and argument to the main hearing.

Normally the procedure before the arbitral tribunal begins with an exchange of pleadings. The claimant will be requested to submit, unless he has done so in connection with his request for arbitration, within a specified period of time a written memorial (Statement of Claim), stating in detail the claims raised, the grounds upon which these claims are based, both in facts and in law, and the relief sought. Frequently the claimant is requested at the same time also to indicate at least what kind of evidence he wishes to invoke in support of his claims. The memorial is then communicated to the respondent together with a request that he submit to the tribunal a counter-memorial (Statement of Defence), stating the respondent's position in regard to the claims raised as well as the grounds, in facts and in law, upon which he bases his position. The respondent is normally also requested to indicate the evidence he wishes to rely upon.

Although counterclaim is not dealt with in the Arbitration Act, it is common practice and generally acceptable that the respondent raises a counterclaim.

However, if such a claim is not covered by the arbitration agreement the tribunal may examine it only if the opposite party gives his consent, or, in other words, if the parties enter before the tribunal into a new and separate arbitration agreement and request the tribunal to examine also the new claim or claims.

Depending on the circumstances a further exchange of one or more written submissions on both sides may follow. If the tribunal finds it appropriate or even necessary, the parties are convened to a preparatory meeting before the tribunal for the purposes of clarifying points on which the position of the parties may yet not be defined, and in particular, to have the parties indicate in precise terms what kind of evidence they wish to invoke and rely upon and what circumstances they wish to establish or ascertain by means of the evidence invoked.

In more complex disputes there may be held two or more preparatory meetings. There may also take place in between such meetings a further exchange of written submissions by the parties.

When the tribunal finds that the case has been sufficiently prepared and that the proceedings should be brought to an end it will normally convene the parties to the main hearing, sometimes also referred to as the final hearing. The main aim of such hearing is to enable the parties to present their case in its entirety, to introduce and lay before the tribunal all the evidence they have invoked, whether in the form of documents or testimonies, and finally to plead their case, in facts and in law.

Powers of the Arbitral Tribunal

While the Arbitration Act presupposes that the tribunal can issue orders to the parties in procedural matters and thus request a party to submit memorials, to hand in documents in its possession which are relevant to the dispute, attend meetings before the arbitrators etc., an arbitral tribunal, as distinct from a court of law, possesses no powers to impose fines or any other means of constraint against a party who does not abide by the tribunal's orders. However, the Arbitration Act offers certain possibilities to the parties and the tribunal to call upon the assistance of the courts of law in matters of this nature.

The lack of powers in this respect also means that the tribunal cannot issue orders for interim security measures for the purpose of securing the claims raised before it. It should be noted, however, that a party who intends to initiate or who has already initiated arbitral proceedings, may under the *Code on Judicial Procedure* apply to the competent court of law for such measures.

The lack of powers of the tribunal to impose means of constraint is also the background to the provisions in the Arbitration Act which, while stating that the arbitral tribunal should give each party sufficient opportunity to present his case, also provide that if a party fails without valid excuse to avail himself of such opportunity, the arbitrators may decide the case on the basis of the existing material. This means that the arbitrators are authorised to proceed *ex parte* if a party fails to appear or to plead without justification. They are not, however, entitled to pronounce a "default judgment", i.e., they may not base their award on the respondent's failure to appear or to plead, but must base their award on the available material. The respondent's failure to appear or to plead may be given evidentiary weight in this connection.

Evidence

There are no restrictions upon the admissibility of evidence. Both written and oral evidence may be submitted. Like judges in court of law the arbitrators are free to evaluate the evidentiary significance of all documents, testimonies, or other circumstances which have been brought before them by the parties.

The production of evidence normally does not give rise to any difficulties. The parties will submit their documentary evidence, and witnesses will attend voluntarily, normally at the mere request of a party and otherwise by order of the arbitrators. The Arbitration Act confers upon the arbitrators a general power to call for production of evidence, and the arbitrators are entitled to assign evidentiary weight to the fact that evidence is not forthcoming.

However, as indicated above, arbitrators are constrained in that they cannot impose fines or invoke other measures to compel attendance of a witness or production of a document. And they cannot administer oaths (as regards witnesses) or truth affirmations (as regards parties), which means that criminal sanctions for perjury are not applicable. But there is a procedure available to compel a witness to attend and give evidence on oath or oblige any person to produce a document. This is by means of the courts. A party who wishes evidence to be taken in this way must ask for approval of the arbitrators. They may refuse their consent if, for example, they believe that sufficient evidence has already been produced on the issue, that the issue is irrelevant, that the costs involved would be exorbitant or that the application has been made solely for some extraneous reason, such as a desire to obtain publicity.

If consent is given, a party will make application to the district court within whose area the witness or the person possessing the document is staying. The party should supply the court with all necessary information, and the arbitrators should aid him if necessary with a brief description of the circumstances and the intended purpose of the evidence. The court possesses the same powers with respect to the application as it would if it were hearing the entire dispute in an ordinary legal action. With the aid of fines, powers of arrest and to some extent imprisonment, it may compel a witness to attend and to testify and may force the production of a document or other object. Sanctions exist for false testimony given under oath or truth affirmation.

If the court grants a party's application to hear a person in court it will so notify the other party and invite him to attend the hearing. The examination of the witness may be conducted by the court itself on the matter described in the application, but, particularly if both parties are present and represented, the examination will normally be conducted by counsel. A transcript of the examination will be transmitted to the arbitrators.

Sweden also has made this procedure available in the case of arbitration taking place abroad.

Experts

The parties may bring expert evidence before the arbitral tribunal. Written reports from such experts are usually submitted to the tribunal and copies given to the other party. Such reports are treated in the same way as documentary evidence in general. Experts are then usually also examined as witnesses before the tribunal.

It is also clear that the tribunal has powers to appoint an expert at their own initiative but at the expense of the parties. However, the tribunal will rarely do so without prior approval of both parties because of the costs involved.

Decision according to Law

Parties may direct the arbitrators to decide the matter *ex aequo et bono* or to act as amiable compositeurs. Unless the parties have expressly given such a direction, however, arbitrators will base their decision on the applicable law.

The autonomy of the parties is the leading principle in Sweden with regard to applicable substantive law. The parties may agree on any law. If the parties have not chosen an applicable substantive law, the arbitrators will make the choice, when necessary.

If the parties have not chosen an applicable substantive law, the arbitrators will make the choice, when necessary.

If the parties have not given any indication of which conflict of laws rules they wish to have applied, arbitrators sitting in Sweden with a Swedish chairman may choose to apply Swedish rules in this respect. They may also apply any other conflict of laws system they deem appropriate in the particular case or may decide the issue of applicable substantive law without having resort to any specific national conflict of laws rules.

The basic choice of law principle in Sweden is that the law which has the closest connection to the contract applies.

Types of Awards

In addition to final awards the arbitrators are entitled to make partial awards. If several claims have been joined, the arbitrators may give a partial award concerning one or some of the claims even if the proceedings are to continue with respect to the other claims so long as "the rights of neither party" are "prejudiced thereby". This possibility is of particular importance in construction disputes which frequently comprise several claims. The arbitrators may also make a partial award based on a respondent's admission of a part of a claim.

The Arbitration Act contains no provisions concerning interlocutory awards disposing of a preliminary question. Clearly, the arbitrators may render such awards with the consent of the parties. However, it seems doubtful whether they have the authority to render interlocutory awards if one of the parties objects. The draft new act, however, includes also provisions on interlocutory awards.

It is generally considered that parties are entitled to have a settlement embodied in an award on agreed terms, although there is no statutory provision on this point either.

Time limit

The parties may lay down a period within which the arbitral award must be given. If they have not done so, a distinction is made in the Arbitration Act, depending on whether the arbitration is international or not.

If at least one of the parties is resident outside Sweden, there is no statutory time limit for making the award. If, however, both parties are resident in Sweden, the award must be made within six months from the request for arbitration. The statutory time limit, but not one fixed by agreement, can be extended by the district court if special reasons are shown and application is made by a party before the expiration of the period.

Voting

The Arbitration Act provides that "all arbitrators must take part in the resolution of a dispute". If the arbitrators have divergent opinions, the Arbitration Act provides for a majority decision, unless the parties have agreed otherwise.

Form and Contents of the Award

The award must be in writing and signed by the arbitrators. If these requirements of form are not observed, the award is void. With respect to signing, the absence of an arbitrator's signature does not cause the award to be void if the majority of the arbitrators have signed the award and certified on the award that the arbitrator whose signature is missing took part in the decision. An arbitrator is thus expected to sign the award even if he does not agree with the decision reached by the majority. It is generally considered, however, that an arbitrator is entitled to express his dissenting opinion, with or without a statement of the reasons therefor. A dissenting opinion is usually annexed to the award.

There is no statutory requirement to give reasons for the award. However, it is the common opinion among practising lawyers in Sweden that arbitrators should give reason for their award, and in practice this is almost invariably done.

Compensation of Arbitrators

The compensation due to the arbitrators may be fixed by agreement between the parties and the arbitrators. This is, however, only rarely done and in the absence of any agreement, the arbitrators may fix the compensation due to them in the final award. They are entitled to reasonable remuneration for their work, and to be reimbursed their expenses.

The parties are jointly and severally liable to pay the compensation specified in the award. A party who is dissatisfied with the decision as to the compensation can bring the question before the district court if he starts an action within sixty days after receipt of the award.

Distribution of costs

Unless otherwise agreed by the parties, the arbitrators are authorised to make an order for the distribution of costs between the parties if either party so requests. The apportionment will in the first place be made in accordance with the agreement of the parties. If no such agreement has been made, the general opinion in Sweden is that the arbitrators should be guided by the rules set out in the *Code of Judicial Procedure*. The basic principle is that the losing party is liable for his own costs as well as those of the successful party.

Deposit

There are no statutory rules as to the payment of any deposit for costs. The general opinion is, however, that arbitrators are entitled to request a prepayment or deposit as a condition of accepting appointment, and in practice this is often done.

Remedies against Arbitral Awards

An arbitral award is final without rights of appeal as to the merits of the case. The remedies that exist apply to irregularities of form or procedure.

The Arbitration Act contains two groups of very narrow procedural grounds, which may be invoked.

Section 20 of the Arbitration Act describes the situation in which an award is void as distinct from challengeable. Section 21 of the Arbitration Act lists a number of grounds on which a party may challenge the validity of an award within a period of sixty days from the receipt of the award. If a successful challenge on one of these grounds is not commenced within this period, the award becomes definitely final and binding. Both sections are severely restricted in scope, since it is a principle of Swedish law to enhance the status of arbitration as an exclusive dispute resolution mechanism.

Void Awards

In rare cases an award may be void. A party need not do anything to have this established, but an action may be brought for a declaratory judgment that the award is void. There is no statutory time limit for bringing such an action. According to section 20 of the Arbitration Act an award is void:

1.   if there was no valid arbitration agreement;
2.   if the award was not in writing and was not signed by at least a majority of the arbitrators;
3.   if the award deals with a non-arbitrable matter;
4.   if the matter was pending in court when the award was given.

Challengeable Awards

An award may be challengeable, i.e., liable to be set aside for one or more of certain specific grounds, if an action is brought in court within 60 days after the date on which the party received a certain copy of the award. Those grounds are set out in section 21 of the Arbitration Act, which lays down that an award may be set aside:

1.   if the arbitrators have exceeded the scope of their mandate, or they have exceeded the time limit for the rendering of the award;
2.   if the arbitration should not have taken place in Sweden;
3.   if an arbitrator was disqualified or not properly appointed, or
4.   if there had been some other irregularity in the conduct of the proceedings that can be assumed to have influenced the outcome and for which the aggrieved party cannot be blamed.

If a ground is substantiated, the court will set aside the award. The court cannot remit the case to the arbitrators or to a new tribunal, nor can it substitute another decision. It can, however, make order which sets aside an award only in part, if such part is severable from the rest of the award.

A party may lose his right to challenge an award by waiving the irregularity. This may be done expressly but is more often done by implication, for instance, "by taking part in the proceedings without objection". But mere conduct will not amount to waiver unless the party has been conscious of the irregularity.

Enforcement

Most arbitral awards are complied with voluntarily. Nevertheless, the possibility of enforcement is of great importance.

As noted above Sweden is a party both to the New York Convention and to the Geneva Convention 1927 on the Execution of Foreign Arbitral Awards. This is important because many States have adopted the New York Convention subject to the reservation that they will only enforce awards made in another Contracting State. The Geneva Convention only applies to such awards. The fact that Sweden is a member of both Conventions therefore ensures the widest possible means of enforcement of arbitral awards rendered in Sweden.

This also means that *foreign* arbitral awards may be enforced in Sweden to a very full extent. Sweden has, in fact, gone further than required by the New York Convention in the direction of recognising foreign arbitral awards. The Convention was ratified by Sweden without making any of the reservations open to the Contracting States, and recognition is thus not limited to awards made in the territories of Contracting States, nor to awards in disputes of a commercial character.

The procedure for enforcement is by application to the Svea Court of Appeal in Stockholm, which has been given exclusive first instance jurisdiction to grant *Exequatur* in respect of any foreign award.

If enforcement of a Swedish arbitral award becomes necessary in Sweden, an application should be made to the competent execution authority. That authority is accordingly charged both with the decision whether to grant execution (*exequatur*) and with the actual enforcement. The execution authority will make a summary check that the award is not void and then invite the other party to state any objections he may have. However, execution will be stayed only if the other party can prove that there is a reasonable chance of success on an action brought to challenge the award. The decision to grant execution is appealable, but an appeal does not by itself result in a stay of execution.

3.2.3   Expert determination

This procedure is occasionally used in Sweden but is not very common. See further 3.1.5 Expert report.

3.2.4   Litigation

Litigation is unusual as a method for resolving conflicts in Swedish construction.

If the parties have concluded a contract stating that disputes between the parties shall be settled by arbitration, this is binding on the court. Thus, the court has to refuse a case if one of the parties ignore the arbitration clause and instead tries to have the dispute subject to court trial.

Specialist judges, courts or procedures for construction cases or for technical cases are not found in our country.

The courts must be considered to have formal possibilities to direct the use of external procedures without the consent of the parties, e.g. mediation, but this does not occur in practice.

The procedures must be said to be adversarial and not inquisitorial, for instance as far as the use of technical experts is concerned.

Disputes referred to court can be tried in three instances: district court, court of appeal and the Supreme Court. Arbitration and written evidence is being used in court as well as in arbitration, and in Sweden "free trial of evidence" is applied. It is possible for the parties to use cross examination, whereas this is not done by the court.

3.2.5   Negotiation

See 2.1.3, 3.1.1 and 3.1.4.

3.2.6   Ombudsman

Sweden being the country of origin of the Ombudsman, the concept is now widely used. There are Ombudsmen for a variety of issues: the Parliamentary Ombudsman, the Ethnic Discrimination Ombudsman, the Equal Opportunities Ombudsman, the Consumer Ombudsman and the Children's Ombudsman. There is, however, no Ombudsman for the construction industry. Nevertheless, the Parliamentary Ombudsman and the Consumer Ombudsman may play a certain part in relation to construction disputes (as in other civil disputes).

The *Parliamentary Ombudsman*, works to ensure that the courts and the civil service enforce the laws and handle their powers properly. In particular, he surveys the handling of issues relating to freedom, security and property of citizens. The Ombudsman is empowered to institute prosecution in court and act against officials who abuse their powers or act illegally.

In most cases, however, the Ombudsman deems it sufficient to issue a statement as to whether, in his opinion, the act or decision of the public servant was lawful or not.

The *Consumer Ombudsman* surveys the application and enforcement of certain consumer protection laws, the most important ones being the Marketing Act 1995 and the Consumer Contract Terms Act 1994. On the basis of these and other acts, the Ombudsman can initiate proceedings in the Market Court against companies which use misleading advertising or unfair contract terms. Such action is preventative only. The Court may order the company to refrain from a certain behaviour and set fines which will become payable in case of further violation. However, as of 1 December 1997, the Ombudsman has been given the additional authority to act as counsel for consumers in individual cases relating to financial services. The Ombudsman can only take such action where the dispute is of general importance for the interpretation of consumer protection rules, or where other special reasons exist. The reason for the limitation to financial services is that this function of the Consumer Ombudsman is experimental - the authorising legislation is temporary and its effects will be evaluated before it is extended in time or scope. Another current experimental function of the Consumer Ombudsman is his power to initiate complaints proceedings before the National Board for Consumer Complaints on behalf of a group of consumers. The Board is an official body which issues non-binding recommendations in consumer disputes.

4.   THE ZEITGEIST

For several years, the Swedish construction market has suffered from a low level of demand. This has led to keen competition among construction companies and, as a result, to low prices. Moreover, the employers have sometimes taken advantage of the situation and reallocated construction risks.

This situation has resulted in more clashes of opinion between the parties to the construction contracts, and the number of disputes has grown. To an ever increasing extent the parties are represented by professional advisers, both engineers and lawyers, and this, too, seems to result in the clashes of opinion becoming more complicated and more difficult to settle.

Previously, it was not very common among specialist sub-contractors to pursue their cases hard against clients and general contractors. Today these groups, too, assert their rights to a higher degree and are nowadays often represented by lawyers when the claims are to be staked out.

Both clients and contractors are of the opinion that, today, it is more difficult to settle clashes of opinion in construction projects. Both parties hesitate, however, before they engage in a court procedure or an arbitration, since they consider it costly, and that the resolution of disputes takes too long.

Many clients and contractors would gladly see that a broad range of alternatives were available for the solution of disputes. It can be seen that parties sometimes try, already at the contract stage, to find other forms than trial to solve clashes of opinion that may arise. A recently signed construction contract includes the following clause:

"The parties are agreed that clashes of opinion that have not been possible to solve by negotiations, shall be referred to a jointly appointed person for a statement before a dispute is referred to judicial trial".

---

1   Support from the Swedish Council for Building Research is gratefully acknowledged.

2   i.e., a member of the Swedish Bar Association.

3   See Swedish Law : A Survey (1994), Tiberg, H. Sterzel, F., Cronhult, P. (Eds) Jurstif rlaget, Stockholm, and An introduction to Swedish Law (1991), Strömholm, S (Ed), Norstedt, Stockholm.

4   Byggandets Knotraktskommitt J (11992), Allmänna bestämmelser f r byggnadsö, anläggningsöoch installationentreprenader, AB 92, Stockholm, Translation, see *General Contract for Building and Civil Engineering Works and Building Services, AB 92.* Distributed by the Swedish Building Centre (AB Svensk Byggtjänst) Stockholm.

SWITZERLAND

Roland Hürlimann
Baur Schumacher & Partner, Zurich, Switzerland

1.   BACKGROUND

1.1   ECONOMIC

For years now, building and construction disputes have amounted to approximately 10 % of all those litigated at the Swiss Federal Supreme Court. Such litigation occurs with astonishing regularity, even though construction generally is in decline[1] and the Federal Supreme Court's workload is, in contrast, sharply on the increase. In fact, the Federal Supreme Court has found there to have been a noticeable rise in the number of construction disputes themselves. The underlying cause of this development would appear to be not so much legal as financial: namely, the fiercer competition which now exists seems invariably to lead to litigation; while at the same time, falling margins appear to be bringing about a greater reluctance to settle disputes subjectively by offering generous (sic) concessions.

Thus, the litigiousness of those involved in the construction industry has increased (and this is not just for financial reasons). At the same time, however, over the last few years, a clear tendency has become discernible to settle those disputes which cannot be avoided out of court, if at all possible. This is particularly true of larger construction projects, the co-operative implementation of which effectively binds participants together for longer periods. This is also true in respect of disputes with public authorities or major institutional clients, i.e. wherever the participants place a high value on remaining on good terms with each other (with an eye to future orders), the desire exists to avoid mutual legal confrontations. One way this is achieved is by reaching agreement on a mandatory pre-litigation dispute settlement procedure.

However, while there are further motives for avoiding public tribunals, arbitration has lost much of its original attractiveness - at least, as far as domestic disputes are concerned. What remains, however, having been promoted by the construction industry with particular vigour in recent years, is the out-of-court dispute-resolution option.

1.2   LEGAL

Judicial organisation and matters of civil procedure: Work contracts are governed by the special provisions of artt. 363-379 of the Swiss Federal Code of Obligations of March 30, 1911 ("CO")[2]. The construction contract is one form of work contract and as such is also subject to the said artt. 363-379 CO[3]. The so called private substantive law is *federal law*[4].

In the field of procedural law, the competence to legislate lies with the states (the so called cantons[5]). Most matters of civil procedure are governed traditionally by *state law*. Judicial organisation is *basically* the prerogative of the Cantons: The Swiss Federal Constitution[6] provides that all the 26 cantons shall organize their court systems and procedural codes *independently*. As a consequence, although Switzerland is a small country[7], there *are 26 different Cantonal civil procedure codes.*[8]

The court system and the procedural code of each Canton differ more or less from those of the others. In discussing the basic principles governing the court and civil procedure systems in Switzerland, it should first be noted that there are *important differences* between the several Cantons (i.e. jurisdictions). Some descriptions given in this report may be true of one Canton, but not of another. Recently, Federal laws have markedly affected the Cantonal civil procedure codes[9]. In addition, there is a *Federal Code of Civil Procedure*[10] applicable to matters reserved for the jurisdiction of the Swiss Supreme Court.

In the international context, court and arbitration proceedings are in many respects governed by the *Swiss Federal Act on Private International Law* of December 18, 1987 ("PIL"). That Act, which took effect on January 1, 1989, overrules the procedural codes of the Cantons in matters coming within its scope (cf. art. 1 PIL)[11].

System of Courts in Construction Law matters: Each Canton has its own court system. In first instance, most construction cases are handled by district courts as they have general jurisdiction. District courts are called *"Bezirksgericht"*, *"Amtsgericht"*, etc. In many matters, judgments of district courts are subject to appeal to a higher court (Cantonal High Court), called *"Obergericht"*, *"Kantonsgericht"*, etc. None of the Cantons has special courts for construction matters. In the Cantons of Zurich, Aargau, Bern and St. Gallen, however, commercial and industrial matters are reserved to commercial courts, usually called *"Handelsgericht"*.

Those commercial courts are vested with limited jurisdiction; only special cases are tried, *e.g.* where the parties (or at least the defendant)[12] are commercial organizations or qualify as merchants[13] and the sum in dispute exceeds a specific amount[14]. In the commercial courts, the case is usually tried by specialized judges having expertise in the construction field. The presiding judge is invariably a lawyer.

Most Cantons have a supreme court hearing appeals. Yet the commercial courts are usually the first and only instance. Some Cantons (*e.g.* Zurich) have a special court (Court of Cassation) to hear points of substantive law and procedure raised on appeal.

Final judgments of Cantonal courts may be appealed to the *Federal Supreme Court* if the interpretation of substantive law is queried and if the sum in dispute exceeds the amount of Sfr. 8,000.--. The Federal Supreme Court will only review Cantonal judgments or orders in applying Federal law. Factual issues cannot be challenged unless the evidence rules have been breached. Points of foreign law are examined only subject to restrictive requirements. As private law is mostly Federal law, an appeal very often lies.

No separate court system exists for points of Federal law or diversity jurisdiction. Parties to a litigation must go through the Cantonal courts irrespectively of whether there is a point of Federal law in dispute, unless they agree to elect the *Federal Court as first and sole instance* (under art. 41/c of the Federal Judicature Act of 16 December 1943). The only requirement is that the sum in dispute should exceed Sfr. 20,000.--. Moreover, the Swiss Federal Court has sole jurisdiction in some special matters, such as cases involving Federal or Cantonal authorities.

The Federal Court has *final jurisdiction* for complaints for violation of constitutional rights of individuals, appeals and complaints for violation of Federal private or criminal law, complaints against the handling of specific matters involving Federal administrative law, etc. The Federal Court also has final jurisdiction in matters concerning international treaties to the extent that such jurisdiction is not reserved to other Federal authorities.

Judicial proceedings in outline: Before a construction dispute is taken by the proper court, the plaintiff is required in most Cantons to institute a conciliation proceeding before a Justice of the Peace *(Friedensrichteramt)*[15]. The purpose of that proceeding is basically to establish whether the parties are really serious about proceeding with the matter in court.[16] As a rule, the parties are required to appear in person rather than to file an answer. However, the J.P. is not authorized to subpoena the defendant or give judgment by default. The parties may not be represented by counsel.[17]

If no settlement is reached, the J.P. grants the plaintiff leave to file his *complaint (statement of claim)* with the court[18]. The complaint must not only set out the facts of the plaintiff's case but also specify the evidence.[19] In construction disputes, the litigants will often file *expert opinions* along with their pleadings. The defendant, if he is to avoid judgment by default, is required to file an *answer (defence)*. Like the complaint, the answer must set out the facts, showing why the relief sought should not be granted, and specify the evidence.[20] The answer may be combined with a counterclaim. A defendant arguing that a third party (e.g. engineering company, subcontractor, etc.) is liable to him for all or part of the plaintiff's claim, may bring in that party as *third party co-defendant*. Other persons affected by the outcome may enter into the action of their own motion as *"intervenors"*.

Usually each party will be given leave to file a second pleading *(replication* and *rejoinder*, respectively).[21] After that exchange of pleadings, the case comes up for trial, at which each party presents the evidence and calls witnesses. The forms of discovery vary from one Canton to another[22]. It is common in most construction disputes to file one or more expert opinions with the pleadings. A majority of construction disputes may be settled before judgment. Where judgment is given, the parties may appeal.

Arbitration procedure in outline: With the enactment of the Swiss Federal Act on Private International Law[23] ("PIL") in 1989, Switzerland adapted its laws with respect to *international arbitration* by embodying chapter 12 of PIL[24]. Unless a written agreement to the contrary exists, these new rules apply automatically to all arbitrations held in Switzerland if at least one of the parties is domiciled abroad[25]. The rules of procedure leave the parties a great deal of freedom of choice in procedural matters in view of the nature of the business involved. Considering that multi-party arbitration is often desirable in construction disputes, the parties should draft the arbitration clause setting out the basic terms for the entire proceedings with due care[26].

For the decision as to *what procedural law* to apply, chapter 12 PIL relies upon the principle of party autonomy and places that decision directly or indirectly into the hands of the parties, subject to the fundamental procedural rules of equal treatment and the right to legal hearing. By art. 182 (1) PIL, the parties may frame the procedural rules themselves, or choose them by reference made to a set of arbitration rules[27]; or the parties may leave that task to the tribunal[28]. It follows from the above that parties and the tribunal are basically free to frame the procedural rules, as long as equal treatment of the parties and the right of the parties to be heard in an adversorial procedure is ensured[29]. Quite often, the parties to arbitration agreements make reference to the procedural code of the canton in which the tribunal has its seat. However, as chapter 12 of PIL defines not a set style of procedure, but a flexible framework, the parties may tailor rules to the particular dispute in hand or include rules of a domestic or international association, a chamber of commerce or organizations having a well-established set of rules for international arbitration. See in detail, chapter 3.2.2 supra.

In domestic arbitration, the Swiss statutory rules of the Intercantonal Arbitration Convention (IAC) of March 27, 1969 apply. In addition, the parties often make reference to SIA-Directive 150 which contains arbitral rules to resolve construction disputes[30].

1.3 Construction Contracts

Standard forms of SIA: Construction contracts are usually complex long-term contracts, differing in several respects from the specimen work contract governed by arts. 363-379 CO. As a matter of fact, the Swiss construction industry is strongly influenced by standard form contracts including general conditions issued by professional associations which, in many respects, overrule the regulations of the codified law. Of outstanding importance for Switzerland are the standard forms of contractual terms issued by the Swiss Society of Engineers and Architects (Schweizerischer Ingenieur- und Architekten-Verein; S.I.A.; Zurich).

- *SIA- Norm 118;* Edition 1977/1992 "General conditions for construction work" ("Allegmeione Bedingungen für Bauarbeiten")

- *LM 95;* Edition 1996 "General conditions for the services and fees of design professionals" ("Leistungsmodell   95:   Planervertrag   / Subplanervertrag / Gesellschaftsvertrag")

- *SIA-Ordnung 102;* Edition 1984 "General conditions for the services and fees of architects" ("Ordnung für Leistungen und Honorare der Architekten")

- *SIA-Ordnung 103;* Edition 1984 "General conditions for the services and fees of engineers" ("Ordnung für Leistungen und Honorare der Bauingenieure")

- *SIA-Ordnung 108;* Edition 1984 "General conditions for mechanical, electrical and other specialised engineers" ("Ordnung für Leistungen und Honorare der Maschinen- und der Elektroingenieure sowie der Fachingenieure für Gebäudeinstallationen")

These standard forms regularly serve as the basis in Swiss construction and/or design (professional) contracts, supplemented *by other standard-forms of SIA* (e.g. *SIA-Nr. 1025/1026* contract form for general contractors; 1982) or/of other professional associations[31].

For special construction, the SIA has created other standard forms such as the *SIA-Norm 198* for "underground construction" (Edition 1993); and the *SIA-Norm 161* for steel construction.

These are just two documents among hundreds that are prepared and issued by SIA's permanent professional staff. Most SIA-documentations also cover *technical aspects of construction*.

For procedural aspects, the SIA has issued the *SIA-Directive 150* "procedure rules for arbitration ("Verfahren vor einem Schiedsgericht, Richtlinie", 1977) *SIA-Directive 152/153* "price competition for design professionals" (Ordnung für Architekturwettbewerbe (1993); Ordnung für Bauingenieurwettbewerbe (1991) or *SIA-Directive 155* "instructions for experts" ("Ausarbeitung von Gutachten, Richtlinie", 1987).

Binding effect of standard forms: The standard forms do not have legal force nor are they binding in the sense of an administrative regulation[32]. To be valid, these general conditions must have been incorporated by the parties into the contractual agreement. If the parties want SIA-Norms to be applicable, they must make explicit *reference* that the relevant SIA-Norm should form part of the agreement. Two exceptions may apply: First even if the parties *did not* incorporate one of the SIA-Norms into their contract, some of its provisions may be relevant as the expression of professional usage or custom and may serve as such to construe and complement the contractual agreement[33]. Second, even if the parties *did* incorporate the SIA-Norm into their agreement, single clauses may be unenforceable as one party (not experienced in construction matters) may submit that an SIA-provision is unfair or impairs their rights[34].

The SIA-Norm 118 in outline: The SIA-Norm 118 is predominant in today's construction practice in Switzerland. It contains general conditions for construction work, primarily governing the conclusion and subject of construction contracts (for both building and civil engineering work). It also deals with the responsibilities of the contractual parties (client and contractor). The SIA-Norm 118 is divided in the following main chapters:

Chapter 1: The building contract in general: This chapter mainly deals with the conclusion of the building contract. According to *art. 3,* the building contract may be concluded in writing, verbally or by tacit agreement. For larger works of construction, the following *procedure* is recommended: The client calls for tenders *(art. 4)* whereupon contractors submit their bids *(art. 15).* The client evaluates the bids *(art. 18).* If he wishes to accept a tender, he expresses acceptance by awarding the building contract *(art. 19).* The client may be *represented by the site supervisor (art. 33;* Bauleitung). Concept, methods and procedure for tender are dealt with in *arts. 4-6; components* and classification of the bidding documents in *arts. 7-14. Art. 21* provides for a *classification of contract components* which is important if some of these components are contradictory. According to *art. 21,* the contractual document signed by both parties has priority over all remaining integral parts of the contract. Construction specifications and bill of quantities have *priority* over plans. According to the classification in *art. 21,* the SIA-Norm 118 *precedes* all other Norms of SIA and/or norms of other trade associations[35]. *Arts. 23-27* contain the *main duties* of the contractual parties, such as loyalty and copyright *(art. 24),* obligation to report and to written warning notices *(art. 25);* contractor's duty to insure *(art. 26),* etc. Another section deals with the (common) case where the construction work is awarded to *several contractors* (joint venture; consortium, *art. 28)* or to subsidiary contractors (Nebenunternehmer) who all have individual building contracts with the client *(art. 30). Art. 29* relates to contracts with subcontractors and rules under which circumstances *subcontracting* is allowed without the owners approval[36]. *Art. 33-36* deal with the so called *supervisor* (Bauleitung) as the representative of the client[37]. *Art. 37* contains a *forum clause* (legal residence of the defendant party) and stipulates, that in the case of disputes neither party may, in violation of the contract, interrupt the construction nor refuse due payments.

Chapter 2: Remuneration of the contractor[38]*:* Chapter 2 deals in detail with the several types of fixed prices, such as unit prices (Einheitspreise; *art. 39),* global prices (Globalpreise; *art. 40)* and lump-sum prices (Pauschalpreise; *art. 41).* The execution of certain works (Regiepreise) as scheduled works instead of fixed prices must be agreed upon in the work contract (see *art. 44-57).* Of particular importance are *arts. 58-61* which entitle the contractor performing the promised work at fixed prices (lump sum or unit prices) to price increases if extraordinary/ unforeseeable circumstances impede the performance of the work. SIA-Norm 118 modifies the legal allocation of duties concerning ground examination: It is the duty of the client (and its project consultants) to determine the sub-soil conditions (see *art. 58 para. 2; art. 5 para. 2).* The contractor is obliged to examine the sub-soil conditions only if the client is not represented by a project consultant (geological expert), if the client is not an expert himself, or if no outside expert is advising him (see *art. 25 para 3* SIA-Norm 118). Consequently, the contractor increases the contract price due to adverse sub-soil conditions. There are many precedents by the Swiss Federal Supreme Court on this issue[39]. The new SIA-Norm 198 "underground construction" (Edition 1993) is very comprehensive with regard to the nature of the rock (loose rock, solid rock) and the level of technology pertaining to the method of tunnelling. It has not yet been established by the courts whether or not this new Norm will cause a further allocation of risks in favour of the contractor.

Chapter 3: Change orders (variations)[40]*: Arts. 84-90* SIA-118 deal with the problems concerning change orders. The consequence of a change is that the contractor may claim an appropriate increase in price if the construction contract is to be performed in a manner other than stipulated, in greater or lesser quantities or if specific performance is to be ommitted.

Chapter 4: Performance of construction[41]*:* Chapter 4 deals with time problems in the performance of the contract, with the requirements of time extensions *(art. 94-96)* as well as with the legal impact of construction schedules *(art. 93)*[42]. An other section describes the technical aspects the contractor is obliged to comply with in the execution of construction, including rules on building site installations *(arts. 123-128)*, the supply of water, electricity and sewage *(arts. 129-135)*. *Arts. 144-152* contain rules for down payments and the right of the owner to ask for security deposits or retention of payments. Especially important are *arts. 153-156* which deal with the final account. According to *art. 154*, the final account must be submitted no later than two months after acceptance, the final account is only due upon the examination report of the owner and is payable within 30 days *(art. 190)*. If the contractor makes no written reservation in the final account, it declares with its submission that it will not issue further bills and waives all further claim to remuneration for services not being accounted up to that time *(art. 156)*.

Chapter 5: Acceptance of performance and liability for defects[43]: *Arts. 157-164* provide for a detailed completion procedure: Once the contractor reports the completion of the work (or part thereof), the client (or the supervisor) and the contractor jointly examine the work within the period of one month *(art. 158)*. If no defects are found during the joint examination, the work is regarded as accepted upon the conclusion of the examination *(art. 159)*. In case of major defects, acceptance is postponed until the contractor has remedied the defects and the work has been re-examined *(art. 161)*. *Art. 165-171* deal with the liability for defects. The contractor is liable for defects without regard to the cause of the defect (e.g. careless work, application of unsuitable material, unauthorised deviation from plans and regulations, faulty behaviour of subcontractor, etc.). According to *art. 169* the client has initially only the right to ask the contractor to remedy the defect. In other words, the contractor has a right of correction before any other remedies are available to the client. In as much as the contractor does not remedy defects within the time period set by the client, the client may execute the correction through another contractor or perform the correction himself; both at the expense of the contractor *(art. 170)*; the client may also make a deduction of the remuneration corresponding to the decrease in value of the work or revoke the contract. In addition, the client has the right to ask for consequential damages in cases where correction, price reduction or revocation are not sufficient remedies at all *(art. 171)*.

According to the SIA-system, the client's rights concerning defects are barred by the statute of limitations five years after the acceptance of the work or of a part thereof *(art. 180)*. The rights regarding defects deliberately concealed by the contractor, however, are barred by the statute of limitations after ten years only. Prerequisite for the contractor's liability is that the client notifies the defects at any time during the guarantee period of two years *(art. 173/174)* or immediately upon discovery after the two-year-guarantee-period *(art. 179)*. The contractor is not liable for hidden defects which could have been discovered during the examination or prior to the expiration of the guarantee period, unless however, the contractor had deliberately concealed the defects. The burden of proof lies with the client if it is questionable as to whether a notified hidden defect is actually a contract deviation and therefore a defect in the sense of Norm SIA 118 *(art. 179)*.

Chapter 6: Termination of the contract and default in payment by the client: With regard to the premature termination of the building contract, SIA-Norm 118 makes reference to the statutory rules of the Code of Obligations. As long as the work is incomplete, the client is entitled to terminate the construction contract at any time, provided he fully compensates the contractor *(art. 184)*. *Arts. 185-189* deal with the legal consequences if the work (or parts thereof) have been destroyed for any reason. In case of accidental destruction or force majeure, the contractor is entitled to total or partial remuneration for services performed before the destruction, even in the absence of agreement *(art. 187)*. According to *art. 190*, the client is obliged to effect due payments within 30 days. Delay of payment by the client entitles the contractor to accumulated interest in addition to compensation if the client is in default *(art. 190)*.

2.   CONFLICT MANAGEMENT

2.1   Non-Binding

2.1.1   Dispute Review Boards/Advisers:

In Switzerland, a working group of the Construction Industry Standing Conference[44] has devised a conciliation procedure appropriate to projects forming part of plans for imminent major infrastructure construction works (NEAT, Alptransit, completion of the motorway network). This procedure can be applied under Swiss law and appears generally to be suitable to Swiss conditions. The Working Group has put forward a type of institutionalized conciliation[45], which (in contrast to previous models) would be operative from the start of the construction. It would also ensure that construction work continues free of interruptions. It achieves this by subjecting disagreements arising during the project to proper, competent quasi-judicial assessment (immediately and not one year later), with a view to bringing about their swift resolution. Hence, this form of conciliation differs from practically all traditional dispute-resolution mechanisms; since the latter only begins to deal with disputes long after they have arisen (if not generally after completion of the entire project).

Under this so called GIB-model, the Disputes Review Board (DRB) is appointed at the start of works and meets regularly until its completion. Its meetings normally take place every 3-4 weeks, regardless of whether or not a specific dispute exists. Members are kept up to date on the progress of works by project participants. Although the conciliators maintain absolute neutrality in relation to projects, they do possess an extraordinary degree of relevant knowledge. Because the conciliators are kept constantly apprised of progress and any difficulties which may arise, project participants can expect them to possess a high level of both competence and fairness. Normally, project participants agree jointly on the appointment of DRB members. They each choose an independent expert they trust, as well as a chairman from among well-known construction lawyers specializing in the area. These conciliators must undertake to remain available for the entire duration of the project.

Under the GIB standard model, the DRB merely issues recommendations, and these are not finally binding on the parties. Nevertheless, the conciliators' opinion will be of critical importance in any subsequent court or arbitration proceedings. The Swiss Association of Master Builders (*Schweizerischer Baumeisterverband*) has emphasized the DRB's preventative effect. The mere existence of a Conciliation Board should serve to encourage project participants to treat disagreements with greater objectivity. It should even encourage them to attempt to resolve minor disputes without involving the Conciliation Unit at all. In this respect, project participants can be said primarily to promise each other that they will strive to improve the working climate on-site, and this in turn promotes improved productivity.

Moreover, the DRB's technical, organizational and contractual knowledge imbues it with a natural authority, thus heightening the reciprocal trust and understanding of the parties. The former are both factors which can help the Conciliation Unit's proposals to achieve a breakthrough. Perhaps the most important advantage of the conciliation procedure, though, is that project participants are virtually forced into submitting for discussion existing disputes all the time (or at least, without too much delay) and, further, are virtually forced into making their demands known. This practice is welcomed by clients too, since it affords them ongoing attention to their unresolved disputes. In the process, it offers them some indications of the success or otherwise of works carried on at a particular site. In turn, this spares clients the nasty surprise (sprung by the contractor) of suddenly finding exorbitant claims for extra payments contained in the project's final account statement.

The Construction Industry Standing Conference Working Party recommends a three-stage conciliation procedure for major infrastructure projects[46]. Accordingly, whenever a relevant dispute arises, the parties must work through the following three stages:

Building Site Decision: The parties must undertake to try initially to resolve disputes between themselves. So as to ease this 'internal' dispute-resolution process, a common negotiating procedure must be gone through. This in turn must follow precisely defined temporal milestones. Right at the outset, such a requirement should have the effect of removing any incentive to take the dispute further. After this, a 'senior management discussion' must take place. This operates as an additional threshold. The discussion is intended to ensure that both parties' senior decision-makers discuss the dispute prior to any intervention by the Conciliation Unit. It is intended also to help the parties avoid a situation where their on-site representatives 'lightly delegate' disputes to the Conciliation Unit.

Conciliation Unit: The contract concluded between the parties provides for a Conciliation Unit, the duty of which is to assist them in resolving disputes. Application to this Conciliation Unit may only be made after the parties' efforts to resolve the matter by way of Building Site Decision (Stage 1) have proven fruitless. Conciliation must be attempted in all cases before a party is entitled to apply to a court for its judgment (Stage 3).

Court Judgment: In view of the fact that the infrastructure projects planned (Rail 2000, Alptransit) are of national significance, the Swiss Construction Industry Standing Conference proposes that the Swiss Federal Tribunal be declared as having sole jurisdiction over litigated disputes. Since the Federal Supreme Court may be invoked at first instance only when disputes involve more than CHF 20,000.-, the Working Group further proposes that the parties should declare the Conciliation Unit to be competent to reach final decisions in disputes involving a lesser sum; and to be capable of functioning as an arbitration tribunal in those cases (Stage 3).

This variant - entrusting the arbitration tribunal with final judgement in small disputes up to a certain threshold value - represents one of several different solutions which the client can, but need not, select. The corresponding arbitral clause may then only be formulated in detail at the time the individual award is made. This is due to the fact that appropriate solutions differ, depending on whether or not foreign companies are involved.

2.1.2   Negotiation:

Until recently, in Switzerland there has been no systematic approach to negotiation. Authorized by the Harvard negotiation project, the consulting firm of Egger, Philips & Partner AG offers training seminars in negotiations, namely in the so-called Harvard-concept. These practical seminars are encouraged and sponsored by the Swiss Bar Association[47]; however, negotiation training is dealt with in a general manner, not focused on construction. In summer 1997, a society for mediation and conflict management was founded to promote mediation in the construction industry. Mediation seminars will start in winter/spring 1998[48].

It should be further mentioned that ADR, including negotiation has been often discussed in conferences dealing among others with construction law issues. Pars pro toto I refer to the cedidac-conference (in collaboration with Vorort and ASA with Marc Blessing[49] or by annual Swiss construction conference (organized by Prof. Gauch and Prof. Tercier) with guest-speaker Michael E. Scheider[50]. Swiss models of negociation and ADR-solutions are subject to international conferences as well: In October 1997, Prof. Peter Gauch was giving a speech on out-of-court dispute resolution procedures in Switzerland and how these models have been implemented in real legal life[51].

2.1.3 Quality matters:

In Switzerland, ISO-standards (SN ISO 9000-9004) were introduced in January 1990[52]. Although quality assurance and quality management has been an issue even before ISO was introduced, quality management in the Swiss construction industry is not fully established. Today however, most major construction firms have passed the certification process. The realization of the ISO-Standards is promoted by the Construction Industry Standing Conference in a so called "Forum Quality Management " to ascertain an unité de doctrine[53]. Since enactment of the new procurement laws (BoeB, VoeB) in Switzerland January 1, 1996, ISO-9000-certification is one criteria of prequalification for public tenders[54].

2.2 Binding

2.2.1 Partnering

Long term commitments between two or more organizations for the purposes of achieving specific business objectives are not known in Switzerland although they may exist[55]

3. DISPUTE RESOLUTION

3.1 Non-Binding

In Switzerland, dispute-resolution is a generic term for various procedures aimed at settling construction disputes without the assistance of either the public courts or arbitration tribunals. To this end, various models have been developed in Switzerland (influenced, however, not least by foreign models). Among these are:

3.1.1   Conciliation

The task of a conciliator is to bring the parties closer together by means of the skilful management of negotiations. The parties may make use of a recognized, neutral expert familiar with the language of the construction industry. Such an individual is therefore in a position to assess the prospects of success of participants' positions, as communicated to him or her during conversations with them. Swiss literature on the subject contains varying accounts of this new (and ambitious) form of dispute-resolution[56]. However, as yet, no statistics are available on how much success this particular method has had.

In practice, the conciliation rules of the Cantonal Chamber of Commerce are used: these are either taken as drafted or amended to suit individual requirements.

Conciliation in a broader sense is also done by a *Justice of the Peace*. According to the civil proceedings rules of most Swiss Cantons, a dispute is first brought before a Justice of the Peace. These judicial authorities have already been discussed above (ch. 1.2). The task of a Justice of the Peace is generally to ascertain the positions of the parties during discussions held with them, and then to attempt by means of mediation to steer such discussions towards a mutual amicable agreement. In most Cantons, appearing before the Justice of the Peace is mandatory. It is only after the latter has certified that an unsuccessful attempt at mediation has been made that the parties may bring an action before the competent District or Commercial Court.

There have been no statistical surveys regarding the extent to which conciliation efforts by Justices of the Peace are successful in construction matters. However, anecdotal experience suggests that Justices of the Peace often fail in their attempts to bring about pre-trial agreement between the parties due to the complexity of the subject matter in dispute. In addition, the parties are frequently represented in court by construction lawyers, who have already sounded out the possibility of out-of-court agreement prior to appearing before the Justice of the Peace. Given this starting point, the decision of whether or not to proceed to full court proceedings has generally been made already; and hence the Justice of the Peace's mediation efforts are superfluous.

In construction matters, it is frequently the case that the both parties petition the *instruction judge* to undertake an attempt at conciliation even before the pleadings have been finished or before evidence procedure has taken place[57].

In Switzerland, construction disputes where there is an *arbitration panel* frequently also begin with conciliation proceedings. This is required by *SIA Guideline 150*, by virtue of which an arbitration tribunal must make an attempt at settlement prior to the submission of written statements. (Art. 21/22). In practice, however, it is often the case that the tribunal panel first demands a legal statement from each of the parties. If settlement is achieved, the arbitration tribunal must document it in full, this record is then signed by the parties and their representatives. In such a case, the arbitration tribunal issues an arbitration decision, declares the proceedings resolved by settlement and the investigation to be closed. Where necessary, a costs decision may be reached at the same time (Art. 22 III).

The *Zurich Chamber of Commerce* also offers its members a conciliation procedure designed to settle differences. A written account of the facts of the dispute must be submitted together with the application to initiate the procedure. If both parties agree to mediation, the Chairman of the Chamber of Commerce appoints a suitable impartial individual as a mediator. The latter is authorised to undertake any actions which appear conducive to amicable settlement of the dispute. The principal advantage in comparison with reconciliation proceedings (*Sühneverfahren*) before a Justice of the Peace is that Zurich Chamber of Commerce mediators often turn out to be individuals mediation with specialist knowledge who also possesses mediation skills[58].

3.1.2   **Executive Tribunal**

This method is in effect a qualified mediation proceeding. Its chief characteristic is that the parties present their case in concentrated proceedings (e.g. best evidence) before a mini-trial panel. This panel is composed of a high-ranking Executive Board member from each of the parties, under the chairmanship of a neutral individual. It is expected of the Executive Board members that they will be in a position to assess the matter in relation to their company or firm with a certain distance and autonomy.

Each of the procedural models described have their respective advantages and disadvantages. The parties will agree upon them either ad hoc or in advance. Such an agreement forms the basis of a 'dispute-resolution arrangement', which from time to time is also pre-formulated in an applicable set of general contractual conditions.

Since 1985, the Zurich Chamber of Commerce has had an independent set of Mini-Trial Rules, modelled along American lines. Until now, however, this procedure has been invoked very infrequently, and none of the cases concerned has involved a construction dispute[59].

3.1.3   **Mediation**

Mediation assumes the most varied forms in Switzerland. All the models employed share in common the fact that the mediator plays a more active role than a conciliator. This dispute-resolution method, whereby the parties systematically isolate individual points at issue with the aid and guidance of a mediator (resulting in a more congenial climate for negotiation and/or settlement), has been used to settle various types of construction dispute in Switzerland.

In accordance with the Swiss understanding of mediation, the mediator is regularly authorised to provide the parties with a settlement proposal. On the other hand, he or she is not authorized to decide the dispute submitted with binding effect; unless, that is, the parties subsequently extend the mediator's brief by appointing him or her as an arbitrator (Med-Arb). Yet this is something which seldom happens in practice. Frequently, in fact, the mediation agreement states that the mediator must cease acting in any further proceedings that may be rendered necessary, whether as an arbitrator or as a witness in court or arbitration proceedings.

How are the mediation models implemented in real life? Swiss construction industry practice is as follows: The Swiss construction industry associations have devised various out-of-court dispute-resolution procedures together with accompanying models. This was done particularly in view of the major infrastructure construction projects now imminent in Switzerland. The one initiative containing specific proposals comes mainly from the above-mentioned Construction Industry Standing Conference and also from the Association of Swiss Highway Professionals (VSS). These associations allude firstly, to the flaws in formal, conventional methods and secondly, to the positive experiences yielded by the use of out-of-court procedures in connection with major foreign construction projects.

Up to now, mediation has been utilised rarely, although where it has been employed, it has taken the form of a simple mediation procedure, generally agreed upon 'ad hoc'. 'Dispute-resolution arrangements' have not formed part of the customary content of construction agreements so far, nor has it been the normal practice of parties to agree upon dispute-resolution once a dispute has broken out. This is true of all branches of the construction industry, irrespective of the size of the project involved. Moreover, the fate of the 'Zurich mini-trial' possesses particular significance for the current status quo. Hence, to summarize mediation or other out-of-court dispute-resolution procedures may in no way be said to amount to an everyday practice of the Swiss construction industry[60]. The *reluctance* which such procedures have encountered in practice has several underlying causes. Although these are not susceptible to scientific investigation, they can nonetheless be guessed at[61]:

-        Many differences are resolved informally by negotiations (often on site) before they have developed into an open dispute between the parties. As regards differences of opinion between the contractor and the client's representatives (i.e. the architect or civil engineer) occurring from time to time, this process can even take place without the client's knowledge.

-        Precautionary agreement upon a simple negotiation procedure or two-stage negotiation may be viewed as otiose, since it is always possible to conduct negotiations using good will and 'without a formal agreement' (even a multiple stage agreement).

-        Both project participants and their legal representatives have little experience in handling out-of-court procedures. With mediation procedures, there is also an occasional lack of suitable mediators possessing the trust and confidence of both parties. Bad accounts as well as good circulate regarding Disputes Review Boards. And were the matter to come to court, there is a fear of the possible prejudicial effect of out-of-court procedures - and of the associated jointly-commissioned expert reports, in particular.

-        In Switzerland, no institutionalized construction industry Conciliation Units exist whatsoever.

-        The representatives of public construction clients sometimes lack either the courage or the authority to consent to an out-of-court settlement.

- Lawyers are trained more for court fights than for situations where the dispute is dealt with out of court. Sometimes they prefer the option of pursuing litigation through one or more courts simply because they are unable to assess the risks involved in out-of-court dispute-resolution.

Abundant literature has been produced on the subject of out-of-court dispute-resolution[62]. The bulk of it, however, is of a descriptive nature or merely deals with the 'negotiating psychology' aspects of the different procedures. Little of it concerns out-of-court dispute-resolution's legal side, despite the fact that many relevant legal issues require clarification.

At the present time, a thesis monography is in course of preparation[63] which in particular should contain answers to the following questions: Under what circumstances may a dispute-resolution arrangement be validly agreed upon? Which duties result from such an arrangement? What are the consequences of a breach of duty? What is the effect of a temporary suspension of the right to call in a judge? What is the mediators' legal position? What use may be made of the events during, or statements and documents from, out-of-court procedures? May the mediator be called as a witness in any subsequent trial? What is the legal effect of a settlement?

Academic commentators have divided opinions about the usefulness and prospects of success of out-of-court dispute-resolution. Professor Gauch has been quite critical:

Mediation and other out-of-court dispute-resolution gives rise to costs as well. It also ties up personnel as well as funds. On these grounds alone, it is no substitute for effective prophylaxis, for dispute avoidance is better than even the best dispute settlement. This maxime is so self-evident that the writer would not even dare to restate it here were it not for the fact that it is so frequently ignored in practice. Poorly drafted agreements, ignorance of their precise content, ignorance of the law and careless contractual performance are fertile ground for unnecessary quarrels.

Over and above this, the climate in which a dispute takes place may be heated further by lack of understanding between the parties. Emotional intelligence is too little taught in schools and colleges, and ultimately this has repercussions on the development of disputes. Kindly note that legal training is no exception to this: although law students learn how to win when situations of conflict arise, they are certainly not taught how to avoid them in the first place.

As far as agreements themselves are concerned, there is much to find fault with. In any event, it is clear that the traditional contractual wordings have been left behind by technological developments. Or do Swiss conditions differ from those in the United States, where complaints have been made that, 'Contract documents...have not matched technological advances in the design and construction of tunnels ... Engineers still copy others' specifications, bid schedules do not give the contractor the incentive to reduce costs, and specifications have too many ambiguities and discrepancies'.

Dr. Blessing takes a far more positive view[64]. He believes that ADR is absolutely necessary in a large number of disputes and he calls for the use of alternative disputes resolution in all construction agreements dealing with infrastructure projects.

3.2.   BINDING

3.2.1   Adjudication:

The method as such is not known in the Swiss Construction industry, Infrequently, it is practised by Dispute Review Boards who are entitled to give binding decision for the duration of a long term contract.

3.2.2   Arbitration

In Switzerland today, arbitration is by far the most well-known alternative option to state jurisdiction over disputes. Arbitration has the major advantage that the parties have the power to appoint an arbitrator themselves, who has a high level of specialist knowledge and authority regarding the issues to be decided[65]. After all, such professional authority has a certain persuasive power. Studies undertaken by the Zurich Chamber of Commerce have shown that Zurich Chamber of Commerce arbitration panels, which are chaired by a professional lawyer with two specialist lay assessors sitting on either side, have a markedly higher incidence of achieving settlement than other arbitration panels consisting of professional lawyers only[66]. The relevant SIA regulations (as between clients and design professionals) or SIA standards (as between clients and builders) provide that the ordinary courts shall have primary jurisdiction over disputes. Agreeing to take the matter to arbitration would normally require a prior written agreement to this effect. In contrast to previous SIA standard agreements, today's SIA model conditions are based on equal status between the ordinary courts and arbitration tribunals, in accordance with SIA guideline 150. At the time of contracting, the parties may opt for one or the other. If neither is opted for, then the jurisdiction of the ordinary courts is deemed to have been selected.

Significance of Switzerland as arbitration place: Switzerland is one of the chief centres of international arbitration[67]. That fact is reflected both in the countless private commercial disputes and in numerous public and international arbitrations for which Switzerland has been elected as venue. The election of Switzerland as venue is prompted by the consideration that Switzerland provides an arbitration-friendly law system and administration of justice ensuring independent and impartial proceedings. Moreover, the parties expect a Swiss arbitration venue to have an open mind for other cultures and the willingness to accommodate parties of different origin, cultural and social background. Finally, there are further reasons for electing Switzerland, not least the country's political neutrality and the justified expectation of finding a well-developed civil, commercial and legal infrastructure[68].

Since the enactment of the *Swiss Federal Act on Private International Law* ("PIL") on January 1, 1989 international arbitration in Switzerland is regulated in the artt. 176-194, the so called chapter 12 of the PIL-act.

The artt. 176-194 of the PIL Act have been favourably received by doctrine and practice. In particular the liberal approach and the wide autonomy allowed to the parties by the rules of its chapter 12 have met with wide approval both in Switzerland and abroad. Specifically, the new rules have brought three main key points of change and improvement: Parties have been given more freedom of choice in procedural matters; the arbitral tribunal has been granted more powers; the ordinary courts' authority to intervenue in the course of proceedings has been reduced.

Scope of application: The provisions on international arbitration are to apply exclusively to arbitral disputes having an *international aspect.* In contrast with other procedural statutes, under chapter 12 PIL-act, the international aspect is not determined by statutory definition. Rather, by art. 176 para. 1, the point depends on two requirements of form (domicile of either party abroad, seat of arbitral tribunal in Switzerland): "The provisions of this Chapter shall apply to any arbitration if the seat of the arbitral tribunal is in Switzerland and if, at the time when the arbitration agreement was concluded, at least one of the parties had neither its domicile nor its habitual residence in Switzerland."

Exclusion clause: Subject to any treaties (which by art. 1 para. 2 PIL take precedence), the provisions of chapter 12 (artt. 176-194 PIL) will invariably and exclusively apply[69]. On the conditions stipulated by art. 176 para. 2 PIL, however, it is open to the parties by their arbitration agreement to exclude application of chapter 12: "The provisions of this chapter shall not apply where the parties have excluded in writing its application and agreed to the exclusive application of the procedural provisions of cantonal law relating to arbitration" (176 para. 2). An exclusion clause made under art. 176 para. 2 PIL must cumulatively satisfy three requirements: It must *expressly exclude* application of PIL; it must *expressly elect* the Cantonal rules of IAC as exclusively applicable; finally, the clause must be *in writing*. If any of those requirements is not satisfied, the exclusion clause will be ineffective, and so chapter 12 PIL will apply[70]. The point whether a valid law-election clause has been made can be submitted to the Federal Court by constitutional appeal. The election of law will not be exclusive if the parties elected IAC merely for appeals from interlocutory decisions[71].

Seat of arbitral tribunal: Art. 176 para. 3 provides: "The seat of the arbitral tribunal shall be determined by the parties, or the arbitration institution designated by them, or, failing both, by the arbitrators." Election of the venue requires a minimum of clarity: The Canton at least is to be named[72]. In case the parties fail to agree on the seat of arbitral tribunal (or other procedural rules), the new regulations give the arbitrators the authority to do so[73].

Arbitrability: Any claim involving assets may be the subject matter of arbitration (see art. 177). The term covers any kind of real and personal property, including tangible and intangible assets[74] (in German: *"jeder vermögensrechtliche Anspruch"*). Para. 2 of art. 177 contains a special provision for the event of a state being a party: "If a party to the arbitration agreement is a state or an enterprise or organisation controlled by it, it cannot rely on its own law in order to contest its capacity to be a party to an arbitation or the arbitrability of a dispute covered by the arbitration agreement."

Form and substance of an arbitration agreement: Art. 178 distinguishes between the formal and the substantive validity of an arbitration clause. As regards its form, an arbitration agreement shall be valid if made in writing, by telegram, telex, telecopier or any other means of communication which permits it to be evidenced by a text (para. 1). As regards its substance, an arbitration agreement shall be valid if it conforms either to the law chosen by the parties, or to the law governing the subject matter of the dispute, in particular the law governing the main contract, or if it conforms to Swiss law (para. 2)[75].

In keeping with that liberal approach, mutual assent does not require any exchange of the declarations of intent in writing; it will be sufficient if the circumstances (e.g. acts of performance) permit a contract (with arbitration clause) to be conclusively inferred[76], or if an arbitration clause (e.g. contained in standard forms) has become part of the contract by express or global reference made thereto[77].

Validity of the arbitration agreement: Art. 178 para. 3 rules out the objection that where the main contract is invalid, the arbitration clause automatically lapses. The provision reads: "The validity of an arbitration agreement cannot be contested on the ground that the main contract may not be valid or that the arbitration agreement concerns disputes which have not yet arisen." Admittedly, scope and effect of the arbitration clause where the main contract is ineffective are controversial in doctrine and in precedent[78]. In default of prior arrangement, the point whether invalidity of the main contract (or non-existence thereof) entails invalidity of the arbitration clause will ultimately be determined in the second place on the hypothetical intention of the parties. Again, it is still a moot point whether under the Swiss principles of law the partial nullity rule of s. 20 para. 2 of the Swiss Code of Obligations also is applicable[79].

Appointment, removal or replacement of arbitrators: The PIL-Act stipulates no special qualifications for arbitrators except independence, want of which constitutes a ground of challenge[80]. Art. 179 reads as follows: "The arbitrators shall be appointed, removed or replaced in accordance with the agreement of the parties" (para. 1). "In the absence of such an agreement, the matter may be referred to the court where the arbitral tribunal has its seat; the court shall apply by analogy the provisions of cantonal law concerning the appointment, removal or replacement of arbitrators" (para. 2). In selecting the arbitrators, the parties as well as the (regular) court must be guided by the principle of *independence* of the arbitral tribunal. The point whether, on the formulation of art. 179 para. 3[81], it is possible, on the conditions mentioned there (existence of arbitration agreement on prima facie evidence), to call on further judicial authorities to assist in the appointment proceeding, is controversial in doctrine and in precedent[82]. Opinion is agreed that, in selecting the arbitrators, the parties as well as the court must be guided by the principle of independence of the arbitral tribunal. It has further been established that no appeal to the Federal Court lies from the appointment of an arbitrator under art. 179 PIL[83].

It is open to the parties to specify by prior agreement the requirements to be met by the arbitrators in personal and professional respects. Even the grounds of challenge and the challenge procedure are amenable to agreement by the parties under art. 180[84]. The only (albeit important) statutory requirement is independence; the *requirement of impartiality* was deliberately dropped from the bill in the Parliamentary debates, not least because impartiality is a subjective, and therefore not readily quantifiable, criterion. In a given case, though, it will invariably be necessary to decide on the circumstances overall whether a "party arbitrator" gives occasion for justified doubts as to his independence[85]. A (positive or negative) decision of the tribunal on a challenge is not amenable to Cantonal appeal; but it will be open to the wronged party - despite the unclear formulation of art. 180 para. 3 PIL - to file constitutional appeal if and insofar as the decision of the challenge breaches constitutional rights[86]. An arbitral tribunal's challenge decision is not an appealable preliminary decision under art. 190 para. 3, but can be appealed from (e.g. on the ground of bias by an arbitrator) by appeal from the final award, on the ground that the tribunal wrongly declared itself to have jurisdiction[87].

The fundamental procedural rules of equal treatment and the right to legal hearing: For the decision regarding what procedural law to apply, chapter 12 relies on the principle of party autonomy and lays that decision directly or indirectly into the hands of the parties, subject to the fundamental procedural rules of equal treatment and the right to legal hearing[88]. By art. 182 para. 1, the parties may frame the procedural rules themselves or elect them by reference made to a set of arbitration rules[89]; or the parties may leave that task to the tribunal[90]. Unlike art. 24 para. 2 IAC, the PIL Act makes no provision for any subsidiary procedural law to be applied in default of party agreement, and so the tribunal is bound in default of agreed procedural rules to set up such rules "to the extent necessary".

On the foregoing, parties and tribunal are basically free in framing the procedural rules. The only reservation is compliance with the mandatory provision of art. 182 para. 3 PIL[91]. If either of those two fundamental procedural rules (equal treatment of parties, right to legal hearing) is breached, there will be a ground of appeal under art. 190 para. 2.d.

Authority to grant interlocutory relief and taking of evidence: In contrast with art. 26 IAC and with arbitration rules in other countries, the arbitral tribunal is newly empowered to order interlocutory relief, unless otherwise agreed by the parties, under art. 183 PIL[92]. The tribunal may basically order *any form* of interlocutory relief, such as safeguards, regulation of terms, acts of performance. The latter form of relief, however, is to be granted with restraint and is not to prejudice the final decision of the dispute[93]. By prevailing doctrine and precedent, the *attachment of assets* (e.g. an alien's Swiss bank accounts) is *not* an admissible form of relief under art. 183 PIL[94]; such interlocutory relief to safeguard a money claim is a matter for the (regular) court and is to be enforced by the (administrative) debt authority of the place where the assets to be attached are located.

Further, the tribunal is empowered to order the *safeguarding of evidence* within an arbitration dispute, yet, there again, the rule is that though the parties may be invited to co-operate, they cannot be compelled without the order of a regular court. All interlocutory relief is subject to the rule that until appointment of the arbitral tribunal, the regular court alone has power to order such relief. Once the dispute is pending, i.e. once the arbitral tribunal is appointed, a party may apply for interlocutory relief *either* to the tribunal or to the court, and in the latter case the opponent will not be able to raise the arbitration objection under art. 7 PIL[95], for, in respect of interlocutory relief, arbitral tribunal and regular court are on an equal footing[96].

Arbitration in most cases involves the *taking of evidence*. As with interlocutory relief, the PIL Act, by art. 184, basically empowers the arbitral tribunal itself to take the evidence. If necessary, application may be made to the regular court for assistance in the taking of evidence[97] as well as in further matters in which the tribunal may require a court's assistance for enforcement purposes[98].

Jurisdiction: If the validity or scope of an arbitration agreement is in dispute, the arbitral tribunal, before entertaining the case, must decide whether it has jurisdiction. Art. 181 PIL[99] defines from what time the case is pending, - a point of importance notably for the interruption of limitation periods or time-bars. Art. 186[100] regulates the tribunal's powers once the arbitration proceeding is pending. The decision as to the tribunal's jurisdiction as a preliminary decision under art. 190 III may (but need not) be appealed from to the proper court within 30 days of notice served.

Arbitral award: The provisions of artt. 187-190 PIL deal with the arbitral award. By art. 187[101], the case is to be decided either *under the law applicable* or - subject to authorization by the parties - *on the principle of equitability.* As regards the parties' election of the law system, the rule of party autonomy largely prevails in Switzerland. The parties are basically free in electing the law to be applied; that is subject to mandatory provisions of (domestic) public law affecting contract law, and in particular also to international mandatory rules[102]. As regards form, contents and validity of the law-election agreement, the tribunal can consult the other chapters of the PIL Act, notably the provisions of debt law[103]. Unless otherwise agreed by the parties, the tribunal may also give *partial awards,* i.e. awards which finally decide some of the claims or which decide some preliminary points of fact or law with the effect of res judicata. The Swiss Federal Court has dealt in two precedents[104] with the problems of "partial awards". By those supreme court precedents, the "partial award" under art. 188 also *includes preliminary and interlocutory decisions.* That is of great relevance, because partial awards have equal status with final awards, notably with respect to the admissibility of appeal, so that under art. 190 an appeal lies to the Federal Court.

From preliminary or interlocutory decisions, however, art. 190 para. 3 merely provides for appeal under art. 190 para. 2.a and para. 2.b. In the two precedents mentioned, the Federal Court opted for greater appeal facility and declared constitutional appeal to be admissible, on the consideration that the terms of "partial" and "intermediate" award were not defined in PIL, wherefore, on the conditions laid down in s. 81 of the Swiss Federal Judicature Act, constitutional appeal is to be admissible not only in national disputes, but also in disputes having an international aspect[105].

By those precedents, since confirmed by the further Federal Court precedent 116 II 80 ff., constitutional appeal lies from any final award; apart from the special case of subjective cumulation of claims, it will lie from a partial award (in the wide sense) only if that award exposes the particular party to an irreparable disadvantage, or if the appellant raises one of the grounds of appeal specified in art. 190 para. 2.a and b. In the latter instance, the premise is that the said grounds of appeal are not obviously inadmissible or obviously unfounded, and that they might not have been raised earlier. Thus, the Federal Court ruled that a partial award ordering an appellant to pay the other party a sum of money, though constituting a partial award, was not amenable to appeal under s. 87 of the Federal Judicature Act because the condition of irreparable disadvantage was not present[106].

The parties are also free to agree terms in respect of procedure and form of the award; in default of such terms, art. 189 provides: "The arbitral award shall be made in conformity with the rules of procedure and the form agreed by the parties. In the absence of such agreement, the award shall be made by a majority decision, or, in the absence of a majority, by the presiding arbitrator alone. It shall be in writing, reasoned, dated and signed. The signature of the presiding arbitrator shall suffice."

Appeal proceedings: Appeal from, and enforcement of, arbitral awards, are dealt with in the provisions of art. 190 para. 2 to art. 194. Art. 190 paras. 2 and 3 specify the grounds of appeal and the decisions amenable to appeal; art. 191 treats of appeal procedure; art. 192 para. 2 defines the conditions of partial or full waiver of appeal from arbitral awards. Finally, art. 192 para. 2 deals with the enforcement of Swiss awards in Switzerland, while art. 194 regulates the enforcement of foreign awards. The provision of art. 193 deals with the registration of awards and certificates, which may be of some importance in connection with enforcement.

Most precedents so far published in respect of chapter 12 of the PIL-act relate to the provisions of art. 190 sqq. The Federal Court has already had occasion to rule on the transitional scope of artt. 190 and 191. By its precedents 115 II 102 and 115 II 288, appeal to the Federal Court lies from all arbitral awards given after 1 January 1989, regardless of whether arbitration was commenced before or after that date[107]. That facility of constitutional appeal to the Federal Court also applies to partial awards, though in that instance it is subject to the condition of s. 87 of the Federal Judicature Act (i.e. the condition that the interlocutory decision entails an irreparable disadvantage for the party affected[108]).

The grounds of appeal: Art. 190 PIL[109] conclusively enumerates the grounds of appeal available against arbitral awards:

a)   *Improper constitution of the arbitral tribunal:* The arbitral tribunal will be inadmissibly constituted (*head a*) inter alia if its appointment was flawed or otherwise contrary to the contract; if an effectively appointed arbitrator failed to attend the proceeding or, conversely, if a person not, or not effectively, appointed attends. A further ground of appeal arises if the appointment of the tribunal runs *against the rule of independence[110]*. Still, an appeal on the ground that the tribunal was not properly constituted will not be admissible if there was occasion and opportunity for challenge at the time of appointment (art. 180 para. 2).

b)   *Improper decision as to jurisdiction:* An *improper decision as to jurisdiction (head b)* will arise inter alia if the tribunal decided without having jurisdiction (e.g. where an arbitration agreement was either wanting or invalid[111]), or if, conversely, the tribunal wrongly denied its jurisdiction[112]. An independent appeal to the (proper regular) court will be admissible even if the party affected failed to object to the tribunal's decision as to jurisdiction, which, being a preliminary decision under art. 190 para. 3, is amenable to appeal.

c)   *Improper decision as to ultra or infra petita:* Para. 190 par 2 (*head c*) relates to arbitral awards which outstep the limits of the claim (ultra petita) or leave some claims (e.g. for set-off) undecided (infra petita)[113]. With regard to the decision of claims for which the tribunal has no jurisdiction by reason of an arbitration agreement being wanting or limited (extra potestatem), the only ground of appeal available is that of art. 190 para. 2.b[114].

An appeal on the ground of a decision ultra or infra petita will not invariably result in the award being set aside merely because the tribunal departed from the motions: A decision which is wrong on the merits does not necessarily involve an exceeding of arbitral decision powers. Conversely, the tribunal has power (under art. 188) to give a partial award. The appellant's right will therefore in many cases be confined to obtaining a supplementary decision of points inadvertently left undecided[115].

d) ***Breach of basic procedural rights:*** By the wording of the Act, a breach of basic procedural rights (*head d*) raising a ground of appeal is to be seen as a breach of the rule of equal treatment and the right to legal hearing.

The right to legal hearing is to give "every party the right to plead facts material to the award, to set forth its position in law, to move material evidence motions, and to attend the hearings. The contentious procedure is to enable either party to examine the other's argument, comment thereon and seek to refute it by its own arguments and evidence"[116]. In the Federal Court's opinion, the arguments that the evidence proceeding ought to have been introduced by an evidence production order, or that the tribunal failed to consider unfounded objections or immaterial evidence, do not constitute adequate grounds of appeal which might be deduced from the rule of equal treatment or the right to legal hearing. Other breaches of procedure can only be raised with a chance of success if they carry weight and in that light qualify as a breach of public policy[117].

e)   ***Breach of public policy:*** By art. 190 para. 2 (*head e*) it is
admissible to appeal from an arbitral award on the sole
(though important) ground that the award involves a
breach of public policy[118].

An award will be contrary to public policy if it is quite
incompatible with the Swiss perception of law, as this will
disregard fundamental provisions of substantive and/or
procedural law[119]. The criteria for and against breach of public
policy must be assessed in each particular case; a generally valid
definition of the term is neither possible nor desirable.

Precedents on public policy as a ground of appeal: The Swiss
Federal Court, in its copious store of precedents on public policy,
has hitherto regularly decided between application of a quasi-
international public policy within the scope of the
acknowledgement and enforcement proceeding on the one hand,
and national public policy (subject to stricter criteria) on the other,
the latter is applied in reviewing an arbitral award for correct
application of the law by the tribunal[120].

What is called "public policy" in art. 190 para. 2.e PIL is entirely
concerned with disputes having an international aspect (art. 176
para. 1). The doctrine is therefore practically agreed that the
Federal Court as appellate court ought to observe restraint in
assessing a breach of public policy, and that *"only flagrant
breaches of the Swiss perception of law"* justify setting aside an
arbitral award[121]. The Federal Court's decisions given since the
taking effect of PIL do indeed reveal the tendency to interpret the
grounds of appeal under art. 190 para. 2.e restrictively, in keeping
with the legislator's intention. The absence of a reasoning in the
award is by itself no longer deemed a breach of public policy (in
contrast with IAC, art. 36 h read with art. 33 para. 1.e)[122]. An
appeal for breach of public policy would have a chance of success
only if fundamental rules of law had been violated. In contrast

clear breaches of the law and obviously wrong findings of fact were by themselves not sufficient to set aside an arbitral award[123].

In the Federal Court's opinion a breach of public policy may arise under the rule of pacta sunt servanda, under the rule against abuse of the law, under the rule of good faith, under the rule against compensationless expropriation, under the rule against discrimination or, e.g., protection of legally incapacitated persons. The censure raised in the particular case, namely, that given the terms of payment agreed (pay-when-paid clause in conjunction with a risk-shifting clause), the arbitral awarding of reward for work done violated the rule of pacta sunt servanda, was held to be unfounded.

Again, the censure that the awarding of reward for work done violated the rule of good faith, because the claimant, under the risk-shifting clause, had fully undertaken the price risk and, moreover, had continued the work although aware of the payment difficulties disclosed, was held to be unfounded under the aspect of a breach of public policy[124].

Appeal procedure: Art. 191[125] appoints the Federal Court as appellate court for the majority of cases, yet the parties may instead agree on the Cantonal court at the seat of the arbitral tribunal. In the latter case, the question arises whether the prorogation under art. 191 para. 2 (despite the unclear wording *"... its decision is final")* has in effect appointed an additional appellate court, in the sense that the Cantonal appeal decision is amenable to constitutional appeal to the Federal Court.

The literature is divided on the point whether the prorogation creates a two-level appeal facility. The Federal Court closely dealt with that difficult problem in its precedent 116 II 721, but was able eventually to leave open the point of law, because the particular appeal was not entertainable for other reasons (want of sufficient legal protection interest). Under art. 191 para. 1, the Federal-law procedural provisions of ss. 84-96 of the Federal

Judicature Act are to apply, and they are to apply even where the parties have elected the proper Cantonal court as appellate court[126]. The time limited for appeal is 30 days from delivery of the arbitral award[127] and cannot be extended[128].

Admissibility of an exclusion agreement: Subject to specified conditions in art. 192, PIL permits the parties to contractually waive in full or in part any appeal from the arbitral award to a regular court. Such an exclusion agreement may be made in advance or in the course of the arbitration. An agreement will be admissible[129] if either party is domiciled or resident in Switzerland[130]. Such a waiver is of considerable importance, as the Federal Court rightly points out. The requirement is an express and unequivocal declaration of the parties' intention; a mere reference made, e.g. to art. 24 ICC, would not suffice[131].

The exclusion of appeals under art. 192 does not mean the loss of all rights of objection. Rather, it will still be open to both arbitration parties to raise, in an enforcement proceeding, the grounds of appeal under art. V of the New York Convention[132] against imminent enforcement, insofar as that Convention is applicable.

The provision made by PIL for waiving an appeal in part or in full where both parties are domiciled abroad, is acclaimed by many Swiss and foreign commentators. Other authors have certain dogmatic and rule-of-law misgivings, because waiver even of extraordinary appeal amounts in their view to a basic infraction of Switzerland's law traditions and might in a given case (e.g. where the arbitral award is flawed in respect of public policy) result in an undue contractual commitment under the personality provisions (s. 27, Swiss Civil Code). However, in the latest Federal Court precedent 116 II 639 officially published that Court had no occasion to decide those difficult points of law conclusively.

3.2.3 Expert Determination

Most of the above-mentioned methods of settling construction disputes may be regarded as flawed inasmuch as the panel's decision is not finally binding: the arbitration decision may proceed further (unless the parties have refrained from appealing the decision) to the superior appeal tribunal. An unwanted expert report is frequently refuted by a contrary one or one of superior status. The goal of efficient dispute settlement can turn into its opposite when a second costly and lengthy proceeding becomes necessary. In order to avoid such (unsatisfactory) situations, it is appropriate in many construction disputes for an arbitral expert or committee of arbitrators to be instructed to make a binding determination on a particular disputed factual or legal issue, with the aid of the parties.

In contrast to an arbitrator, the arbitral expert (= Schiedsgutachten) is not authorised to order a party to do, tolerate or omit to do, anything. An arbitral expert's instructions are limited to determination of disputed factual and/or individual legal issues; e.g. the technical defectiveness of works, the legal timeliness of a complaint or the fault of a design professional[133]. An arbitral expert's report differs from an ordinary expert report by being of binding effect. Both parties as well as any judge are bound by the former, unless the loser proves - whether by formal suit or otherwise submitted objections - that it was clearly incorrectly produced (substantive flaws) or breached a fundamental rule of procedure (procedural flaws)[134].

Hence, an arbitral expert's report has both preventative and dispute-resolution functions. In practice, it is not always easy to distinguish between an arbitral expert's report, an arbitral decision and an ordinary expert report, especially as the designations used by the parties in the arbitration agreement are often unclear. The will of the contracting parties and content of the arbitration agreement are the decisive factors in all cases.

An arbitral expert is very often invited onto Swiss building sites to clear up individual factual or individual issues. Especially where a number of different parties are involved (several different planners and sub-contractors, for example), an arbitral expert is generally extremely useful. This is because he can deliver a binding decision to all participants, allowing them to resume working without great loss of time.

The arbitral expert must have a very high level of professional competence, though, since the parties undertake to be bound by his decisions. The Swiss experience in this area is certainly on the positive side: all participants are aware that the report of a court expert used before the court or arbitration tribunal often forms the basis for deciding a construction dispute and thus the parties may just as well accept the decision of an arbitral expert.

3.2.4   Litigation

Disclosure and preliminary taking of evidence: Unlike Anglo-American practice, pre-trial discovery by the contracting parties is very restricted in Switzerland[135]. Generally, the parties are *not* required to disclose any evidence before the case goes to trial.[136] Once the case is pending, each party may call the other participants as witnesses and require them to produce documents. A party may be privileged and may in certain circumstances refuse to answer a question, *e.g.* if he would expose himself to criminal prosecution. The same applies to the inspection of documents.

Where the *SIA-Norm 118* is applicable, the parties are *contractually* bound to give notice of the certain relevant facts, such as:

-    Written notice of any supplements and alterations to the contract as well as of any unilateral alteration to orders (cf. art. 27 and artt. 84-91 SIA-Norm 118).

- Written notice of any "discrepancies or other defects recognized by the contractor during the execution of his work, or other situations endangering proper or timely execution of the work" (art. 365 head 3 CO; art. 25 SIA-Norm 118).

- Further obligations to report or to give written notice are provided by art. 30 heads 4 and 5 with respect to subcontractors, and by art. 136 heads 2 and 3 with respect to building material supplied by the owner.

- Written notice of extraordinary circumstances that entitle the contractor to additional remuneration (cf. art. 373 head 2 CO; art. 59 SIA-Norm 118[137].

- Notice of all dimensional documents, if measurements can no longer be taken during the progress of work (art. 142 SIA-Norm 118).

- Completion of work and the test results thereof summarized in minutes (art. 158 head 3 SIA-Norm 118).

- Expert examination and notarized findings after completion of work (art. 367 head 2 CO)[138].

In most of the provisions cited, the SIA-Norm 118 requires the parties to keep written record of those facts. By art. 25 head 2 SIA-Norm 118, "reports shall be in writing; oral reports shall be recorded".

Pre-trial discovery by the court is admitted where evidence might be destroyed or more difficult to obtain later on. In such situations, a party may be allowed to produce such evidence in a special proceeding even before the case is pending. However, the proceeding is not left to the parties. Only a court can compel a party to give evidence relevant to a future action[139]. We speak here of pre-writ discovery in the sense that the case for which evidence is admitted is not yet pending. The preliminary taking of evidence is provided for by all Cantonal civil procedure codes, with the object of securing evidence before writ. Such pre-writ discovery may be granted by the court if the petitioner can show with a reasonable probability *(with prima facie-evidence)* that the evidence will no longer be available at the trial or, if at all, only under difficult circumstances. The respondent may object on the ground that the petitioner does not have a valid interest in securing or preserving such evidence[140].

In construction disputes, the test of the usefulness or admissibility of evidence is *not a strict one,* because on construction sites the pre-writ taking of evidence is often the only way of securing any evidence for the prospective action. In urgent situations, where no expert opinion is required, a party may apply to a magistrate or a notary public to secure or preserve certain facts for the *"probatio ad perpetuam rei memoriam".* Such official reports may qualify as *public deeds.* They are usually deemed conclusive evidence unless there is proof to the contrary[141].

Interlocutory proceedings: Cantonal civil procedure codes provide not only for pre-trial discovery. On application by a party, many other orders may be obtained in *summary proceedings.* As earlier mentioned, a Cantonal court may in a proper case exercise jurisdiction founded solely on the *attachment* of the defendant's assets in the Canton, unless the defendant can argue the constitutional right to be sued in the court of his domicile. However, the plaintiff must follow up the attachment by filing an action *within ten days.* The defendant may recover damages for loss incurred through the attachment.

An interlocutory injunction may be granted if the plaintiff can show that the defendant is doing or about to do an act which a subsequent or pending suit is to prohibit permanently, and that continuance of such conduct would damage the plaintiff. If the plaintiff has not yet filed the action for a final injunction, the court will set him a time limit for doing so. The civil procedure codes do not prohibit a defendant from selling or assigning the assets in dispute. So, to secure enforcement of a judgment, it may be essential to obtain an *injunction* or a *sequestration* in order to prevent the transfer or sale of such assets (chattels, real estate, securities, etc.). Most Cantonal jurisdictions provide for orders available even before action is brought. However, there are no effective safeguards against deterioration of the defendant's financial situation even when an action is pending. Money claims can be secured only by an attachment order against specific assets of the defendant.

Under the prerequisites of art. 837 subsection 3 Swiss Civil Code *("mechanics lien")*, contractors hold a *lien on the property* for the event of not being paid by the party who has promised to pay for their labour and materials[142]. Prime contractors as well as subcontractors, suppliers, etc., are entitled to a mechanic's lien for their labour and materials that went into property improvement. In the past, lien rights have been denied to design professionals[143]. Plaintiffs suing on a mechanic's lien are subject to *many restrictive requirements*. The most important one: The contractor must file the action *within three months* of completing his work or delivering the materials (see art. 839 subsection 2 CC). Failure to comply with these requirements will invalidate the mechanic's lien.

Multi-party and multi-claim litigation: Most Cantonal procedure statutes have similar provisions on the *joinder of several parties* (plaintiffs or defendants)[144]. Plaintiffs or defendants *may* join or be joined if there is a common point of law or of fact affecting each of the parties to be joined, and if the common point arises from the same transaction or series of transactions. Moreover, the court must have a basis for jurisdiction over the particular parties. Substantive private law (e.g. CO) - rather than the Cantonal procedural codes - describes in what circumstances joinder of parties is *not at the plaintiff's option* (optional joinder), rather than what parties *must* be joined (compulsory joinder). Generally, the whole action must be dropped if joinder of parties is later found to be necessary. It will be dismissed if an omitted party is required by substantive law. Joinder has been held necessary in cases involving partnerships, joint tenancies and joint heirs. A court having jurisdiction over one party will also have jurisdiction over the necessary parties to be joined. None of the Cantons admits a class action; but several plaintiffs may file similar claims against the same defendant (e.g. the tortfeasor). However, it is in the discretion of the court whether or not to hear the actions together[145].

Third party practice differs from canton to canton. Generally, the defendant may request a third party to support him in court. The request is by mere notice, and the third party is neither required nor expected to serve an answer. He may not raise a counterclaim or cross-claims. His function is solely to support the defendant. The third party may or may not join the defendant. In any event, the court will give judgment against the defendant only. A fresh action will be necessary between defendant and third party. However, if the defendant served the third party with timely notice, a judgment against the defendant will be binding in an action between defendant and third party.

Generally, a party asserting a claim may join the plaintiff and may raise any other claim he wishes against the defendant. Additional claims usually raise no problems of jurisdiction as to subject matter, as the joinder of claims merely affects the sum in dispute. Most civil procedure codes (as well as art. 8 PIL "in international matters") permit a counterclaim to be raised on any claim "arising on the contract or transaction set forth in the complaint or connected with the subject matter of the action". Most codes still follow the rule that only claims related to the plaintiff's claim may be raised in counterclaim by the defendant[146]. It is usually within the defendant's discretion whether or not to raise a counterclaim. However, if the counterclaim arises on the same transaction or occurrence as the plaintiff's claim, failure to raise it may result in loss of the claim in future litigation. A later suit may be barred by the *rules of res judicata.*

Trial: Commercial and other civil law matters - such as construction disputes - will *not* be tried before a *jury* except in some jurisdictions if the case involves criminal aspects. In the district or commercial courts, cases will typically be heard before several (usually three or five) judges, viz. one (the presiding judge) *trained in law,* and two or four lay judges having *specialist qualifications* (*e.g.* in construction). The court has power to try all the issues. Summary proceedings are usually held before a single judge.

Judgment may be found by voting. Each (professional or lay) judge has one vote. Usually, one judge will sum up the case before the court.[147] The same judge will also give the reasoning. However, he will not sign the reasoning, nor will the other judges write a dissenting opinion. At the end of the judgment document, there is invariably a summary of the judgment in the form of numbered court orders (termed "Dispositiv"); that summary is most important in several respects (*e.g.* enforcement, res judicata).

Res judicata and collateral estoppel ("barring of issue") apply when there has been a final judgment on the merits by a court of proper jurisdiction in an earlier action between the same parties on the same cause of action. However, collateral estoppel will only apply if the recalled issue was decided in the *judgment summary ("Dispositiv")*. The mere fact that an issue was argued in the reasoning will not estop a party from having it retried.

Once an issue is decided in the judgment summary (*e.g.* that defendant pay a specified sum), the court may adduce the reasoning to the earlier judgment to ascertain whether the same issue is being retried. Whether or not the issue is based on the same cause of action is considered a point of Federal law and may therefore be heard by the Federal Court on appeal.

Venue: In construction, the parties often elect their venue *by pre-suit agreement.* The forum clause will generally apply. Yet such an agreement will not be enforced if the parties have no discretion on the point, or if statute law stipulates a specific venue or rules out any pre-suit agreement[148].

Forum clauses naming the jurisdiction of a Swiss court are very common. In an international context, non-Swiss contracting parties should bear in mind that their agreement includes clauses electing *Swiss jurisdiction as well as Swiss law:* By art. 5 PIL, Swiss courts cannot refuse a suit filed by a non-Swiss party for want of jurisdiction, provided that the dispute is subject to Swiss law, or that Swiss law has been elected by the contracting parties[149].

Art. 5 PIL requires that a pre-suit agreement electing the forum must be in writing to be enforceable. If the forum clause occurs in a standard-form contract, it should be suitably highlighted, and if the defendant has no or little experience of business, the plaintiff must have explained the meaning of the forum clause to him.[150]

Those new provisions (of PIL and/or Lugano Convention) now overrule in many respects the relevant provisions of the Cantonal and Federal procedure codes. As a result, a contractual election of Swiss law is sufficient ground for accepting a Swiss forum clause, *e.g.* even if the construction site is in Eastern Europe or in South America and the non-Swiss parties have no connections with Switzerland.

This contractual choice of jurisdiction and of law - i.e. to elect a Swiss forum and to include a governing-law clause that provides for the applicability of Swiss law - may be an interesting alternative under given circumstances, namely to contractors. However, it repeatedly occurred that foreign contract parties chose Swiss law not having exact knowledge of some (in their view rather surprising) statutory provision such as art. 373, section 2 CO, which under certain (extraordinary) circumstances may give the contractor legal basis to increase the price or to terminate the contract although there were stipulated fixed prices (lump sum or unit prices) in the contract[151].

Motion practice: In general, Swiss courts are not authorized to dismiss a case by summary judgment. In contrast with Anglo-American practice, Swiss courts are regularly bound to try a case *on the merits.* However, there are several motions by which a defendant, by way of preliminary answer, may raise one or more grounds for dismissing the case without trial, viz. -

-    *Want of jurisdiction (in rem or quasi in rem jurisdiction).* This objection should be raised before filing the full answer.

-    *Lis pendens or res judicata.* These objections that another court is or was trying a case with identical issues should likewise be raised before filing the full answer.

-    *Want of jurisdiction as to subject matter* (case for commercial court or arbitration instead of district court).

-    *Motion to answer on restricted points,* such as whether or not the action is barred by the *statute of limitations,* etc.

Even in jurisdictions where such (or other) motions are admissible, there are only few occasions for a preliminary answer.[152] In some Cantons the defendant must file a full answer and argue such motions therein. Persuant to Procedure Codes in other Cantons, courts may narrow the pleadings to certain controversial issues[153]

Recognition and Enforcement of judgments: By art. 61 of the Swiss Federal Constitution, *judgments of one Canton* are enforceable in all the others and have res judicata-effect if the judgment is *final* and has been rendered by a *competent court.* Treaties, if any, will settle the point whether *foreign judgments* in private law matters are recognized and enforceable[154]. If there is no international treaty, Federal law rather than Cantonal law will govern the recognition and the enforcement of such judgments[155]. The relevant provisions of the PIL (art. 25-27) require that the judgment is *final,* that the foreign court had *jurisdiction* to render the judgment, that the defendant was *properly served* with process, that enforcement does not violate the *"ordre public",* and that the foreign country *likewise* enforces Swiss judgments.

Abbreviations

ASA	Swiss Arbitration Association
BGE	Decisions of the Swiss Federal Court (Official Reports)
CC	Swiss Civil Code of December 10, 1907 (as amended)
CO	Swis Code of Obligations of March 30, 1911 (as amended)
Fed. Ct	Federal Court
IAC	Inter-Cantonal Arbitrtion convention of March 27, 1969 (as amended)
PIL	Swiss Federal Act on Private International Law of December 18, 1987
Pra	Praxis of Decisions of the Swiss Federal Court (Helbling & Lichthahn

1   Building and construction accounts for some 15 % of gross
       domestic product (GDP) making them a major factor in the
       economy as a whole. Around 400,000 people are employed in the
       sectors of building, planning, and suppliers. Construction activity
       also has a considerable influence on other parts of the economy
       significantly affecting the living standard of the Swiss population.

2   To work contracts in general: GAUCH, Der Werkvertrag, 4th
       Edition Zurich 1995; GAUTSCHI, Berner Kommentar, Der
       Werkvertrag, Kommentar zu Art. 363 - 379 OR;
       HONSELL/ZINDEL/PULVER, Komm. zum Schweiz. Privatrecht,
       Der Werkvertrag, 2nd Edition, Zurich 1996; KOLLER, Das
       Nachbesserungsrecht im Werkvertrag, Zürich 1995; PEDRAZZINI,
       Der Werkvertrag; TERCIER, La partie spéciale, pp. 230 seqq.

3   See esp. GAUCH, Der Werkvertrag. Contracts with architects and
       engineers may be governed by the provision of artt. 363 - 379 CO
       as well: See: SCHUMACHER, Die Haftung des Architekten aus
       Vertrag, pp. 105 seqq.

4   Cf. art. 64 I and II of Swiss Federal Constitution.

5   Switzerland's constituent states are called Cantons. There are 23
       Cantons, three of which are subdivided into "Half-Cantons".

6   Cf. art. 64 (III) of Swiss Federal Constitution of May 29, 1874
       (as amended).

7   Switzerland's population is about 6.5 million.

8   The fact that the Cantons are very small (some having only a few
       thousand inhabitants) may be one main reason why law firms in
       Switzerland are generally small. A law firm of five to ten
       attorneys is considered a large one (in the Zurich area, a midsized
       one). Only in the Zurich and Geneva areas will some law firms
       have twenty and more attorneys.

9   See a survey in: VOGEL, Zivilprozessrecht, 2 N 11 f.; WALDER-
       BOHNER, pp. 27 seqq.

10   Swiss Federal Code of Civil Procedure of December 4, 1947 (as
       amended).

11   BERTI, Zum Einfluss ungeschriebenem Bundesrechts, pp. 12
       seqq.

12   In this case the plaintiff usually has the choice between the district
      court and the commercial court.

13   The meaning of merchant in this context is very broad. A person
      is considered a merchant if registered in the commercial register.
      As construction companies are usually registered and the sum
      involved in construction suits often exceeds the jurisdictional
      minimum, many construction disputes are handled by the
      commercial courts.

14   See the specific requirements: VOGEL, Zivilprozessrecht, 3 N 65
      seqq.; WALDER-BOHNER, pp. 75 seqq.; see the specific
      requirements for all proceedings at Swiss Federal Court: VOGEL,
      Zivilprozessrecht, 13 N 105 seqq.

15   Service of process is not in private hands, but effected by the
      courts.

16   If the commercial courts have jurisdiction as to subject matter, in
      some cantons, the plaintiff files the complaint directly with the
      court (e.g. in the Canton of Aargau; but not in the Canton of
      Zurich).

17   Several exceptions apply, e.g. when a party's domicile is not
      within the jurisdiction.

18   Once the case is pending in the district court, in most cantons a
      case is sub judice and must be decided by a court unless the
      parties settle the dispute. The notice granting leave must be filed
      within three months after its issuance.

19   It is very important to note that the plaintiff cannot commence an
      action by summons with notice or the like. The plaintiff must have
      the facts and evidence ready before going to court. To the civil
      procedure rules in the Canton of Zurich: STRÄULI/MESSMER,
      ZPO; WALDER-BOHNER, Zivilprozessrecht.

20   A mere entry of appearance would not be considered an answer.

21   Hence the plaintiff can file a replication ("Replik"); the defendant
      a rejoinder ("Duplik").

22   See for example the discovery rules of the Canton of Zurich,
      supra 3.2.4.

23   Swiss Federal Act on Private International Law of December 18,
      1987. Previously, the Swiss statutory rules applicable to

international arbitrations were those of the *Intercantonal Arbitration Convention (IAC)* of March 27, 1969 (as amended), which is primarily designed to the needs of domestic arbitration but is often applied to international disputes, too.

24   See esp. WALTER / BOSCH / BRÖNNIMANN, Internationale Schiedsgerichtsbarkeit in der Schweiz, pp. 32 seqq. and all the authors cited therein; BLESSING, The New International Arbitration Law in Switzerland, pp. 5 seqq.; HÜRLIMANN, International Arbitration in Switzerland, pp. 8 seqq.

25   Art. 167 PIL describes the *scope of application* as follows: (1) The provisions of this chapter shall apply to arbitrations if the seat of the arbitral tribunal is in Switzerland and if at least one of the parties at the time the arbitration agreement was concluded was neither domiciled nor habitually resident in Switzerland. (2) The provisions of this chapter shall not apply if the parties have excluded its application in writing and agreed to the exclusive application of the cantonal rules of procedures concerning arbitration. (3) The arbitrators shall determine the seat of the arbitral tribunal if the parties or the arbitration institution designated by them fail to do so.

26   See: ASA Special Series N° December 1994, The Arbitration Agreement, its multifold critical aspects.

27   Art. 182 para. 1: *"The parties may, directly or by reference to arbitration rules, determine the arbitral procedure; they may also submit it to a procedural law of their choice."*

28   Art. 182 para. 2: *"Where the parties have not determined the procedure, the arbitral tribunal shall determine it to the extent necessary, either directly or by reference to a law or to arbitration rules."* Unlike art. 24 (2) of IAC, PIL makes no provision for any subsidiary procedural law to be applied in default of party agreement, and so the tribunal is bound in default of agreed procedural rules to set up such rules "to the extent necessary".

29   HÜRLIMANN, International Arbitration in Switzerland, p. 13.

30   The "SIA-Richtlinie 150" on arbitration ("Richtlinie für das Verfahren vor einem Schiedsgericht") is issued by the Swiss

Society of Architects and Engineers. This directive is based on the IAC and the Federal Code of Civil Procedure. See: INDERKUM, Zur Schiedsgerichtsbarkeit des SIA, pp. 187 seqq.

31 See FELIX HUBER, Der GU-Vertrag, Zurich 1996.

32 BGE 117 IV 168.

33 GAUCH, Der Werkvertrag, N 283 seq.; TERCIER, La loi, les normes et leur complément, in: Baurecht 1983, pp. 63.

34 BGE 109 II 217 and BGE 109 II 456 ff.; JÄGGI/GAUCH, Zürcher Kommentar, Zurich 1986, N 472 zu Art. 18 OR.

35 See GAUCH, Der Werkvertrag, Nr. 300 seqq.

36 See HÜRLIMANN, Subcontracting in Switzerland - national report, 8 ICLR, 1991, pp. 152 seqq.

37 See SCHWAGER, Die Vollmacht des Architekten, in: Das Architektenrecht, 3rd Edition, Zürich 1995, pp. 253-294.

38 See in general: GAUCH, Kommentar zur SIA-Norm 118, Zurich 1992, pp. 41 seqq.

39 See BGE 113 II 513; 109 II 335; 104 II 315; see also: GAUCH, Price increases or contract termination in the event of extraordinary circumstances, 4 ICLR 1987, pp. 261 seqq.; MERONI, Sub-Surface ground conditions - risks and pitfalls for project participants, 7 ICLR 1990, pp. 198 seqq.

40 See GAUCH/EGLI, Kommentar zur SIA-Norm 118, Zurich 1992, pp. 208 seqq.

41 See in general: GAUCH/SCHUMACHER, Kommentar zur SIA-Norm 118, pp. 252 seqq.

42 See: WERZ, Delay in construction contracts, Fribourg 1994.

43 See in general: GAUCH, Kommentar zur SIA-Norm 118, Zürich 1991, pp. 33 seqq.

44 Arbeitsgruppe der Schweiz. Bauwirtschaftskonferenz = GIB.

45 These documents ("Streitschlichtung") are available at Schweizerischer Baumeisterverband, Weinbergstrasse 49, 8035 Zürich, Fax 01/258 83 35.

46 A similar project - based on the GIB-model - was worked out by Swiss Association of Highway Professionals (= Vereinigung Schweizerischer Strassenfachleute). Address: VSS, Seefeldstrasse 9, 8008 Zürich, Fax ++ 41 1 252 31 30,

47   Schweizerischer Anwaltsverband, Bollwerk 21, 3001 Bern, Fax:
        ++41 31 312 31 03.

48   More informations can be received by Mediartis AG, Society for
        Mediation and Conflict Management, Fraumünsterstrasse 19,
        8001 Zürich, Fax: ++41 1 218 60 33.

49   Partner with Bär & Karrer, Zurich/Lugano, former president of
        ASA. This Report was published in ASA 2/1996, pp. 123-189:
        "Streitbeilegung durch ADR und pro-aktive
        Verhandlungsführung".

50   Partner with Lalive & Partners, Geneva, correspondent of the
        International Construction Law Review. This report was
        published by the Institut for Swiss and International Construction
        Law, Fribourg 1997: "La résolution des littiges dans la
        construction", pp. 28-56.

51   Prof. Dr. Peter Gauch is professor in civil, commercial and
        construction law at Fribourg Law School, University of Fribourg.
        Peter Gauch is president of the Swiss Society of Construction
        Law and is heading the editorial committee of the Institute of the
        Swiss Construction Law Journal (Baurecht/Droit de la
        Construction), Address: Chemin des Grenadiers 2, 1700
        Fribourg, Fax: ++41 26 300 97 20.

52   See HENNINGER, Die Qualitätssicherung beim Bauen,
        Baurechtstagung 1995, pp. 45-70.

53   See introduction to SIA-2007 and SIA-documentation D0102.
        HENNINGER, loc. cit.

54   See criteria of prequalification under VoeB (Verordnung für
        öffentliches Beschaffungswesen of December 11, 1995,
        especially appendix 3, para. 10 (certificate of quality management
        system). To the public procurement laws in Switzerland in
        general:HÄUSSLER/HÜRLIMANN/MEYER-MARSILIUS, Oeffentliche
        Bauaufträge in der Schweiz: Einfluss des EU-Rechts, Zürich
        1997; GALLI/LEHMANN/RECHSTEINER, Das öffentliche
        Beschaffungswesen in der Schweiz, Zürich 1996.

55   See BLESSING, Einführung in die Schiedsgerichtsbarkeit, p. 139.

56   See BLESSING, Streitbeilegung durch ADR und pro-aktive Verhandlungsführung ASA 2/1996, pp. 123 seqq., esp. pp. 150 seq.

57   The Commercial Court of Zurich (Handelsgericht) is famous for their quite successful attempts to bring the parties to an agreement. In the so called "Referentenaudienz" - a court hearing in the presence of the parties and their counsels - the Handelsgericht settles approximatly 70 % - 80 % of all pending cases.

58   According to research by DERENDINGER (p. 175), such proceedings are used relatively infrequently.

59   See: GAUCH, Aussergerichtliche Streiterledigung, p. 3 FN 6; BLESSING, Streitbeilegung durch ADR und pro-aktive Verhandlungsführung, p. 144.

60   See EIHOLZER, Die Streitbeilegungsklausel.

61   See GAUCH, Aussergerichtliche Streiterledigung, p.3/4.

62   See DERENDINGER, Alternative Methoden zur Beilegung von Baurechtsstreitigkeiten, pp. 155 and cited authors therein.

63   EIHOLZER, Die Streitbeilegungsabrede. Ein Beitrag zu alternativen Formen der Streitbeilegung, namentlich Mediation, thesis Fribourg 1997.

64   BLESSING, Streitbeilegung durch ADR und pro-aktive Verhandlungsführung, ASA 2/1996, pp. 123-189.

65   In July 1994, the *Swiss Arbitration Association* ("ASA") issued a brochure introducing all ASA members with biographical information. Address: ASA secretariat, St. Alban-Graben 8, 4001 Basel. Fax: ++41 61 272 80 60. ASA is a non-profit association consisting of more than 650 individual members (where of about 125 being resident outside of Switzerland) and about 30 corporate members. ASA is not an arbitral institution hosting or administrating international arbitral proceedings. On purpose ASA has not promulgated arbitration rules of its own. ASA-membership is open to all those interested in international arbitration, irrespective of nationality. ASA publishes a quarterly bulletin containing current information on arbitration matters,

materials, arbitral awards, court decisions rendered in the field of arbitration.

66   BLESSING, loc. cit.

67   When the particular publications mentioned are reviewed, Switzerland emerges as the preferred venue at any rate for ICC arbitrations (along with Paris) (cf. SIGVART JARVIN, "From the Practice of the ICC Arbitral Tribunal", published in KARL-HEINZ BÖCKSTIEGEL, "Law and Practice of the International Chamber of Commerce as Arbitral Tribunal (1986)", pp. 7 ff. On the frequency of arbitration in Switzerland as compared with other countries, see BUCHER, "Switzerland as Traditional Venue", pp. 119 f.

68   Switzerland's reputation for being a traditionally "ideal" venue for arbitration had suffered but little from the fact that the organization of its courts and the judicial and arbitral procedures are basically a matter not of the Federal administration, but rather of the Cantons, under the country's constitutional division of powers. However, on 27 March 1969, with a view to harmonizing procedural law, a supra-Cantonal arrangement was reached, the *Inter-Cantonal Arbitration Convention ("IAC")*.

69   WALTER/BOSCH/BRÖNNIMANN, pp. 39 ff.

70   See the requirements in BGE 115 II 390 ff. and in 116 II 721; further, LALIVE/POUDRET/REYMOND, p. 300; VISCHER, in: IPRG-Kommentar, p. 1496.

71   See BGE 116 II 721.

72   WALTER/BOSCH/BRÖNNIMANN, p. 40. An arbitration clause such as "In case of dispute arbitration in Switzerland" is best avoided (cf. BLESSING, Das neue Int. Schiedsgerichtsrecht, p. 29).

73   BUCHER E., Die Schweiz als traditioneller Sitzort, p. 122.

74   See BGE 118 II 353. HOFFET, Fragen der subj. und obj. Schiedsfähigkeit, pp. 252.

75   On the importance and practical significance of that collision-law solution, see WENGER, Schiedsvereinbarung und schiedsgerichtliche Zuständigkeit, pp. 225; VOLKEN, in: IPRG-Kommentar, pp. 1511; BLESSING, Das neue Int. Schiedsgerichtsrecht, p. 40. Drafting the arbitration clause, it is

generally speaking, "the simpler the clause the better" (KAUFMANN/BYRNE-SUTTON/SCHENKER, p. 64; Multi-party arbitrations - as they often take place in construction law cases - may require a more detailled agreement (see: KARRER, Multi-party and complex Arbitration under the Zurich Rules, Festgabe Juristentag 1994, Aspekte des Wirtschaftsrechts, pp. 261-271; WIRTH, Die Reglemente der massgeblichen Schiedsinstitutionen, pp. 197. See BGE 119 II 380: Validity of an arbitration clause agreed upon under the influence of bribery.).

76   BLESSING, p. 39; further, Fed Ct in BGE 111 II 253 ff. and 110 II 54 ff.

77   Applicable are the General Provisions of Contract Law on tender and acceptance (cf. JÄGGI/GAUCH, nn. 469 ff.; GAUCH/SCHLUEP, nn. 1132 ff.).

78   Fed Ct in Pra 1990, p. 536; BLESSING, pp. 42 ff.; WALTER/BOSCH/BRÖNNIMANN, pp. 71 ff. According to those authors, by s. 27 para. 2 of the Swiss Civil Code the arbitration clause may relate to future disputes as well, insofar as the underlying contract is sufficiently defined.

79   See HÜRLIMANN, Teilnichtigkeit von Schuldverträgen, No. 242, p. 70; RÜEDE/HADENFELDT, pp. 71 ff. On the scope of arbitral jurisdiction and on the point whether the scope of an arbitration agreement should be interpreted liberally (e.g. to include set-off claims or a counterclaim, cf. Fed Ct in BGE Pra 1990, p. 536; WALTER/BOSCH/BRÖNNIMANN, pp. 73 ff.; further, BUCHER E., pp. 136 ff.).

80   See art. 190 para 2 head a PIL (ch. 3/a supra). Fed Ct in Pra 1995, Nr. 258: An arbitration clause in which a minimum of two arbitrators are to be appointed, is valid but needs interpretation.

81   Art. 179 para. 3: "Where a court is called upon to appoint an arbitrator, it shall make the appointment, unless a summary examination shows that no arbitration agreement exists between the parties."

82   Pro: BLESSING, Das neue Internationale Schiedsgerichtsrecht, p. 45; contra: WALTER/BOSCH/BRÖNNIMANN, p. 106; further WENGER, 117/118; WALDER, Internationale

Schiedsgerichtsbarkeit und Konkordat, pp. 376; see further BGE 115 II 294; 118 Ia 20.

83   BGE 115 II 294: That restriction does not adversely affect the parties, as both parties may raise any irregularities of the appointment before the arbitral tribunal, and may thereafter appeal from the tribunal's decision under art. 190 para. 2.a and b, and para. 3.

84   Art. 180: "An arbitrator may be challenged if he does not meet the requirements agreed by the parties, or if the arbitration rules agreed by the parties provide a ground for challenge, or if circumstances exist that give rise to justifiable doubts as to his independence.
A party may challenge an arbitrator whom it has appointed or in whose appointment it has participated only on grounds of which it became aware after such appointment. The ground for challenge must be notified to the arbitral tribunal and to the other party without delay.
In the event of a dispute and to the extent to which the parties have not determined the procedure for the challenge, the court of the seat of the arbitral tribunal shall decide; its decision is final."

85   See A. BUCHER, nn. 167-170.

86   WALTER/BOSCH/BRÖNNIMANN, p. 111.

87   WALTER, p. 169, referring to art. 190 para 2 head b.

88   See in detail: SCHNEIDER, in HONSELL/VOGT/SCHNYDER, pp. 1504.

89   Art. 182 para. 1: "The parties may, directly or by reference to arbitration rules, determine the arbitral procedure; they may also submit it to a procedural law of their choice."

90   Art. 182 para. 2: "Where the parties have not determined the procedure, the arbitral tribunal shall determine it to the extent necessary, either directly or by reference to a law or to arbitration rules."

91   Art. 182 para. 3: "Whatever procedure is chosen, the arbitral tribunal shall ensure equal treatment of the parties and the right of the parties to be heard in an adversarial procedure."

92   Art. 183: "Unless the parties have agreed otherwise, the arbitral tribunal may, at the request of a party, order provisional or protective measures.
      If the party so ordered does not comply therewith voluntarily, the arbitral tribunal may request the assistance of the competent court. Such court shall apply its own law.
      The arbitral tribunal or the court may make the granting of provisional or protective measures subject to the provision of appropriate security."

93   VOGEL, p. 95; WALTER/BOSCH/BRÖNNIMANN, p. 130, BERTI, in HONSELL/VOGT/SCHNYDER, pp. 1537.

94   BLESSING, Das neue Int. Schiedsgerichtsrecht, p. 56; BUCHER, n. 24; HABSCHEID, p. 135; WALTER, p. 239; WALTER/BOSCH/BRÖNNIMANN, p. 130.

95   WALTER/BOSCH/BRÖNNIMANN, pp. 144 ff.

96   According to LALIVE/POUDRET/REYMOND, No. 9 to art. 183, that parallel power follows from art. 10 PIL. The coordination between arbitrators and ordinary court's authorities is best described in: REYMOND, pp. 113 ff.

97   Art. 184: "The arbitral tribunal shall itself take the evidence. Where the assistance of state authorities is needed for taking evidence, the arbitral tribunal or a party with the consent of the arbitral tribunal may request the assistance of the court of the seat of the arbitral tribunal. Such court shall apply its own law."

98   Art. 185: "For any further judicial assistance the court of the seat of the arbitral tribunal shall have jurisdiction."

99   Art. 181: "The arbitral proceedings shall be pending from the time when one of the parties submits its request to the arbitrator or arbitrators designated in the arbitration agreement or, in the absence of such designation, from the time when one of the parties initiates the procedure for the constitution of the arbitral tribunal."

100   Art. 186: "The arbitral tribunal shall decide on its own jurisdiction.
      Any objection to its jurisdiction must be raised prior to any defence on the merits.

The arbitral tribunal shall, in general, decide on its jurisdiction by a preliminary decision."

101   Art. 187: "The arbitral tribunal shall decide the dispute according to the rules of law chosen by the parties or, in the absence of such a choice, according to the rules of law with which the case has the closest connection.
The parties may authorize the arbitral tribunal to decide ex aequo et bono."

102   See WALTER/BOSCH/BRÖNNIMANN, pp. 185 ff., and authors there in cited; cf. also BLESSING, Das neue Int. Schiedsgerichtsrecht pp. 61 ff. and KARRER, in HONSELL/VOGT/SCHNYDER, pp. 1602.

103   SCHNYDER, IPR-Gesetz, p. 147.

104   Fed Ct in BGE 116 II 80 and 115 II 282.

105   Fed Ct in BGE 115 II 292, and reference made to LALIVE/POUDRET/REYMOND, p. 431; BLESSING, p. 77.

106   Fed Ct in BGE 116 II 85/86.

107   SCHNYDER, Intertemporalrecht, pp. 60 ff.

108   See: BGE 115 II 288 and 116 II 80.

109   Art. 190: (The award is final from the time when it is communicated.) Proceedings for setting aside the award may only be initiated: a. Where the sole arbitrator has been incorrectly appointed or where the arbitral tribunal has been incorrectly constituted; b. Where the arbitral tribunal has wrongly declared itself to have or not to have jurisdiction; c. Where the award has gone beyond the claims submitted to the arbitral tribunal, or failed to decide one of the claims; d. Where the principle of equal treatment of the parties or their right to be heard in adversarial procedure had not been observed; e. Where the award is incompatible with public policy.
As regards preliminary decisions, setting aside proceedings can only be initiated on the grounds of the above paragraphs 2 (a) and 2 (b); the time-limit runs from the communication of the decision."

110   See BGE 118 II 359.

111   cf. A. BUCHER, N 340; BRINER, p. 104;
       WALTER/BOSCH/BRÖNNIMANN, pp. 215 ff.; HEINI, in: IPRG-
       Kommentar, pp. 1580.

112   See: BGE 117 II 94; BGE 120 II 155. Further 118 II 193.

113   BGE 116 II 639; 120 II 172. WALTER/BOSCH/BRÖNNIMANN, pp.
       219 ff.

114   As was ruled by the Federal Court in BGE 116 II 639 in pointing
       out slight discrepancies of wording between the German- and
       Italian-language versions of the Act and the French version

115   e.g. want of order as to costs, non-decision of motions for interest;
       cf. Federal Court precedent 116 II 373

116   BGE 116 II 643; 119 II 386.

117   See BGE 116 II 643; 118 II 357; 119 II 386. In BGE 117 II 346,
       the Federal Court held that a violation of procedural rules to
       which the parties agreed upon in their "règlement d'arbitrage" is
       not sufficient ground to challenge an arbitral award as long as the
       rule of equal treatment and the right of legal hearing is breached.

118   The ground under art. 36 IAC which was regularly raised prior to
       PIL taking effect, namely, that the award was *arbitrary* because it
       rested on findings obviously contrary to the record, or because it
       involved an obvious violation of law (or of equitability), is not
       now relevant under art. 190 para. 2 PIL. In omitting the ground of
       arbitrariness, the legislator deliberately sought to restrict the
       scope for appeal.

119   See BGE 120 II 155; HABSCHEID, pp. 575 ff.; BRINER, p. 105.

120   See BGE 109 Ib 235; 103 Ia 204.

121   cf. Briner, p. 105; WALTER/BOSCH/BRÖNNIMANN, pp. 225 ff.;
       BERTI, Rechtsmittel gegen Schiedsentscheide nach IPRG, pp.349;
       BERTI/SCHNYDER, in: HONSELL/VOGT/SCHNYDER, pp. 1679.
       Further Fed Ct in BGE 116 II 373 and 116 II 634. Finally: BGE
       119 II 380.

122   BGE 116 II 373

123   BGE 116 II 634.

124   BGE 116 II 634.

125   Art. 191: "Setting aside proceedings may only be brought before
       the Federal Supreme Court. The procedure is governed by the

provisions of the Federal Judicial Organisation Act relating to public law appeals.

However, the parties may agree that the court of the seat of the arbitral tribunal shall decide in lieu of the Federal Supreme Court; its decision is final. For this purpose the cantons shall designate a sole cantonal court."

126   cf. VOGEL, p. 315; WALTER, p. 250; critically, BRINER, p. 108; BLESSING, p.83.

127   s. 89, Judicature Act; art. 190 paras. 1 and 3 PIL.

128   s. 33, Judicature Act.

129   Art. 192: Where none of the parties has its domicile, its habitual residence, or a business establishment in Switzerland, they may, by an express statement in the arbitration agreement or by a subsequent agreement in writing, exclude all setting aside proceedings, or they may limit such proceedings to one or several of the grounds listed in Article 190, paragraph 2.

Where the parties have excluded all setting aside proceedings and where the award are to be enforced in Switzerland, the New York Convention of 10th June, 1958 on the Recognition and Enforcement of Foreign Arbitral Awards shall apply by analogy.

130   cf. BLESSING, Das neue Int. Schiesgerichtsrecht, p. 83; WALTER/BOSCH/BRÖNNIMANN, pp. 253 ff.; SIEHR, in: IPRG-Kommentar, pp. 1601.

131   BGE 116 II 639; BLESSING, p. 84

132   Art. 193 (XI. Deposit and Certificate of Enforceability) Each party may at its own expense deposit a copy of the award with the Swiss court of the seat of the arbitral tribunal. At the request of a party, that court shall certify the enforceability of the award. At the request of a party, the arbitral tribunal shall certify that the award has been made in conformity with the provisions of this Act; such certivicate has the same effect as the deposit of the award.

Art. 194   (XII. Foreign Arbitral Awards) The Recognition and enforcement of a foreign arbitral award is governed by the New York Convention of 10th June, 1958 on the Recognition and Enforcement of Foreign Arbitral Awards.

133   See HÜRLIMANN, Das Schiedsgutachten als Weg zur
       aussergerichtlichen Beilegung von Baustreitigkeiten, Baurecht
       1992, pp. 108; see the characteristics described by Fd. Court in
       BGE 117 Ia 365.

134   See § 258 Code of Civil Procedure of Canton Zurich.

135   See: HÜRLIMANN, Der Experte - Schlüsselfigur des
       Bauprozesses, p. 105; HÜRLIMANN, Der Architekt als Experte,
       pp. 429. The same is true for most European countries:
       HABSCHEID/BERTI, pp. 48 seqq.

136   There are only very few exceptions provided by statute.

137   GAUCH, Price increases or contract termination in the event of
       extraordinary circumstances 4 ICLR 1987, pp. 265 seqq.;
       MERONI, Sub-surface ground conditions, risks and pitfalls for
       project participants; 7 ICLR 1990, pp. 198 seqq.

138   VOGEL, Dispute and settlement of disputes, Baurechtstagung
       1985, Tagungsunterlagen 1, pp. 71 seqq.; HÜRLIMANN, Der
       Experte, pp. 105 seqq.

139   The duty to produce documents or to testify may also be on a
       third party.

140   HÜRLIMANN, Der Experte, p. 106 seqq.; SCHUMACHER,
       Beweisprobleme im Bauprozess, pp. 203 seqq.

141   Art. 9 Civil Code; HÜRLIMANN, loc. cit. p. 107.

142   See art. 837 subsection 3 CC; as to the mechanic's lien under
       Swiss Law: SCHUMACHER, Das Bauhandwerkerpfandrecht, 2nd
       ed., Zürich 1982.

143   SCHUMACHER, loc. cit., N 271 seqq.

144   See: VOGEL, Zivilprozessrecht, 5 N 45 seqq.; GULDENER, p. 297;
       STRÄULI/MESSMER, § 46, N 1 seqq.

145   See: § 58 Civil Procedure of Canton of Zurich.

146   VOGEL, Zivilprozessrecht, 7 N 53 seqq. See § 60 of Civil Procede
       Code of Canton of Zurich.

147   Normally the case is deliberated and decided in camera. Only
       appeals may be deliberated and decided in the presence of the
       parties.

148   See REISER, Gerichtsstandsvereinbarungen nach dem IPR-
       Gesetz, pp. 97 seqq.; HÜRLIMANN, Gerichtsstandsklauseln unter

Berücksichtigung des neuen IPRG, pp. 94 seqq. *Art. 59* of the
Swiss Federal Constitution provides that generally a defendant
domiciled or permanently resident in Switzerland may only be
sued in actions in personam in a court at his domicile. In
accordance with this jurisdictional rule, art. 112 PIL provides that
the courts of the defendant's domicile or other places of business
(location of a branch office, etc.) in the forum Canton have
jurisdiction for an action in contract. There are many exceptions
to this rule; thus, art. 59 does not apply when the action is in rem,
etc.

On the other hand, the mere fact that a defendant has no residence
in Switzerland does not mean that he can be sued in any court. As
art. 59 is confined to limiting the jurisdiction of the Cantons, it
does not entitle the plaintiff to elect the venue arbitrarily. Yet the
Cantons do not offer so wide a basis for in personam jurisdiction
as might be found in countries with a "long-arm" facility of this
jurisdiction.

Construction companies having no domicile in Switzerland are
not amenable to this process simply for doing business in
Switzerland unless there are some legally relevant connections.
For instance, if the defendant has neither a domicile nor a branch
office in Switzerland, the plaintiff may sue at the Swiss place of
performance (cf. art. 113 PIL). For actions in tort, the venue is
generally the defendant's domicile or place of business (art. 129
PIL); if the tortfeasor has no domicile or place of business in
Switzerland, the plaintiff may bring the action at the place where
the tortious act took place or took effect (art. 129 head 2 PIL).

If the defendant is not protected by *art. 59 of the Swiss Federal
Constitution* (*e.g.* not resident in Switzerland), he may only be
sued upon attachment of his assets within the jurisdiction (cf. art.
4 PIL). Thus, a creditor can sue a debtor who has no domicile in
Switzerland148 but whose assets have been attached, even if the
action has no direct connection with Switzerland. In other words,
unless a specific jurisdiction has been elected by the parties,
attachment may create a venue for claims against non-resident
debtors (termed "forum arresti"). The creditor may sue at the

forum of attachment only if PIL does not stipulate some other forum in Switzerland. Within the scope of the *Lugano Convention,* the forum of attachment is abolished with respect to defendants having their domicile in member countries.

149   HÜRLIMANN, Gerichtsstandsklauseln unter Berücksichtigung des neuen IPRG, pp. 102 ff.; VOGEL, Zivilprozessrecht, 4 N 85; REISER, Gerichtsstandsvereinbarungen nach dem IPR-Gesetz, pp. 97 seqq.

150   To the formal requirements: BGE 109 Ia 57 seqq.; BGE 104 Ia 278. As by such an agreement, the defendant may waive his constitutional right to be sued in the court at his domicile, the Swiss Federal Court held that the agreement must be precise on its face.

151   See GAUCH, "Price Increases or Contract Termination", 4 ICLR 1987, pp. 265.

152   See for the Canton of Zurich: STRÄULI/MESSMER, § 108 N 1 seqq.; § 111, N 1 seqq.

153   See art. 189 Civil Procedure Code of the Canton of Zurich.

154   There are numerous international treaties.

155   There are few exeptions: SCHNYDER, Das neue IPR-Gesetz, pp. 36 seqq.

U.S.A.

Stuart Nash
Freshfields, London, England
Douglas L Patin
Spriggs & Hollingsworth,
Washington DC, USA
Ralph C Nash
Professor Emeritus of Law,
The George Washington University, Washington DC, USA

1.   BACKGROUND

1.1   Economic

The total "value of new construction put in place" of the national construction industry in 1996 as measured by the US Department of Commerce, Bureau of the Census was $585 billion.[1] The "value of new construction put in place" is a measure of the value of construction installed or erected at the site during a given period. The Gross Domestic Product (GDP) in 1996 as measured by the Department of Commerce, Bureau of Economic Analysis was $7,636 billion.[2] Thus, the percentage the value of construction put in place to the GDP was approximately 7.7%. The Census Bureau has estimated that the value of construction put in place for 1997 will be $605.5 billion. This figure represents a 4% increase over the 1996 rate.

The value of construction put in place is subdivided into two components: private construction and public construction. Private construction in 1997 is estimated at $466 billion. Private construction spending is broken down into the following broad categories: $265 billion for new residential building construction and improvements on residential buildings, $160 billion for industrial, office, hotel, private hospital and other commercial buildings construction, and $40 billion for railroad, electric light and power, and gas and petroleum pipeline construction.

Public construction in 1997 is estimated at $139 billion. Public construction spending is broken down into the following categories: $25 billion for education buildings, $5 billion for housing and redevelopment, $5 billion for hospitals, $25 billion for other public buildings, $39 billion for highways and streets, $10 billion for sewer systems, $6 billion for water supply facilities, and the remainder for miscellaneous public construction.[3]

1.2   Legal

1.2.1   General

The United States legal system is a dual system: a federal system of law overlying individual state systems. The structure of the US legal system is defined by the Constitution. Article III of the Constitution provides that "The judicial power of the United States shall be vested in one Supreme Court, and in such inferior courts as the Congress may from time to time ordain and establish." Article IV of the Constitution provides that the "full faith and credit shall be given in each state to the judicial proceedings of every other state and the congress may by general laws prescribe the manner in which such proceedings shall be proved and the effect thereof". Thus, the federal system of law has priority over the state systems involving constitutional rights and peculiar federal interests, such as interstate commerce.

Although the federal courts are courts of limited jurisdiction, their jurisdiction over construction disputes is extensive. The federal courts exercise jurisdiction over (i) "federal question" cases arising under the US constitution, treaties, federal statutes (e.g. the Federal Arbitration Act, see section 3.2.2), and the Miller Act (actions by unpaid subcontractors and suppliers on public construction bonds) and (ii) "diversity" cases involving controversies between citizens of different states where the amount in controversy exceeds $75,000.

The state courts in the state system of law are courts of general jurisdiction. Each state has its own state law and has authority over the property, contracts and most commercial matters related to the state. Consequently, an understanding of both the federal and state legal systems is necessary to fully understand the methods of managing conflicts under construction contracts in the US.

The federal legal system and each state legal system, except Louisiana (because of its early ties to France) is based on common law. All US legal systems are adversarial. Federal and state courts are both civil and administrative law courts. A US lawyer's relationship with his or her client is a contractual relationship with state bar imposed ethical obligations. As a professional, the lawyer owes his or her client a duty of care and confidentiality.

1.2.2   Rights of Audience

Admission to a state bar allows a lawyer to practice anywhere within the state and to engage in all types of practice, in a law firm, or corporation or in the government as a defence attorney, prosecuting attorney, or judge. There is no distinct judicial or prosecutional career; and there is no division between barristers and solicitors as in other common law countries. Any lawyer of a state has rights of audience in that state's courts. To regularly practice in another state (i.e., appear in court) a lawyer normally must either take its bar examination or, if he or she has practised for five years, apply for admission on motion. If a lawyer is not a member of a state court, it is very common practice for the out of state lawyer to retain local counsel to comply with the rules regarding rights of audience and yet exert complete control over the litigation, including the trial of the case, by admission *pro hac vice* (i.e. for purposes of the case only).

1.2.3   Size of Profession

There are approximately 320 lawyers per 100,000 population. Therefore, with the population of the US over 265 million there are now almost 1,000,000 US lawyers. More lawyers are in the pipeline. In 1996 approximately 128,000 students were enrolled in the 180 accredited US law schools.[4]

1.2.4   Courts

First instance courts in the federal system are United States district courts. There are 94 district courts in the 50 states, the District of Columbia, the Commonwealth of Puerto Rico, and the territories at Guam, the US Virgin Islands and the Northern Mariana Islands. Currently there are 494 magistrate judges and 649 federal district court judges presiding over district court cases. Appeals from these district courts are heard by 12 regional courts of appeals plus one specialist court of appeal, with 179 appellate court judges. The Supreme Court is composed of 9 Justices sitting in Washington DC.[5]

Over 250,000 cases were filed in US federal district courts in 1996. It is difficult to determine how many of these cases addressed construction disputes. However, it can be estimated that at least 10% involved contract disputes and, of these, perhaps up to 20%, involve construction disputes.[6]

1.2.5   Congressional and Specialist Courts

One specialist federal court that has jurisdiction over construction disputes is the US Court of Federal Claims. This court hears cases involving claims against the United States government including disputes between contractors and the government. Approximately 200 construction contract cases are filed each year in the US Court of Federal Claims and approximately 50 decisions are reported each year. The US Court of Federal Claims has been in existence for over 100 years and it is estimated that the court has reported over 3000 decisions during its existence.

In addition to the judicial branch there are executive agency board of contract appeals. These boards of contract appeals decide construction contract claims filed by contractors against the US government. They have concurrent jurisdiction with the US Court of Federal Claims. The three most important boards of contract appeals are: (1) the Corps of Engineers Board of Contract Appeals which decides civil construction cases, (2) the Armed Services Board of Contract Appeals which decides military construction cases, and (3) the General Services Board of Contract Appeals which decides construction disputes with the General Services Administration, the agency charged with the responsibility for building and maintaining most office buildings housing the US government. These boards decide approximately 600 cases a year of which approximately 30% are construction disputes. Executive Agency Boards of Contract Appeals have been publishing their decisions since 1956. Between 1956 and 1997 it is estimated that 20,000 decisions concerning construction disputes have been published by these boards.

The US Court of Appeals for the Federal Circuit decides all federal construction law appeals from both the US Court of Federal Claims and all federal agency boards of contract appeals. The US Supreme Court rarely will entertain an appeal from this appellate court involving substantive construction law issues.

1.3   Construction Law

1.3.1   General

There is an informal, but recognised construction law bar in the United States. There is no official state or federal bar association limited to construction attorneys, but the American Bar Association (ABA) has several sections that focus on construction law. The Section of Public Contract Law of the ABA has a specific Construction Division devoted to federal, state and local public construction law, with subcommittees dealing with Architectural and Engineering Services, Bonding and Insurance, Construction Claims, Construction Contract, Dispute Resolution, and several other construction related topics. A separate ABA group of attorneys is called the "Forum on the Construction Industry." This group has a quarterly publication entitled "The Construction Lawyer". In addition, many state bar associations have Construction Law divisions. Many large private law firms have construction law departments, and many smaller to medium size firms specialise in construction law. Thus, all major metropolitan areas have private attorneys specialising in construction law. Many attorneys representing state and federal government agencies have construction law expertise. In addition, major construction companies have in-house counsel specialising in construction law matters.

The sources of construction law in the United States are dramatically different for private commercial construction contracting and public construction contracting. For private commercial construction contracting the major source of construction law are (1) the terms of the contract, and (2) the general common law developed in construction contracts in that state, or if no such law, the federal common law, or common law developed in other states. Except in the area of mechanic's lien laws (which provide limited payment protection for contractors, subcontractors and suppliers on private projects), the state legislature rarely impose statutory requirements on private construction contracting. Some state courts have, as a matter of public policy, declared void construction contract indemnity clauses which impose liability on contractors and subcontractors for personal injury or property damages caused by a third party's sole negligence.

The sources of federal public construction law are (1) a complex set of procurement statutes and regulations, (2) contract clauses, and (3) federal common law embodied in the reported decisions. State and local public construction law is made up of the same three sources, with each state having somewhat different statutory and regulatory provisions. The complex statutes and regulations principally govern contract formation, and dispute resolution, while the contract clauses, and common law generally govern contract performance issues.

1.3.2   Contract Formation Procedures

Private construction contracts traditionally have been awarded after competitive bidding against a set of detailed drawings and specifications created by an architect/engineer under a separate contract. In recent years, this process has become known as the design-bid-build process. On more complex projects, in an effort to ensure that the design can be built within the project budget, owners have commonly contracted with construction managers to work with the architect/engineer on the design phase and to build the project. In most such cases, the construction manager agrees to a guaranteed maximum price before the construction work is begun. Another process being used more frequently is the design-build process where the owner contracts with a single entity for both the design and construction of the project. In this case, the design-build contractor assumes far more risk for the accuracy of the drawings and specifications than is the case under the design-bid-build arrangement.

The federal government traditionally used the design-bid-build process under sealed bidding procedures (with public opening of the bids). However, in 1984 the statutes were changed to permit negotiation of all contracts.[7] Since that time there has been a trend towards using negotiation procedures instead of sealed bidding to award construction contracts. This permits government agencies to negotiate with offerors after receipt of proposals and to award the contract to the offeror that is perceived to have offered the "best value" to the government (permitting award to a higher priced offer if the offeror has greater capability or offers superior work). Federal agencies have also used construction managers for some projects and have adopted design-build as a contracting technique for many recent projects.[8]

State and local governments are more likely to use traditional public bidding procedures, following the design-bid-build process. Generally, this is required by state statutes with a variety of exceptions for different states. Under these exceptions, state and local government have also used negotiation procedures on major projects and there is also a trend to use construction managers and design-build contracts for such projects.

1.3.3   Construction Contracts

Private construction contract forms generally are developed by national organisations involved in the construction process. Each set of contract forms may reflect terms and conditions favouring the constituents forming that organisation, i.e., such as architects, engineers, or contractors. However, these forms tend to contain provisions that balance the risk between the owner and the contractor (generally containing clauses making the owner responsible for differing site conditions and owner-caused delays). The principal private construction forms utilised are the American Institute of Architect (AIA), 1735 New York Avenue, NW, Washington, DC 20006[9], the Associated General Contractors of America (AGC), 1957 E Street, NW, Washington, DC 20006[10], and the Engineers Joint Contract Documents Committee (consisting of American Consulting Engineers Council, National Society of Professional Engineers, and American Society of Civil Engineers), 1420 King Street, Alexandria, VA 22314[11].

Most large private owners will have their standard contract forms which are tendered to general contractors for negotiation, which may allocate more risks to the contractor than the standard forms. Most general contractors have standard subcontract forms and, to a lesser degree, standard purchase order forms for equipment and materials. The AIA and AGC organisations also have standard subcontract forms.

The federal government has a complex set of mandatory and non-mandatory contract clauses set out in the Federal Acquisition Regulations (FAR). The FAR is published by the federal government at 48 C.F.R.[12] These federal forms balance the risk between the government and the contractor with the government bearing the risk of differing site conditions, defective specifications and government-caused delays.

Many state and local public contracting authorities also have their standard contract forms. If not, they will typically rely heavily upon forms from the organisations referenced above. These forms are very diverse in the assignment of risk between the government and the contractor, with some state and local government contracts still containing clauses that assign the risk of differing site conditions or defective specifications to the contractor and relieving the government of liability for damages for its delays.

In 1989, the ABA Section of State and Local Government Law issued the Model Procurement Code for State and Local Governments. The Model Procurement Code contains proposed statutes and regulations for the procurement of goods and services, including provisions regarding source selection, modification and termination of contracts, mechanisms for dispute resolution, socio-economic policies, and ethical standards. At least 15 states and 25 localities have adopted legislation based on the Model Procurement Code.[13]

The ABA also recently published a volume entitled Annotations to the Model Procurement Code (3d ed. 1996) which contains analyses of cases and state legislation based on the model code.

1.3.4   Dispute Resolution Clauses

Dispute resolution on private construction contracts has generally been assigned to the architect with a requirement for arbitration if the architect cannot resolve the dispute. However, there is a distinct trend to provide for mediation as the second step in resolving disputes. Thus, the 1997 edition of the construction contract forms of the AIA calls for mediation if the architect cannot resolve the dispute, followed by arbitration if mediation fails. Major private owners use a variety of techniques - usually either mediation or arbitration. As arbitration has become more expensive, there is a distinct trend toward greater use of mediation.

Federal dispute resolution is governed by the federal Disputes Clause which provides a contractor with the option of pursuing a dispute in the less formal setting of an agency board of contract appeals (where contractors may appear without attorneys, and less rigid rules of evidence apply) or the more formal US Court of Federal Claims (where corporate contractors must be represented by counsel and formal rules of evidence apply). In addition, the current trend in the federal construction contracting arena is to utilise many forms of ADR (e.g., mediation, mini-trials, arbitration) and several other non-judicial means of resolving disputes, even after a formal complaint is filed in court or at a board of contract appeals.

Dispute resolution in state and local procurements may require litigation in state court, or an informal state board of contract appeals, or they may specify arbitration. The current trend is also towards mediating large construction disputes, provided this is permitted by the state's procurement statutes or regulations.

2.   CONFLICT MANAGEMENT

There is a long tradition in the United States of the parties resolving conflict in the course of performance of a construction contract. Thus, each party will generally appoint an employee with the authority to resolve conflicts as they arise. In the case of government agencies, this employee is called a "contracting officer" and contractors must exercise care to ensure that agreements reached during performance of the work are signed by this official.

2.1   Non-binding

2.1.1   Dispute Review Boards

Dispute review boards have been used on major projects to assist the parties in resolving conflicts when they fail to resolve them through normal negotiations. The US Army Corps of Engineers has used such boards on a few projects with some success. These boards are generally composed of three people capable of giving an opinion on the technical issues that may arise during contract performance. They meet either on a regular basis or on call.

Examples of dispute review boards used on major state projects are found in Massachusetts which has boards in operation on the Central Artery Tunnel project and the Boston Harbour cleanup contracts.

2.1.2   Dispute Review Advisers

There is no reported use of dispute review advisors in the United States.

2.1.3   Negotiation

Negotiation between the parties is the major technique for conflict management. In most construction organisations, negotiation is a skill that is learned through personal experience on the job. However, there are numerous courses on negotiation in the United States, and some employees attend such courses.

Many employees of construction organisations become skilled at the art of negotiating to a point where the conflict is resolved to the satisfaction of both parties (the so-called "win-win" position). This is particularly true of more senior members of the organisation. However, there is a tendency for employees at the working level to use strong negotiating tactics, taking a firm position and refusing to compromise. This is a major cause of the large number of disputes that arise in construction contracting in the United States.

2.1.4   Quality Matters

The standard approach to quality control is governed by contract clauses. The ISO 9000 is not applied. Quality control procedures usually involve inspection by an owner's representative (which is often an outside architect/engineer firm), and contractor's representatives. Disputes over quality control are governed by the disputes clauses of the contracts.

The substantive quality control standards applied are generally found in the technical specifications, and are either expressly contained in the specifications, or the specifications incorporate them by reference to quality standards developed by industry groups. Typical of such groups are American Society for Testing and Materials (ASTM), American Concrete Institute (ACI), American Institute of Steel Construction (AISC), American Institute of Timber Construction (AITC), American Welding Institute (AWI), American Society of Mechanical Engineers (ASME), and National Electric Code Association (NECA).

2.1.5   Partnering

Partnering has become a major form of conflict management in the last decade. It was pioneered in the governmental arena by the US Army Corps of Engineers in the 1980's. It has subsequently been adopted by most federal construction agencies and many state and local government agencies. It is also being used, to a lesser degree, by private organisations.

Partnering, as used at the governmental level, is a quite formal procedure where the parties to the construction contract meet at the inception of the contract to agree on common goals and establish procedures to implement those goals. These common goals are often written into a "charter". The charter is not a contract and does not affect the parties contractual responsibilities. Generally, the parties agree to meet regularly throughout contract performance (weekly in many instances) to review all problems encountered and assign responsibility for resolving those problems. There is a general consensus that partnering reduces the number of disputes to a significant degree.

2.2   Binding

2.2.1   Partnering

There is no reported instance of partnering being used as a binding conflict management technique.

3.   DISPUTE RESOLUTION

Although litigation has been the traditional form of dispute resolution, there has been a growing awareness in many jurisdictions that this is no longer an effective way to resolve disputes. Hence, numerous surveys and studies document the increasing use and effectiveness of alternative dispute resolution (ADR) procedures as an alternative to litigation. A survey conducted through the ABA 1990-91 Forum on the Construction Industry examined the experiences and views of construction attorneys on the use of various ADR procedures.[14] Arbitration proved to be the most frequently used form of ADR in construction disputes, with 81.5% of those surveyed having experience with the procedure and 72.5% having used arbitration in the last two years. Many participants also reported use of mediation, with 64.2% of the respondents having some experience with it and 58.3% having mediated a dispute in the last two years. Other ADR procedures were less familiar to those surveyed. The survey also showed that the use of ADR resulted in full settlement in 57.4% of cases, and partial settlement resulted in another 8.4% of cases.

A smaller but more recent survey confirms the growing use of ADR in government contracts disputes. In 1993, the ABA Section of Public Contract Law, Alternative Dispute Resolution Committee surveyed section members to evaluate familiarity with and use of ADR in government contract disputes.[15] Binding arbitration remained the ADR technique most familiar to and most often used by those surveyed, followed by mediation. Other forms of ADR, such as mini-trial, early neutral evaluation, dispute review boards, fact-finding, non-binding arbitration and partnering, are less familiar to practitioners. Participants were also asked to evaluate various forms of ADR for effectiveness in terms of saving time, cutting costs, providing realistic evaluations of cases, and overall effectiveness. All approaches received high ratings, with mediation proving to be the most favourable.

Statistics from the American Arbitration Association (AAA) also reflect the widespread use of ADR. The AAA reported 70,516 case filings in 1996, including 4,114 construction arbitration and mediation filings. These figures are up from 62,423 and 3,991 filings, respectively, in 1995[16].

The growing popularity of ADR extends beyond the realm of public contracts and construction. A survey of 530 of the largest corporations in the United States conducted by Cornell University and Price Waterhouse, LLP confirms the widespread use of ADR[17]. Of the corporations surveyed, 88% have used mediation and 79% have used arbitration. Mediation again appeared to be the most favoured ADR approach. In another recent survey of large and mid-size companies conducted by the Centre for Public Resources, 24 companies reported estimated savings of $24 million in legal costs through the use of ADR[18].

The impressive results gained through the use of ADR also extend to mandatory court-ordered ADR programs. A mandatory program carried out by the Commercial Division of the New York County Supreme Court reported complete settlement in 52% of cases[19]. In another 16% of the cases, the use of ADR was reported to have contributed to a settlement reached shortly thereafter. Mandatory programs at the United States District Courts for the Northern District of California and the Western District of Missouri have also reported high user satisfaction and substantial cost savings[20].

3.1 Non-binding

3.1.1 Conciliation

There is no reported instance of the use of conciliation as a technique for resolving construction disputes in the United States.

3.1.2 Executive Tribunal

This technique is known in the United States as the "mini-trial." It was pioneered by the US Army Corps of Engineers beginning in 1985 and has been used effectively by that agency and other government agencies to resolve construction disputes. There is little reported use of mini-trials by private organisations.

The mini-trial is a structured settlement technique where the parties each appoint a principal with the authority to settle the dispute and a neutral advisor to assist the principals ascertain the facts and negotiate a settlement. The principal is usually an executive of the organisation who has not been involved in the dispute. Each party is given a period of time (frequently one day) to present its case. The presentation may be made by an attorney but, more commonly, they are made by officials of the parties that worked on the project (frequently using slides and graphic material). No rules of evidence are applied and the witnesses are not sworn. During these presentations, the principals and the neutral advisor are permitted to ask questions. Rebuttal presentations are sometimes scheduled but cross examination of witnesses is usually precluded.

After the presentation of the dispute, the principals and the neutral advisor meet to discuss the issues and negotiate a settlement. During this negotiation, the principals are free to obtain the neutral advisor's view of the facts and recommendations as to the settlement. However, the negotiation of the settlement is their sole responsibility and they are free to break off negotiations at any time.

Mini-trials have resulted in the parties reaching settlement in most disputes using this procedure but their use has declined as contractors and agencies have become more familiar with mediation as an alternate procedure. However, it is now common for the parties to blend mini-trial procedures with mediation procedures to achieve the procedure that they regard as the most efficient manner of resolving the dispute.

In construction contracts with the federal government, both the US Court of Federal Claims and the agency boards of contract appeals provide that they will conduct a mini-trial at the request of the parties. However, this procedure has been elected infrequently.

3.1.3   Mediation

In the past decade, mediation has become the ADR method of choice for resolving construction disputes. However, mediation is conducted in many forms as determined by the parties at the time they establish the procedure.

Traditionally, mediation involved the appointment of a mediator who would meet with each party to ascertain the bounds of their case and would then discuss settlement with each party, attempting to reach a common point by continual evaluation of their positions. This process is still used in many instances, but it has also become common for the parties to conduct all or part of the mediation process together. Most commonly, the presentation of the case will be done with both parties attending, either in a structured form such as the mini-trial or in a less structured form such as a round-table discussion of the issues. Sometimes, the negotiation of the settlement will be conducted by a principal for each party with the mediator assisting (similar to the mini-trial). The procedures can be varied to meet the circumstances and can be changed as the mediation progresses.

Mediation has been highly successful in resolving construction disputes at a fraction of the time and expense required for litigation.   It has proven to be most effective when used immediately after the parties have determined that conflict management techniques have failed. However, it is also used, in some instances, immediately before trial.   This occurs most frequently in states where the courts require mediation prior to litigation of construction cases (a requirement that is becoming increasingly more prevalent).

Occasionally parties to a construction dispute will provide for the mediator to issue a binding decision if the parties cannot agree to a settlement (Med-Arb). There are no federal laws in the United States prohibiting the use of such a procedure. The relevant statutes in each state would need to be inspected to determine whether state law would prohibit the use of such a procedure.

3.1.4   Negotiation

Negotiation can be and is used to resolve disputes before litigation commences and during litigation. Indeed, most judges urge the parties to resolve their disputes by negotiation and are pleased to see a negotiated settlement at any time during the pre-trial or trial proceedings.

3.2   Binding

3.2.1   Adjudication

There is no reported instance of the use of adjudication as a binding technique for resolving construction disputes in the United States.

3.2.2   Arbitration - Domestic

Arbitration has been the traditional technique for resolving private construction disputes and some state and local government contract disputes. Most standard forms for private construction contracts have designated arbitration as the dispute resolution procedure.   It has been generally successful as an economical and timely method of resolving small to medium sized disputes. Arbitrator's may apply "equity" principals to resolve the dispute. Expensive discovery is generally avoided and small to medium size cases are generally quickly disposed of. Generally arbitrators are experienced professionals in the industry and "care" about the proper resolution of the dispute. Many parties who have onerous contract provisions written in their favour will attempt to avoid arbitration since arbitrators are not bound to strictly enforce contract clauses or common law precedent.

Many practitioners have become dissatisfied with arbitrating large construction disputes. It is not uncommon for long arbitration hearings to be conducted in a non-continuous fashion over many months (e.g., alternating weeks, or several days a week) and are entirely dependent on the arbitrators' and parties' schedules, unlike court settings where trial schedules are generally conducted in a continuous fashion. In addition, some states, such as California, allow arbitrators to impose expensive discovery procedures, such as depositions. The summary nature of the arbitration award generally leaves the parties without information as to the reasoning of the arbitrators in reaching their result. Furthermore, abuse of the arbitration process is difficult to challenge.

The American Arbitration Association is the major institution that administers construction arbitrations. Other arbitration associations include ENDISPUTE and Judicial Arbitration and Mediation Scheme (JAMS). Many standard arbitration clauses provide that the arbitration shall be conducted in accordance with the AAA Rules. These rules provide that the arbitration award will not be supported by a decision giving the rationale for the arbitrator's finding. In addition, arbitration awards are rarely published.

The number of construction arbitration's administered by the AAA in the last ten years are listed below:

1987	4,582	1992	4,387
1988	4,940	1993	4,094
1989	5,132	1994	3,564
1990	5,440	1995	3,991
1991	5,189	1996	4,114

All 50 states plus the District of Columbia and Puerto Rico have enacted general arbitration statutes. Thirty four of the states arbitration statute and the District of Columbia arbitration statute are based on the Uniform Arbitration Act. The remainder of the state statutes and the Puerto Rico statute are unique. Four states have based their statutes on the UNCITRAL Model Law on International Commercial Arbitration: California, Connecticut, Oregon and Texas[21]. Two states Alabama and West Virginia do not have modern arbitration statutes. These state statutes only enforce the arbitration of existing disputes. The full text of all state statutes can be obtained from the internet.[22]

The parties to a construction contract may avoid the jurisdiction of the federal court by agreeing to arbitrate their dispute. The Federal Arbitration Act, 9 USC §§ 1 et seq. (FAA), applies to cases where there is federal question jurisdiction, i.e. the contract evidences a transaction involving interstate commerce.[23] The FAA strongly favours arbitration as an alternative dispute resolution forum. In fact, Congress adopted the FAA in 1925 to overrule the federal judiciary's long-standing refusal to enforce arbitration agreements.[24]

Whether the parties can avoid state court jurisdiction by agreeing to arbitration will vary on the specific state statutes and rules across the 50 jurisdictions. Generally, a policy favouring arbitration has been adopted by the state courts.[25]

Arbitration awards will be enforced by converting the award into a civil judgment and enforcing the civil judgment in the same manner as any other civil judgment entered by a US court (see section 3.2.5). Following the completion of an arbitration, the prevailing party will file a petition with the appropriate court to confirm the arbitration award. The process by which the court confirms the arbitration award does not involve judicial review of the merits. Although each state court will have its own statutory scheme, the standard grounds for setting aside an arbitration award are limited to where the award was procured by corruption, the arbitrators engaged in misconduct, or the arbitrators exceeded their powers.[26]

The doctrine of Kompetenz-Kompetenz, the arbitral tribunal has the competence to rule on its own jurisdiction, is traditionally an issue that will be decided as a preliminary matter by the court rather than by the arbitral tribunal.[27] The more modern rule, however, is that if the arbitration clause expressly provides for arbitral Kompetenz-Kompetenz, then the arbitral tribunal can rule upon its own jurisdiction. Upon review by the court, the arbitral tribunal's decision will be subject to a deferential standard of review.[28]

Generally, the state statute or the state law provide that the arbitrators have the same immunity of a judicial officer from civil liability when acting in the capacity of an arbitrator.[29]

3.2.3   Arbitration - International

International arbitration is used by US companies and international arbitration awards are enforced in the US. In fact the international branch of the AAA administers more international arbitrations than any other body except for the ICC.

The number of international arbitration's administered by the AAA in the last ten years are listed below:

1987	103	1992	204
1988	139	1993	207
1989	190	1994	187
1990	205	1995	180
1991	262	1996	226

AAA's international arbitration rules are based on the UNCITRAL Rules and the AAA will administer cases under either its rules, the UNCITRAL rules or specialised industry rules.

The United States is a party to the convention on the Recognition and Enforcement of Foreign Arbitral Awards (The New York Convention).   Since its adoption in 1958, 112 counties have become a party to the convention.   A list of the current member states can be found at AAA's internet site.[30]

The United States is not a party to any treaty either bilateral or multilateral with respect to enforcement of foreign judgments. Therefore, the enforcement of any foreign judgment in a US Court will first require the judgment creditor to obtain a judicial declaration recognising the foreign country judgment.   By contrast, an arbitral award made in the US is enforceable, by virtue of the New York Convention, both in the US courts and the courts of the other signatory states.   Therefore, a foreign party when contracting with a US party should ensure that the contract contains an arbitration provision if the foreign party desires a quick, reliable, and comprehensive enforcement mechanism.

The following journals publish international arbitral awards: The Arbitration and Dispute Resolution Law Journal, Lloyd's at London Press Ltd., The ICC Yearbook, Kluwer Law International[31], and Mealey's International Arbitration Report, Mealey Publications, Inc.[32]

3.2.4   Expert Determination

The Administrative Dispute Resolution Act of 1996 provides for a procedure described as "fact-finding" where parties to a dispute appoint an expert to make a binding finding on a specific issue[33]. This form of expert determination is thus available in resolving construction disputes but there are few reported instances of its use.

3.2.5   Litigation

As noted earlier, construction disputes are litigated in either the federal courts, state courts, or governmental boards of contract appeals. However, both the courts and the boards of contract appeals have become inefficient forums for the resolution of disputes.   Many courts have such crowded dockets that construction disputes may languish for years awaiting trial. Boards of contract appeals have also become slower in recent years.   Thus, litigation is generally not highly favoured by contractors.   The federal boards of contract appeals have expedited procedures which work well for claims less than $100,000. However, it is unusual for a large construction dispute to be resolved in less than two years, in most court or board of contract appeal settings, and such litigation may easily extend to five years or more, if subsequent appeals are taken.   For these reasons, ADR procedures are highly favoured, and a "cottage industry" of mediators has developed to meet the demand of parties dissatisfied with courts, boards of contract appeals, and arbitration.

Discovery in US courts is time consuming and expensive.   Jury trials often yield arbitrary results from jurors who have no understanding of the construction process, and do not have a keen interest in the proper result. The same can be said for many state and federal judges who are not interested in complex construction cases that clog their docket of cases. It is common for private or public opponents to utilise "the system" to "wear out" the small to medium sized construction firms through litigation to either (1) settle the case "cheaply" or (2) run them out of business. In a disturbing trend, large federal agencies aggressively use threats of criminal and/or civil prosecution for False Claims Act violations as a means to force settlement of claims. While such threats are appropriate for real substantive violations, frequently standard construction disputes are improperly converted into weak or baseless False Claims Act[34] violations to enhance negotiation leverage. For a civil or criminal judgment can result in the debarrment of that company from any further federal (and possibly state work as well) work for three years, which is a "death sentence" for a firm heavily dependent on public construction contracting, to say nothing of the prospect of criminal confinement for the company's principals.

Most states have enacted the Uniform Enforcement of Foreign Judgments Act, 13 U.L.A. 149 (1995) which provides that any final judgment, order or decree of a court of the United States is entitled to full faith and credit in the enacting state.[35] To enforce an out of state judgment, a party need only file the judgment with a court in the enacting state. Thus, a creditor wishing to enforce such a judgment need not go through the formalities of commencing a new action to enforce the judgment. Once filed, the judgment is then treated as a judgment of the enacting state. In other words, the judgment has the same force as a judgment issued by a court of that state, and is likewise subject to the same procedures and defences[36].

The federal and state courts have strict rules on the admissibility of evidence. For the most part, documents generated during the performance of the contract are admitted as business records, as an exception to the hearsay rule. Fact witnesses are generally limited to testifying based upon their personal perceptions and involvement. Expert evidence is generally presented orally and is admitted as long as acceptable expert qualifications are presented.

Public boards of contract appeals have less rigorous standards for admission of oral testimony. Some board judges nevertheless lose sight of the informal nature of the process and impose court like standards.

By comparison, arbitrators generally are noted for the admission of evidence that would not be admissible in either state or federal court. Competent counsel generally avoid the type of evidence to enhance the credibility of their cases. Some arbitrators will be more vigorous. But the arbitration process can be easily abused by the less competent or by those who simply have weak cases.

Courts have minimal filing fees. Public boards of contract appeals have none. Neither forum charges for judges or jurors. The compensation or arbitrators can be very expensive for large cases. Most experienced arbitrators now charge at least $1,000 per day. With panels of three typical for large cases, these costs can arise quickly. Filing fees are also expensive. For example, the AAA charges a minimum of $2,000 and up to $7,000 to file an arbitration for any case having three arbitrators. The filing fees, the arbitrators' compensation costs, and expert fees may be awarded entirely against one party, or apportioned equally, or in some other matter. Unless allowed by contract, attorneys' fees are not awarded in court or arbitration.

3.2.6   Negotiation

There is no reported use of negotiation as a binding dispute resolution technique in the United States.

3.2.7 Ombudsman

The use of ombudsmen is a very new development in the United States. The federal National Aeronautics and Space Administration has appointed an ombudsman with broad powers to hear any complaints registered by contractors and presumably this person could attempt to resolve a construction dispute. A few other federal and state agencies have also appointed ombudsmen. There is no reported instance of such a person addressing or resolving a construction dispute.

4.   THE ZEITGEIST

Litigation has become so costly and slow in the United States that it is no longer regarded as an effective means of resolving most construction disputes. However, the adversarial nature of construction contracts has resulted in a large number of instances where conflict management techniques have not been effective in preventing such disputes. Construction contractors, private owners and public agencies have addressed this problem in two ways. First, they have adopted conflict management techniques, such as partnering, to improve their abilities to manage conflict more effectively. Second, they have used ADR techniques more frequently to resolve disputes early. The result of these efforts is a slow reduction in the amount of litigation of construction disputes.

The federal government has taken the lead in this movement. The US Army Corps of Engineers has pioneered in the use of conflict management techniques. The availability of ADR to resolve disputes with the federal government has increased with the help of initiatives from Congress, the courts and the President. In 1990, Congress passed the Administrative Disputes Resolution Act of 1990[37], The Act authorises agencies to use ADR "for the resolution of an issue in controversy that relates to an administrative program, if the parties agree to such a proceeding".[38]   Agencies are also to consider not using ADR in certain proceedings, such as where a definitive resolution is needed for precedential value.[39]   The Act also amends the Contract Disputes Act of 1978 to authorise the use of ADR in government contract disputes.[40]   The statute was amended and permanently re-authorised last year by the Administrative Disputes Resolution Act of 1996[41], which further strengthened the federal policy favouring ADR.   Among other things, the amendments require the President to designate an agency or interagency committee to "facilitate and encourage agency use of dispute resolution under this subchapter".[42]

ADR in federal district courts has been strengthened by the passage of the Civil Justice Reform Act of 1990.[43]   This Act required each federal district court to develop a civil justice cost and delay reduction plan.   Of the 94 district courts 80 have implemented some form of ADR.

Courts and boards with jurisdiction over government contract disputes have also implemented policies promoting the use of ADR. In response to "rising litigation costs and the delay often inherent in the traditional judicial resolution of complex legal claims," the United States Court of Federal Claims adopted Amended General Order No. 13 on November 8, 1996[44].   The order implements three ADR procedures to be used in appropriate cases: (1) the use of "settlement judges" who are neutral advisors brought in to provide "frank, in-depth discussions of the strengths and weaknesses of each party's case," (2) "mini-trials" wherein each party provides a brief presentation of their case to a neutral advisor who then assists the parties in reaching settlement, and (3) the use of "third-party neutrals" who meet with the parties in an attempt to help resolve the dispute.

The promotion of ADR has also spread to the federal Boards of Contract Appeals. Ten of the boards have generated a "sharing agreement" pursuant to which a judge from one board may serve as a "neutral" in facilitating the use of ADR in disputes before other boards[45].

State courts have also resorted to ADR to help reduce the backlog of cases. For example in New York the chief judge announced in October 1997 that a program of ADR initiatives ranging from neutral evaluation to binding arbitration will be introduced in counties and city courts in the NY state court system.[46]

Finally, on February 5, 1997, the President signed Executive Order 12988 entitled "Civil Justice Reform"[47]. The order provides that government attorneys should suggest the use of ADR techniques to parties where appropriate. The order also states that government attorneys should be trained in ADR techniques to facilitate their use. In response to the order, the Department of Justice issued interim ADR guidelines on July 22, 1997.[48] The guidelines state that "litigation counsel should consider ADR as soon as adequate information is available to evaluate the litigation and settlement, as well as throughout the course of the litigation."[49]

---

1   See "www.census.gov".

2   See "www.bea.doc.gov" to obtain information regarding the "national economic accounts".

3   A Department of Commerce publication titled "Statistical Abstract of the United States" contains a vast quantity of information regarding US economic data. This publication is prepared by the Census Bureau and is known as the national data book. See "www.census.gov" to obtain additional economic information concerning the US construction industry.

4   See generally "www.abanet.org".

5   See generally "www.uscourts.gov".

6   See "www.uscourts.gov".

7   The Competition in Contracting Act, Pub. L. 98-369.

8   Section 4105 of the Clinger-Cohen Act of 1996, Pub. L. 104-106, which permits "two-step" design-build contracting - implemented in FAR Subpart 36.3.

9   See "www.aia.org".

10 Telephone: 202-393-2040.

11 See "www.asce.org".

12 Contact the US Government Printing Office, Superintendent of Documents at Mail Stop: SSOP, Washington, DC   20402, telephone 202-512-1800 to purchase the FAR. The FAR is also published by CCH, Inc., 4025 W. Peterson Avenue, Chicago, IL 60646, "www.cch.com".

13 See *O'Hare Trucking Serv., Inc. v. City of Northlake*, 116 S.Ct. 2353, 135 L.Ed.2d 874, 1996 US LEXIS 4263 at *37-38 (1996); *Board of County Commissioners v. Umbehr*, 116 S.Ct. 2342, 135 L.Ed.2d 843, 1996 US LEXIS 4262 at *42-43 (1996). A copy of the Model Procurement Code can be obtained from the

ABA by telephone at 312-988-5522, by mail order to ABA Publications Orders, P.O. Box 10892, Chicago, IL 30310-0892, or on the Internet at "www.abanet.org."

[14] See Thomas J. Stipanowich & Douglas A. Henderson, *Settling Construction Disputes with Mediation, Mini-Trial, and Other Processes--The ABA Forum Survey*, Construction Lawyer (April 1992).

[15] See Thomas J. Stipanowich, *The Quiet Revolution in Government Contracting: Dispute Avoidance and Resolution* 30 Procurement Lawyer 3 (1994).

[16] See "www.adr.org".

[17] See David B. Lipsky & Ronald L. Seeber, *The Use of ADR in U.S. Corporations: Executive Summary* (issued by Cornell University School of Industrial and Labour Relations) "www.ilr.cornell.edu"; and Rinat Fried, *Corporations Backing ADR, Study Says*, American Lawyer Media, Apr. 29, 1997, p.4.

[18] *CPR Survey in Variety of Industries Confirms ADR Cost Savings*, Metropolitan Corporate Counsel, Inc., Oct. 1997, p.25.

[19] Robert C. Meade, *Commercial Division ADR: A Survey of Participants*, New York Law Journal, Oct. 17, 1997, p.1.

[20] See Donna Stienstra et al., *Report of the Judicial Conference Committee on Court Administration and Case Management: A Study of the Five Demonstration Programs Established Under the Civil Justice Reform Act of 1990*, (Federal Judicial Center 1997); and Joshua D. Rosenberg & H. Jay Folberg, *Symposium on Civil Justice Reform: Alternative Dispute Resolution: An Empirical Analysis*, 46 Stan. L. Rev. 1487 (July 1994).

[21] See "www.un.or.at/unicitral".

[22] See "www.adr.org"

[23] *Barbier v. Shearson Lehman Hutton Inc.*, 948 F2d 117 (2d Cir. 1991).

24   *Collins & Aikman Products Co. v. Building Systems, Inc.*, 58 F3d 16 (2d Cir. 1995).

25   *Guadano v. Long Island Plastic Surgical Group, PC*, 607 F.Supp. 136 (NY 1982) (Public policy of NY favours arbitration, in part, due to its virtue of conserving financial resources of the contracting parties).

26   See e.g. NY Civ. Prac. Law section 7511(b), Cal. Civ. Prac. Code section 1286.2, and generally 9 U.S.C. section 10.

27   *Steelworkers v. American Manufacturing Co.*, 363 U.S. 564, 80 S. Ct. 1343 (1960); *Steelworkers v. Warrior and Gulf Navigation Co.*, 363 U.S. 574, 80 S. Ct. 1347 (1960); and *Steelworkers v. Enterprise Wheel & Car Co.*, 363 U.S. 593, 80 S. Ct. 1424 (1960).

28   *First Options of Chicago v. Kaplan*, 514 U.S. 938, 115 S. Ct. 1920 (1995).

29   See Cal. Civ. Proc. Code section 1280.1 and *Morelli v. Provenzano*, 115 NYS 2d 489.

30   See "www.adr.org".

31   See generally "www.iccwbo.org/arb".

32   See "www.mealeys.com".

33   Public Law No. 104-320, for US laws, see generally "www.thomas.loc.gov".

34   31 USC § 3729 and 18 USC § 287.

35   The states which have adopted the Act are Alabama, Alaska, Arizona, Arkansas, Colorado, Connecticut, Delaware, District of Columbia, Florida, Georgia, Hawaii, Idaho, Illinois, Iowa, Kansas, Kentucky, Louisiana, Maine, Maryland, Minnesota, Mississippi, Missouri, Montana, Nebraska, Nevada, New Hampshire, New Mexico, New York, North Carolina, North Dakota, Ohio, Oklahoma, Oregon, Pennsylvania, Rhode Island, South Carolina, South Dakota, Tennessee, Texas, Utah, Virgin

Islands, Virginia, Washington, West Virginia, Wisconsin and Wyoming.

36   See Sara L. Johnson, *Validity, Construction and Application of Uniform Enforcement of Foreign Judgments Act*, 31 A.L.R. 4th 706.

37   Pub. L. No. 101-552, 104 Stat. 2736 (Nov. 15, 1990).

38   5 U.S.C. § 572(a) (1996).

39   Id. § 572(b).

40   41 U.S.C. § 605(d).

41   Pub. L. No. 104-320, 110 Stat. 3870 (Oct. 19, 1996).

42   5 U.S.C. § 573(c) (Supp. 1997).

43   Pub. L. No. 101-650, (Dec. 1, 1990), see generally "www.uscourts.gov".

44   Rules of the Court of Federal Claims, Appendix L.

45   See Melanie Dooley, *Boards of Contract of Appeals Agree to Sharing Arrangement to Promote Use of ADR*, Fed. Cont. Rep., Apr. 14, 1997, at 428-29.

46   See "www.ljextra.com".

47   Exec. Order No. 12988, 3 C.F.R. 157 (1997).

48   Office of the Attorney General; Memorandum of Guidance on Implementation of the Litigation Reforms of Executive Order No. 12988, 62 Fed. Reg. 39,250 (July 22, 1997).

49   Id. at 39,253.

APPENDIX

Conseil International du Bâtiment pour la Recherche , l'Étude et la Documentation (CIB)

The International Council for Building Research, Studies and Documentation

International Task Group TG15
Construction conflict: management and dispute resolution

CIB TG 15

Monographs of Practice and Procedure In Member Countries

Guidelines for authors

CIB TG15
Construction Conflict: Management and Resolution

Monographs of Practice and Procedure In Member Countries

Background

TG15 was formulated to encourage the study of construction conflict and disputes by the collection of data with a stated objective of empirical data rather than anecdote.   Such data, it is proposed, will allow comparison between countries and informed management decision making.   Where structured data is available please include it if you can;   graphs may serve well.

Contextual Statement

Authors of the national monographs are asked to follow a consistent outline. While this may be strictly impossible the intention is that two major advantages will flow from consistency:

1.   Monographs will be consistent and rigorous.

2.   Comparisons of monographs will allow meaningful comparisons of national practice and procedure.

   The comparative analysis of national legal systems is an academic discipline in its own right.   Compiling an analysis of legal matters arising from different sources would require the utmost care.   Laws establish or define rules chosen according to the particular social, historic, economic or judicial context in which they are applied and according to the common usage to which they are put.   This collection of national monographs **is not** a comparison of legal systems, it is a comparison of the techniques for managing conflict and resolving disputes on construction projects in different countries. Of course any such analysis necessarily deals with legal systems; but this should not be seen as a restricting factor since any such analysis also deals with social, economic and cultural systems.

In addition to the national monographs authors are asked to contribute to a taxonomy, system of classification, and to an annotated vocabulary. The collected national monographs will be published together with an editor's report on the taxonomy and vocabulary.

CIB TG15
Construction Conflict: Management and Resolution

Monographs of Practice and Procedure In Member Countries

Guidelines for authors

These guidelines are not intended to be restrictive and all contributions will be gratefully received; however it may help if the papers follow a common format.

Length
There is no requirement, or restriction, on length.

Submission
The papers will be made available on the Internet, it would help if all submissions were made electronically as detailed below via the Internet to PFENN@UMIST.AC.UK or on floppy disk. **Deadline 1 December 1997.**

Publication
The collected monographs will be published: as a formal book; and electronically via the Internet, copyright issues will be decided when the book contract has been finalised. My previous experience of this is that you will retain copyright of your submission but that the copyright of the edited collection will be held by the editor (Peter Fenn probably). Your work remains yours the concept of the book as a whole is the copyright of the editor. There will be some royalties to the editor but I submit that they will be very small; if you are doing this for the money forget it, I suggest that you would be better paid pumping gas. The royalties could be split between the editor and the monograph authors or they could be donated to a charity, I favour the latter. It would be fun if a book about construction disputes

created a copyright dispute; it would at least make someone (probably a lawyer) happy.

Format

Detailed guidelines on format are given below: under the headings of format and content. Like all matters of consensus and compromise the guidelines for content are bound to reflect the experiences and culture of the main authors. I reiterate that the intention is to collect empirical data in some kind of structured sense; if however you are restricted in any way by the guidelines please deviate from them to describe your country's experience of practice and procedure.

Sub-contracting

Of course it is expected that specialist contributions will be invited, where necessary. If you can involve anyone else; please do so, all contributions will be acknowledged.

CIB TG15
Construction Conflict: Management and Resolution

Monographs of Practice and Procedure In Member Countries

Guidelines for monograph format

The national monographs are intended to be a source of reference; they will be published electronically via the TG15 World Wide Web page and formally as a book. To ease the passage to the two publication media please try to comply with these guidelines.

Electronic media
Each hard copy of the monograph **must** be accompanied by an electronic copy of the full text of the monograph. The electronic copy can be sent on disk or via the Internet as a file. The following word-processing packages are acceptable.

WordPerfect 5.1
WordPerfect for Windows 6.0a.
Word for Windows 6.0

I cannot cope with Apple Mac I am afraid; if you really must: produce it in Apple Mac and export to DOS ASCII but remember to take care with any special features, eg footnotes, which may not export.

Paper Size
The document shall be formatted for A4 paper 210 x 297 mm, with 25 mm borders all round.

Fonts

The text shall be in 12 point Times New Roman. Full justification, single spaced. No indentation at first line of paragraphs; leave a single line between paragraphs.

Sections

The text of the monograph shall follow the headings shown in the guidelines for content. Section headings shall be as follows:

1 LEVEL ONE (UPPERCASE, BOLD FACE).

1.2 Level two (initial letter only in upper case, bold face).

1.2.1 Level three (initial letter only in upper case).

CIB TG15
Construction Conflict: Management and Resolution

Monographs of Practice and Procedure In Member Countries

Guidelines for monograph format

<u>Page numbering</u>
The word-processed file should **not** be numbered but each page of the hard copy **must** be numbered in pencil on the reverse of each page.

<u>Explanatory notes and citations</u>
Explanatory notes and citations shall be placed in footnotes to each page numbered consecutively from the beginning of the document, starting at footnote number 1.   Explanatory notes should use the Harvard system of referencing, citations of legal cases must give the full citation and reference (no abbreviations).   Whilst being reluctant to provide an exemplar, I hope that this document complies with my own instructions.

CIB TG15
Construction Conflict: Management and Resolution

Monographs of Practice and Procedure In Member Countries

Guidelines for monograph content

Please use these guidelines as a template for your monograph. Any problems with compliance should be included at the end.

1.   BACKGROUND

1.1   Economic
1.1.1   Trends

1.2   Legal
1.2.1   Construction Law
1.2.2   Construction Contracts

2.   CONFLICT MANAGEMENT

2.1   Non-binding
2.1.1   Dispute Review Boards
2.1.2   Dispute Review Advisers
2.1.3   Negotiation
2.1.4   Quality Matters
2.1.5   Partnering

CIB TG15
Construction Conflict: Management and Resolution

Monographs of Practice and Procedure In Member Countries

Guidelines for monograph content

1.   BACKGROUND

1.1   Economic
Describe the main features of the national construction industries: in terms of output; and if possible in relation to Gross Domestic Product (GDP). Where possible, but do not be intimidated by the need, provide any useful statistics available

1.1.1   Trends
If possible describe any fluctuations in the economic features over time, particularly recent history.

1.2   Legal
Describe in outline the legal system in which the national industries operate. Common Law or Civil; adversarial or inquisitorial. One overall legal system, or federal, or many jurisdictions within one system. Describe the court system: if appropriate annotate any special construction courts and cross-reference to later (see section 3.2.4). Outline the arrangement or system of legal profession(s); is advocacy separated; is it a distinct profession. Describe the rights of audience within the court system. Describe the system, if it exists, of consumer protection; do the courts (and or the legal professions) differentiate between individuals and corporate concerns.

Describe, if it exists, any special relationship which may exist between the clients and professional advisors. In English law a special contractual relationship exists under the broad term of agency used to describe the relationship between two parties. The agent acts on behalf of the other party the principal.

1.2.1   Construction Law

To save time in the body of the monograph describe here matters of general relevance to construction law. The first point may be that there is no recognised legal specialism of construction law. Describe the major sources of construction law. List any specialist reports and journals. If appropriate discuss the professions with specialist expertise. Outline the major players.

1.2.2   Construction contracts

Describe the nature of construction contracts: bespoke contracts or standard forms: national or international. If the contracts used are published internationally it is expected that the reader of these monographs will be able to obtain copies. However if the contracts are national or bespoke some explanation and list of sources will help. List the standard forms and their use, this may centre on the use of specific procurement systems see 2.1.4. Outline the dispute resolution clauses contained within those standard forms e.g. arbitration clauses.

2.   CONFLICT MANAGEMENT

Describe the systems of conflict management (an alternative approach might centre on the avoidance of disputes). Where possible include data on use and abuse, frequency if appropriate. Opinion of the author on national perceptions is welcomed. Please address all the sections below; even if only to say not used in this country. Include extra techniques and therefore sections as required.

The taxonomy of conflict and dispute provoked much comment in this group.   The conceptual differences are blurred and the literature reflects this.[2]   For the purposes of this group I suggest we use conflict as the 'softer' term. Conflict exist where there is an incompatibility of interest.   When a conflict becomes irreconcilable and the mechanisms for managing it are exhausted, or inadequate, techniques for resolving the dispute are required.   Interestingly, to me anyway, dispute resolution may also be identified by third party intervention.   If a national perspective on conflict versus dispute is pertinent please debate this under
5 Taxonomy and List of Terms.

The taxonomy of binding and non-binding on the other hand provoked total silence.   If a national perspective on this exists please debate it under 5 Taxonomy and List of Terms.

2.1   Non-binding

2.1.1   Dispute Review Boards
         See the list of terms[3]

2.1.2   Dispute Review Advisers
         See the list of terms[4]

2.1.3   Negotiation
         See the list of terms. If possible describe your country's approach to negotiation.   What are the appropriate theories; how is negotiation training dealt with, is it formally taught, or is it a practical skill learnt pragmatically by personal experience.

2.1.4   Quality Matters
         Please describe here the techniques which centre on management control: Total Quality Management; Quality Assurance. What is the

approach to standard control; is ISO 9000 applied, what are the national standards. Are there any initiatives to improve quality by information control, e.g. Co-ordinated Project Information.

If possible outline the major Procurement Systems used for construction projects and any data on usage, perhaps as a percentage of: value of construction work, or number of contracts let.

2.1.5   Partnering
See the list of terms. In particular discuss the nature of long term commitments: formal or informal. Suggest that informal goes here and formal in 2.2.1

2.2   Binding

2.2.1   Partnering
See 2.1.5

3.   DISPUTE RESOLUTION
Describe the systems of dispute resolution. Where possible include data on use and abuse, frequency if appropriate. Opinion of the author on national perceptions is welcomed. Please address all the sections below; even if only to say not used in this country. Include extra techniques and therefore sections as required.

3.1   Non-binding
3.1.1   Conciliation
See the list of terms. Is conciliation recognised as distinct from mediation. If so highlight and refer to 3.1.3. Comment on the use and adoption of UNCITRAL rules

3.1.2   Executive Tribunal
See the list of terms.

3.1.3   Mediation
See the list of terms and refer to 3.1.1.   Is the process where a mediator can be asked to provide a binding decision permitted (Med-Arb)?

3.1.4   Negotiation See 2.1.3.

3.2   Binding
It is anticipated that this section will concentrate on the tribunals for resolving disputes.   It may be necessary to refer to national legal systems: how is evidence admitted/presented; does the particular tribunal have to follow any rules of evidence or procedure.   Any systems for the delivery of expert witness or witness of opinion. How do tribunals apportion costs, can advisers work on a contingency fee basis, are there any limits on extent.

3.2.1   Adjudication
The intention here is to examine the specialised meaning of the term, as found, in my experience in the United Kingdom construction industry.   This meaning is defined in the list of terms, it is accepted that this may not translate to other countries; but of course that is what TG15 seeks to find out.   Refer to 3.2.3

3.2.2   Arbitration
It is anticipated that this will be a major part of the monograph (since it is assumed that most countries have adopted arbitration as the default dispute resolution technique).   Once again if international arbitration is used or if international arbitration uses the country please discuss.   The following are areas where particular comment is required.

Equity clauses; ex aequo et bono; amiable compositeur. Separability, Kompetenz-Kompetenz. Immunity of Arbitrators, ditto appointing bodies. The arbitrator's powers. The arbitrator's ability to use his own knowledge (and to give himself evidence).

Are arbitration awards published.
Has the country ratified the New York Convention 1958.
Has the country adopted the UNCITRAL Model Law for Arbitration.

3.2.3   Expert Determination
The intention here is to examine the specialised meaning of the term, as found, in my experience in the United Kingdom construction industry. This meaning is defined in the list of terms, again it is accepted that this may not translate to other countries; but of course that is what TG15 seeks to find out. Refer to 3.2.1

3.2.4   Litigation
What is meant here is the resolution of a dispute by an institution usually called a court or a tribunal set up and organised by (if not controlled by) the state. The Court usually has the power to conduct such resolution under the general law of the state rather than by the agreement of the parties and the power will include the ability in most cases to enforce its procedure and decisions compulsorily on the parties whether or not they agree or submit to its jurisdiction.

Of particular interest is (in no particular order):

(a)   Whether the parties can avoid the jurisdiction of the court (eg by a legally binding agreement to arbitrate) in a way that the Court will recognise and enforce by preventing cases proceeding where the parties have (in their contract) entered in to an agreement to arbitrate which one party seeks to ignore.

(b)   Whether there are specialist Judges, courts or procedures for construction cases or for a range of technical cases including construction cases. For example in England and Wales the Official Referees are specialist Judges who sit in separate court buildings with specialised procedures designed for "technical" cases such as construction and computer disputes.

(c)   Whether the Courts have the power to direct the use of external procedures without the consent of the parties (eg) mediation.

(d)   Whether the procedures are adversarial or inquisitorial such as the use of court appointed technical experts.

(e)   What appeals systems are available including the Court's ability to review the decisions made in other dispute resolution methods (eg) appeals from arbitrators' awards

(f)   What legal systems are used (eg) will the Court apply foreign law to a contract said to be subject to a foreign legal system.

(g)   What procedures are used where one or more of the parties is not resident or based in the state in which the Court is based particularly in respect of enforcement and, vice versa, will the Court enforce awards made by foreign courts or arbitrators?

(h)   Will the Courts prevent parties from giving evidence as to statements made in negotiations (eg) is there a concept of "without prejudice"?

(j)   To what extent is evidence given orally or in writing and what opportunities exist for cross examination by the parties and/or the Court?

3.2.5   Negotiation
See 2.1.3 and 3.1.4.

3.2.6   Ombudsman
This is an area where the group would welcome any experiences, or even definitions, it may be a suitable place to describe innovative techniques in member countries. English definitions[5] centre on officials to investigate complaints against public bodies. However there is also a development in the UK of Ombudsman operating as consumer protection eg insurance, banking and legal services[6]. Perhaps you have a system of Ombudsman in construction?

4.   THE ZEITGEIST

What is the spirit of the times for the nation on which you are reporting. In the UK the legal zeitgeist may be summarised by the aims of the review of the current rules and procedure of the civil courts in England and Wales carried out by The Right Honourable the Lord Woolf: the aims of the review were:
to improve access to justice and to reduce the cost of litigation;
to reduce the complexity of the rules and modernise terminology;
to remove unnecessary distinctions of practice and procedure.[7]

The economic zeitgeist is depressing, the recession of the late eighties still lingers output is low and construction prices lag behind costs.[8]

5.   TAXONOMY AND LIST OF TERMS

Whilst the taxonomy, that has been developed by the group, has formed the structure for the monographs it is inevitable that anomalies will emerge, and misclassification will arise through error or be forced due the rigidity of the system.   Please comment on the taxonomy and suggest improvements.

The list of terms is perhaps the most corrupted of the guidelines it is inevitably influenced by the experience of the proposer.   Some terms will have posed special problems and the guidelines have attempted to deal with this; please annotate the list of terms with your suggestion/alternatives.

The editor will collect all the submissions and provide a report.

5.1   Taxonomy

Conflict Management

Non-binding	Binding
Dispute Review Boards	Partnering
Dispute Review Advisers	
Negotiation	
Quality Matters	

- Total Quality Management
- Procurement Systems
- Co-ordinated Project Information
- Quality Assurance
- Partnering

Dispute Resolution

Non-binding	Binding
Conciliation	Adjudication
Executive Tribunal	Arbitration
Mediation	Expert Determination
Negotiation	Litigation
	Negotiation

5.2   List of terms

Adjudication

Adjudication is where a third neutral third party gives a decision which can be binding on the parties in dispute unless they wish to proceed to formal arbitration. In most instances this decision is binding for the duration of the contract (Construction Industry Council).

The procedure where, by contract, a summary and interim decision making power in respect of disputes is vested in a third party individual (McGaw).

Arbitration

Arbitration in law is a process, subject to statutory controls in England and Wales, whereby formal disputes are determined by a private tribunal of the party's choosings. The decision awarded is final, except for a few safeguards, and enforceable at law.

Conciliation

The terms mediation and conciliation are often interposed. Some authors believe that conciliation is more at the adjudicative end of the spectrum of ADR techniques.

Disputes Review Adviser

A neutral third party who simply advises on a problem or potential dispute which requires clarification as to the best methods of reaching a settlement. Sometimes described as early settlement advisor (CIC).

Disputes Review Board

Disputes Review Board process is an expedited non binding ADR procedure whereby an independent board, usually of three persons is established to evaluate disputes and make settlement recommendations to the parties. The Board members become knowledgeable about the project by periodically visiting the site.

Executive Tribunal

Executive tribunals are a formalised method of ADR consisting of one executive from each side or party in dispute and a neutral. The executive must have power to settle. This process is sometimes termed "mini-trial".

Expert Determination

Expert determination is a means by which the parties to a contract jointly instruct a third party to decide an issue.

Litigation

Procedure of taking a dispute to courts for settlement at law.

5.2   List of terms

Mediation

A voluntary, non-binding private dispute resolution process in which a neutral person helps the parties to try to reach a negotiated settlement (CEDR).

Negotiation

Negotiation is a process in which individuals or groups seek to reach goals by making agreements with others (Johnson)

Negotiation is a process for resolving conflict between two or more parties whereby both or all modify their demands to achieve a mutually acceptable compromise (Kennedy, Benson and McMillan).

Negotiation is a basic means of getting what you want from others (Fisher and Ury).

Partnering

Partnering is a long term commitment between two   or mor organisations for the purposes of achieving specific business objectives. By maximising the effectiveness of each participant's resources (NEDC).

---

[1]D. Dagenais, (1993), *Law and Contract*, in Post Construction Liability and Insurance, Knocke, J., (Ed), Spon, London, p59.

[2]See Brown and Marriot (1993), *ADR Principles and Practice*, Sweet and Maxwell, London, p5 for a review.

[3]Further definitions may help; e.g. Wall, C., (1992) *The Dispute Resolution Adviser,* in Construction Conflict Management and Resolution, Fenn and Gameson (Eds), Spon , London, p328.

[4]See Wall, C,. (1994) *The Dispute Resolution Advisor System in Practice*, in Construction Conflict Management and Resolution, Fenn, P., (Ed), CIB Publication No 171, CIB, The Netherlands.

[5]See for example The Little Oxford and Collins Concise

[6]See Brown and Marriot (1993), *ADR Principles and Practice*, Sweet and Maxwell, London, pages 279-281.

[7]The Woolf Review *Access to Justice*, Woolf Enquiry Team, Room 438, Southside, 105 Victoria Street, London.

[8]For an excellent discussion of the economy see: Hutton (1995), *The State We're In*, Cape, London.

Index

This index includes only those terms listed under monograph sections 2 and 3 in pages 842–8 of the Appendix.